Science, Understanding, and Justice

Science, Understanding, and Justice

The Philosophical Essays of
Martin Eger

Edited by
Abner Shimony

OPEN COURT
Chicago and La Salle, Illinois

To order books from Open Court, call 1–800–815–2280,
or visit our website at www.opencourtbooks.com.

Open Court Publishing Company is a division of Carus Publishing Company

Printed and bound in the United States of America

Library of Congress Cataloging-in-Publication Data

Eger, Martin, 1936-2002.
  Science, understanding, and justice : the philosophical essays of Martin Eger / edited by
Abner Shimony.
     p. cm.
  Summary: "A collection of essays by the late Professor Martin Eger applying the
hermeneutic approach to issues of science, education, and ethics. Discusses how concepts of
science are conveyed to the general public through popularizations, and includes exchanges
between Eger and other leading philosophers"--Provided by publisher.
  Includes bibliographical references and index.
  ISBN-13: 978-0-8126-9461-1 (trade paper : alk. paper)
  ISBN-10: 0-8126-9461-9 (trade paper : alk. paper) 1. Hermeneutics. 2.
Science--Philosophy. 3. Education--Philosophy. 4. Ethics. I. Shimony,
Abner. II. Title.
  BD241.E35 2006
  121'.68--dc22
                                                                    2006021990

Contents

v

PART TWO

PHILOSOPHY AND EDUCATION

Preface

Martin Eger was trained as a physicist and for almost the whole of his professional career served as a professor of physics at the City University of New York, College of Staten Island. He also became an accomplished philosopher, essentially by self-education, which consisted of assiduous reading, attendance at lectures, visits to departments of philosophy during sabbaticals (where he sought and found opportunities for intense discussion and dialogue), and sustained independent thinking.

Most of Martin's completed philosophical writings are collected in the present volume. Omitted are drafts of articles and variants of finished articles which were presented on different occasions. Almost all of the papers in this volume are divided into three groups—Hermeneutics and Natural Science, Philosophy and Education, and Morality; two works, however, Martin's "Letter to a Friend Regarding the Farm" and the "Dialogue" with the editor Abner Shimony, cut across the philosophical boundaries. Within each group the papers are arranged in roughly chronological order, but the editor has often departed from this order for the purposes of exhibiting continuity of ideas from one paper to another and of showing the deepening and enrichment of Martin's thought. The reader who is interested in one of the groups is urged to read the papers of that group sequentially in order to see the remarkable maturation of the thought of an independent, disciplined, and self-critical philosopher. The editor's Introduction is intended to help the reader follow the development of Martin's thinking. The Dialogue between Martin and the editor, written in the last six months of Martin's life, contains indications of directions in which his research would have proceeded had a longer life been granted to him.

A list of Martin's published and unpublished papers is given at the end of the volume. None of the purely scientific articles listed there is reprinted in the body of the volume, but the reader should be aware of the scope of these articles, because problems within physics were both stimuli and resources in his philosophical investigations.

The editor wishes to express his appreciation to the publishers of books and journals who gave permission to reprint Martin Eger's published papers.

Permission has been granted by Kluwer Academic Publishers to reprint the following essays by Martin Eger:

"Hermeneutics and Science Education: An Introduction," first printed in *Science and Education: Contributions from History, Philosophy and Sociology of Science and Mathematics* 1, No. 4. (1992), 337–348. ©1992 Kluwer Academic Publishers.

"Hermeneutics as an Approach to Science: Part I," first printed in *Science and Education: Contributions from History, Philosophy and Sociology of Science and Mathematics* 2, No. 1 (1993), 1–29. ©1993 Kluwer Academic Publishers.

"Hermeneutics as an Approach to Science: Part II," first printed in *Science and Education: Contributions from History, Philosophy and Sociology of Science and Mathematics* 2, No. 4 (1993), 303–328. ©1993 Kluwer Academic Publishers.

"Achievements of the Hermeneutic-Phenomenological Approach to Natural Science: A Comparison with Constructivist Sociology," first printed in *Man and World* 30 (1997), 343–367. ©1997 Kluwer Academic Publishers.

"Language and the Double Hermeneutic in Natural Science," first printed in *Hermeneutics and Science*, ed. M. Feher, O. Kiss, and L. Ropolyi (Boston Studies in the Philosophy of Science 206, Dordrecht: Kluwer, 1999), 265–280. ©1999 Kluwer Academic Publishers.

"Alternative Interpretations, History, and Experiment: Reply to Cushing, Crease, Bevilacqua, and Giannetto," first printed in *Science and Education: Contributions from History, Philosophy and Sociology of Science and Mathematics* 4, No. 2 (1995), 173–188. ©1995 Kluwer Academic Publishers.

"The 'Interests' of Science and the Problems of Education," first printed in *Synthese* 80 (1989), 81–106. ©1989 Kluwer Academic Publishers.

Permission has been granted by Blackwell Publishing to reprint the following essays by Martin Eger:

"A Tale of Two Controversies: Dissonance in the Theory and Practice of Rationality," first printed in *Zygon* 23, No. 3 (1988), 291–325. © 1988 Joint Publication Board of *Zygon*.

"Reply to Criticisms", first printed in *Zygon* 23, No. 3 (1988), 363–68. ©1988 Joint Publication Board of *Zygon*.

Permission has been granted by Blackwell Publishing and Dr. Mary Hesse to reprint "'Rationality' in Science and Morals," first printed in *Zygon* 23, No. 3 (1988), 327–332. ©1988 Joint Publication Board of *Zygon*.

Permission has been granted by Blackwell Publishing and Dr. Abner Shimony to reprint "On Martin Eger's 'A Tale of Two Controversies,'" first printed in *Zygon* 23, No. 3 (1988), 333–340. ©1988 Joint Publication Board of *Zygon*.

Permission has been granted by Blackwell Publishing and Dr. Thomas F. Green to reprint "A Tale of Two Controversies: Comment," first printed in *Zygon* 23, No. 3 (1988), 341–46. ©1988 Joint Publication Board of *Zygon*.

Permission has been granted by Blackwell Publishing to reprint "Positivists v. Positivism," first printed in *The Center Magazine* 6, No. 4 (1973), 61–66. ©1973 The Fund for the Republic, Inc.

Permission has been granted by the American Association of Physics Teachers to reprint the following papers:

"Physics and Philosophy: A Problem for Education Today," first printed in *American Journal of Physics* 40 (1972), 404–415. ©1972 American Association of Physics Teachers.

Review of V. Weisskopf's *Knowledge and Wonder,* first printed in *American Journal of Physics* 49 (1981), 605. ©1981 American Association of Physics Teachers.

Permission has been granted by the University of Georgia Press to reprint "Hermeneutics and the New Epic of Science," first printed in *The Literature of Science: Perspectives on Popular Scientific Writing*, ed. M.W. McRae (Athens GA and London UK, 1993), 186–209. ©1993 University of Georgia Press.

"The Price of Collaboration," first printed in the *Bulletin of the Atomic Scientists* (December 1978), 55–56, is reprinted by permission of the *Bulletin of the Atomic Scientists*, ©2005 *Bulletin of the Atomic Scientists*, 6042 South Kimbark, Chicago, Illinois 60637, USA.

Fig. 1 of "Meaning and Contexts in Physics: A Case Study" is reprinted by permission of John Wiley & Sons, Inc. from *The Elements and Structure of Physical Science*, 2nd ed., by J.A. Ripley and R.C. Whitten (New York, Wiley, 1969), 203. ©1969, John Wiley & Sons, Inc.

Three graduate students in the Philosophy Department of Boston University—Dimitri Constant, William Devlin, and Matthew Shaffer—were assiduous assistants in the preparation of a typescript and an electronic copy of this volume, with serious interest in its philosophical content.

Samantha Varner brought expert knowledge and skills to the final copy-editing and proofreading of the manuscript.

André Carus of Open Court Publishing Company encouraged the collection of Martin Eger's philosophical papers and offered valuable advice on the format of the volume.

Prof. Thomas Lickona of The Center for 4th and 5th Rs, SUNY Cortland, provided valuable information about moral education programs in the public schools of New York State.

Most of all I want to thank Martin Eger's widow, Judith Eger, who was an indispensable partner in editing this volume. She provided biographical information and insights that could not have been obtained elsewhere, she clarified obscurities in the texts, and she offered illuminating suggestions about ideas which Martin did not have time to develop. Without her enthusiasm and thoughtful supervision this volume would not have been completed.

Finally, I wish to express my indebtedness to Martin Eger, for his warm and generous friendship, the serious and enlightening philosophical conversations which we had during his lifetime, and the illuminating instruction which his papers gave me during the process of editing them.

ABNER SHIMONY

Introduction

by Abner Shimony

1. Some Biographical Information

From conversations with Martin, from his papers and letters, and from the biographical material compiled by Judith Eger, I have constructed a survey of Martin Eger's philosophical achievements. In spite of its deficiencies of details and possible errors, the survey should help the reader to link the papers collected in this volume and to understand Martin's transformation of a vast reading into an original and coherent philosophy. His philosophy is clearly based upon the hermeneutic tradition of Jürgen Habermas, Martin Heidegger, and Hans-Georg Gadamer and followers like Patrick Heelan, Mary Hesse, and Don Ihde, but is modified (and in my opinion enriched and corrected) by his careful analyses, his professional expertise in physics, his observation of and reading about social phenomena, and his dedication to objective truth in scientific research and objective justice in practice.

It is beyond my knowledge and psychological insight to exhibit sources of Martin Eger's mature thought in the painful events of his early life. He was born in Lvov, Poland in 1936 into a comfortable bourgeois Jewish family. His father was arrested and killed by the Nazis after their conquest of Poland, but he and his mother escaped by traveling to a part of the country where they were not known, pretending to be Polish peasants and surviving by her work as a maid and cook for a German officer's wife. The reader must infer as well as possible the intellectual consequences of this harrowing experience, taking into account a remarkable statement by Judith Eger: *"In all the grief and constant danger of these years, Martin was*

The italicized passages have been taken from unpublished biographical material about Martin Eger prepared by Martin's widow, Judith Eger.

able, both at the time and in retrospect, to rescue and retain those aspects of life that were good. That meant, in particular, the land."

When Martin came to New York at age ten he was placed in a public school with no knowledge of English and no instruction in the language, but learned it in the street. He did well in school, though it was uninspiring, and only learned that books could be interesting when he worked at the Wenner-Gren Foundation (of anthropology) and borrowed books from its library. Expecting to become an engineer he entered Brooklyn Polytech for one year, where his greatest intellectual stimulus came from excellent philosophical discussions led by the very well-read president of his fraternity; this was his introduction to the fascination and discipline of dialogue. Martin transferred after one year to M.I.T., where he became a physics major and the beneficiary of profound and illuminating classes by great physicists, among them Victor Weisskopf and Laszlo Tisza. Graduate school in physics at Brandeis University emphasized problem-solving, but his interest in conceptual questions and the philosophical side of physics had already been aroused by classes at MIT and textbooks like Max Born's *Atomic Physics* and David Bohm's *Quantum Theory.*

Martin's interests developed further during his postdoctoral visit to Lawrence Livermore Laboratory, 1965-1967. Although Martin appreciated the enthusiasm and technical skill of the staff at the Laboratory he was not satisfied with the work from a human/dialogical point of view. He was, however, sympathetic with the concerns of Berkeley students *"about the nature of their education expressed by the original protesters, especially their perception of the university as a 'machine' and the belief that the current modes of teaching and learning devalued the human being. . . . [H]e thought that the times might yet be conducive to some genuine reform or revitalization of university education. . . . It was against this background that he made the decision to accept an appointment (in physics) at the new Richmond College of the City University of New York. . . . The practical realities of the college (the initial political/cultural clashes between segments of the faculty, and between the campus and the surrounding community; the sixties-heightened and ongoing human and organizational challenges; and above all, the interaction with his varied student constituencies . . .) turned out to form not only a backdrop to but a contributing element of his philosophical thinking. During virtually his whole teaching career, spent at this single but changing institution, he was interacting simultaneously or in close succession with groups of students who stood in differing relationships to the corpus and enterprise of science. Thus, they challenged and nourished his growing insight into the related philosophic questions. And in this sometimes roiling organizational setting, he himself had to make, and could see others making, choices based on a self-understanding that depends in great part on one's cognitive/moral framework."*

Until 1970 Martin's publications were all technical articles in theoretical physics, concerned with many-body systems and mainly written in collaboration with his Ph.D. adviser Eugene Gross. Thereafter his publications were primarily philosophical, historical, and pedagogical. Martin's paper "Physics and Philosophy: A Problem for Education Today" (Eger 1972, this volume, 167–185) argues that college students were more seriously interested in the philosophical aspects of science than in earlier years, discusses the achievements of a year-long course in physics which he offered at Richmond College (later called the College of Staten Island) focusing on the 'process of science,' and gives a bibliography of the wide-ranging readings on the history and philosophy of science distributed to students along with specially prepared notes on the technical aspects of physics.

Noteworthy in this bibliography are works by Thomas Kuhn, Adolph Baker, Herbert Marcuse, Lewis Mumford, and Hannah Arendt that express skepticism about claims that science, and physics in particular, can achieve good approximations to the objective truth about the constitution of the universe, and argue against the 'relevance' of science to basic human needs and problems. The inclusion of these works shows attention to and respect for questions raised by many of the students, but the students' concerns are sharpened and tightened.

2. Hermeneutic Philosophy and Natural Science

The summer after submitting "Physics and Philosophy," Martin spent in the Max Planck Institute in Starnberg (near Munich) where he studied and discussed intensely the work of contemporary German philosophers, including Heidegger and the hermeneuticist Habermas. In Martin's later work the version of hermeneutics which he followed most was that of Gadamer, but with modifications suggested by Hesse, Heelan, Ihde, and Richard Bernstein and also important innovations of his own. *"I note the fact that he did not become an enthusiast or disciple; he thought that some of the categories, insights, modes of thought, and emphases of the hermeneuticists were valuable, especially as correctives to various excesses and omissions in the self-understanding, teaching, and social (normative) role of the natural sciences; but he certainly was also alert to the comparable (or worse) excesses that could be perpetrated in the opposite, anti-objectivist direction. As always, he was a cognitive/personal counterweight not only to one excess or the other but to the bit-between-the-teeth mode itself."*

Hermeneutics began in the nineteenth century as a technique for understanding texts in ancient languages. The meaning of a difficult passage was conjectured in the light of a preconception of the text as a whole, but the resulting translation of the passage could induce a correction of this preconception, which in turn would necessitate a revision of the translation of the passage. Thus the scholar was caught in a "hermeneutic circle," for which the remedy was successive corrections in the hope of eventual convergence of the local and the global interpretations. The process tacitly and optimistically assumed that there was an objective meaning to be extracted by the process of successive corrections. Philosophical hermeneutics generalized this philological procedure by applying it to any domain in which the scholar—classicist or historian or anthropologist—studies an alien or temporally remote culture, and, more importantly, it radicalized the procedure by doubting that initially erroneous preconceptions could be overcome by successive corrections. Hence, "meaning *emerges* from the interpretation. Interpretations become, at least in part, *constructive*, not *reconstructive*" ("Hermeneutics as an Approach to Science: Part I", Eger 1993a, this volume, 10). And the preconceptions are embedded more deeply in the language and culture of the interpreter than the philological hermeneuticists had realized—buried in practices of the life-world, in the concepts and perspectives of life, and in language (ibid., this volume, 7–14). Heidegger's and Gadamer's version of the hermeneutic circle is devastating for the ideal of objectivity not only in interpreting written texts but elsewhere in human activities where traditionally objectivity has been prized and sought. Martin writes, "By relating knowledge first to the pre-scientific world of practice, to the depths of human existence itself, then to the inherited language, philosophical hermeneutics closed off for itself any hope of objectivity in the absolute sense. . . . No longer may preconceptions be regarded as something merely negative, since in effect they are *part* of the 'seeing' (interpreting) apparatus without which nothing at all can be seen" (ibid., this volume, 10).

Those philosophers who look for objectivity in the natural sciences might take comfort in the distinction that Habermas made in his early work between the natural sciences which are "empirical-analytical," dealing with entities which do not have a spoken or written language that needs interpretation and furthermore aim not at understanding but at prediction and technical control, and the human sciences, which are "historical-hermeneutic" because they deal with agents who use language (cf., *Knowledge and Human Interests*, translated by Jeremy Shapiro [Boston: Beacon Press, 1971; German publication date: 1968]). But this comfort evaporated under the critiques of scientific investigation by Kuhn and Michael Polanyi, who insisted upon the essential role of "tacit knowledge" that is presupposed in all measuring, classifying, and reasoning in the natural sciences. They "elevated the role of such preconceptions to a level of importance in science comparable to that given them by Heidegger in hermeneutics. On the one hand we cannot completely rise above such bias;

on the other, because of its *positive* role, it is now clear that we cannot wish to do so" (Eger 1993a, this volume, 12).

Martin now finds another—surprising, and to my mind quite puzzling—consolation in Gadamer's writing for the loss of objectivism in the natural as well as the human sciences: an 'ontological turn,' whereby "hermeneutics moves 'beyond' the subject/object dichotomy" (Eger 1993a, this volume, 14). Martin cites, with apparent approbation, Gadamer's concept of the 'horizon':

> ". . . the range of vision that includes everything that can be seen from a particular vantage point" (Gadamer 1975, 269). . . . But—and here we really see the two-sidedness of this theory—the text we are trying to interpret also has its horizon—a limit to all those meanings to which a text of this sort, employing a language of this sort, possibly could give rise—which includes, of course, those specific possibilities I have called the 'cloud' hovering about it. . . . All this . . . is a description of dialogue, or *being in* dialogue. . . . It is this ontology of dialogue that saves hermeneutics from relativism and subjectivism. (Eger 1993a, this volume, 17, 18)

Martin's presentation of Gadamer's hermeneutics with an ontological turn is a series of metaphors which I find difficult to translate into literal discourse, but Martin comes close to achieving this desideratum by providing case histories of scientific investigation: the biologist Barbara McClintock, the physicist and astronomer George Darwin, the cosmologist Fred Hoyle, and the various investigators of the neutrinos generated in the interior of the sun. Of these only McClintock testified to a transformation of her mentality during research that suggests to my ears what Gadamer calls a "fusing of horizons":

> "I found that the more I worked with them [chromosomes] the bigger and bigger they got, and when I was really working with them, *I wasn't outside, I was down there. I was part of the system.* . . . I even was able to see the internal parts of the chromosomes." (Eger 1993a, this volume, 20, quoting from Keller [emphasis Eger's])

There is surely something mystical in this introspective report, but even a reader with a mundane frame of mind cannot help being impressed by it to the point of granting that such ecstasy could be heuristically valuable in the process of discovery. But such a reader would also wish to quote the conclusion of the passage: "And you forget yourself. The main thing is that you forget yourself" (ibid., this volume, 20). One could argue that scientific objectivity is enhanced when the ego is prevented from screening the objective fact from the investigating subject. Martin describes her state of mind as "listening" (ibid., this volume, 22).

George Darwin was obsessed with finding the mechanisms for the persistence of certain physical and biological structures and the susceptibility of others to variation, thereby attempting to fill in crucial missing details in his father's theory of the origin of species. He also took seriously Lord Kelvin's physical arguments that the age of the earth is much too short to accommodate the process of biological evolution. And Darwin was a serious and accomplished master of fluid dynamics, not satisfied with the impressionistic and hand-waving explanations (like those of Herbert Spencer current in the late nineteenth century), and undertook with pencil and paper extraordinarily demanding calculations on the instabilities and bifurcations of rotating viscous bodies of fluid. Although Darwin did not succeed in his grand project of establishing the common ground of the origins of physical and biological species, his dedicated work helped to establish the modern theories of fluid dynamics and their application to astrophysics ("Meaning and Contexts in Physics: A Case Study," Eger 1992a, this volume, 101–114).

Hoyle was one of the founders of the "steady-state" cosmology, for many years the main rival of the "big bang theory of creation." He was often accused of self-interest that made him impervious to evidence and selectively attentive to data. Martin presents two important steps in Hoyle's scientific career which contradict this accusation, and incidentally reveals Martin's understanding of the "ontological transformation" that according to Habermas and Gadamer emerges during a struggle with the hermeneutic circle ("Rationality and Objectivity in a Historical Approach," Eger 1989a, this volume, 121–131). First, in order to show that the production of carbon and heavier elements could occur in the interior of stars, without the supposition of a yet hotter origin (as the big bang would be), Hoyle postulated the existence of a 7.68 MeV resonance of the carbon nucleus which would speed the process of carbon creation. When nuclear experimenters searched for this resonance they found it at the energy that Hoyle had postulated on the basis of astronomical observations. Secondly, Hoyle later performed a calculation showing that all the energy emitted by fusing hydrogen into helium in all galaxies since their formation would be too small by a factor of ten to account for the abundance of helium in the observable universe. Since this calculation ran counter to the requirements of the steady-state theory it provided evidence that the self-interest of defending his own theory did not control his scientific labor. Without any mysticism whatever Hoyle's behavior exemplifies McClintock's dictum, "You forget yourself."

The most striking of Martin's case histories is that of solar neutrinos. The synthesis of heavier elements from lighter ones in the sun was expected by standard elementary particle theory to produce a strong flux of neutrinos as a by-product. J. Bahcall convinced an experimentalist who specialized in neutrino detection to measure the detection rate of solar neutrinos in a large chlorine tank deep in a mine. The rate was much lower than Bahcall had calculated, giving rise to a discrepancy between theory and experiment. This state of affairs was publicized in a book by two British sociologists of science, H. M. Collins and T. Pinch (*The Golem* [Cambridge: Cambridge University Press, 1993]), who concluded that "science is politics," with consensus reached by "negotiation"—science "un-made" and replaced by "construction." Although there are similarities between this conclusion and the criticism of objectivism by hermeneutic philosophers, Martin argues that there is a radical difference between the spirits of the two critiques. The later stage of the story of the solar neutrinos contradicts the cynical account of the constructivists. According to them the "experimenter's regress" can be broken only by nonscientific tactics, which they sometimes call "antics" ("Achievements of the Hermeneutic-Phenomenological Approach to Natural Science," Eger 1997, this volume, 53–71), whereas hermeneutics recognizes "the long-term change in our self-understanding as humans" (ibid., this volume, 66).

The long-term change consists partly of a radical change of physical theory—namely, that neutrinos are not massless, that they come in three varieties (one associated with electrons, one with muons, and one with tau particles), and that the interactions of neutrinos with a conjectured but as yet unobserved field causes transformations from one variety to another of neutrinos, the so-called "neutrino oscillations" (hence a lowered rate of detection of the solar neutrinos associated with electrons)—and partly of an increased open-mindedness and humility of investigators in the face of experimental anomalies.

What do these case histories reveal about Martin's apparent acquiescence in Gadamer's radical ontologization of the "hermeneutic circle"? The case of Barbara McClintock, with her ecstatic reports of entering into the plants which she studied, comes close to a literal acceptance of Gadamer's metaphorical language. But others of Martin's comments are closer to translations of metaphors into prose, though elevated prose: that one must "listen" to what the material is saying (Eger 1993a, this volume, 22); that "it is this ontology of dialogue that saves hermeneutics from relativism and subjectivism" (ibid., this volume, 18); and that serious dialogue genuinely transforms its participants—"It suggests an intensity of participation that evokes from the other 'partner' *assertive* meaning" (ibid., this volume, 18). These comments of Martin's apply even to the case of McClintock, but more clearly to the other three case histories just summarized. I shall strengthen this conclusion by saying that Martin has substantially modified hermeneutic philosophy, however much it has inspired him, by playing down the obscurities in its *ontological* terminology and replacing it by a more straightforward and comprehensible terminology of *dialogue*. In the case of scientific investigations the interlocutors of the dialogue clearly include other scientists with differing hypotheses and conflicting experimental data. But the individual scientist is also involved in dialogue with himself or herself because of evolving ideas. And Martin's quotations about "listening" to the systems under investigation suggests that these systems too are involved in the dialogue (though, admittedly, saying this risks a reversion to metaphor). Most important, there is a moral element in genuine dialogue. Martin contrasts scientific dialogue with the instrumental and adversarial rationality that is used in politics and law cases:

> [I]nvariably the sort of reason-giving and treatment of evidence in these practices excludes precisely the element of self-limiting inquiry in opposition to the cherished commitment. . . . As is the case with morals, whether religiously-based or not, self-limitation, the distinctive mark, points to that transcendent element which, in both realms, makes for something quasi-religious: a corrective, even a prophetic note, in our modern culture of self-expanding institutions and self-assertive individuals. (Eger 1989a, this volume, 130)

There is a deep and dark historical reason for applauding his transformation of the 'ontological turn' in hermeneutic philosophy, which I have not yet mentioned but deserves attention. One of the founders of modern hermeneutic philosophy was Heidegger, whose terminology is notoriously obscure and whose political sentiments—disguised or misled or somehow twisted by the very obscurity of his philosophy—led him into active collaboration with the Nazi regime. The dialogue that Martin insists upon in his restatement of the hermeneuticists' ontology requires an understanding of the point of view of others, a self-limitation, a truthfulness, and a morality, all at the far extreme from the evil that Heidegger abetted.

In general, Martin's philosophical analyses became more original, complex, and nuanced with the passage of time. A reader of this collection who has not time to read it in its entirety is therefore encouraged to study the later papers on any topic of special interest.

"Academic Dialogues and the Post-Modern Mood" (Eger 1995b, this volume, 161–64) is not a published article or talk to a learned society, but is rather a memo to Martin's colleagues. It is the consequence of disagreements at the College of Staten Island regarding educational policies, and one can discern in it some of the practical issues in dispute. But the memo itself concerns the 'meta' question of how disputes can be conducted fruitfully. In the memo one sees Martin applying the philosophical ideas that he has crystallized from his epistemological studies, in particular his conviction resulting from reflection upon the literature of hermeneutics that serious dialogue is the optimum instrument of rationality available to human beings. The memo is an eloquent plea to his colleagues for engagement in constructive dialogue. Near the conclusion of the memo is a passage which ought to be heeded not only in academic disagreements but in the increasingly rancorous political disputes of our nation and the world:

> The very notions tending to make genuine dialogue increasingly difficult are now taught, formally and informally, in colleges throughout the country. The only thing I can suggest by way of antidote is increased consciousness of precisely this situation. For if we are aware of the pressures, then those of us so inclined can make the needed corrections. Liberals who think that most conservative arguments are 'garbage' might try to find the best of the conservative thinkers (e.g., Sowell, Gadamer) to keep themselves from believing all true insights lie on the left. Conservatives might likewise seek out those on the left whom they can respect (Charles Taylor, Richard Bernstein, etc.). This will seldom change our commitments but it can substitute a genuine hermeneutic consciousness for the bogus kind of which we hear so much. (Eger 1995, this volume, 164)

3. Hermeneutics and Science Education

In the early stages of hermeneutic philosophy, the natural sciences were not subjected to hermeneutic analysis because their subject-matter was nature itself, whereas historical-hermeneutic analysis was applied to the human sciences, whose subject-matter was users of language. As Martin pointed out, however, later hermeneuticists broke this dichotomy, by emphasizing the preconceptions of ordinary life and ordinary language from which investigators of nature can never completely escape but which can be understood to some extent by analysis similar to that applied in the human sciences. In his articles on science education, Martin proposes that there is an important and more straightforward reason for recognizing continuity between the human and the natural sciences: that the study of science by students and by a non-professional general public inevitably confronts a language, namely the language that professionals in the natural sciences use in textbooks, lectures, and popularizations to convey their discoveries and theories. By turning to the educational situation,

> . . . we avoid for the time being the difficult philosophical question of whether the activity of the researcher in natural science is itself hermeneutical. Whatever the answer to that may be, regarding the position of the student (including students of the natural sciences) we are on much more solid ground: Whenever a strange language is encountered, does it not need interpreting? Whenever there is interpretation, does it not entail hermeneutics? . . . [W]hat the human being faces is not really the phenomena of nature themselves, but various forms of written and spoken *text*. ("Hermeneutics and Science Education: An Introduction," Eger 1992b, this volume, 235)

As a prelude to counseling about the study of science Martin points out that there are roughly three different kinds of interests that students may have in natural science: "technical, methodological, cosmological" ("The 'Interests' of Science and the Problems of Education," Eger 1989b, this volume, 211–230). The educational system today is focused almost entirely on the first two, and he grants that this focus does no harm for those students who aim at technological professions. But the belief that a methodological focus will fascinate and serve students with non-technological professional aims, using the rationalization that the reasoning and the puzzle-solving techniques of science are applicable to the problems arising in non-scientific activities, is erroneous, in large part for the reason that was noticed above—that the reasoning in such endeavors as law, politics, and business is adversarial, and that of the natural sciences is not. Martin argues that the interest that in fact does, and that can motivate a large part of the student body which does not have technological professional aims is the cosmological interest: the implications of the natural sciences for *Weltanschauung*, overlapping with and perhaps in competition with myth and religion. One kind of evidence for this assertion is the type of question which students ask—when they are permitted and encouraged to do so—about the "relevance" of science to their lives (Eger 1972, this volume, 167–185). Further evidence is the immense popularity of high quality, difficult books, different from both professional texts and from popular science, which make serious presentations of bodies of scientific material along with serious discussions of the wider significance of this material—books like Jacques Monod's *Chance and Necessity,* Ilya Prigogine's and Isabelle Stengers's *Order Out of Chaos,* Douglas Hofstadter's *Gödel, Escher, Bach,* and E. O. Wilson's *On Human Nature* (Eger 1989b, this volume, 214). Martin concludes that this literature provides a cultural bridge among various sciences and also between professional science and interested non-scientists.

Martin's conclusion for teaching the natural sciences is to emphasize programs that serve the cosmological interest of students, which is pervasively present. And he gives outlines of courses designed to do so, which have been tried out at the College of Staten Island. (See here Eger 1972 [this volume, 167–185], and Eger 1987 [this volume, 195–210].) There is much historical material in these courses, for instance following the sequence of mechanical theories from the scholastics to Galileo to Newton to Einstein, with similar sequences in biology. An abundance of scientific data is presented to the students in order to show what is the evidence that the rival theories attempt to explain. The confrontations and comparisons of the various theories are presented as dialogues, in the spirit of a hermeneutic analysis that attempts to make the interlocutors cognizant of their preconceptions. And the relevance of these dialogues to philosophical questions like the place of man in the universe—questions raised over the centuries as well as by contemporary students—is made explicit.

This proposal of a kind of science course that could satisfy the cosmological interest of students has a profound implication for the training of teachers. Instead of designing curricula that are intended to be 'teacher-proof,' which is all too common in state educational systems, Martin urges a "new relation between 'the teacher and what is taught.' I propose a shift of focus: from the teacher as a means of transmitting prepackaged curricula, as a trainer of puzzle solvers for the science-system, to the teacher as carrying out primarily the hermeneutic-communicative action of science. In so doing, he or she does not neglect the other interests, but puts them in perspective. Above all, the teacher must *embody* that perspective" (Eger 1989b, this volume, 211–230). I cannot resist at this point an expression of great admiration: Martin's achievement in drawing suggestions for teacher training from his reflections on hermeneutic philosophy is an inspiring model of the utility of philosophy!

4. Moral Education

Martin's essay "The Conflict in Moral Education: An Informal Case Study" (Eger 1981b, this volume, 317–334) has many fascinating aspects. It is a philosophical reflection concerning the grounds of moral judgment (in spite of strenuous efforts on the part of many of the central personages in the story to avoid being drawn into controversy on this fundamental question). It is a meta-ethical study of *reasons* for moral choices, which modulates in the hands of some participants in the story into a surrogate for a system of morality. It is, above all, a deep study of the possibility and the desiderata of moral education. It is also a vivid sociological study of a largely rural but partially professional community that is subjected to a particular educational crisis. This occurred when the local school board introduced into the middle school curriculum a technique called "Values Clarification" in compliance with a list of goals and responsibilities mandated by the New York State Department of Education. Martin's analysis of this crisis also is a general sociological study of the way in which the structure of a society depends upon shared values. And finally, by quotations from the personages involved, it is a revelation of psychological conflicts within families, between families and the school system, and between teachers and educational theorists. Martin dramatically and painfully says that "To disenchanted parents . . . the relation between mental growth and this type of values program . . . deepens their despair over the intellectual quicksands into which the schools have now sunk, where each step taken to solve a problem seems to make things worse" (ibid., this volume, 324).

No summary can do justice to the richness of Martin's case study. It may be useful to the reader, however, to compare the "Values Clarification" technique taught in the local middle school (which Martin strongly criticizes) with the superficially similar dialogical aspect of hermeneutic philosophy, to which he is strongly attached.

The designers of the Values Clarification technique claim that it satisfies desiderata that prima facie pull in opposing directions. It recognizes the importance of values in human life and therefore is an antidote to being silent and negligent about morality. But it also recognizes diversity in our culture and the obligation of a public school to respect diversity. Hence the technique should not be biased in favor of one system of values, e.g., those of a particular religion or of a particular secular viewpoint. But abstention from bias, carried to an extreme, runs the danger of propagating ethical relativism or subjectivism—that one system is as good as any other, or that the choice of values is an individual's personal option—and these nihilistic consequences must somehow be avoided. The reconciliation of these different desiderata is achieved, according to the technique of Values Clarification, by focusing upon the *process* of valuing. The concern of the Values Clarification technique, and of the sciences of genetic and humanistic psychology and of decision theory which underlie it, is "with the 'how' of valuing, with the structure of decisions rather than their outcome. One uses 'is' to derive not the 'ought' of morals themselves, but another 'ought': the manner of thinking about morals, and the way of evaluating that manner" (ibid., this volume, 318–19).

This rationalization of the Values Clarification technique is reminiscent of philosophical hermeneutics in several respects. Philosophical hermeneutics is also concerned with intellectual process, and it relinquishes objectivism, recognizing that the successive corrections of the hermeneutic circle may not converge to a unique point. Both activities aim at interpretations of the subject's intellectual starting points and at increase of self-understanding, and both claim some degree of success in achieving these aims. Both are dialogical, hermeneutics envisaging an "ontology of dialogue" (Eger 1993a., this volume, 18), and Values Clarification proceeding by confronting the student with questionnaires about personal preferences and staging dialogues between teacher and student concerning the answers submitted by the latter.

The similarities just listed are coarse, however. There are great differences in the details between philosophical hermeneutics and Values Clarification—differences which account for Martin's attraction in one case and revulsion in the other. First of all, there is an immense difference between the personnel for whom the two programs are intended. Philosophical hermeneutics is intended for mature thoughtful people, professional philosophers or adults deeply concerned with problems of epistemology, ontology, and self-knowledge, and acquainted with the literature on these subjects; by contrast, the Values Clarification program is imposed by the decision of a school board on children in middle school, around thirteen years old, who have not requested a study of values and in many cases are bored by the subject. There is a fundamental idea of philosophical hermeneutics that makes the level of maturity particularly important: namely, the insistence by the hermeneuticists upon the cultural and linguistic preconceptions that constitute the background of any scientific or humanistic investigations. In the case of the school children who are supposed to engage in the dialogues of Values Clarification, the background of preconceptions is incomplete and confused, constituted by family indoctrination, religious training, television and other entertainment, and the pressure of contemporaries. One could charitably say that both groups generically share a need for self-understanding, but the contents of the selves that are to be understood are radically different in character, and the intellectual tools that the middle school children possess for the enterprise of examination are much weaker than those of professional and amateur philosophers engaged in hermeneutic analysis. There are good reasons for Plato's recommendations (*Republic*, Book III and Book VII) of a proper sequence of studies, starting with gymnastics and music for the very young, to instill strength and harmony in the body and the soul, then proceeding to mathematics between the ages of twenty and thirty, and finally (for those properly endowed) intensively studying Dialectic between ages thirty and thirty-five with emphasis on the principles of morality. Socrates warns against premature exposure to Dialectic:

> You must have seen how youngsters, when they get their first taste of it, treat argument as a form of sport solely for purposes of contradiction. When someone has proved them wrong, they copy his methods to confute others, delighting like puppies in tugging and tearing at anyone who comes near them; . . . and the result is that in the eyes of the world they discredit, not themselves only, but the whole business of philosophy. (*Republic*, Book VII, 539, Cornford's translation)

The great difference in maturity of practitioners of philosophical hermeneutics and middle school students implies very different demands upon the teachers of the two groups. The teachers of the middle school children who lead the classes in Values Clarification ought to be prepared to compensate for the ignorance, inexperience, and confusion of their students. Martin strongly criticizes the program of Values Clarification on this matter:

> Can the average teacher with the average training be a "teacher"—or "facilitator"—in such deep waters without comprehensive additional education in ethics, moral history, theory of knowledge, and the like? . . . Values

Clarification, which recommends itself as "easy to get started," is especially culpable in this respect; to become a practitioner, the novice is assured, "a relatively short training period, as short as a few hours, will suffice." (Eger 1981b,
this volume, 330)

In the four case histories of scientific discoveries presented by Martin
of the ontological transformation achieved in philosophical hermeneutics,
one feature common to all of them is self-limitation ("You forget yourself"
in McClintock's words). But the technique of Values Clarification works
against this purpose in two respects. First, the reasons which the students
are led to proposing as justifications for taking stands are characteristically
related to their hedonistic preferences in the present or else to considerations of self-realization in the future. Second, the entire process of clarification questions the derivation of moral rules from authorities, notably parents; but surely for most children the internalization of concern for others
than one's self occurs in the family under the supervision of parents. Hence
there is a danger that the technique of Values Clarification will conflict with
philosophical hermeneutics by increasing rather than limiting self-centered-
ness.

The complexity of the problem of teaching morality is shown by
Martin's paper "A Tale of Two Controversies: Dissonance in the Theory and
Practice of Rationality" (Eger 1988b, this volume, 335–362) which had
appeared in *Zygon*, followed by commentaries by Mary Hesse, Abner
Shimony, Thomas F. Green, Holmes Rolston III, and Daniel R. DeNicola
(the first three of which are reprinted in this volume, 363–381), then followed by Martin's "Reply to Criticisms" (Eger 1988d, this volume,
383–88).

The reader may be interested to know that the New York State
Department of Education did not continue to recommend the Values
Clarification program to local school boards as strongly as during the period studied by Martin. The Department continued to require some type of
teaching of morality, particularly the responsibilities of citizens in a democratic society; and it continued to be concerned to abstain from teaching a
preferred system of ethics that would be disturbing in a culturally diverse
society, and nevertheless not to lapse into ethical relativism and subjectivism. A favored program from the 1980s onward has been Character
Education. The shift of emphasis was summarized by James Leming: "By
the middle of the 1980s, values clarification, due to the shifting political climate in the nation and devastating scholarly critiques, disappeared from
schools. . . . During the past two decades we find that the idea of foster-
ing character in children, rather than morals or values, has once again
become an important focus in schools."[1]

5. The Egers's Farm

This Introduction would be incomplete without some information about the farm in upstate New York, where the Egers moved in 1979. Although Martin maintained a pied-a-terre near the College of Staten Island, it was their only home, their principal physical and spiritual recreation, and a uniquely idyllic and harmonious locus for Martin's reflections. The farm is 140 acres in extent, the entire working farm of its previous owner, and it remains a working farm, not just metaphorically as the locus for serious intellectual labor, but in the ordinary agricultural sense. Neighbors who raise beef cattle have leased fields where they plant, mow, and bale forage crops. But more important, the farm has been transformed to Martin and Judith's aesthetic ideals by their sensitivity to the environment, careful planning, landscaping experience and skill, and strenuous physical labor.

Although I had more horticultural knowledge and practical experience, Martin very much took the lead in learning, rescuing, and developing the land. The first thing we did was to explore and chart it, no small job: We kept a small notebook which we carried with us into the fields, woods, and old orchards, making maps and locating the fruit trees and berry thickets that we discovered. We did a number of things quite soon, as we became oriented: We began making trails so that we could get around and visit the deeper places; we started to transplant waist- or shoulder-high spruces and hemlocks to make a belt of trees along the road from our house toward our neighbors, increasing our privacy; and we devoted a lot of labor to the rescue of an old orchard, in which the great, gnarled apple trees—some of their trunks lying along the ground but still blooming and bearing—were being overgrown by other trees of all kinds. We decided that we could not hope to prune or otherwise help the apple trees directly; but we could and would free them from encroachment. It was a tremendous effort with a marvelous result: In spring the forest floor under the apple-trees is covered with violets and filtering petals of the apple-blossoms, in summer with blackberries, in fall with aromatic, fermenting apples, and in winter with frozen apples, preserved and available for the deer and the rabbits and the raccoons. . . .

I have told you about making the big pond—the shallow-water wildlife area, as the USDA calls it. It was certainly an instance of the coming-together of the cognitive, the moral, and the aesthetic: We learned as much as we could about pond and wetland ecology in order to participate optimally in the attempt to reverse the decline of the frog population and address other problems facing wildlife; to increase the beauty and interest of our land for ourselves; and to provide more sustenance and gladness-of-heart to the creatures that live here with us. And we saw many wonderful things there: the frogs that appeared overnight in the test-holes; the fox that caught and ate an unfortunate animal along the sloping bank and then came down and washed his paws and face in the water; great blue herons that we learned to recognize as individuals. . . .

Oh, Abner, I could go on. Martin was so respectful of the land, and so unsentimental. . . . He once looked out from the breakfast table and beyond the small pond to the field where our neighbors were baling and taking away the hay. He watched for a while and then he said, "Persons of intellect have prints on their walls of van Gogh's painting of a hay wagon. A Japanese billionaire has the actual van Gogh painting of a hay wagon. But we—we have the hay wagon!"

In the "Letter to a Friend Regarding the Farm" (Eger 1994b, this volume, 405–409), Martin describes the physical appearance of the farm and his and Judith's activities in caring for it and being nurtured by it. One sees from this letter that the farm not only gave Martin the serenity and harmony which he needed for working out his philosophical ideas but also provided some of the content of his ideas—for instance, his interpretation of Gadamer's cryptic metaphor of "the fusion of horizons." (See Eger 1993a, this volume, 3–27.)

And that is where we're sitting now, this Saturday, looking out over the water at the four geese by the cattail reeds; they've been with us from the start, two pairs of Africans, with black beaks and knobs on their heads, now all lined up in a row—still, perfectly reflected in the pond's mirror surface. We're looking out on a scene that's not alien. It is not "a world I never made." Of course, neither is it a world "I made." I intervene to dig large holes. Then, a whole sequence of processes takes over to change the shore line, the pond floor, and the life in the water. I act—nature reacts. Other times, nature acts and I react. And the scene outside is the outcome of this dialogue. It is not quite "nature" and not quite "man-made." It is something between nature and us—shared with nature, and among us. (Eger 1994b, this volume, 409)

6. The Concluding Dialogue

Following Martin's papers in this volume is a Dialogue between Martin and Abner. To some extent it is a continuation of their exchange in *Zygon* (1988c, d, this volume 363–388), but it is much more wide-ranging—discussing scientific methodology, the history of quantum mechanics, and related topics of interest to philosophically inclined physicists, but also wandering onto remote topics like the current state of the medical profession (illustrated by Martin's and Judith's experiences) and the prospects for peaceful coexistence of Israel and the Arabs.

The Dialogue was not planned. At the end of October 2001 Abner visited the Egers' farm in upstate New York, and for two days had strenuous discussions with Martin. These were so interesting that both decided to write them out while memory of them remained fresh. Thereafter there were four more exchanges in the Dialogue by email, the last being Abner's communication on March 6, 2002. Martin wished to continue, but was prevented by fatigue due to leukemia, from which he died on March 24, 2002. Martin's decision to continue the Dialogue during the last four months of his life entailed a sacrifice. Judith wrote, *"he was inclined to do it, knowing . . . that if he chose the Dialogue, he would be giving up the possibility of finishing the 'realism/instrumentalism' book."*

Both Martin and Abner argued strenuously throughout the Dialogue, because both were serious about their respective convictions and wished to convince the other. The clarity and vigor of Martin's argumentation were triumphs of character, for he was suffering from pain and fatigue during the entire time of the Dialogue. Also noteworthy was the fact that, despite the differences between their philosophical starting points and between the bodies of literature that influenced them, they came increasingly close to agreement as the Dialogue proceeded. This outcome was in part due to their deep friendship, but it also was a vindication of Martin's commitment to the idea that dialogue, of which listening and open-mindedness are essential components, is a good vehicle for approaching the truth.

PART ONE

Hermeneutics and Natural Science

[1]

Hermeneutics as an Approach to Science: Part I

A broadly conceived attempt is initiated here to extend hermeneutic philosophy to the understanding of natural science, with a special interest in the significance of this for education. After a brief discussion of several current problems to be addressed, a general outline is given of the evolution of hermeneutics in comparison with the evolution of the philosophy of science. The main thrust of this paper, however, is to show how the 'ontological' version of hermeneutics, developed in various ways by Hans-Georg Gadamer and others, describes rather closely features of scientific activity neglected by most philosophies of science and of education. To this end, the paper focuses especially on the biologist Barbara McClintock. [—A.S.]

Introduction

From the works of Jürgen Habermas, Hans-Georg Gadamer, Mary Hesse and others, and from the lucid analyses of Richard Bernstein, a new awareness of the pervasive role of hermeneutics has emerged, affecting radically a whole cluster of disciplines.[1] The shift in focus from *determin*ation to *interpret*ation can have such an impact because today this shift bears directly on key issues under debate. By questioning the putative neutrality of the reading of a text, hermeneutics in effect supports constructivist theories of knowledge and communication. By criticizing overdependence on formal methods, it highlights the linguistic component of the sciences—their shared, social element. By emphasizing history, tradition, and cultural perspectives, and by placing meaning at the center of inquiry, it aims to shore up in all scholarly activity an apparently declining human dimension. We can understand, therefore, that in educational practice too a hermeneutic approach has been advocated—to oppose "decontextualization,"* for

---

*The reader is reminded that double quotes are used either to indicate metalinguistic usage or to indicate a usage introduced by a known author. This paper appeared in *Science and Education: Contributions from History, Philosophy and Sociology of Science and Mathematics* 2, No. 1 (1993), 1–29.

example (Young 1990), and to foster critical thinking (Hostetler 1991).

Yet for the most part, whether in research or in teaching, a conscious-ly hermeneutic orientation is recommended only for the social or human sciences. Gadamer and Habermas make this restriction explicit.[2] Others, especially those dealing with education, imply much the same thing, or, by speaking very generally, merge their hermeneutic views with the broad stream of humanistically based educational critique.

It is appropriate to ask, therefore, to what extent hermeneutics is rel-evant also to *natural* science and its study. So far, in this realm, it has received little attention. Despite the efforts of a few lonely philosophers, and friendly nods from scientists here and there, the very thought of bring-ing hermeneutics into something like physics seems both pointless and dis-tasteful.[3] Historically, after all, hermeneutics was supposed to *distinguish* between the two main types of research: The idea was (and still is) that while the humanities and social sciences may need a language-oriented, interpretive approach, the natural sciences do not[4]; and the reason they do not is the very thing that makes them successful. Sociologists might give alternative yet valid interpretations of the same phenomenon because the questions they study often are 'undecidable.' In natural science, however, since meanings do not lie in purely verbal usage, differences *are settled* sooner or later, even when major paradigms clash. It seems reasonable then that science education should try to get students to understand what has been thus settled through hard and detailed work, not to train them in arguing all plausible interpretations of phenomena.

Nevertheless, I would like to present programmatically the case for hermeneutics in the *appropriation of natural science*—that is, in every kind of presentation, study, and understanding of what a particular science is say-ing to us. The argument is based on three major points: first, that despite deeply rooted opposition to such an idea, the mutual resemblance and par-allel evolution of historical-literary hermeneutics on the one hand and phi-losophy of science on the other is striking enough to suggest that, in some general but important sense, science itself is a form of hermeneutics; sec-ond, that the branch of hermeneutics called "ontological" seems especially appropriate to science because it makes possible the formulation of the most basic questions concerning the relation of the student to the objects of study, questions that lie at the center of at least two problem areas where controversies are now taking place—constructivism and the feminist cri-tique; finally, that whatever may be the place of hermeneutics in profession-al research (science as inquiry), in the cultural-educative realm (science as knowledge) the issue is fundamentally different and the case here much stronger; for in education it is not nature itself but a *language of nature* that one encounters initially.

The first two of these points are taken up in the next two sections; the last, though briefly dealt with in a previous paper (Eger 1992b), remains for the sequel. Although hermeneutics is essentially a wide-ranging philosophical orientation and will be discussed here in general terms, one goal of the present treatment will be to show that, nevertheless, several areas of current concern are immediately addressed by it. Constructivism, already mentioned above, is addressed because it entered hermeneutics more or less the same way, and more or less at the same time, as it entered the philosophy of science. One of the things that emerges in tracing out this correspondence is the possibility of a critique of certain extreme forms of the constructivist philosophy: Precisely because the viewpoint of hermeneutics is so broad, we can see clearly how constructivism is both embraced and limited by it; and this in itself, if accepted, would be an outcome of no small value. The feminist critique is addressed because that critique, in some of its best formulations, involves crucially the question of the objectivity of science, and of the 'detachment' practiced by the scientist and student—questions, as we shall see, that stand at the very center of modern, philosophical hermeneutics. The whole area of 'misconception' research is addressed, as previously indicated (Eger 1992b), because the idea of *pre*-conception is a key factor in hermeneutical thought. Among other issues addressed, though barely touched upon here, is the very serious problem of the 'commoditization' of science, a problem now increasingly coming to the fore as technology and changing social factors impinge more strongly on both the practice and the study of science.[5]

In short, what is suggested here is a general theoretical framework with possibilities for a consolidating as well as an innovating function. Clearly, this is a long term project. For the moment the immediate task is: by focusing only on the basic ideas, to work out the sense in which hermeneutics is relevant to science, and then to indicate with a few examples what difference it makes.

However, there are dangers on this path. With its concern for meaning and understanding, with its dialogical bent, hermeneutics calls attention not only to conceptual difficulties, but to language as such; and through language we enter onto the ground of rhetoric. Already scholarly works have appeared, suggesting it is the 'rhetoric of science' that accounts for the settling of differences noted above, rather than 'contact with reality,' the 'scientific method,' or even Kuhn's scientific "values." From anthropology, from sociology, from literature, from "deconstruction" and the "strong program" as practiced in Britain and France, and from the "archaeologies" and "genealogies" of knowledge, a renewed relativism—some say nihilism—emanates, and penetrates every field of learning today.[6] In regard to natural science, it is (still!) the Kuhnian outlook, widened, adapted, and pressed to its limit, that represents this mood best.

Yet an extreme form of relativism is inimical to the study of science. By implying, in effect, that laws and theories depend more on the scientist than on nature, it undercuts a major motivation for such study, the cosmological interest—a need for synthesis, a desire to understand what *kind* of world this is and how, in it, things hang together (Toulmin 1982; Eger 1989). If we say, for example, that "modern people . . . invent 'scientific' laws of conservation [of energy] to persuade themselves that substance is permanently with us" (Oliver and Greshman 1989, 145), then on what grounds do we recommend the study of such laws? If the "statements of science . . . can be *nothing more* than constructions of the humans who use them" (Bazerman 1988, 313, emphasis mine), then the study of science, like the study of non-representational art, is about human creativity and the 'worlds' it constructs, not about *the* world, common to us all. The critiques of 'objectivism,' of 'foundationalist philosophies,' and of 'grand narratives' have for years practiced a kind of intellectual devaluation of the natural sciences by separating the technological success of these sciences from their significance concerning human beings and the world.[7] The success is accepted, albeit with some trepidation; the significance is ruefully declined or denied. In exchange for the benefits of the former, it is said, the latter foists upon us the 'Western metaphysics' of an 'alienated' objectivity, suitable mainly for 'domination.'[8]

These critiques, offered frequently on the authority of well-known philosophers, though sometimes in adulterated form, are increasingly influential in the educational world as a whole, and may be expected increasingly to exert an effect on teachers' and students' understanding of what science is about.[9] The separation mentioned above between the efficacy of science and its meaning constitutes *de facto* an 'instrumentalization' of our understanding of scientific ideas (quite aside from instrumentalist philosophies of science as such), since the only meaning this leaves for the ideas is indeed instrumental—their *use* in 'predicting and manipulating' nature.[10] So understood, it is hard to see what interest science might have for people not pursuing a scientific career, and even for these people the interest is considerably reduced. Therefore, to the extent that a social-philosophical pressure for relativization and instrumentalization does exist, science teaching is confronted with yet another problem.

Because hermeneutics can indeed relativize the meaning of the content of science *to an extent,* because in some writings it *is* associated with devaluations of science, it could, in the absence of any checks on these features, undermine the very idea of science study—and that is why I emphasize that the approach has its dangers. But I will claim that in fact hermeneutics itself includes appropriate checks, and more: that despite its use by some writers to redefine science as just another "genre of literature" (Rorty 1980; 1987, 61), this is by no means inevitable. On the other hand, if we simply ignore the evidence for interpretive freedom emanating from the histories, philosophies, and sociologies of science, and from cognitive psychology,[11] if we present scientific findings unreflectively yet authoritatively, in the manner of naive realism, then science—its teaching and its works—is undermined in *other* ways: The more science takes on a cut-and-dried appearance, a learn-it-and-use-it format, and an aura which announces that nothing profound, mysterious, or specifically human is to be found here, nothing requiring judgment or subtlety, *the more* vulnerable it becomes to just those criticisms that deny its anchor in reality. For if science is both distasteful *and* unenlightening (about the world), its value can lie only in professional ambition.[12] In response to this growing problem, I would like to show that hermeneutics rather supports the cognitive seriousness of science than endangers it. In addition to the various expected benefits already mentioned, it has still this feature to recommend it: By elucidating the types of interpretive freedom inherent in the study of science, and delineating its bounds, *while at the same time* using it to enhance the *encounter* with science, a hermeneutic approach may show more convincingly than our present one that science is after all about the world, the common world in which we live, not about its own ways.

Hermeneutics Old and New

The perennial urge to take a literary task as the model for science is embodied in the classic book-metaphor—nature as a 'book' to be deciphered (Jones 1989). Even before Galileo 'discovered' that mathematics was the language of this book, before Francis Bacon's *interpretatio naturae,* the late middle ages were familiar with the metaphor. In the early part of the nineteenth century, it still stood, as, for example, in Ralph Waldo Emerson's famous essay *Nature*. That in our time this way of looking at things seems archaic is perfectly understandable: Increasing emphasis on the hypothetico-deductive process and, more recently, on post-empiricist philosophies of all sorts has raised to prominence the role of the *human* contribution. Doing science now seems less like reading nature, more like "dressing her up," taking her measurements for a "garb of ideas" (Husserl 1970, 51; Gregory 1988, 186).

But the decline of the old metaphor, as well as its present revival, has much to do with how the word "reading" is understood. As passive reception it became increasingly discordant with the flavor of contemporary science on the one hand, with historical and philosophical studies on the other. Yet in the modern, more constructivist sense of the word, an *interpretive* reading as a new "garb of ideas" is anything but passive. We can understand, therefore, why Mary Hesse, for example, and a number of other writers have returned to the older conception in which learning the 'language,' 'reading,' and 'interpreting' are once again central.[13] It is this combination—the old metaphor of nature as a book, *and* the newer understanding of 'reading' as an *act,* with its own degree of freedom—that gives impetus to the whole idea of hermeneutics in natural science. Let us look at this idea more closely.

In its classic, nineteenth century form, hermeneutics was a demanding art but philosophically straightforward. To understand a text written in an ancient language, one had to recover not only the meaning of expressions long out of use, but the meaning of actions, situations, customs, institutions—entire ways of life, in effect. Of course it was assumed that the text *has* a meaning. This might be the contents of the author's mind or the consensus reading appropriate to the author's milieu. But no one doubted that this meaning *exists* independently of the interpreter, and prior to the interpretation. To find the *true* interpretation, one would have to purify oneself, shedding the prejudices of one's own culture, and enter mentally into the culture of the text. One would often fall short, naturally, but that is true of any skill. What distinguished the master interpreter is just the ability to do better and get closer to the 'thing itself,' what the text is really talking about.

The great difficulty in all interpretations, the famous "hermeneutic circle," is so crucial to our discussion that I would like now to take it up it with some care. As an instance of a general epistemological mode, it links the two realms before us—text interpretation and natural science. Since the goal of this paper is to transfer, from the former to the latter, key concepts related to this mode, we will be in a better position to do so if in each realm we trace out how "the circle" is pictured and how the relevant philosophies have dealt with it.

When facing a difficult book, we know that the meaning of a phrase depends on the paragraph, on the whole text, and on the context; yet, to understand the text, we must of course understand the individual phrase. In all such situations (especially well known to science), the obvious method to use is that of 'iteration' or 'successive approximations': Start with some original "projection" or "preconception" for the text as a whole—a guess, a hunch, a prejudice carried over from previous readings— use that to make sense of the smaller parts of the text, and then see whether all these partial meanings, taken together, reinforce the original projection in a consistent way. It is important to note that although preconceptions may be deliberate, often they are unconsciously or "tacitly" adopted. This is likely to be so especially when it seems that the individual parts are clear enough, and no special assumptions need be made. In any case, one usually finds that the parts so interpreted *do not* in fact add up harmoniously; some discrepancy remains. Using this 'remainder' to correct the original projection, we start the cycle over again, and then again, and continue in this way, in a series of back-and-forth movements between the part and the whole (Gadamer 1975, 236, 261; 1988).

Implicit in this procedure (though not usually expressed as such) is the idea of *convergence*. With attention focused on a particular phrase or on the text as a whole, and noting what happens after each cycle, one should see the changes getting smaller, the circle shrinking. In the classic view, the *true* meaning of the text is being approached as, step by step, we purge our interpretations of all spurious and subjective factors. In practice, the process may be interminable, of course, but every cycle brings us closer to the goal. What is important is that we do not go *drastically* astray or oscillate forever between opposing poles—we converge. As an idealization, this convergence is to be understood in two senses: Not only does the circle shrink to a point, so to speak, in any particular attempt; but in any other attempt too, starting from a different projection, it should approach the *same* point. Clearly, this latter feature of convergence suggests *objectivity*. For a skilled interpreter, a prejudice is an obstacle but not a determining factor of the end result. The text determines the result.

It is precisely on this issue that the twentieth century radicalized hermeneutics. The transformation occurred more or less in three phases: an explosive widening of scope, then a drastic *reinterpretation of interpretation* as constructivist, accompanied by a shift from the epistemological to the ontological point of view. If one looks first at the question of scope, one sees immediately what really was at stake in the later reinterpretations— what is at stake today. Already in the nineteenth century, it was noticed that the problem of understanding encountered in historical or theological studies has its counterparts in other areas. In anthropology too one faces large cultural gaps, strange languages; and this, of course, is true in any situation where we try to communicate across linguistic or behavioral frontiers. In time, it became clear that similar situations occur even within one's own culture, in regard to art for example, or whenever one is trying to understand something new. Finally, in Gadamer's words, it encompasses "the whole human experience of the world. I call this experience hermeneutical, for the process we are describing is repeated continually throughout our familiar experience. There is always a world already interpreted, already organized in its basic relations, into which experience steps . . . upsetting what has led our expectations and undergoing reorganization itself in the upheaval" (Gadamer 1976, 15).

It is in this form, as *philosophical hermeneutics,* that it confronts us today. But it does so only after taking its second step—the shift toward constructivism. Now this line of thought, begun by Martin Heidegger (1962), and developed in depth by Gadamer, parallels so closely the later course of the philosophy of science that in comparing these two developments—changing ideas on how a text is read, and changing ideas on how nature 'is read'—we will find both the differences and similarities illuminating.

Hermeneutics was transformed when it was denied that prejudices, or fore-meanings, or preconceptions, as they are called, could be *essentially* overcome. For if they cannot, then the original idea of recovering the pre-existing meaning of a text collapses as meaning ceases to be the fixed target toward which interpretation is merely a path. Instead, meaning *emerges* from the interpretation. Interpretations become, at least in part, *constructive,* not *reconstructive.* This implies that fore-meanings, which now acquire a new significance, cannot just be cleared away but must themselves become a focus of inquiry; their several kinds can then be examined somewhat in the manner of Bacon's idols but for a different purpose—not so much to purge them as to understand their effect. What results is a set of distinctions and descriptions which, when taken together, make up the hermeneutic 'fore-structure' of knowledge.

There is first of all that 'primordial,' most deeply rooted kind, called "fore-having" (*Vorhabe*), which stems from the common practices of the "life-world." We already handle unreflectively, as part of routine experience, the things we might later contemplate; and this 'having,' or handling, or encountering, constitutes—tacitly—the primary layer of 'knowledge' of the things around us. A second layer, closer to consciousness, is "fore-sight" (*Vorsicht*), the whole category of concepts and perspectives inherent in our specific situation—'glasses' for the mind's eye, so to speak. This provides the means with which to 'see' what has already been handled in experience. Most important here is the suggestion that language, a basic element in "fore-sight," ought not be understood merely as an instrument of communication but as a collective field, a tradition, from *within which* we see whatever we see. As may be expected of a philosophy of interpretation, language plays in it a key role, an exalted role one might almost say; but that is a matter to which I will return. The last layer of fore-meanings consists of "incipient ideas" (*Vorgriffe*) that serve as 'gripping' tools, in effect, allowing one to handle cognitively what has already been handled physically. These may be consciously held, communicated and revised.[14]

By relating knowledge first to the pre-scientific world of practice, to the depths of human existence itself, then to the inherited language, philosophical hermeneutics closed off for itself any hope of objectivity in the absolute sense. Hermeneutic thought does not relax its effort to identify prejudices and take them into account, but getting at the truth, at 'things themselves,' is now a more complex, problematic enterprise. It must be done from *within* the fore-structure and from within "the circle" of understanding to which that leads. No longer may preconceptions be regarded as something merely negative, since in effect they are *part* of the 'seeing' (interpreting) apparatus without which nothing at all can be seen.[15]

One easily recognizes throughout this account the recent history of the philosophy of science, with its paradigms, its conceptual frameworks, its drift away from objectivism.[16] Significantly, that resemblance holds not just for the general trend but for important particulars. Throughout most of the nineteenth century, science aimed at truth in its unique, pre-existing sense just as did the hermeneutics of that time. Hypotheses, as we know, had been under suspicion from the very beginning of the scientific revolution, when Newton said he did not use any at all. If one did use them, they had to be justified directly. It had to be shown, in other words, that whatever projecting or assuming or *pre*-conceiving was done did actually hit the mark. With time, however, a hypothesis began to be regarded less as a basis for the final solution to a problem, more as a step in a process. After the Second World War, when Popper's falsificationist philosophy became better known, this understanding of conscious preconceptions as *means,* as the foil against which nature is to have her say, prevailed (Popper 1968, Ch. 1). In this form it is taken for granted today; and in this form, it matches well the minimal, cautious hermeneutic attitude:

> Anomaly appears only against a background provided by the paradigm. The more precise and far-reaching the paradigm is, the more sensitive an indicator it provides for *anomaly and hence of an occasion for paradigm change.* (Kuhn 1970, 65, emphasis mine)

> The important thing is to be aware of one's own bias, so that the text may present itself in all its newness and thus be able to *assert its own truth against one's own fore-meanings.* (Gadamer 1975, 238, emphasis mine)

In both realms, notice, 'prejudgment' appears as a *probe,* with which an object is approached, allowing that object to be 'felt,' in a sense, as it reacts on the probe. In both realms, thereafter, this leads to "the circle"—an alteration of the probe, and another turn in the movement: "Coulomb had to employ electrical theory to determine how his equipment should be built. The consequence of his measurement was a refinement of that theory" (Kuhn 1970, 33–34). Similarly, in the back-and-forth movement of reading a book, the interpreter 'works out' the "fore-project, which is constantly revised in terms of what emerges as he penetrates into the meaning" (Gadamer 1975, 236).

All this working out, revising, penetrating, still suggests that we are approaching a goal; but today this has to be understood in a new light. The minimal, cautious formulations above are not yet the whole story. For science, the 'other shoe' fell not long after Popper's main impact, when a number of writers showed independently that preconceptions were far more involved in all phases of inquiry than had been previously thought.

Not just Popper's explicit, deliberate conjectures, not only biases that could be exposed, but deeper, more 'primordial' commitments are at work.

In the theories of Michael Polanyi and of Kuhn the counterparts of "fore-having" are certain kinds of 'tacit knowledge.' They are the most firmly held part of the paradigm or disciplinary matrix, lodged in a tradition of laboratory procedures, in methods of handling the material physically and mentally, in ways of classifying, analyzing, measuring, recording. Superimposed on this, the *language* of the paradigm, the counterpart of "fore-sight," *pre-structures the seeing* of what goes on in a problem or on the workbench, and the manner of talking about what is seen (Kuhn 1970, 187–191; Polanyi 1958, 59; 1969, 188). A famous instance of the latter was the use, before Einstein, of the word 'simultaneous,' without reference to coordinate frame. Just as interesting, however, from our perspective, is the recent research on student preconceptions—that motion requires force, for example; that moving bodies 'have' force, and so on. Since these appear as a recapitulation of beliefs widespread at the very beginning of the scientific revolution, we may suppose them to be the most primitive, unexamined effects of a "fore-having" in the life-world.[17]

The point is that, by elucidating the deeply rooted, pervasive, and *indispensable* nature of preconceptions of the tacit sort, Polanyi, Kuhn, and others elevated the role of such preconceptions to a level of importance in science comparable to that given them by Heidegger in hermeneutics. On the one hand we cannot completely rise above such bias; on the other, because of its *positive* role, it is now clear that we cannot wish to do so.

This exactly is the place where modern radicalism enters—in science just as in textual interpretation. Deepening analysis of preconceptions gives rise to a most far-reaching and controversial development: the attack on convergence. In philosophical hermeneutics and textual analysis, it is called 'undecidability' (Connolly and Keutner, 55–58). Whatever knowledge is contained in the fore-having, in fore-sight or in language, will in most circumstances *not* be revised to insignificance. The hermeneutic circle may not shrink to a point, may not even shrink to the same neighborhood for different interpreters, and may not therefore be expected to converge to a unique truth. Successive projections may differ *drastically,* not just marginally, because the starting point matters after all and enters the outcome. Consequently, multiple interpretations can coexist, and no test can really *decide* among them. The first result of this 'reinterpretation of interpretation' was to regard the interpreter as *part* of the interpretation.[18]

So also in the more radical post-empiricist views of science: Here the counterpart of "undecidability" is "conceptual nonconvergence."[19] Within a paradigm, in times of 'normal' research, one may expect refinement of theory to approach some stable limit; but revolutions, because they affect *meanings,* can result in wild deviations. Just as historical revisions change the significance of major political events, though the events themselves are not altered, so in the scientific realm successive paradigms give increasingly accurate *numerical* results, yet differ essentially on the meaning of concepts. In this view, the theory of relativity did not merely extend classical dynamics by correcting and explaining it in a wider realm: When it changed the meanings of 'simultaneity' and 'mass,' it changed the very language of dynamics (Kuhn 1970, 98–103). What is being said here, in a most emphatic way, is that numerical and conceptual convergence are *decoupled:* The former no longer entails the latter. The succession of paradigms from Ptolemy to Newton to Einstein improves calculation, but does not approach ever closer to any picture we can discern and *call truth* (Kuhn 1970, 99–102; Laudan 1981, 250). Conceptually, that sequence oscillates.

No one should underestimate the genuine sense of loss, of sadness, among scholars like Israel Scheffler, for example, who at an earlier stage of this trend rose to defend *objectivity* in science (Scheffler 1967). The ensuing debate was not just academic in the usual sense. Far more is at stake here than professional theories: Objectivity is the *morality* of science. Its relation to the moral life in general rests on its demand for transcendence—the overcoming of our many parochialisms. It is an *effort* to get at truth or whatever is the closest thing available, arising not from personal detachment or superhuman neutrality, nor from contempt of feeling or of values, but from the need for *common ground*—solidity beneath our feet—in the tragic human situation. Scheffler's (and Popper's) call for objectivity as "responsible belief" (1967, 4) reflects, I think, a recognition that most injustices, large and small, have *un*truths of one size or another at their foundation.

It is this that should be remembered by all of us who now wish to 'humanize' science in some sense, fearing that objectivity, as 'objectivism' and 'detachment,' is perhaps doing more harm than good. It should be remembered especially by those who see in the methods of science a form of 'domination' and would therefore replace our Western, 'value-free,' 'hierarchical,' 'reductionist' way of knowing with ways sensitive to wholeness, indeterminacy, and the personal factor (Polanyi 1958; Gadamer 1970; Keller 1985; Harding 1991; Bohm 1987). Such critiques are prominent among the hermeneutic philosophers of Europe whose ideas I wish to use; but the approach of this paper is to deny to neither side the very serious and valid concerns which it undoubtedly has.

Various roads in both directions are still open. The shift of the philosophy of science toward subjectivity and relativism has been severely criticized and is by no means universally accepted.[20] In hermeneutics, on the other hand, the ontological turn, yet another aspect of the radicalization, allows us to see interpretation and the 'thing interpreted' in a far less subjective light than has been generally admitted. The next section takes up this point more thoroughly by examining the *modes of being* of the knower and the known, in order to show how hermeneutics moves 'beyond' the subject/object dichotomy.

The present section was preliminary in nature. Its goal was to discuss three independent yet closely related aspects of our topic. First, as background for various discussions yet to be presented, I tried to exhibit the close correspondence and parallel development of certain concepts in hermeneutics and in the philosophy of science. This effort is only begun here and will continue. Second, the juxtaposition of scientific and textual research helps to frame a major question of this inquiry: *What is the role of interpretation in the understanding of science?* Finally, by following the increasing constructivist and relativist trend in these two disciplines, I am preparing to suggest that the well-known problem this poses for the *philosophy* of science is reflected as well in the *study* of science; and, moreover, that in hermeneutics itself may lie part of the answer to this very problem.

The Ontological Turn

We are ready now to consider what I take to be most distinctive aspect of that branch of hermeneutics developed by Heidegger and Gadamer, and also most relevant to our inquiry. It is the feature that limits, counterbalances, and places in proper perspective that subjectivism and relativism with which hermeneutics is often associated. It is this feature also that opens up new perspectives on a whole complex of problems by inviting us to reconsider fundamentally the text-teacher-student relation in the sciences. The point is that the foregoing sketch of the reading of a text, of "the circle," and of the fore-structure of knowledge, must not be taken as a description of *knowing* by a subject but as a description of *being* in a certain mode that is peculiarly human. In Gadamer's words, "the circle of understanding is not a 'methodological' circle, but describes an *ontological element in understanding*" (Gadamer 1975, 261, emphasis mine). Here we have perhaps the most surprising thrust in the modern reinterpretation of interpretation.

In the first instance it means that, far from being optional, hermeneutics is the universal *way* of being in the world, for us humans, whether we face a written text, an art work, or each other. The sequence is not *first* to get to know something, *then* to interpret it. Interpretation and understanding are not two separate cognitive acts performed by the subject, as previously envisaged; rather, "understanding is always an interpretation, and hence interpretation is the explicit form of understanding" (Gadamer 1975, 274). To interpret, however, is to "enter" the hermeneutic circle, to project, and to remain thus in motion between the text (or text-equivalent) and the projected fore-meanings. Therefore, a fundamental mode of human *being* is precisely *being in* this circle in which the mind is not with itself but is *drawn* by its own projections to that which it attempts to understand. The reason for describing this as a state or a mode of being, rather than simply an activity among other possible activities, is precisely to identify this mode as 'primordial'—prior to all constructions—one without which the human, *qua* human, cannot be understood.[21] That such a mode is indeed part of research, and is explicitly recognized as a *state,* will be shown in the next section when we take up the case of Barbara McClintock.

Here, however, let us clarify the special sense in which the word "text" and "student" are used. In general, "text" may refer not only to expressions coded in natural or specialized languages but to all objects of study and hermeneutical concern. In the context of science, this may include such things as graphical representations, models, and experiments; it may also refer to nature itself, in the sense of the classic metaphor. Normally, the context will determine specific meaning, but sometimes the term "text equivalent" will be used as a reminder. Just as important, "student" is not to be identified exclusively with the usual image of a youth who is the object of educational activity; it will be used here in a much broader sense, as a characteristic mode of being in our time. Anyone who faces nontrivial "texts" with an intention of reading and understanding them is to that extent a "student," obliged to interpret consciously and systematically. My contention will be that today this is a more fundamental mode than "consumer," "producer," or even "worker." Without passing through an extended period (and to some extent continuing) in the "student" mode, one cannot become a member of modern society.

This much, then, in its minimal form, is the ontological claim about the interpreter. Concomitantly, there is an ontological claim about the thing interpreted: The meaning of a text, a work of art, a history, must not be looked upon as 'existing' or standing alone prior to its understanding. Meaning *arises* in the interpretation itself. The subject/object cut does not lie between the interpreter on one side and the text with its meaning on the other (objectivism). Neither does it lie between the text alone on one side and the meaning *inside the interpreter's mind* on the other (subjectivism). As a fixed boundary, the cut is just not there; meaning is not disjoint either from the text or from the interpreter. Rather, a 'bare' text is to be thought of as an ontological core, around which *potential* meanings hover, so to speak, in a space of all possible meanings. A human consciousness, in its understanding-seeking, interpreting mode, interacts with this text by means of its fore-structure (preconceptions), and brings into being one or another of the potential meanings. Only in the interpretation is the *being* of the text fully realized because only then does the thing it talks about come to light. Bare or 'in itself,' a text is incomplete.[22] Thus, interpretation must be viewed as an event, a *performance* of the text, like the interpretation on stage of a play or musical composition: In script form, the text of a theatrical play serves as the core that structures interpretations; but only the *act* of a performance (necessarily interpretive) allows the text to achieve its *realization,* its being.[23]

In this way, by contradicting radically the classic notion of meaning as something *conveyed* by a text, or *extracted* from it, contemporary hermeneutics shows its constructivist side. But it is important to keep in mind that this is indeed only one of its sides. Thus far, our discussion of projections and pre-understanding might imply that since all initiative lies with the interpreter, the resulting interpretations are achievements of the interpreter alone; and from here the road seems wide open to subjectivism and relativism on the largest scale. In fact, that road is not open—in testimony of which we have for example Gadamer's own lengthy critiques of what he calls "hermeneutic nihilism"; his insistence that the text "speaks to us," "asserts its own truth"; and many similar locutions (Gadamer 1975, 85, 442–43). What precisely does all this mean?

It means, first of all, that while potential interpretations surround a text (or text-equivalent) and the "distance" between any two of these may be considerable, nevertheless they are not running around all over the yard to be scooped up wherever one pleases. Where they may be found and what sort of "space" they occupy depends on the core, the text itself. Interpretation is *not* invention; there is something *there* to interpret. We approach that something with the pre-understanding of our fore-structure (*our* way of being), but within or against this fore-structure it "addresses us" with its own message. This is what Gadamer means when he says that "the constant task of understanding is to work out the proper, *objectively* appropriate projections . . . " (Gadamer 1988, 72, emphasis mine). If, in other words, the projections are appropriate to the objects of the text, and the interpreter is "open," listening, then an encounter with meanings—with the *text's* meanings—becomes increasingly probable. What the interpreter constructs by means of his projections, and how much, is then guided by clues in the text, confirmations or rejections by the text, a stream of responses from the text—all of which, together, evoke progressively an imposing *presence* that can, in the manner of a "voice," begin at some point to reveal the *un*anticipated. Of this phenomenon, one can say, indeed, the text "speaks."

Of course, the text can speak only in its *own* language. Yet if it speaks *to us* it must do so in a language (in concepts) *we* understand, which means that the text's language must to some extent overlap our own or be understandable from within our own; in short, the two languages must be "commensurable." Recall here the great debate provoked by Kuhn and

Feyerabend on just this question of what happens when a scientific language belonging to one paradigm *cannot* be translated into that of another paradigm—the question of '*in*commensurability' (Kuhn 1970; Feyerabend 1988 [1975). Kuhn's original answer, that a kind of psychological or religious conversion occurs, apparently outside the realm of reason—a 'leap of faith'—was widely criticized precisely on account of its subjectivist, relativist character.

Hermeneutics does not take this route. To the contrary, it repudiates outright any suggestion of totally closed-off worlds. Because of the projecting, anticipating character of all interpretation, dialogue is possible even between radically different languages. To shed light on how, in general, communication finally does occur even when at first there is little common ground, Gadamer employs the concept of "horizon."[24]

The horizon, "the range of vision that includes everything that can be seen from a particular vantage point" (Gadamer 1975, 269), encloses the field of meanings that our various pre-conceptions, tacit and otherwise, allow us at any given time. It defines, in other words, both our grasp and our limitations. But—and here we really see the two-sidedness of this theory—the text we are trying to interpret also has its horizon: a limit to all those meanings to which a text of this sort, employing a language of this sort, possibly could give rise—which includes, of course, those specific possibilities I have called the 'cloud' hovering about it. If our own horizon and the text's horizon do not overlap at all, there is no way for our projections to fall within the realm of the text's potential meanings, and the attempt to reach understanding fails—incommensurability.

If, on the other hand, an overlap does exist, the back-and-forth movement of interpretation permits us to traverse the horizon of the text and move deeper into *its* language domain. Then, by coming within the range of the text's own potentialities, we allow it to 'speak to us' in the sense indicated above, and thereby to *expand* our field of vision to a point where the two horizons can be said to have "fused." When this occurs, we have the supreme moment of understanding (Gadamer 1975, 269–273; Linge 1976, xix).

In the "fusion of horizons," a central metaphor, Gadamer offers an alternative to incommensurability: The human being is *not* rooted in his standpoint or trapped within his language game; nor is he forced to jump over a chasm, blind to what lies ahead. The process of interpretation itself, in which construction of concepts plays its part, accompanied by "listening," has the capacity to move or expand horizons. This simple but powerful formulation is just as applicable to the "texts of science," it would seem, as it is to other sorts of texts; and we will exploit it. Significantly, years after Kuhn offered his original thesis he explicitly accepted the hermeneutic viewpoint, referring to well-known papers by K.-O. Apel

(1972) and Charles Taylor (1971): Yes, he said, even when one paradigm cannot be *translated* into another, it may still be *interpreted* because interpretation is much more than just the replacement of some groups of words by other groups of words (Kuhn 1983, 671–77).

"Fusion of horizons" embodies the idea of interpretation as a state of motion (in cognitive space), and of human beings who enter into this state as participants in the motion. The cause of the motion is the interaction itself—between the text and the interpreter—which is attractive and made possible by some initial overlap of horizons. All this constitutes a phenomenological description of an *encounter with meaning* that is *not one's own,* but toward which one has deliberately oriented oneself.[25] It is a description of dialogue, or *being in* dialogue; and it is in this sense that hermeneutic consciousness is understood as "dialogical."

The projections toward the text and the potentialities of the text typically have the character of questions, answers, counter-questions; and typically this is the description offered by those involved in such an encounter. What questions are asked depends on the horizons involved; what answers are given, or potential meanings realized, depends on the questions.[26] But here again, the ontological perspective gives to the outcome a more objective status: The realized possibilities, a product of the interaction of text and interpreter, exist in the world no less than do the text and the interpreter. The performed play exists no less than the script. It is this *ontology of dialogue* that saves hermeneutics from relativism and subjectivism, even as it allows 'play' for interpretation; and it is this feature that must be developed in relation to science.

That one can actually have dialogue with things like books and paintings and history is a familiar enough idea if taken figuratively. Hermeneutics, however, suggests it in a stronger sense. It suggests for the human partner in the dialogue an intensity of participation that evokes from the other 'partner' *assertive* meaning. Pursuing this line one step further, we come in sight of its truly radical implications: 'Things' are, in this sense, 'not dead.' They put forth, or present, or 'have' a meaning that is *theirs,* a part of *their being* (in relation to us). Ultimately this suggests—if the hermeneutic viewpoint is indeed universal—that the 'things' so understood could well include not just human creations, stamped with the human imprint, but equally the natural things like trees, molecules, or stars. By this route, we arrive once again in the realm of the natural sciences, and we arrive there accompanied by a challenging thought.

Because the notion that 'things speak' is so provocative and contrary to prevailing philosophies, it will be worthwhile to show that nevertheless it is a notion held by at least some outstanding scientists. In research, it testifies to the dialogical quality of the interaction between the scientist and the object of investigation. At another level, in the *study* of science, the corresponding dialogue is between the student and the scientific tradition ('texts'). It is the latter, more obviously linguistic level that will concern us in the long run. But first it is important to focus on what sorts of things happen when scientists face nature 'in the raw,' and when the interaction is at its strongest. That some researchers describe this interaction precisely as does ontological hermeneutics is a major clue: You must have the patience to "hear what the material has to say to you," claims the biologist Barbara McClintock, and the openness to "let it come to you." She is talking about corn (Keller 1983, 198).

In her fine biography of this Nobel laureate, Evelyn Fox Keller emphasizes the contrast between McClintock's attitude and the better known, more 'aggressive' style of the "bold hypothesis," made famous by such stories as that of Watson and Crick. In that style, according to McClintock, preconceptions—or "tacit assumptions" in her words—are given so much weight that they become over-restrictive, blinding. Scientists who for years ignored her major work in genetics, for example, "didn't know they were bound to a model, and you couldn't show them . . . even if you made an effort" (McClintock quoted in Keller 1983, 178). More generally, she believes

> the work is done because one wants to impose an answer on it. . . . They have the answer ready, and they [know what they] want the material to tell them. [Anything else it tells them] they don't really recognize as there, or they think it's a mistake and throw it out. . . . *If you'd only just let the material tell you.* (Keller 1983, 179)

From our hermeneutic perspective, McClintock's complaint is that in the aggressive, overly constructive style, dialogue between the scientist and nature turns into a *monologue:* The scientist speaks, nature hardly murmurs. By contrast, McClintock describes her own way as follows:

> I start with the seedling, and I don't want to leave it. I don't feel I really know the story if I don't watch the plant all the way along. So I know every plant in the field. I know them intimately, and I find it a great pleasure to know them. . . . I have learned so much about the corn plant that when I see things, I can interpret [them] right away. (Keller 1983, 198)

This degree of attunement—what Keller calls a "feeling for the organism," an "extended vision"—is precisely the entrance point to the state of *being in* that which one is interpreting. I have been at pains throughout this section to emphasize that the ontological claim of hermeneutics must be taken most seriously. As confirmation of the validity of this claim, evidence that it reflects more than arm-chair philosophizing, here now, in the words of McClintock herself, is a description that outdoes even the theorists' most daring formulations:

> I found that the more I worked with them [chromosomes] the bigger and bigger they got, and when I was really working with them *I wasn't outside, I was down there. I was part of the system.* . . . I even was able to see the internal parts of the chromosomes. . . . As you look at these things, they become part of you. And you forget yourself. The main thing is that you forget yourself. (Keller 1983, 117, emphasis mine)

McClintock's ecstatic language is to be understood as precisely that, a report of *ekstasis*—a being outside oneself. In no way should this be viewed as strange or mystical, though some people have characterized it so. Rather, let us note first the similarity of her *experience* to Gadamer's purely theoretical description:

> To be *present* . . . has the character of being outside oneself. . . . In fact, being outside oneself has the positive possibility of being *wholly with something else.* This kind of being present is a self-forgetfulness. . . . Self-forgetfulness here is anything but a primitive condition, for it arises from the attention to the object. . . . (Gadamer 1975, 111, emphasis mine)

"Being outside oneself" or "self-forgetfulness" or *ekstasis* are used above in the technical sense. Far from being a mystical experience, what is described here is simultaneously a state of cognitive detachment and attachment: detachment from the everyday self ('forgetfulness') required for an attachment to the object or to the instrument that brings one closer to the object—in McClintock's case, the microscope. Both the attachment and the detachment constitute an extension of one's perceptive and cognitive organs beyond one's own body, and can be given a rigorous formulation. This has been done by Patrick Heelan (1983) and Don Ihde (1979):

> whenever familiar scientific instruments function in the cognitive way described above as extensions of the neurophysiological organism for the purpose of perception, they become part of a perceiving subject's Body: the subject is then said to be embodied in the instrument . . . conjoined with the instrument . . . as one coordinated operating system in which in some way are represented . . . the possible states of the environment as 'felt' or probed. . . . (Heelan 1983, 206–207)

Nor is this "embodiment" limited to instruments of the sort we call hardware. Here, in regard to *cognitive* instruments, is the same idea in the words of Polanyi: "When we accept a certain set of presuppositions, and use them as our *interpretive framework,* we may be said to dwell in them as we do in our own body" (Polanyi 1958, emphasis added).

To *inhabit,* as it were, the structured world consisting partly of experimental hardware, partly of the software of scientific tradition, partly of one's own emerging addition to that software—and from the center of such an interpreting system ('extended vision') to find oneself 'seeing' or 'touching' something 'out there'—can be taken as an *ontological* state in which the subject, no longer the human alone, is the interpretive system as a whole, *tuned* to something 'other' than itself.

> Empirical procedures in science are not pieces of Nature, pure and simple, but each is a humanly contrived phenomenon in which Nature is made to write in conventional symbols a text through which it 'speaks'. . . . (Heelan 1983, 224)[27]

It is surely not beside the point that independent, theoretical descriptions of such ecstatic states match to a surprising degree the kind of report given by McClintock. What she experiences as "being down there," Heelan and Ihde describe as "embodiment," and Polanyi calls "indwelling." I have referred to the same phenomenon as "inhabiting," and its implications are important:

> [A]ll understanding is achieved by indwelling. The idea developed by Dilthey and Lipps, that we can know human beings and works of art only by indwelling, can thus be justified. But we see now also that these authors were mistaken in distinguishing indwelling from observation as practiced in the natural sciences. The difference is only a matter of degree. . . . (Polanyi 1969, 160)[28]

For a long time, Polanyi (himself also a distinguished chemist) had been developing the idea of "personal" and "tacit" knowledge as a deliberate antidote to the over-formalized, theory-centered, and logic-driven philosophies of science that dominated the first part of this century. Although the word 'hermeneutics' can hardly be found in his writings, Polanyi was strongly influenced by this school. Taking to heart their basic problem, the alienating quality of modern, 'rationalized' life, and believing—as do so many others—that this problem originates in science, he became perhaps the first philosopher to go beyond the leaders themselves and apply extensively a hermeneutic viewpoint to natural science. In the quotation above, he challenges the perennial belief that the scientific attitude is based on *detachment*—a key point in hermeneutics, and a key point for this inquiry as it moves on to the issues of education. Instead of detachment, followed by active construction, he posits *attachment* and 'sharing.' Instead of the traditional emphasis on inductive and deductive logic, he calls attention to "tacit" integration of clues.

Keller has made it a point to discuss this contrast between the active and passive styles of doing science—the dominating and the receptive—in the context of the gender question. Although she rejects any idea that the passive is intrinsically feminine or feminist, there is more than a suggestion in her writings that the *active,* if not 'masculine' in the proprietary sense, is nevertheless unduly permeated by an attitude of objectivizing, controlling, manipulating, 'dominating'; and this attitude, she believes, is indeed part of a "commitment of scientists to the masculinity of their profession" (Keller 1985, 175). By contrast, for McClintock (as for Polanyi),

> science has a different goal: not prediction per se, but *understanding;* not the power to manipulate, but. . .the kind of power that results from an understanding of the world around us, that simultaneously reflects . . . our *connection* to that world. (Keller 1985, 166, emphasis added)

In this revealing passage, McClintock shows how pervasive in our thinking is the dichotomy between predicting and understanding, linked, as is often the case, to the one between detachment and attachment. The classic hermeneutic tradition, by separating science from understanding, unfortunately legitimates these dichotomies. McClintock bridges them in her work, even as she acknowledges them with her words. But in this paper enough has been said already to make clear that our *extension* of the hermeneutic tradition—essentially in agreement with Hesse, Heelan, Polanyi, and McClintock—moves in the direction of the *dissolution* of the dichotomies, at least in the sense of their exclusiveness: A dialogue must have its speaking phase and its listening phase—the one being the necessary complement of the other. When speaking, one is often detached and imposing; when listening, taking it in, one must of course be attached, attuned, and receptive. An overwhelming emphasis on either phase, though possibly appropriate in particular situations, will on the whole be distortive. Even McClintock was not always taking it in, but projecting like everyone else: "Her theory," says Keller, was "a hierarchy of hypotheses, each more abstract and further removed from objects of perception than the one before" (Keller 1983, 126).

Interim Summary

Let me summarize the argument to this point and indicate where it is leading. First, we followed the parallel evolution of hermeneutics and the philosophy of science—reflection on the interpretation of texts, and reflection on the interpretation of nature—showing how the two histories, starting from a position of unquestioned objectivism, moved, more or less in step, toward ever increasing constructivism or freedom of interpretation. In the course of this evolution, the meanings of both texts and scientific theories were, in varying degrees, relativized—which implies, of course, that the meanings were put in doubt. For the study of science, the danger in this trend is that, ultimately, *no meaning worthy of study may be left*—only techniques for professional puzzle-solvers and 'manipulators' of nature. On the positive side, however, the same evolution has raised consciousness of the historical, philosophical, and social dimension of science, and of its relation to other disciplines such as the interpretation of texts. At this point, my suggestion was that if the parallel between literature and science is extended, the *further* development of hermeneutics offers insights for the understanding of science; and the pursuit of this development should repay the effort.

Therefore, in the following section, we focus on the ontologization of hermeneutics—something for which mainstream philosophy of science shows *no* parallel. Not only Kuhn, but the great majority of theorists have stayed within the epistemological approach in which the object/subject distinction is kept clear. In these theories, the individual scientist or the community of scientists as a whole is the subject, seeking knowledge about the objects of this world, and the big question is: What is the *character,* the import, and the cognitive status of knowledge so obtained in relation to the subject and to the objects investigated? Is it a mirror-like image of the objects? Is it a map, a model, a metaphor? Is it an instrument, a Darwinian adaptation of the subject? In education, traditionally, the answer of choice was the 'mirror-like image'; increasingly now it is becoming 'the model,' with an emphasis on *construction* of the model by science and by the student. But whatever answers are given by the various schools, the question is essentially the same.

In contrast, philosophical hermeneutics poses the question in an entirely different way. It asks, first of all, What is understanding of any kind? It asks then, in particular, What sort of thing is a historical text in relation to understanding? What sort of thing is a work of art in relation to understanding? And what is the mode of being of a human being *trying to understand* a text, a work of art, and so on—to which, a first-level answer under all circumstances is: Understanding is the emergence ('coming into being') of meaning that occurs in *interpretation,* which is a dialogical process. This locates meaning not in the mind of the interpreter—as epistemology does—but in the interaction or dialogue between "embodied" interpreters (in the sense of Heelan or Polanyi) and what is interpreted.[29]

If we are to follow this pattern of questioning, then in a completely analogous way, we will have to ask: What sort of thing is a *scientific text* in relation to understanding? What sort of thing is a *work of science* (experiment, theory, etc.) in relation to understanding? And what is the mode of being of a human being trying to understand such a text or such a work? It may seem inordinate, even presumptuous, to raise elemental questions of that sort; after all, people have been dealing with textbooks, experiments, and the like for a long time and have some fairly adequate conceptions of what these things are. However, my point is, first, that such conceptions are by no means fixed, as recent history and philosophy of science show; and, second, that contemporary problems centering on the issue of meaning do raise doubts about the appropriateness of many of the conceptions current now.

For example, even before we pursue these big questions—about the mode of being of science, let us say, in relation to the student—it is clear enough that the current, *de facto* answer, as reflected in educational practice, stands in sharp opposition to the hermeneutic attitude. It is characterized by an image of scientific knowledge as *transferable commodity.* In this view, knowledge, or information, is *produced* in the laboratories, observatories, computer centers, and on researchers' desks; it is then recorded in papers, in monographs, and in books at various levels of expertise; finally, through lectures, courses, popularizations, and the work of teachers, that *same* knowledge (information) is *transmitted* to other scientists, to students, and to the public. This is the *sender-channel-receiver* model of information theory. It gives us the research scientist as producer-sender, the textbook and teacher as channels, and the student as receiver. When it leaves the hands of the scientist-producer, the commodity (scientific knowledge) is fixed and finished; thereafter, the only way to measure the quality of the transmission is to compare that commodity, as the student receives it, with its form as the scientist sends it.

We know that such views of science, of study, and of understanding have been criticized recently in a number of independent efforts, among which constructivism is perhaps the most prominent. The constructivist critique of education has convinced many people that students—normally placed at the 'receiving' end of the channel—ought not be cast in such a passive role because in order to *understand,* to attain meaning, they themselves must do some "building up" (von Glasersfeld 1989, 135). It has shown, further, that this activity on the students' part involves more than a mere decoding of the scientific text, since, in the language of Piaget, it often requires "accommodation" of already existing concepts to new situations (Piaget 1971). In this way, psychologically-based constructivism makes contact with ideas about 'conceptual change' emanating from the history and philosophy of science (Nersessian 1989; Garrison and Bentley 1990).[30]

However, in its desire to combat the commodity model—to press home the message that even the 'hard,' 'durable' information sent out by the physical sciences is not to be merely *received*—constructivism perhaps creates problems as large as those it tries to address. And this precisely is one area where the value of a hermeneutic orientation comes to the fore. From the viewpoint of this paper, radical constructivism as a philosophy of education can be criticized on two broad fronts: in its implications for both the *objects* and the *subjects* of the study of science.[31]

Because, implicitly, such a constructivism assigns primary status ('privilege') to the subject only, not the object of knowledge, it reduces drastically the horizon of meanings for anyone facing science in the student mode. Yet historically, the excluded meanings (realist interpretations) have been part of both science and its philosophy, as they still are today, even though their status has for a long time been under debate.[32] Whether a philosophy of education can take it upon itself to adjudicate this classic debate, and to so orient both teacher and students that *what* they study is *a priori* to be understood as "nothing but the cognitive structures they themselves have put together,"[33] is a question that certainly deserves more discussion. Under some circumstances, a hermeneutically conscious student of science might be willing to accept this as a starting point, but only with the understanding that here we have a preconception *par excellence*.

Regarding the subjects of study, and their attitudes, we have another kind of problem. Constructivism has been highly effective in directing attention to the *active* role of the 'readers'—the scientists *vis-à-vis* nature as they construct theories, the students *vis-à-vis* science as, under guidance, they reconstruct these theories in their own minds. But from our present perspective, all this construction is only the 'speaking' phase of the interpretive dialogue. Missing, and disturbingly conspicuous by its absence, is the *listening* phase. Where, in this picture, is that experience of attunement, of being in touch, of *entering into the knowledge of things,* prominent in the thought of Polanyi, McClintock, Heelan, and most other scientists perhaps?[34]

Of course, constructivism by no means excludes listening from its purview. The more radical it is, the more it insists that listening itself is a constructing, and without constructing ("building up") there is no listening that does any good (yields understanding). However, it was not the purpose of the section above, dealing with McClintock, to imply that in the experience she regards as "letting it come to you" we have the pure reception of an incoming message. Rather, her contrast of what happens when she listens with what happens when people "impose" suggests, first, that this constructing must vary in an important way both quantitatively and qualitatively, and, second, that possibly something else may be involved in listening besides constructing.[35]

Consider for a moment the mode McClintock calls "imposing": Does it not resemble just that popular image of the scientist as 'master builder,' in bondage to his own inner imperatives,[36] alone, out of touch with the ground on which he builds, erecting doomed houses? But this is the sort of image constructivism might well encourage, if pushed to extremes, by overextending and artificially raising in consciousness the building aspect of scientific thought. What McClintock and Polanyi provide is evidence that the *phenomenon* (experience) of listening is there. Even radical constructivism does not deny the phenomenon, but by interpreting *this too* as construction, as 'nothing but' construction—in addition to outright construction as such—it, in effect, transforms that phenomenon (reductively!) into its opposite. Listening becomes speaking: No 'voice' other than one's own is now heard, and no more of the original phenomenon is left for other kinds of interpretation.

My brief discussion of these things is simply a reminder that the phenomenon is indeed there, and must neither be forgotten nor interpreted away. Nor is it to be confused with the old question of whether theorizing or observing is primary; what is at issue here is a *special kind* of observing, taken phenomenologically as a possible state of a human being in the world. How we ought to interpret it, and what role that might play in education, is just the sort of problem that hermeneutics is about, though not one that can be pursued here.

The further development of this theme—hermeneutics as critique of constructivism in the context of education—is a project in itself and one that should certainly be undertaken. In earlier parts of this paper, we saw how, in the twentieth century, hermeneutics itself had turned toward constructivism; and the value of this is now clear. The critique I envisage, therefore, is a friendly one: Among other things, it should clarify limitations, greatly extend the discussion of interpretation, and situate constructivism within a larger philosophy of science education. A major feature of such a philosophy would be to restore to the dialogical process of the study of science the ontological dimension of an essential mode of being in the world, and thus to avoid substituting for the image of the student as 'receiver' another image just as doubtful: a cybernetic mechanism, connected to a 'black box,' seeking its own 'equilibration.'[37]

It should be clear by now in what sense the word 'ontological' is used here. For the most part, it has little to do with classic problems about the existence of such objects as electrons, quarks, or black holes. It is not primarily a matter of the ontology of the *constructs* of science, but of science as such, its various modes of being in relation to nature, to the scientist, to the student, and the corresponding modes of the human being in dialogical interaction with these modes of science—possibly, with nature itself. To this, in particular, the sequel must address itself.

Thus far, and *inter alia,* I hope to have shown how quickly and naturally this highly general approach to understanding—a 'universal' approach, according to its founders—engages with issues both current and traditional, both philosophical and pedagogical. Constructivism, misconceptions and the feminist critique have all now been brought into the arena of discussion, albeit in a preliminary way. Another issue on the agenda mentioned here briefly is the matter of *commoditization,* which plays such a large role in the 'post-modernist' critique of Jean-François Lyotard.[38] The decline of interest in scientific careers among young people in certain industrialized countries is now receiving a great deal of attention.

Usually, however, this decline is addressed as a problem of the 'packaging' and 'selling' of science. It is addressed, in other words, largely within the framework of the commodity model of knowledge. But is it not possible that this model itself—described so well by Lyotard, yet cheerfully accepted by him—is actually a *cause* of the trends we are seeing, and is unlikely therefore to promote an understanding of these trends, much less their reversal? If so, then here again is a place where hermeneutics may help shed light.

[2]

Hermeneutics as an Approach to Science: Part II

ABSTRACT: This paper continues the hermeneutic-phenomenological investigation of natural science, in which understanding plays a role comparable to creative construction (see "Hermeneutics as an Approach to Science: Part I" [Eger 1993a])[1]. The first issue treated is that of language: Is the language of science part of the equipment of the scientist, part of the subject, or part of the object itself—nature already linguistically encased? This issue, arising from the so-called argument of "the double hermeneutic," relates the general question of the role of the subject in natural science to the role of interpretation. Examples of major interpretive developments in physics are discussed. The inquiry suggests that the role of interpretation and hermeneutics is tied to the educative or 'study-mode' of science; and that this mode can, apparently, be found at all levels and stages of science. The nature of this interpretive mode, and its relation to the creative mode, is then analyzed on the model of Gadamer's description of the interpretation of art. [—M.E.]

Introduction

Toward the close of the nineteenth century, a widespread intellectual movement, with currents in science as well as philosophy, came to the fore in Europe bearing the name "phenomenology." It was in part a reaction to what was happening then in the theoretical sciences, and in society itself under the impact of those sciences. The sudden growth in abstraction, as the mechanical picture of nature collapsed, accompanied by accelerating fragmentation of disciplines, brought about for many people a sudden 'loss of meaning' in science and in knowledge itself, just at a time when similar

This paper appeared in *Science and Education: Contributions from History, Philosophy and Sociology of Science and Mathematics* 2, No. 4 (1993), 303–328. Research for the paper was carried out, in part, under a City University of New York Scholar Incentive Award, and a version of it was read to the Philosophy Department, University of Auckland, New Zealand, in October 1992.

forms of fragmentation and meaninglessness emerged as well in their working and private lives.

Phenomenology, in its broadest sense, is that approach in science and philosophy which tries to stay close to the phenomena by avoiding as much as possible all abstraction and imposition of constructs, and by always relating the object of study to the experiences of the subject who does the studying. In philosophy especially, this effort to recapture the original, to return to the 'primordial,' was an effort to preserve the *wholeness* of human experience, whether this be in the day-to-day "life-world," in the laboratory, or in our theoretical imagination. By turning back to "the things themselves," it was thought, one might avoid or at least diminish the loss that follows upon reductive analysis; one might envisage a world from which the human being, as subject, is not always excluded. In the early part of this century, it was Edmund Husserl (a mathematician) who led the development of this approach, introducing a number of distinctive concepts and procedures (Husserl 1962 [1913]; 1970).

One of these procedures is the so-called "variation of profiles." The object of attention is viewed by the subject from different angles, so to speak, against different backgrounds and in different contexts. Each viewpoint affords the subject a different "profile" of the object; and it is only by examining many such profiles that a conception of the essence of the object is attained. The purpose of the procedure is to shake off, as much as possible, preconceptions inherent in any conventional or habitual viewpoint, while preserving all the particular views so obtained, which together constitute the object itself.

Hermeneutics, as self-conscious *interpretation,* becomes relevant to this entire approach in the following way: For each of the viewpoints selected in each specific context, an interpretation of the object within that context is called for. It is interpretation that is called for, not hypothetico-deductive analysis, precisely because of the phenomenological imperative to stay close to "the things themselves." One "interprets" the meaning of the thing from such-and-such a viewpoint in such-and-such a context, by relating it to the background, to its "horizon"; one does not postulate pure concepts or invisible entities in terms of which the thing may be understood. Joined together like this, as hermeneutical phenomenology (or phenomenological hermeneutics), it appears full-blown now in the work, for example, of Hans-Georg Gadamer (1975; 1976; 1988) and Patrick Heelan (1977; 1983; 1988).

Language enters fundamentally into hermeneutics because in a given language we already have, to an extent, a selection of viewpoints. Although such selections do allow for an enormous number of different perspectives, this freedom is bounded even in natural language, and far more so in the special languages of science. But language enters also in a more positive sense. The need to interpret—in terms of contexts, of experiences, and of specific standpoints—calls for metaphorical and analogical descriptions. Since new or unusual contexts cannot always be physically or visually inserted, it turns out often that language alone, and only language, can perform this function. Thus, in his major work, Gadamer (1975) focuses first on the structure of play, of games, to exhibit the reality of certain kinds of experiences; he then transfers the description of the same sorts of experiences to the context of art, and later again switches to consider them in the context of historical study.

Beginning with the next section, we will attempt the same kind of interpretation in regard to natural science. One thing to emerge from this, very much in the spirit of phenomenology, will be the close relation between science's quest for knowledge and the *study* or *understanding* of that knowledge—that is, between science 'itself' and education. Normally, as we know, the analyses of these two domains are kept quite separate: There are on the one hand 'real science' and its philosophy, and on the other, the philosophy of education and pedagogy. Insights from the former may be transferred to the latter, as has been done, for example, with Kuhn and Piaget. But usually that is an afterthought. A view of science that *includes* education as a central theme does not exist at present, as far as I know, though probably the treatments of Polanyi and Kuhn come closest to it.

The hermeneutic-phenomenological approach to natural science is not new. In previous papers I indicated the relevance to the present study of the work of Polanyi, Hesse, Heelan and Ihde (Eger 1992a; Eger 1993a). On the broadest theoretical scale, it is probably Mary Hesse who is most responsible for making at least somewhat familiar the idea that, various objections notwithstanding, hermeneutics in natural science is not an oxymoron. She has done so by calling upon the "new philosophy of science," and showing the extent to which, in one sense at least, that philosophy projects a view of science that is already hermeneutical (Hesse 1980, 167–186; Bernstein 1983, 31–33). In response, a number of philosophers—defending the traditional separation between the human and the natural sciences—have argued that actually hermeneutics is the relevant *distinguishing* feature, for it properly belongs only in the former sciences, not in the latter.

These debates have not settled the issue yet; but because they take up the question of the essential nature of the various sciences, there is no better place to begin the present discussion. To one such debate, therefore, let us now turn.

The Double Hermeneutic in Natural Science

If with Mary Hesse we take the view that, regarding the role of hermeneutics, all the sciences lie on a continuum, then the difference between physics and anthropology is not that the latter only is hermeneutical, by virtue of some inherent interest (as Habermas at one time claimed). Rather, the difference is a matter of degree: Anthropology is located on that side of the continuum where hermeneutics is prominent, while physics is on the other side, where it is less so. But in this view, even physics, the most determining and exact science, involves interpretation in some aspects of its basic activity. Now, in reply to this, Habermas has countered with an argument—attributed to Anthony Giddens—that is so suggestive and goes so deeply into the heart of our concern that it will be worthwhile to devote to it close attention. It epitomizes a whole cluster of such arguments, and therefore, as a genre, it is called the argument of "the double hermeneutic."[2]

In outline, the reasoning is this: Of course, the anthropologist trying to understand the behavior of some remote culture and the physicist trying to understand the remote aspects of nature are to this extent in the same position: Both will have to *interpret* what they observe, and construct theories. But that is only one stage of what goes on. Actually, the anthropologist's situation is different because, unlike the physicist, he finds a pre-interpreted world and a *language,* or symbol system, already in being when he comes on the scene. And this is the language to which he must—hermeneutically—gain access *before* any theoretical interpretation can begin about the world of interest, the world inhabited by the people who speak this language. That world cannot be directly encountered by the investigator as can the world of natural phenomena. And so, Habermas concludes:

> If the paradigm-dependent theoretical description of data calls for stage 1 of interpretation that confronts all sciences with structurally similar tasks, then we can demonstrate for the social sciences an unavoidable stage 0 of interpretation, at which there arises a *further* problem. . . . The social-scientific observer . . . has to make use of the language encountered in the object domain. (Habermas 1984, 110)

The argument is, then, that in natural science there is a *single* hermeneutic (stage 1) which lies in the effort to understand phenomena by means of theory: But we have a *double* hermeneutic in all those sciences that deal with a language, because there this language itself needs to be interpreted (stage 0), in addition to whatever phenomena the language describes. Apparently, by directing attention to whether or not a *language* is included in the *object* of study, this argument by Giddens and Habermas again cuts the sciences neatly in two; and at first it seems impressive.

Yet from our point of view, precisely this focus on language reveals a glaring omission: It is implied in all such treatments that whenever a natural scientist comes on the scene to work on a new project, he finds no pre-interpreted world, *no language* there already in being. What social philosophers have in mind, when contrasting natural and human sciences in this way, is an imaginary situation in which the physicist, say, always faces the phenomenon of nature *ab ovo* and directly, unmediated by any symbol system other than that of the life-world. But since the seventeenth century at least such a thing has rarely happened. Of course the scientist finds a language already in being—he or she finds the language of the particular science within which the new project belongs. Initially, the scientist is always a student of symbols created by others.[3]

In general, if we take the perspective of the one who *appropriates,* rather than the pioneer creator—a student, a scientist entering a new field, a scholar from another discipline altogether, a layman—anyone who for some professional or personal reason wishes to understand a particular domain of nature, then surely, not just the phenomena themselves but the scientific language that confronts such a person must be interpreted. Don't we have then in this case also a "double hermeneutic"? Even when this appropriator—let us call him a 'student' in the generic sense (Eger 1993a, this volume 3–27)—is ready to go into the laboratory and observe personally, he can hardly do so without first interpreting the language used in such laboratories.[4]

At this point we face of course a major objection: Are we not confusing two completely different types of activity, two separate domains? Even though one grants that physicists must learn the language of the previous generations of their own kind before tackling a genuine research problem, yet isn't that a matter of studying—of being a *student,* as indicated earlier, rather than a *scientist*—or, more generally, of 'preparing oneself' for the task? The task itself, *science itself* (research) begins only *after* this preparation has been completed. By contrast, the anthropologist desiring, let us say, to study an Amazon tribe, is in a different situation: The language of the tribe is part of the research task itself; it is faced out there in the field, not studied in class, with textbooks.

The issue raised in this objection is fundamental and (I would like to suggest) at bottom an ontological one. What *is* science? Where does it begin or end? What actions or circumstances constitute its *being*, as distinct from other modes of being? It is a basic aim of this paper to argue that the usual answers to these questions, or, rather, the usually unarticulated preconceptions in regard to such questions (as implied by the hypothetical objection before us), represent a dubious ontological stance—yet a stance with significant consequences for education, for the understanding of science's role in society, and perhaps for science itself.

Let me begin by showing an alternative to this stance. One way of doing so is to use again the concept of "embodiment" discussed in Part I of this study (Eger 1993a). In that paper, it was said that the hardware and software of a scientist's profession—experimental equipment and cognitive tools like theories and models (language)—by extending his perceptual reach, in effect function in the manner of an artificial body. By mastering these tools, making himself 'at home' in them, the scientist comes to "inhabit" them or "be embodied" in them in a specific sense that has been explicated by various authors (Polanyi 1958; Heelan 1977 and 1983; Ihde 1991); and this embodiment is precisely the 'mode of being' of the scientist *qua* scientist. In fact, according to these authors, the modern scientist cannot normally face nature "naked," without "inhabiting" this sort of "body," because only by means of it can the scientific view of nature come into focus, and scientific interpretations begin.

In the case of deep-sea divers and astronauts, the special suits literally worn and inhabited—the additional, artificial bodies—are not incidental to the mode of human existence in such alien environments, but belong to that mode. This follows from the very nature of human beings and of the environments involved. If, now, the scientist's extended body is a part of the being of science itself, as a space suit is part of astronautics, then the 'entering into it' by each individual, the *process of embodiment* going on all the time—this too must be a part of science. For it is not a matter of just *once* learning to use a suit. There is one kind of suit for the ship's interior, another for the moon, a different one for empty space; and new, improved suits are tested all the time, guided by feedback from the users. And so too with science.

But this issue of the suit or extended body—of putting it on and learning to live in it—is just the issue of education, of 'being a student,' in the more inclusive and generic use of that phrase. Thus, strangely enough, the question of whether natural science is or is not hermeneutical in some deeper sense than Giddens's and Habermas's "stage 1" is bound up with the question of whether the 'student mode' of being is or is not included in the being of science itself. And this, in turn, is related to one aspect of the question of objectivity, the question of the subject/object 'cut'—where the subject ends and the object begins: Is the suit, and putting on the suit, on the subject side of the cut, or on the object side? When we talk about the language of science, the apparatus of experiment, and the mastering of these instruments, are we talking about the investigator only or about the *investigated* as well?

To say that the astronaut's suit is only a means for the investigation of space, or of the moon, is to overlook the fact that that suit already embodies a partial knowledge of whatever is still to be investigated. A suit designed for the moon takes into account the gravity of the moon, the pressure there, the temperature and variations in temperature, the consistency of the surface, solar radiations, and so on. At the start of the astronautical project, the suit itself had to be the focal concern, the object; but as that problem was 'solved,' and the astronaut 'entered in,' the suit was joined to his body, became more or less peripheral to awareness (as our natural bodies are), and could be viewed thereafter as part of the subject.[5] What happened is that the subject/object cut *shifted* in the course of the enterprise.

This shift, this movement of the boundary between subject and object, Heelan takes to be characteristic of hermeneutics (Heelan 1977, 12; 1983, 206); and in this sense, the subject 'enters into' the object of study. Consequently, all arguments based on a fixed subject/object distinction become suspect, and, in particular, those that categorically separate the 'study' mode (entering the body) from the 'research' mode of natural science.

The Two Books of Science

Let us return now from the relatively new metaphor of the body to one that is much older—science as a reading of the "book of nature." The body metaphor, as developed from Merleau-Ponty (1962) to Heelan and Ihde, is powerful because it is so comprehensive, taking into account, as extensions of the human perceptual system, physical tools as well as cognitive ones. For some purposes, however, just because it is so comprehensive, it becomes unwieldy: Specifically linguistic features of the situation cannot easily be portrayed in terms of a body, yet it is these features that must now be discussed. The book metaphor will allow us to confront from yet another side the argument of the double hermeneutic, and to do so more explicitly in regard to natural science itself. For if we say that science is a reading of the book of nature, we raise again the whole issue of whether the language of that book is found there (to be deciphered, as Galileo thought), or whether it is brought to the book by the reader as part of the interpretive effort.

Taking advantage of the insights of the embodiment discussion, we are not likely now to opt for either of these alternatives alone, suspecting again some sort of a boundary shift in the very process of interpretation. From the student's standpoint, as was noted before, the language is already in being. From the researcher's standpoint, it is (in part at least) yet to be created. The body metaphor led to a solution of this problem by means of a separation in *time*—first one situation obtains, *then* the other. If now we wish to consider a student and a researcher, side by side, each trying to understand some sort of 'book,' this solution is not available, although the scene seems realistic enough. Perhaps we should say, rather, that the two are not facing the same book. Perhaps, in some sense at least, there really are *two* books of nature, not one.

On this view, we have the following picture: There is the primary book of phenomena, the one Francis Bacon told us to interpret; and then there is the secondary book, the book of interpretations started, more or less, in Bacon's day—to which we have been adding ever since at an increasingly furious pace. We could call the first book, the "book of nature," and the second, "the book of science." This second book is, of course, a sprawling, rambling, collective work, with its own gradations as to depth and importance. In Karl Popper's language, it is "world 3"—the sum total of recorded theories, reports of experiments, problems, solutions, and so on. Moreover, as Popper points out, it is this book with which even the mature scientist interacts most of the time.[6]

Recently, all sorts of people have been suggesting that the second book *as such* deserves more analysis. Not only have the historians and sociologists of science flourished by telling us how it gets written, but a new field has now arisen, the *literature* of science (promoted by a new organization), where even the rhetoric of this book receives the specialist's attention.[7] The increased scrutiny seems warranted because it has become clear that most of what society knows about nature comes from the *second* book, not the first—and that society includes scientists themselves.

The question arises (with Giddens and Habermas in mind): If reading the first book, the book of nature, is admittedly interpretive, is reading the *second* book also interpretive? Is there a second tier of interpretation? The most common answer, implied usually rather than spoken, is "no": Understanding scientific results *after* they have become established, or verified, or constructed is a matter of teaching and learning, which is in the realm of education or popularization. Certainly interpreting observations and experiments is part of science, but this belongs to the reading of the *first* book, or the *writing* of the second. It does not belong to the *reading* of the second book—except in the common pedagogical sense in which the content of textbooks, monographs, reports, must be properly interpreted by the student; and this, according to the prevailing view, is not constructive interpretation in the *serious* sense. All real construction, the kind that 'makes a difference,' takes place in the 'reading of,' or interpreting, the original book of nature.

Again we come upon that familiar separation, which, for natural science, posits only a single hermeneutic; and it is this understanding of things that we are questioning here. The line between 'science itself,' as the *construction* of particular interpretations of nature (of phenomena) and the *study* of science as the *reception* of these interpretations—this line is too sharply drawn: On the one side is the writing—or, rather, the *composing*—

of the second book, the book of science, to which attaches all the creativity associated with the writing of original books; on the other, is a *reading* of this book, regarded as a matter of 'getting it right' but not essentially interpretive, and therefore uncreative.

To see that this picture misses important aspects of what actually takes place will require a two-step argument: First, one must at least make plausible the claim that acts of creative interpretation and acts of 'mere' appropriation (understanding) cannot be so neatly separated; that the composing and the reading of the second book are not always disjoint because both—in principle, and often in fact—involve interpretation of the serious sort. Second, if these two aspects are indeed intertwined, then it has to be shown why, in what way, and at which stages of the enterprise this occurs. Let us turn now to the first of these points.

The idea that true interpretation inheres only in the composing of the book of science, not in its reading, derives from the common belief that although raw data does need interpreting, by the time results reach the textbook all interpretation must surely be over. Famous 'exceptions' serve only to reinforce this belief. Today, even outsiders know that a major debate is still going on about the meaning or interpretation of quantum theory, long after that theory has become established. But invariably, this is attributed to the uniqueness of quantum theory itself. Except for scholars in this area, nearly everyone tends to forget all the previous interpretation debates, such as those about Newton's laws, about the statistical interpretation of the second law of thermodynamics, or about Maxwell's equations—all taking place *after* the laws themselves had entered the book of science.[8]

Should these cases be thought of rather as *continuing* research, and therefore clearly not in the realm of the 'study' or 'reception' or 'understanding the results'? Perhaps not. In the end they often *appear* as research but, on closer examination, many instances cannot easily be placed within the normal paradigm, in which acknowledged problems are attacked with experiment or calculation, and some new result is obtained. Frequently the impulse to interpret arises from a *vague unease,* a lack of *understanding.* Moreover, the unease tends to occur precisely at the moment of the *reading* of the book of science—sometimes literally a textbook.

Consider two examples. In the 1890s, the physicist Heinrich Hertz undertook a fundamental reinterpretation of mechanics, despite the fact that at that time no crisis, not even any serious anomaly, troubled this basic science. He did so, as he says, "Solely in order to rid myself of the oppressive feeling that to me its elements were not free from things obscure and unintelligible. What I have sought is not the only image of mechanics, nor yet the best image; I have only sought to find an *intelligible* image" (Hertz 1956, 33, emphasis mine).

To share with us his concern, he reviews, in the lengthy introduction, all the existing formulations of the subject—the old Newtonian version based on the concept of force and used then in every introductory textbook, as it is now; and the more advanced versions of Lagrange and Hamilton, based on energy. In each of these he finds much to bemoan. For example, here is what Hertz thinks of the Newtonian approach:

> We see a piece of iron resting upon a table, and we accordingly imagine that no causes of motion—no forces—are there present. Physics, which is based upon the mechanics considered here and necessarily determined by this basis, teaches us otherwise. Through the force of gravitation, every atom of the iron is attracted by every other atom in the universe. But every atom of the iron is magnetic, and is thus connected by fresh forces to every other magnetic atom in the universe. Again, bodies in the universe contain electricity in motion, and this latter exerts further complicated forces . . . and in addition to these again various kinds of molecular forces. Some of these forces are not small: if only a part of these forces were effective, this part would suffice to tear the iron to pieces. But, in fact, all the forces are so adjusted amongst each other that the effect of the whole lot is zero; that in spite of a thousand existing causes of motion, no motion takes place; that the iron remains at rest. Now if we place these conceptions before unprejudiced persons, who will believe us? (Hertz 1956, 13)

We know, of course, who believes such things, more or less—average undergraduate students (or do they?). But the *unusual* student, the very bright one, probably has the same difficulty as did Hertz. I must emphasize that I am not quoting here some eccentric. Not only was Hertz the discoverer of radio waves, and co-developer, with Maxwell and Lorentz, of the classical theory of electromagnetism, but he was probably the only person ever to achieve fame as an experimentalist, a theoretician, and a philosopher of science. Not even the brightest undergraduate is likely to do what Hertz did then, however unintelligible the textbooks seem. In the body of his monograph, Hertz presents us with his own interpretation of mechanics—mechanics without forces.

It did not cause a revolution. Because it was complicated, and unwieldy, no one used it or followed up on the idea. To this day, it remains of theoretical and philosophical value only; but that does not mean it had no effect. Not only did it demonstrate convincingly the degree of interpretive freedom at the higher theoretical levels, but its influence on a later, more famous critique was direct and significant, as we learn from Einstein's autobiographical notes and other sources (Einstein 1949, 31). Hertz's suspicion of constructions by the scientist himself prepared the way for general relativity.

I take as my second example the work of a contemporary physicist whose lectures many of us enjoyed and still remember: Richard Feynman's reinterpretation of non-relativistic quantum mechanics—the well known "path-integral" method. Feynman too was puzzled by the standard formulation of his day. To the consternation of colleagues, even near the end of his life he insisted that "no one really understands" it. But back in the 1940s, as he tried to do so, using the famous textbook by Dirac, he came upon a discussion there of the quantum analog of the classical action principle. It was a brief excursion by the author into a Lagrangian perspective on quantum theory, in place of the usual Hamiltonian perspective. This is precisely what Feynman needed to translate his ideas about classical electromagnetism into quantum theory, solve the problem he was working on, and obtain his Ph.D.

It turned out, however, that he could not *understand* what he was reading. Dirac had said that a certain quantum transition probability was "analogous" to a certain exponential function of the classical action.[9] But, asked Feynman, "What does he mean they are analogous; what does that mean, *analogous*?" Having put this question to a visiting physicist who tried to help, and having received no satisfactory answer, Feynman started to interpret for himself—that is, he began a kind of dialogue with Dirac's text that we can recognize immediately as a *hermeneutic circle:* Assuming first that "analogous" stood for simple equality, he checked the consequences against other parts of the text—standard theory, the Schrödinger Equation—and found a discrepancy. Next, he changed the assumption slightly to a proportionality and went through the same calculation. On this second try he found that indeed the consequences were consistent: "I turned to Professor Jehle, not really understanding, and said, 'Well, you see, Professor Dirac meant they were proportional'" (Feynman 1966, 37).[10]

This was the beginning of a new way of thinking in quantum theory—its *reinterpretation* on the basis of entire histories of particles (paths) rather than just their states at a particular time.[11] For Feynman, though, still officially a 'student,' it was already his *second* notable reinterpretation; the first had been a recasting of classical electromagnetism into a form that dispensed with fields (much as Hertz had dispensed with forces). This already reformulated electromagnetism then led Feynman to reinterpret quantum mechanics in terms of paths. The third reinterpretation—of quantum *electrodynamics*—is the one that brought the Nobel Prize. We could, with justification, call this man the 'master reinterpreter' of the physics of our time.

An interesting question comes up now regarding that crucial dialogue with the text—Feynman's attempt to construct for himself the meaning of Dirac's use of the word "analogous": Was Feynman's construction a *correct* interpretation of Dirac, or was it a *creative* reinterpretation? In other words, did Dirac *know* (and was he actually trying to say) that the two mathematical expressions in question were proportional, but chose, for some reason, not to develop this fact as did Feynman; or, was he (and everyone) unaware of it until Feynman came along? Feynman himself said, "I thought I was finding out what Dirac meant but, as a matter of fact, had made the *discovery* that what Dirac thought was analogous was in fact equal" (emphasis mine).[12] On the other hand, Julian Schwinger, fellow Nobelist with Feynman, thinks "Dirac surely knew" (Schwinger 1989, 45). Clearly, the question was of some importance to Feynman because on one occasion he tried to find out directly. James Gleick describes what happened when, at a 1946 Princeton conference, Feynman came up to Dirac and asked, point blank, whether "the great man had known all along that the two quantities were proportional": "'Are they?' Dirac said. Feynman said yes, they were. After a silence he walked away" (Gleick 1992, 226).

The question remains undecided. I have raised it here only to illustrate the point often made by constructivists and hermeneutic philosophers that the line between receptive and creative interpretation—between *finding* the meaning and *constructing* meaning—cannot always be drawn.

Although Feynman's reinterpretation of quantum mechanics is far better known than Hertz's of classical mechanics, it too caused no revolution—since it revealed no new laws of nature, predicted no new effects. For this reason, some physicists have belittled it, and even Feynman himself voiced disappointment. Yet on the whole it *is* appreciated because it affords an entirely new way of conceptualizing phenomena. The thinking it arouses has contributed not only to Feynman's third, prize-winning reformulation but also, as a technique, to a number of other areas in physics. And even in the strictest sense of achieving something really new, no one can say with finality that the Feynman path-integral formulation is a 'mere' alternative; recently, the possibility has arisen that in cosmology it may play a unique role (Gell-Mann 1989).

My aim in both these examples was to raise the following question: Shall we regard these initial frustrations, interpretations, and groupings—of a Hertz or a Feynman—as activities of research or activities of *trying to understand the results of research* (i.e., study)? Were these physicists interpreting the book of nature or the book of science? A good case can be made either way, but the second alternative is surely not excluded. If we accept it, even in part, then we see that the process of *studying established science*, of trying to fathom it, occurs at the highest levels of competence and involves (perhaps more frequently than we think) problems of understanding not different from those faced by the average student at various stages of education. The examples show also how such problems lead one to reinterpret. For hermeneutics, this close tie between understanding and interpretation is a major feature, as I have emphasized throughout (Part I[13]; Gadamer 1975, 274); and it was a feature also in the scientific life of Richard Feynman, a feature of which he was especially conscious. In a letter to a friend, written in the same year in which the path-integral method was being finalized, Feynman expressed a view of science rarely encountered in the literature. Here are some excerpts:

> I find physics is a wonderful subject. We know so very much and then we subsume it into so very few equations. . . . Then we think we have *the* physical picture with which to interpret the equations. But there are so very few equations that I have found that many physical pictures can give the same equations. So I am spending my time in study—in seeing how many new viewpoints I can take of what is known. . . . I dislike this talk [of there] not being a picture possible but we only need know how to go about calculating. The power of mathematics is terrifying—and too many physicists finding they have correct equations without understanding them have been so terrified they give up trying to understand them (Feynman, quoted in Schweber 1986, 472).

But do such episodes represent anything more than an interesting sidelight on the thinking of outstanding scientists? Or do they represent hermeneutical difficulties, and impulses to reinterpret, that permeate science throughout? Although more stories like the ones above can easily be given (and some will be), a confident answer to this question can of course come only from extensive historical research. Nevertheless, I do suspect it is so; and, for the sake of discussion, I assume so in what follows. What I will argue, and try to show eventually, is that the process of science involves, in effect, a "cascade of interpretations," from the highest levels of the kind just mentioned, all the way to teacher and student at the various stages of education. Interpretation, I will maintain, is a fundamental and pervasive event that, potentially, ties the enterprise together; and it is for this reason that subdividing science into its 'educative' and 'creative' modes is problematic and, in an important sense, inappropriate.

It remains now to take up the second of the two tasks projected earlier: If we are willing to entertain the idea that interpretation does occur as a matter of course in the 'study of results,' then, still, we need a clearer picture of how this interpretation comes about and what place it occupies at various stages, and in science as a whole. For this purpose, I would like once more to turn to the work of Gadamer. Although, for the most part, Gadamer himself opposed the extension of hermeneutics to natural science, a certain component of his analysis seems to me particularly appropriate to the task at hand. Surprisingly, perhaps, that component is his treatment of art.

Interpretations

In the first part of his *Truth and Method,* Gadamer gives a phenomenological description of the arts, particularly the performing arts, as a model for similar descriptions of history, of language, and of the social sciences. His goal is to exhibit universal features of the encounter—the interpretive encounter—between humans and the cognitive cultural objects they create. He chooses the performing arts for special attention, since in this realm, where interpretation is *physically* enacted, the desired features will be easier to see. Despite the habitual contrast, repeatedly drawn in our society, between art and science, Gadamer's treatment is relevant here because the role it assigns to interpretation in the very *being* of art is just what needs to be emphasized in science. If we consider art not in the obvious ways it differs from scientific inquiry, but in the *manner of its being* and of its appropriation, then the analogy will turn out less far-fetched than at first might appear. As in art so in science, interpretation occurs at all levels whether or not it is acknowledged—and, I will maintain, as in art, the *being* of science depends upon it. Gadamer is not alone in taking up, as a central question, the ontology of the art object. The distinguished philosopher of art Roman Ingarden (1989 [1961]) has done so too, as have others. Since the problems of hermeneutics are usually regarded as epistemological rather than ontological, the question of being may seem out of place in discussions of interpretation. However, as emphasized in the previous paper (Part I), in Gadamer's work (and in Ingarden's) we have a hermeneutics with an ontological turn; and this precisely is the direction to take if we wish to examine the sense in which the *works* of science *exist* for the one who encounters them and wishes to appropriate them—our generic student.[14] By "works" we shall not mean here theoretical constructs—electrons, genes, or black holes—but rather the theories, experiments, and texts that make up the 'book' of science. These may *refer* to electrons or black holes, but it is not the latter that 'students' actually encounter when they read the book of science. First they encounter the 'works,' and these nearly always in the framework of language.

For art, *being* becomes a problem if one asks, for example, what *is* the Shakespearean play *Julius Caesar* or Vivaldi's musical composition *The Four Seasons*. It is a problem because we realize that what is contained in the script or in the score cannot really be the answer to this question, but only one extreme of a spectrum of possible answers. One cannot say that a piece of music is ontologically identical to the musical notation read by the performers, since the notation itself *is* not music. Similarly, the lines of *Julius Caesar,* as read, *are* not in themselves what comes into being when the play is properly staged. Nor can we say that the story, the plot, constitutes the play. We realize soon enough, if we pursue the question, that whatever a musical work or a theatrical play *is* has vitally to do with *performance*. The score or the script are crucial, of course, yet only in the performance do such works come *fully* into being. We arrive then at the hypothesis that if these art objects must be performed to actually *be there,* then their very *being* may *include* their performance.

Of course, a performance is interpretive. There is much missing in the script that is supplied by the director, the actors, the stage designers, even the audience; and this is why the play is more than the script. Consequently, the being of such works of art depends on their interpretation at least in so far as each must be interpreted to be performed, and must be performed to be *there* at all. However, the role of interpretation is more pervasive still. Because the same script can be interpreted differently by different directors, different groups of actors, and different audiences, what we have if we view the work as a total phenomenon—from creation to appropriation—is in fact a *cascade of interpretations*: First, a scripted play like *Julius Caesar* may be the author's interpretation of real events; then a director makes an overall interpretation for the staging; this is followed by specific interpretations by the actors, resulting in a concrete performance; finally, all members of the audience interpret what they behold.

As in any cascade, we see in this description a sequence of movements alternating with a sequence of relative stabilizations, of concretizations. Phenomenologically, we are encouraged to contemplate the total *event of art,* spread over time as well as space. By contrast, in the normal way of looking at things, one notices only the stabilizations—scripts, particular stagings, particular performances—not the movements, the interpretations that lead to these stable structures.

The spectrum of possible answers to the question of being is generated when different weights are assigned to the different levels of this cascade. At the most objective end of the spectrum, one would insist that all, or practically all, of the work of art *is* the arrangement of words or notes, as understood by the author or composer; at the other extreme, one says that it is the reader or the viewer or the listener who *determines* what the work finally is. Thus, the cascade and the associated spectrum of views define also the dimension of objectivity—or objectivism versus subjectivism—in the ontology of art: The being of the work may reside wholly in the 'schematic formation' originally created and handed down to us, or it may reside wholly in the perception of the beholder.

This subject/object distinction—in every kind of encounter, whether among human beings or between humans and their creations—is a special concern for Gadamer. It was his attempt to avoid that distinction altogether that took him on another detour, even before the discussion of art, to focus on *play,* or games in general. In a game, he points out, the subject/object dichotomy disappears as the player becomes part of the action: One can say not only that the player plays the game, but, just as well, that the game plays the player. One 'enters into' a game and 'forgets oneself' or 'loses oneself' in it; one is forced, or obliged, to act in a certain way by the state of the game.[15] Treating all forms of art as play, or as games in this sense, Gadamer places the being of the work of art more or less in the middle of the scale of objectivity: All the levels of interpretation contribute significantly.

The being of *Julius Caesar*, then, lies certainly in the script of that name, but only when taken together with the whole ensemble of possible or acceptable interpretations. It consists, in other words, of a formal or schematic core (the script, in this case) surrounded by *potential* presentations of this core, possible performances. In the language of phenomenology, *Julius Caesar* is not a set of symbols that can be read off, but an *intentional object* constituted, in any particular instance, by the whole cascade of interpretive perceptions—which are not arbitrary since they are part of a game of interaction with the core (the script), a game that imposes broad but powerful criteria.[16] This is the essence, and the distinctive feature, of the theories of art of both Gadamer and Ingarden.

One implication of these theories is that the viewer, the listener, or the beholder of a work of art becomes part of a larger subject whose boundary with the object is fuzzy. Or, put another way, at every level, each interpreter enters the circle of understanding—the hermeneutic circle—which is the game (dialogue!) between himself and the formal core of the work; and this too implies that the boundary between the interpreter and what he interprets is neither sharp nor immovable. A second implication is that without the later interpretive stages, a work of art is *incomplete*. If art is a kind of game, a play within a structure, then a material object as such—a script, a canvas—cannot, by itself, be an instance of art. What are usually considered 'receptive' interpretations, as contrasted with the acknowledged 'creative' interpretations, are crucially part of this game, and therefore part of the *being* of art.

The foregoing is of course a rough and drastic condensation of what is in the original an elaborate, subtle, in-depth treatment involving several authors and many works. We must take into account also that although the performing arts are chosen as prime examples, the same kind of analyses have been carried out in the case of other arts as well, where performance in the literal sense does not enter. For literature and painting especially, this has been done in great detail by Ingarden (1973; 1989), with no substantial difference in the role of interpretation.

Science and the Cascade of Interpretations

With this as background, then, let us return to the main problem: Were such a theory applied to *natural science,* what would it look like? Of course, whatever is said here can be no more than a suggestion and the beginning of a development; but it seems worthwhile to present this suggestion, even as only a bare skeleton, to indicate the outline of the idea as a whole. My procedure will be to transfer, modify, and adapt to science any element of this theory that seems appropriate—from Gadamer, from Ingarden, from Husserl, and from other hermeneuticists and phenomenologists—*unless* it can be shown why this cannot or should not be done. In effect, this assumes an underlying structural similarity between all symbolic cultural creations, among which art and science are just two.[17]

We begin with what corresponds in science to a work of art. I have already begun to call this a "work of science." A work of science can be an experiment, a law, a model, a theory, and so on. Every such work becomes at some point a paragraph or a section or a chapter in the second of the two books discussed earlier; this includes experiments as well, which are, of course, written up, recorded, and read, as is every work of science. In the 'writing up,' however, much is necessarily left out. As a playwright cannot specify, along with the script, the gestures, the facial expressions of the actors, or the entire stage setting, so a scientist cannot exhibit every aspect of his experiment, every nuance or possible visualization of his theory. Recall Feynman: "We know so very much and then we subsume it into so very few equations."

What happens then as the equations are 'read'? Just as a performed play is inconceivable without *some* gestures and *some* kind of staging, so the *understanding* of an experiment or theory is impossible without *some* definite interpretation, which, in effect, supplies what has been left out. It follows that interpretation does not stop after the works have been produced, but enters a new stage; and what we should look for now is the rest of the cascade. So far, we have discussed only the higher levels of this cascade, the kind of interpretation done by the leading scientists themselves, especially when the meaning of theoretical terms or claims is not clear. Now the hermeneutic imperative tells us to look further down; and this, of course, is where the strongest objections come up. They come especially as we realize what this procedure implies in regard to the ontological question—the question of the *being* of the work of science.

The first thing it implies, of course, is that the being of such a work— an experiment or a theory, for example—does not consist wholly of what is written about that work in the book of science. What is written and found in any authoritative segment of that book corresponds to the script of a play or the score of a musical composition, which, ontologically speaking, is merely the *core* of the work. This core, taken in a particular context, has the potential for particular interpretations. The context may or may not lie wholly in the book of science, but, in any case, that book is very large, so a unique context does not come unambiguously tied to the work. It remains precisely for successive interpreters to present some profile of the work within a chosen context, to constitute its meaning and so to bring the work more fully into being. Feynman, for example, saw in textbooks a profile of quantum mechanics in the context of a Hamiltonian picture; what he wanted was its profile in the context of a Lagrangian picture—which led to the path integrals.

One way of seeing more clearly what is implied in the view of science being sketched here is to compare it with Karl Popper's objectivistic philosophy of the three worlds: Recall that in Popper's framework, in addition to "world 1" (matter) and "world 2" (the subjective realm of feelings and thoughts), there is also a third realm—one of ideas, theories, problems, and so on. All three "worlds" exist fully, according to Popper. The point is that "world 3," as he calls it, is ontologically distinct and independent even though it is humanly constructed. In this scheme, both theories and problems can be *found* in world 3, essentially complete; they are not constituted in the dialogical activity of *interpreting* that world.[18]

By contrast, in our view here, neither quantum mechanics nor the theory of relativity (to choose another example), nor any of the problems raised by these theories, is fully *there* until the cascade of interpretations runs its course, from the book of nature to the understanding of individual human beings. As part of this cascade, performances, presentations, and therefore interpretations take place, having much in common with those in the arts, and blurring therefore the subject/object distinction even in the most exact disciplines. This element has been largely ignored in the philosophy of science, though there are signs that in the future it may receive more attention.[19] If it does, then yet another departure will have begun from the treatments of science to which we have become accustomed. Let us examine briefly some aspects of this performative and presentational element.

If we focus on the *experiment* as a "work of science," the idea of performance, repeated interpretive performance, is surely not far-fetched. It can almost be said that a new experiment is *designed* to be performed—again and again—as is a dramatic play. It is performed more than once by its creators, to bring it to perfection; then it is repeated, with variations, by others who try to improve it even more; later, it is repeated with students at various levels, to give them a chance to participate in and witness the same experiment. To this day, in undergraduate and even high school laboratories, we 'perform' experiments first done, for example, by Galileo and Atwood in mechanics, by Young in optics, by Joule on heat, by Michelson in regard to relativity, and so on. That at many stages reinterpretation takes place is by now well known. At the higher levels, we have such examples as the reinterpretation by Lavoisier of the experiments of Priestley, which resulted in the identification of oxygen (Kuhn 1970, 56).

The audiences at these performances vary. It has not been unusual for only one person to be present, trying to verify for himself or herself what had been reported. Some occasions are designed to convince or impress peers, and some to demonstrate the phenomena for large groups of all sorts. At the Royal Institution in London, we know, figures like Faraday took the matter of presentation seriously enough to study it methodically, write on it, and practice it to widespread acclaim. The resemblance of all this to theater has not gone unnoticed by historians of science. One of them, L.P. Williams, compares Faraday's advice on demonstrating to Hamlet's speech to the players.[20]

Can one say something similar about the non-experimental works—the theories, models, and so on? I think the answer is yes, though in these cases, *presentation* naturally takes a different form. From here on, therefore, I will use this term in a more general sense, in which it may refer to spoken or written presentations without dramatic aspects as such.

Consider again the special theory of relativity: What is often called its discovery or formulation or construction by Einstein was, in fact, a reconstruction—a reinterpretation of an already existing theory of H.A. Lorentz and others. Regrettably, many students and laymen today believe that this pre-existing theory could not explain some of the optical or other experiments then at issue, and that only with Einstein's theory was physics at last able to do so completely. But this is a misconception, abetted unfortunately by lack of regard for history.[21] Not only were most of the distinctive "relativistic" effects—length contraction, time dilation, inertia of energy—already known, but so were the basic equations of relativity (Lorentz transformations) which explain these effects.[22] The achievement of Einstein's 1905 reinterpretation was to place all this on a theoretically simpler, more unified, more general basis, by discarding the concept of the *ether*. In so doing, Einstein gave what is perhaps the outstanding example of a hermeneutic approach to the book of science: He questioned fundamental preconceptions regarding time and simultaneity—preconceptions implicit in all previous theories.[23]

Yet this was only one step in the cascade—the second, let us say. The third followed quickly—Hermann Minkowski's (1908) reinterpretation of the same equations and the same characteristic effects in terms of a 4-dimensional space. Since that time, numerous presentations have been made re-creating or re-deriving or re-casting that theory on a different basis or in a different form—even *in terms of the ether* (a 'backward' reinterpretation!).[24] One of these, the so called "K-calculus" of Hermann Bondi, which minimizes the mathematics of the theory, was designed exclusively as *presentation,* and is used in lectures and books emphasizing ideas rather than formalism.[25] Concerning its value, David Bohm says in his book on relativity, "The Lorentz transformation approach and the K-calculus approach complement one another, in the sense that each provides insights that are not readily obtainable in the other" (1965, 144–45).

From Lorentz and Einstein at the top, through Minkowski, to Bondi's presentational interpretation, we see an example of a *cascade in science*. I have not described it completely, for we have not followed here many side branches, some of which can certainly be taken as responses of individuals, the final steps in the cascade. But enough has been said, perhaps, to make plausible the idea that even as far 'down' as Bondi's little book for the public (1964), serious rethinking is still taking place—inviting, challenging, stimulating the reader to do likewise.

Again someone may object: Does it really make sense to blur the difference between a major theoretical innovation like Minkowski's with a purely presentational idea like Bondi's, just because both were, in a strict sense, reinterpretations? The former was clearly an extension of the theory, recognized as such by all physicists, the latter a contribution to education. But this is precisely my point: This line between the writing of the book of science and its interpretive appropriation is drawn too sharply; and usually it is drawn in retrospect. As in Feynman's case, the value of Minkowski's reinterpretation was not so clear in the beginning, not even to Einstein, who called it superfluous learnedness ("*überflüssige Gelehrsamkeit*").[26] Some interpretations simplify and minimalize, like Bondi's; others complexify and maximize, like Minkowski's. Some develop into whole new areas of research; others become accepted as just another way of looking at the same thing. But even in this last-mentioned case, as Feynman pointed out, alternative views still have an important role:

> Different physical ideas may be equivalent in all their predictions, and are hence scientifically indistinguishable. However, they are not psychologically identical. . . . For different views suggest different kinds of modifications . . . and hence are not equivalent in the hypotheses one generates from them in one's attempt to understand what is not yet understood. (Feynman 1966, 44, emphasis mine)

The common factor in all reinterpretations is the *effort to understand* by looking at the thing in whatever way seems best to the interpreter—whether the interpreter is doing it for himself or for others. As in art, interpretation takes place all along the line from creation to appropriation; and in science this interpretive activity reveals a *continuity* usually obscured—a continuity that deserves more notice.

Another way of looking at the matter is to recall Reichenbach's distinction between the "context of discovery" and the "context of justification." It was offered originally to differentiate between the contingent and possibly irrational ways in which discoveries are made, and the rational ways in which they must be justified, established. By implication, it leads one to assume that discovery and justification are the only modes of science as such, and its only *goals;* and that all scientific activity must be seen as belonging to one or the other of these contexts. But if so, it is hard to see how one classifies a major reinterpretation like Minkowski's, or Feynman's path integral formulation. These contributions did not justify the already existing theories, nor were they a discovery of anything new about nature; yet, compared with their predecessor formulations, both offer alternative or expanded views of the world. They are interpretations. One could say therefore that there really are *three* contexts, not two: discovery, justification, and *interpretation.*[27]

In the normal view of things, both discovery and justification go into the composition or construction of the book of science. The book thus composed becomes the object of understanding for its readers or students, the subjects. Since the subject/object cut is taken to be sharp, the subjects at this level are not part of the *being* of the object or of science, regarded as independent of the subject's existence. Whatever interpreting they do is aimed only at 'getting straight' the meaning of what they are reading, assumed to be determined and contained in the book itself. But close historical study immediately points to a difficulty here: We have already seen that sometimes the "book of science" offers more than one interpretation of phenomena; and this is true not only when major reinterpretations are undertaken but in commonly used textbooks as well, even in fields considered 'settled' long ago.

Fabio Bevilacqua (1983) has analyzed comprehensively and in depth the differences of interpretation in 20th century presentations of classical electromagnetism by leading physicists; and he has traced these differences, in great detail, to the 19th century debates that went into the making of this science. That such differences are not noticed by many teachers and students is due to the fact, apparently, that in science today we are so focused on the 'bottom line' that many kinds of meaning escape us.

Still another problem with the normal view is that textbooks written by major contributors to the field, like the ones by Dirac or Heisenberg on quantum mechanics, are clearly part of the real book of science, the object; but those churned out by the hundreds each year, to facilitate course work—these are more like digests, or books *about* the book. The writing of these texts is regarded as a kind of journalism, not true *composing* or constructing. So the cut is positioned somewhere on the spectrum of all books, and the average textbook is either on the subject's side as a mere tool, an aid to reading the real book, or on the cut itself, as a channel of communication.

In contrast to all this, if presentation and interpretation are taken as a basic context of science itself, more or less on a par with discovery and justification, then the line between composing and reading the book of science does indeed blur, because interpretation takes place on both sides of any such line. In this view, *all* presentations and all interpretations, however far down the cascade, are part of the *being* of science as a whole. In regard to any individual work, presentation and interpretation belong to the work of science as they belong to the work of art. The textbook, even a hack textbook, is part of the interpretation, and thus acquires increased importance for better or worse; it is one of the *faces* of science, the face encountered by whoever reads that book.

Does this mean that bad books about relativity or evolution detract from the very being of these theories? That seems absurd, but let us consider it for a moment. Does a bad performance of *Julius Caesar* diminish that work? Not ordinarily, most people would say, if they have seen a good performance of it, or read it seriously. Potential interpretations that present a 'true' *Julius Caesar* can still be visualized. But should a time ever come when no one really understands what this play was about—because our own concerns and way of life have become so different—and few are interested any more either in seeing or performing it, then of course the script might still be there deep in the vaults of electronic memory banks, but what could one say about the *existence* of *Julius Caesar*, as a work of art?

By no means should this be construed as a form of relativism. Relativism is no more the direction of this study than it is of hermeneutics as practiced by its foremost exponents.[28] The understanding of science is now affected by extreme subjectivism *and* extreme objectivism, all at once; and I suspect that hermeneutics can counteract both extremes—in science as it does in art. Subjectivism comes, for example, from sociologists of science, some of whom say boldly that "The natural world has a *small or nonexistent* role in the construction of scientific knowledge."[29] Objectivism, on the other hand, is standard in almost all formal science education, which takes for granted the subject/object cut in the way I have discussed.

Clearly, the hermeneutic perspective bears on this realm (education) directly, because it challenges there a highly entrenched view that is embodied in the 'commodity model' of knowledge: Scientific knowledge, in this model, regarded as a *transferable commodity*, is *produced* by researchers, *stored* in papers and technical monographs, *distributed* by teachers using textbooks and popularizations, and finally *received*—for *use* in some manner—by students and the public at large. Thus, the teacher is portrayed as the *retailer*, and the student as the *consumer* of science's cognitive products; and thus they see themselves as *outsiders* to science 'itself,' identified with the production process. The subject/object distinction is not only taken for granted in this view, it *reifies* everything in sight. In contrast, hermeneutics fosters a vision of science somewhat closer to our vision of art (though, of course, not identical to it), in which the researcher, the teacher, the student, all *participate* in the *being* of a science by interpreting at each stage the objects of that science.

Yet equally, the hermeneutic perspective challenges the *relativist* trends now expanding in influence, for it takes science *seriously* as a structure in human existence, and the "world of science" as part of that structure. It takes seriously also the *works* of the "world of science"—the theories and models and experimental observations—not as *unique* determinations of reality, nor yet as pure constructions of the mind, but as *profiles and interpretations of reality from a human viewpoint.* It does not reduce science to rhetoric.

This paper has moved rapidly over a number of ideas that are not generally familiar in analyses of natural science. Its purpose is to serve as an overview, orienting and situating the more concentrated studies that are, of course, still needed. Specific consequences, in education for example, have not been examined here although these should be substantial. It could hardly be otherwise in a portrayal that highlights *understanding,* and posits 'student' as a basic mode of *being* toward science—a mode taken as the standpoint for the portrayal itself. But the actual discussion of such consequences must be left for another occasion.

[3]

Achievements of the Hermeneutic-Phenomenological Approach to Natural Science: A Comparison with Constructivist Sociology

ABSTRACT: The hermeneutic-phenomenological approach to the natural sciences has a special interest in the interpretive phases of these sciences and in the circumstances, cognitive and social, that lead to divergent as well as convergent interpretations. It tries to ascertain the role of the hermeneutic circle in research; and to this end it has developed, over the past three decades or so, a number of adaptations of hermeneutic and phenomenological concepts to processes of experimentation and theory-making. The purpose of the present essay is to show how appropriate these concepts are to an important current research program (solar neutrinos) and thus to point out what difference they make to our understanding of science as a whole. This goal is pursued by means of comparison. The program of social constructivism in natural science has produced alternative but parallel concepts, embodied in an alternative and parallel vocabulary. The contrast between this vocabulary and that of hermeneutics and phenomenology reveals, so I argue, the advantages of the latter. But actually it does more: It reveals as well the 'pre-understanding' or 'prejudgment' of science embedded in each approach. [—M.E.]

Introduction and Background

In the past, as we know, most continental philosophers denied that hermeneutics has any place at all in natural science.[1] For many of them, it actually served as a demarcation criterion between the social and natural realms. I, on the other hand, am convinced that even though the hermeneutic-phenomenological approach to the understanding of the study of nature is still young, it has already made important contribu-

This paper was presented at the fourth annual meeting of the International Society for Hermeneutics and Science (ISHS) held in 1996 at Stony Brook, New York. It later appeared with a selection of papers from that meeting in a special issue of *Man and World* (30: 343–367, 1997).

tions—achievements, if you will—though it is true that these achievements have not yet had the influence they should have.

To discuss some of the achievements by comparing them with certain rival formulations is, therefore, my first goal here. The second is to indicate why, nevertheless, they have not yet had the desired impact, and to suggest along the way what might be done about that. I have chosen to argue by way of comparison because, as in science itself, no approach is ideal or comprehensive; we are always obliged to pick out the best, or the best part of 'what's available.' Yet in doing so, in making explicit comparisons, we gain not only confidence in our decision, but greater clarity about what each of 'the available' really offers, what *the field* is like. And this, of course, is of value whatever the outcome.

When I say "rival formulations," I will not mean here the various analytic, objectivist or positivistic philosophies of science. The difference between these and hermeneutics is well-enough known within groups like the one at this conference.[2] Rather, I have in mind a certain segment of the sociology and sociologically-oriented history of science that flourishes now and includes such writers as Harry Collins, Trevor Pinch, Steve Shapin, Simon Schaffer, Karin Knorr Cetina, David Bloor, Steve Woolgar and Bruno Latour. For brevity, I will call this group "the constructivists."

The contrast between hermeneutics and *this* type of sociological work is rarely mentioned because, in fact, the two share some common ground, borrow from each other, and are often perceived as moving in the same direction. Yet, as I want to show, there are important differences that should not be played down, even in the interest of presenting a united front against objectivism and scientism.[3]

The common ground is the critique that both hermeneutic philosophy and constructivist sociology direct against several key features of the dominant analytic view of science: at the small role ascribed to interpretation, at the way objectivity is portrayed, and at the claim to universality. Clearly, all these are affected by the extent to which the scientist engages in construction. Both the hermeneuticists and the sociologists are convinced that construction plays a larger role than analytic philosophy of science allows. Yet the spirit of these two critiques, their import, and many of the specific criticisms differ in a way that is of real significance to various groups of people inside and outside the academic world.

To see the difference fully, we would have to take up one at a time at least the three features I mentioned—interpretation, objectivity, universality—and exhibit side by side the hermeneutic and the constructivist approach to each. This I would certainly like to do, and it remains my long-term plan. But the subject is vast. Every one of these features alone can take up a whole talk or paper; so today I confine myself to the first—the matter of interpretation.

When we ask what *kinds* of interpretation there are, or what *stages* of science call for interpretation, one crude scheme I like to use, that has emerged from previous discussions, is this:

Stage 0: Interpreting the received heritage or tradition of a science as a whole:

   a. reading the "book of science."

   b. practicing routine procedures of science.

Stage 1: Interpreting at the level of the research experiment:

   a. interpreting data.

   b. interpreting phenomena in terms of high level theory (or, interpreting theory by 'performing' it experimentally).

Stage 2: Interpreting high-level theories in alternative ways.

The numbering (starting with 0) is something I have taken over from Anthony Giddens and Jürgen Habermas, from their discussions of the "double hermeneutic." Having in mind recent work in the history of science, Giddens and Habermas do not deny even to physics a certain hermeneutic aspect. But they allow for natural science only one stage at which interpretation plays a serious role—stage 1. By contrast, in the human sciences, there is in their view a prior stage—stage 0—where the investigator must first come to understand the *language* of the people being studied, just to get *access* to the data of stage 1.[4]

In opposition to this, a number of people, myself included, have argued that in the natural sciences, as in the human, interpretation occurs at several stages. Despite the fact that the objects of natural science—stars and atoms—do not speak, there is here too a prior problem of 'language' learning, and often a further problem at a much higher level that I have called "stage 2."[5]

Be that as it may, there is little doubt that it is stage 1 that has been receiving most of the attention, that it is here that the greatest concessions have been made, like those by Habermas, and rightly so. For it is here that science becomes "a 'reading' of the 'book of nature,' requiring circular interpretations between theory and observation and also theory and theory, and also requiring a 'dialogue' about the meaning of theoretical language within the scientific community" (Arbib and Hesse 1986, 181). It is this stage, therefore, that will concern us today.

Interpretation and "Negotiation"

Perhaps the most important achievement of hermeneutic thinking has been to call attention to the role of interpretation precisely at *this* stage, the stage of "observation." Here we always knew it existed but, until recently, since we thought it unproblematic, it had gone largely undiscussed. Yet, arriving at a conclusion as to what the experiment shows—or, to use Robert Crease's terminology, *recognizing* the phenomenon—is, as nearly everyone now admits, far from straightforward.[6]

Awareness of this has been growing at least since the 1950s when Polanyi's writings started coming out. It reached a watershed with Kuhn.[7] Since then more and more explicitly hermeneutic and phenomenological treatments have made their appearance, including of course by several contributors at this conference, as well as by others such as Mary Hesse (1980), Marjorie Grene (1985), and Joseph Rouse (1987).[8]

At this point, however, concurrent with the rise of hermeneutic thinking, and influenced by it no doubt, enter the new breed of sociologists and anthropologists of science—the ones I've mentioned—with their ever-more-detailed studies and with conclusions ever more shocking to the uninitiated. They pounce upon the problematic aspect of experiment like detectives who think they see a smoking gun right under the suspect's bed. They treat it as the great exposé of our times.

Collins, for example, tells us that "truth, rationality, success and progress are not found to be the driving forces of science" because, he says, "it is not the regularity of the world that imposes itself on the senses but the regularity of our institutionalized beliefs that forces itself on the world" Collins (1985, 185, 148). Or, as Latour puts it, "a stable interpretation . . . of microbes is provided by bacteriology . . . exactly like . . . a stable state of society is produced by the multifarious administrative sciences. . . . No more no less" (Latour 1987, 256).

I will not multiply these quotations, which everyone has read by now. My goal is to focus on the key mechanism used by these constructivists to portray the way science works in cases of disagreement, how the problematics of experiment and the relations of theory to experiment are handled. This mechanism they call—"negotiation."

I was surprised to read in a review of the field by the sociologist Joseph Ben David that he found this word—"negotiation"—*fitting* to the situations to which it was applied *despite* the fact that he criticized its use severely (Ben David 1981, 45, 48–51). Just how fitting it is we will examine in a moment, but first I want to explain why it interests me enough, why I think it is important enough, to take up a good part of this talk.

Let us begin by noting that although this word *derives* from hermeneutics, as I will show, it is not a part of the hermeneutical vocabulary with which we are familiar. That this is so is perfectly understandable in view of the fact that phenomenology and hermeneutics have several other well-known terms which, in application to natural science, denote the same activities as does "negotiation" but say something entirely different.

It is in fact these terms that I would describe as *fitting* in their adaptations to natural science, and it is these adaptations, already developed in some depth, that I would count among the truly impressive achievements of the hermeneutic-phenomenological approach. Yet it is "negotiation" that has carried the day, that pervades the literature and can be heard in STS departments from coast to coast in America and Britain.[9] Why this is so is in itself an interesting sociological question, but one I will not pursue here.

What I do want to pursue is the contrast between the way in which the word "negotiation" is used and the meanings of its hermeneutic counterparts when applied to the same situations. This will lead us to several related concepts that we can also compare in the same way: the "experimenter's regress," the "black box," and the "un-doing" or "deconstruction" of science. My aim is to exhibit the cumulative effect of this new vocabulary that—sometimes *in the name of* hermeneutics—embodies a substantially different philosophy and a substantially different stance in relation to science.

The power of the word "negotiation" in our context stems from the fact that, in the Western world at least, it has a fairly well-defined meaning; it does not signify *any* sort of a discussion that people may have. What it requires, first of all, is parties with *opposing interests,* such as one finds in the political and economic arenas. When such parties attempt to reach agreement by *trading* on their interests, by giving up something here to gain something there, each party trying to maximize its own advantage, frequently at the expense of the other—this we all recognize as negotiation, and something like this always comes to mind whenever the word is used.

But what does this word denote when used by the constructivist sociologists of science?

Just about anything, it seems—from what was once routinely called critical discussion to the sort of bargaining I have just described. At the same time, *all* negotiation is characterized as "funnelling in" social interests, turning them into nonscientific negotiating tactics, using them to "manufacture certified knowledge," and finally "rendering them invisible" by "laundering" (Collins 1985, 143–144). One effect of this terminology is that it assimilates very different kinds of activity under the same rubric and labels them all "social."

For example, in one of the many stories Latour tells, when Elmer Sperry, early in this century, needed money to develop his idea of the gyroscopic compass, he started *negotiations* with the Navy to fund the project; he convinced them that out of it would emerge the instrument they were looking for to replace the magnetic compass on iron ships (Latour 1987, 112).

We nod our heads—that's scientists negotiating all right. But when physicists criticized Joseph Weber's experiment to detect gravity waves and convinced him to adopt their standard of antenna calibration—that too, according to Collins, was "negotiation." In fact, it was badly conducted negotiation on Weber's part; in his own interest he should have stuck to his guns and refused to accept his critics' ideas (Collins 1985, 103).

"Negotiations" over calibration come up again in the story—told by Shapin and Schaffer—of conflicting results obtained by different experimenters working with the vacuum in the 17th century. This is a major theme of the highly-regarded and much-cited book *Leviathan and the Air Pump,* chock-full of impressive scholarship, but tendentious, trying to convince us that in the argument over experimental method—the argument between Boyle and Hobbes—it was in the end Hobbes who was right: Experiment proves nothing because it is always possible to *negotiate your way around* any result (Shapin and Schaffer 1985, 226, 282, 344).

However, the most revealing such story that I have recently come across, and the most worthy of scrutiny, is that of the search for solar neutrinos, recounted in the popular book by Collins and Pinch, *The Golem.*[10] In this case, it will be worthwhile to go over a few of the scientific ideas, before we consider it philosophically.

A Contemporary Example—Solar Neutrinos

A crucial part of stellar evolution theory is the set of nuclear reactions believed to be going on in the cores of stars, some of which result in the production of neutrinos—the massless or nearly massless particles whose weak interactions with matter make them extremely hard to detect. Since the 1960s, a major effort has been building to try to measure the flux of neutrinos from the sun using large underground containers of various fluids in which a fraction of the neutrinos heading our way should be trapped and detectable.

The original motivating idea was that if the predicted number of neutrinos is found, then, in effect, we would be 'looking into' the core of the sun, verifying that the supposed synthesis of heavier elements from lighter ones is really taking place and that the evolution of stars and elements proceeds just the way we theorize. However, for over a quarter century now, the number of neutrinos found is considerably lower than that predicted, and, moreover, the results from the different experimental groups do not cohere properly.

Starting at the beginning, Collins and Pinch tell a dramatic tale of how Ray Davis of the Brookhaven Laboratory devoted his career to building the first detector—a giant tank of chlorine, deep in the Homestake gold mine; how a young theorist, John Bahcall, started calculating the expected neutrino flux; then, how Bahcall *negotiated* with Maurice Goldhaber, Director of Brookhaven, to convince him that the experiment was worth doing, and how—somehow—he got the prediction of the neutrino flux to be *high* enough to convince funding agencies to provide the money; how depressed Bahcall became after the first disappointing results were in; how he started fiddling with his models, *negotiating downward* the predicted figure for the neutrino flux so as to meet the low flux detected by Davis; how another theorist, Icko Iben, who didn't believe the new figures, applied pressure to Bahcall in his own negotiations with him, getting Bahcall to switch sides "dramatically" and declare that a real discrepancy does after all exist between theory and experiment; how other physicists criticized Davis's procedure but succeeded by *negotiation* to get him to perform highly-demanding tests of his equipment; how new and more-sensitive detectors are now beginning to produce data—though, as might be expected, these too are in disagreement. And how, nevertheless, despite this lack of success, Bahcall "managed to . . . make a career out of solar neutrinos" by stressing the importance of the problem, thus landing a prestigious professorship at the Institute for Advanced Study in Princeton. In the very last sentence of this chapter, ending the story, the authors tell us how the problem is being resolved: "Negotiations are in progress!"[11]

What is the upshot of all these detailed narratives? By the time the reader has wended his way through the 15th or 16th such example he may well be ready to conclude, as so many have done, that negotiations about the very substance of science, about what is to count as fact and be accepted as such by the public, are, as Latour and Woolgar say, "no more or less disorderly than any argument between lawyers or politicians" (1979, 237).

In short, *science is politics.* The effect of the term "negotiation" is to habituate us to *this* understanding as the *only alternative* to objectivism in science and to scientism in the culture.

In a section they call "Science Unmade," Collins and Pinch describe how, in trying to account for the neutrino discrepancy, theorists were willing to throw overboard the most secure and trusted elements of the theoretical edifice: Energy conservation, stellar evolution, models of the interior structure of the sun, even the famous reactions of nucleosynthesis, supposedly the source of our energy, all seemed up for grabs in the negotiations—*science un-made.* The idea is to show that since scientists *con*-structed all these things in the first place, they can *de*-struct them if they so choose.

But is it *true* (in the simple "life-world" sense of that word) that were we to enter a lab or attend a conference on solar neutrinos, what we would see taking place is the *un-making* of science—a re-negotiation of laws and principles of physics as lawyers and politicians might do it? It happens that, recently, I was present at one of these "negotiation sessions," so I can bring you personal witness of it—and some reflections on how fitting these vocabularies are to "science-in-the-making," as it is now called.

In March of last year, Cornell University held a celebration to mark the 60th anniversary of its most distinguished scientist's—Hans Bethe's—association with that campus. As might be expected, the proceedings included scientific talks by physicists working on topics in which Bethe has been involved. Since it was Bethe who proposed the particular nuclear reactions that produce the neutrinos for which the search is now going on, it was no surprise that solar neutrinos were on the agenda and that leaders such as John Bahcall (the same Bahcall whose "career" maneuvers Collins and Pinch described) were on hand to give the latest thinking on that. *Physics Today* called it a "Bethe Fest."

Well, what went on? How did the negotiations proceed?

One of the most interesting talks, I thought, was by Bethe himself, eighty-eight years old at that time and still at it. His was a general review of the solar neutrino situation to date.

As he talked—slowly but very clearly—putting on the screen various graphical depictions of the match and mismatch of theories, models, and data flowing in from Japan, Italy, Russia; and as the questions and discussion continued throughout the meeting, what came to my mind was a passage from Bob Crease's recent book, *The Play of Nature*: "an interpretive process, which we can understand thanks to Heidegger's conception of the hermeneutic circle . . . a making explicit of what I understand, a developing, deepening, and enriching one's involvements and expectations—so that the eventual moment of confidence occurs in the form of recognition of the presence of something that is already familiar" (Crease 1993, 150). Something like this is what I would say was going on at the Bethe Fest, and what has been going on with this whole project for over a quarter of a century. But let me try to be a little more specific.

One reason that hermeneutic accounts fit better is because their conceptual repertoire is richer. They avoid, for example, the flat-footed conflation of the *production process* involved in an experiment with the *performance* of the experiment. Again, I am alluding to Crease's book, in which this distinction is treated at length, separate chapters being devoted to production and to performance. A complex experiment, like a theatrical performance, does require the marshalling of resources and of talent. This is its production aspect. Here, negotiation of the usual sort takes place—as everyone has always known. But performance—the experimental procedure and its results—is something else.[12]

When Bahcall tried to get Goldhaber to allow Davis to proceed with the experiment, this might well have been negotiation of some sort, but it dealt with the production of the experiment (see Bahcall 1996b). When Davis accepted suggestions from others on how to check his neutrino-detection procedure, this had to do with performance. To place the two in the same category, in order to show linkage between production and performance, is to mix up second order interactions with first order distinctions.

Regrettably, what the public is getting in such accounts as *The Golem*'s neutrino chapter is nothing more than a story of experiment contradicting theory and of scientists trying to negotiate their way around that fact. (The book won a prize!) It is to be regretted because the real story is much richer and more interesting than that, and no more difficult to tell to non-scientists than the one Collins and Pinch have told. Moreover, it is a story that becomes more understandable when regarded with an eye for its truly hermeneutic aspects.

That we are not getting the expected number of neutrinos—Collins and Pinch's main plot line—is, in any case, no longer the real story. That was the story twenty years ago. But during those twenty years scientists have been circling, enriching their involvements and expectations, as Crease says, shifting the subject/object cut, as Heelan says, to include more and better instruments on the subject side—until today the issues are wonderfully different from what they were at the start.[13]

The upshot is that science has been getting more and clearer profiles of the phenomenon it is trying to bring to presence, and in the process something unexpected seems to be occurring: A new phenomenon, not the one sought at the start, is beginning to make its presence felt.[14] To explain this, let me use a chart showing a simplified version of the *phenomenon originally sought*—the presumed reactions in the core of the sun (Fig. 1). Essentially, the whole process is the fusion of hydrogen into heavier elements, ending in helium. Most of the reactions are left out of this picture so as to focus on the ones we can actually 'see,' the ones involving neutrinos. All the other particles produced are in various ways trapped inside the sun, but to neutrinos the sun is nearly transparent.

Now, I have used the word "see" deliberately; and it certainly is a fair question to ask just what it can possibly mean to say that we are "seeing" into the core of a star.[15]

It is one of the achievements of hermeneutic phenomenology to have addressed this question and offered an interesting answer—in the embodiment theory that Heelan has introduced and Ihde has elaborated. According to this theory, we can "see" or otherwise "perceive" phenomena

Nuclear Reactions in the Sun's Core

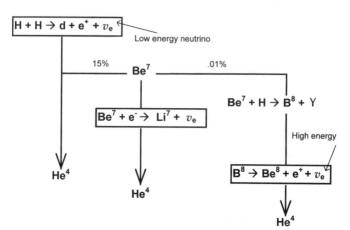

Figure 1. Nuclear Reactions in the Sun's Core

by means of instruments, once these instruments—in effect—become part of the subject and form an extension of his perceptual organs. The "seeing" here occurs in a way analogous to that of seeing a solid object in space: From any one position, we see only one profile. But when we change the viewing angle and perceive an invariant form as the source of the various profiles thus obtained, the solid itself comes to presence. The analogy depends on our decision to regard perfected scientific instruments as artificial sense organs of an extended subject; and to regard different instrumental arrangements oriented to the same phenomenon as, in effect, different viewing positions.[16]

Constructivist sociology has developed a parallel concept to embodiment—the "black box"—with an entirely different connotation. A black box is a standardized measuring device, a commodity, "a piece of furniture" to be borrowed or bought.[17] Both the black box and the extended body can be used without attention to their functioning, as an eye or an ear is used. Both yield "inscriptions" that are *read*. But while the theory of "the body" implies that "perception" is direct, that readings are accepted when taken as responses of the object of study (an epistemic reason), in the case of the black box acceptance is achieved by extortion, essentially, based on the *expense* resistance would require (a sociological reason). The power of black boxes to coerce assent is proportional to the cost of challenging their testimony, which is proportional to the costs of comparable boxes in the hands of challengers (Latour and Woolgar 1979, 242).

The difference between the two concepts could hardly be sharper: One gives us a picture of a subject *accreting*, as it were, new perceptual organs that open up new realms of being. The other points to scientific careerists with ample funds, acquiring or "enrolling" as many black boxes as they can, to present as formidable a defense as possible against challengers and competitors. In the solar neutrino research, the large detectors are as yet neither black boxes nor embodiment, since they are still far from unproblematic. We do not yet "see" the nuclear reactions in the core with confidence. But the process of embodiment ("black-boxing," in the other language) is certainly under way.

Looking with our present equipment at the total neutrino flux, we get less of it than predicted, and that was the original puzzle. But note that, according to the theory, neutrinos are expected from three different branches of the phenomenon, and in each branch their energy distributions should be different. Using this presumption, detectors have been designed to be sensitive to neutrinos coming from the different branches. In effect, we have today several viewpoints, yielding several different profiles of the phenomenon. And now a totally unexpected thing has emerged. You recall that, according to Collins and Pinch, in order to account for the neutrino deficit, physicists were busy "un-making" science, negotiating to sacrifice crucial parts of the theory, like our models of the sun.

However, from the viewpoint I am taking, what they were really doing is quite different. They were "playing" (Crease 1993). They were *circling* hermeneutically. They were changing something in one or another part of the theory to see how that would affect predictions ("what-if" games familiar to every business student). They were trying new equipment. They were simulating different evolutionary scenarios on computer, feeling their way about, making explicit what they understood. And out of this "play" emerged the so-called "beryllium/boron anomaly"—which is what the story today is about, and was the sort of thing discussed at the Bethe Fest (Raghavan 1995; Bahcall 1996a).

The beryllium/boron anomaly refers to a comparison of the number of neutrinos coming out of the beryllium branch (Fig. 1, middle) with those from the boron branch (right end). It turns out that regardless of solar models and nuclear reaction rates, the experimental results show a big deficit in the beryllium we "see," as compared to the boron we "see." But the beryllium is a component in the synthesis of the boron. How can it be missing if the boron is seen?[18]

This is the really meaningful outcome of all the playing with theory—not the "un-making of science" but the realization that the problem goes deeper than had been suspected, that it is largely independent of models of the structure of the sun but strikes at the nature of the phenomenon itself. If this phenomenon (the reactions) is real, and if our understanding of our situation is correct, then regardless of anything else, the phenomenon itself does not have the expected shape, and no amount of "negotiation" with models will change that.

Where then do we stand in this research now, and how does the hermeneutic-phenomenological approach help us understand it?

Theodore Kisiel, who has paid close attention to the discovery process, has a fitting description for this kind of situation: "a transitional experience [which] transposes us from a disintegrated context and directs us toward another integration" (Kisiel 1973, 406). What it means here is that we stand possibly at the threshold of the emergence of an entirely new phenomenon with enormous significance.

This is so because there appears to be only one promising way now to understand the beryllium/boron anomaly and related problems. It has to do with the fact that there exist other kinds of neutrinos than the one for which we have been searching, the electron-neutrino expected from the sun. According to the standard theory, all neutrinos are supposed to have zero mass; but if actually neutrinos do have a small mass, then quantum mechanics tells us that a fraction of the electron neutrinos would convert to the other kinds (called "flavors") while en route to Earth. Since our detectors "see" only the electron neutrinos, we should then see fewer of them—as in fact happens.

Needless to say, part of the "what-if" games is to calculate what we *should* see *if* neutrinos have various masses and therefore convert at various rates. This has been done, and it shows that the deficits, including the beryllium/boron ratio, could be accounted for if some of the neutrinos have a mass of about 3 milli-electron-volts. If they do have mass, however, there would be a fundamental change in particle physics, and possibly in cosmology too, since the mass-carrying neutrinos might account for some of the missing mass of the universe—another puzzle, that at first had nothing to do with this whole project.[19]

The Hermeneutic Circle and Other Vocabularies

Notice the truly hermeneutic feature here: We start with the sun's core as our object, while neutrinos are to be the medium carrying the information. We rely on our pre-understanding of these particles and of our apparatus (our embodiment) and of the entire background of practices that goes along with all that. But in the course of interpretation, we find that this pre-understanding fails, and the neutrinos become part of the object under investigation. To understand the sun's core we must understand neutrinos; but to understand the neutrinos, we must (so it seems) understand the sun's core—we must know the flux of electron-neutrinos coming out of there to know whether any of them are converting while en route. We are in "the circle."

Now at this point, a clarification is in order. Several formulations of the term "hermeneutic circle" are known, so it is important to indicate the sense in which I use that concept here. We have, first of all, the classic notion of the "circle of understanding" as the back-and-forth movement of thought from a part of the object of investigation to the whole, in which each new understanding of the latter modifies the understanding of the former, and vice versa. An enlargement of scope leads to the more dialogical

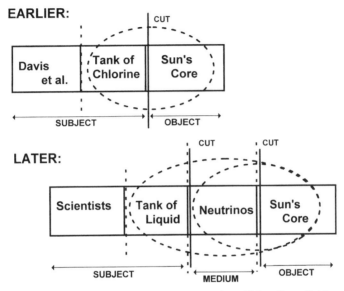

Figure 2. Movement of the Boundary Separating Object from Subject.

and ontological version I associate with H.-G. Gadamer: Here the circling is between a global pre-understanding of the subject's own world and the response of the object—again with the latter modifying the former and vice versa.[20]

In Heelan's adaptation to natural science, this mutual influence results in a movement of the boundary separating the object from the subject. The movement is *away from* the subject when progress has the result that more and more of the object is incorporated into the apparatus of the subject, becoming "transparent" and, in effect, part of the subject's sensory system. The movement is *toward* the subject when troublesome responses from the object cause the gaze of the experimenter to veer closer to home—to 'question his senses,' as it were.

Thus, at first, our pre-understanding of our equipment and of the sun—the flux of neutrinos from it, their energies—led not only to the construction of a "sensing" apparatus but to its *integration* into the routine practice of the experimenters—a movement of the boundary away from the scientist in whose mind the idea of such an experiment first arose. Later, the low response from the sun forced the objectivation of the chlorine tank and the neutrinos, in a move back toward the scientist (Fig. 2, top diagram).[21]

Today, what we have is not just a dichotomy but a trichotomy—subject/medium/object (lower diagram). The old circle between the sun and the detector still involves us; but in addition there is another one between the sun and the neutrinos. As the neutrinos become suspect, the second cut moves back and forth; the neutrinos are understood sometimes as medium, sometimes object (Fig. 2, lower diagram).

Note, please, that this at least partially answers the objection of some philosophers that the circles now finally acknowledged in natural science are only 'theoretical' or 'methodological' but not 'practical' or 'ontological,' and therefore not truly hermeneutic.[22] The argument is that the circling between hypothesis, experiment, modified hypothesis, and so on, involves changes in the scientists' *theory* of the object without changing their 'practical' engagement in the world, their understanding of being. In Heidegger's terms, such circling involves only the *Vorgriff*, maybe the *Vorsicht*, but not the *Vorhabe*—the tacit knowledge of background that comes with being a particular kind of human in a particular place.[23]

This criticism loses much of its force when applied to the formulation I have been discussing because here interpretation is understood more universally: It includes not only the classic idea of the methodological circle, but the post-Heideggerian ontological concept as well. Here not only models but performances are affected. Changes take place not just in the structure of theories but in scientists' *embodiment* and in their total *problem situation*. The circle that *cannot be diagrammed* is the one involving the scientists' *understanding* of this whole world—of equipment, of problems, and of practice—in which Fig. 2 is contained.[24] Ultimately, this larger circle touches also the long-term change in our self-understanding as humans, a change that results from high-level evolutionary and cosmological theories whose development is affected by research such as that on solar neutrinos.

There is more to be said about these circles, but let me emphasize at this point how important it is now that stories like the one about solar neutrinos be told with a sensitivity to the genuinely hermeneutic features they include, rather than in a way that shoe-horns events into a far-fetched, ill-fitting mold borrowed from politics.[25] The conceptual enrichment in the approach to science that hermeneutics has already achieved makes possible such alternative stories. That it is the social constructivists who are doing most of the storytelling today, rather than hermeneutic writers, is a circumstance that perhaps some of us may want to do something about.

Not that the constructivists avoid altogether these hermeneutic features, but they describe them differently, using their own vocabulary, and conveying therefore an entirely different meaning. Thus, for a certain kind of circle, dear to his heart, Collins has coined his own term—"the experimenter's regress." It is formulated usually as an interdependence between the object of an experiment and the adequacy of the equipment. To obtain the right outcome concerning the object, we must have properly functioning equipment. But if, as Collins and Pinch say, there is no independent check on the equipment in an experiment such as this, then, in practice, only the right outcome gives confidence in the equipment. But we don't know what the right outcome is and so *ad infinitum—the experimenter's regress* (Collins 1985, 84).

No doubt, Collins is aware of the large literature on the hermeneutic circle. Why then has he invented this new name, one that suggests a bottomless pit, epistemically speaking—a defeat? Actually, Collins is right. The name he chose is fitting indeed for the view *he* holds of what he is describing, since for him the circling is vicious in the classic sense and therefore certainly a defeat. It is the hidden secret of science. According to Collins, progress can only be made when the circle is broken; and the circle can only be broken when *nonscientific* tactics—"antics" he sometimes calls them—enter the negotiations and restrict interpretation in some way (Collins 1985, 103, 106, 143).

Hermeneutic thinkers, on the other hand, believe that circling is precisely what *has* to take place. It is the *means* to understanding, because rarely are our preconceptions corroborated in every respect. It is not escape from the circle by restriction of dialogue that brings about closure, but a continuing effort (more of the same!), sometimes for decades, that shrinks the circle and stabilizes the boundary between subject and object. The aim, in a situation like this, is for theory and experiment to validate each other *simultaneously*. Since, in the case of the neutrinos, they don't validate each other yet, the people working on it have to keep on circling from nuclear physics to solar models to neutrino theory—to see if they can make the circle shrink; or find a new entrance into it; or reach some new standpoint from which the whole circle looks radically different.

For hermeneutically-oriented scholars, it is hardly surprising that, in circles like these, even the most stable interpretations may be challenged, possibly altered. Such scholars see no "un-making" here but precisely the dialogue between scientists and their tradition, their history. With Gadamer, they say that taking one's tradition seriously is not the same as slavery to tradition. To take it seriously is precisely to come to grips with it. One entertains the thought that the solar models are wrong in some respect, but not lightly. One raises the possibility that neutrinos may be converting to different kinds, but this leads to a whole new series of difficult, very large scale experiments on terrestrial neutrinos—a veering of the dialogue in a new direction that might actually break some circles to some extent by providing the independent checks whose possibility Collins denies.[26]

But even if one day theory and experiment do converge, even if some circles are broken, we will not have escaped all hermeneutic circularity—as scientists or as human beings. Theory and experiment are science's way of interpreting the world, but they always are interpretations carried out from the inside. The chance that the whole web of theory and experiment is wrong may diminish but it is never zero. The largest circle is never broken. Closure, or temporary closure, comes when we have a workable interpretation that rests on stable invariants under variation of both instrumental and theoretical standpoints. If this is not God's truth, neither is it a bargain struck by politicians in the back room.[27]

I would say that the interaction between current science and its tradition should certainly be one of the great themes of a hermeneutic approach to science. Where some sociologists, with a smile and a wink, point to scientists un-making the past, engaging in something suspicious, something contrary to our understanding of how it is all supposed to work, *there* hermeneutics *recognizes* one of the truly familiar features of the enterprise of seeking understanding.

Does it make a difference? Does it make a difference whether we call it "science un-made" or "dialogue with tradition"? Whether we say "hermeneutic circle" or "the experimenter's regress"? Whether we use the words "negotiation" or "interpretation," "black box" or "embodiment"? Do particular expressions count that much?

Yes, I think so. Because a whole understanding of what science is about and how it works is carried by these words, as we see; and their constant use, their repetition, like repetition in commercials, has its reinforcing effect on the public and on all of us. I think the vocabulary of hermeneutics is more fitting, more capable of doing justice to what goes on, even in natural science. For those who doubt it, let it be juxtaposed to competing vocabularies and tested against the things at issue—*die Sachen selbst*—like solar-neutrino research. This is one of the things, it seems to me, that should be done more vigorously than in the past.

An Unscientific Genealogical Postscript

As a final point, and because questions of origin are often enlightening, let me return now to that key word "negotiation" that has led us to this whole issue of fitting and unfitting languages. And let me ask just how and when did this particular word find its way into the sociological descriptions of science to displace what most scientists, most philosophers, and even the average person would be inclined to call "interpretation." How did this come about? I confess that I did not do a rigorous historical search, and I don't know whether what I am about to discuss is really the first instance of the use of this word in our context. But, at any rate, it claims to be that; it is itself an interesting and revealing case; and it does come early in the history of the constructivist sociology of science.

In the 1976 book by David Bloor, *Knowledge and Social Imagery* (a revered and ancient work by now), the author tells us at the beginning of Chapter Seven that he wishes at this time to introduce "an entirely new process, which I shall call 'negotiation'" (133).

The issue he is dealing with is the role of informal thought in logic and mathematics. In this connection, following John Stuart Mill, he denies that the syllogism is an entirely deductive process, because the relation between general principles and the cases that fall under them is not inferential but interpretive. At this point, he offers the following quotation from Mill: "This is a question, as the Germans express it, of *hermeneutics*." So, Bloor concludes, the informal really has primacy because in this space, between the general and the particular, the informal may always "seek to criticize, evade, outwit or circumvent formal principles . . . [and] this negotiation is what Mill referred to as an interpretive or hermeneutic process" (133).

We see, first of all, that the word "negotiation" is here quite openly *substituted* for what Mill and the Germans called interpretation and

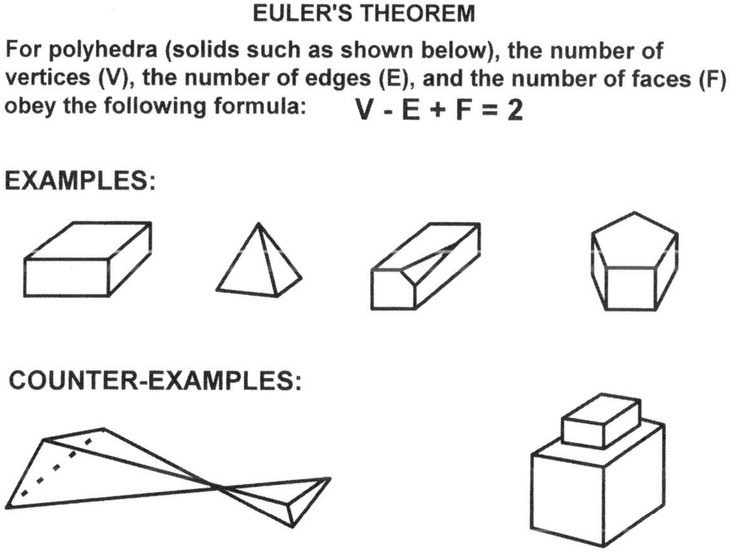

EULER'S THEOREM

For polyhedra (solids such as shown below), the number of vertices (V), the number of edges (E), and the number of faces (F) obey the following formula: $V - E + F = 2$

EXAMPLES:

COUNTER-EXAMPLES:

Figure 3. Euler's Theorem

hermeneutics. It is just an outright replacement, without any argument as to whether or to what extent the two concepts actually overlap. In place of argument, Bloor gives us examples, histories.

I would not ordinarily want to go into these examples, but it happens that the main one was already discussed by this group at the Holland meeting last year, when Olga Kiss gave her paper (Kiss 1995), and this circumstance affords us yet another revealing comparison. The example is the nineteenth century debate over Euler's theorem, as dramatized by Imre Lakatos (Fig. 3).

The theorem is a simple relation between the number of vertices, sides and faces of a polyhedron; and for the ordinary sorts of polyhedra it can be proven. The trouble started when mathematicians began to think up the uncommon types (lower part of the diagram) for which the theorem does not hold. For each new counter-example to the proof, the validity of the theorem would be preserved by limiting its scope, by amending the definition of "polyhedron" to exclude the troublemaker. And that is mostly what the debate was about (Lakatos 1976).

From this, Bloor concludes: "The meaning of the term 'polyhedron' was in need of decision. . . . It had to be created or *negotiated*." There are two points now to be made regarding this use of the history of Euler's theorem.

First, I want to pinpoint how the idea of negotiation enters descriptions of mathematics and natural science: Examples are found of something that *could* be called negotiation, such as arguments over the appropriateness of *definitions* (which usually are conventional without prejudice to results); or dealings about the *production* of experiments, which of course involves money and careers. *These examples are then generalized to all of science, without discussion as to whether or when such generalization is warranted.*[28]

Second, I would like to call attention to the difference between the way David Bloor and Olga Kiss treat this very same piece of historical writing by Lakatos. In Bloor's story, the prominent part is the thrust and counterthrust of conjecture and refutation, the battle of wits. Just to give reasons for a claim is, as he says, to "open up a front along which they may be attacked" (Bloor 1991, 148). And this indeed is the picture drawn by most of the sociologists to whom I refer. As Lyotard put it, science is *agon,* and the strategy of science is "*agonistics*."[29]

Here, it seems to me, is the deeper significance of the term "negotiation." It is not just that science involves competition which often gets hot, but that conflict—between interests and individual antagonists—is of the *essence* of science. For only where such conflict exists is negotiation really needed. Criticism of the Popperian sort does not necessarily make for conflict. But if it is true that "nature is mute," then science is an affair between scientists only, and *in that* case, negotiating skills are indeed of the essence.

Olga Kiss, on the other hand, takes up this Eulerian example for altogether different reasons—the title of her paper was "Hermeneutic Roads to Commensurability" (1995).

In her treatment what comes to the fore is not battle but *dialogue*—dialogue between mathematicians engaged in interpretation of a received piece of mathematics; dialogue in their encounter with their tradition; dialogue arising from the need to make explicit the tacit pre-understandings that cause trouble. What in Bloor's story was the lunge and parry of a duel appears in hers as an extension of horizons, first in one direction, then in another.

Therefore, Kiss never uses words such as "negotiation." From her, we get an essentially Gadamerian interpretation of the same Lakatosian text. In Bloor's story, human will *dominates* logic and ideas. In that of Kiss, as in Gadamer's, knowledge and understanding do not arise at the conclusion of some *agon*, a prize won in a war with fellow scientists. Rather, it is the outcome of a dialogue or play in which *die Sache selbst*—not the players—is dominant; and to which the players, scientists or mathematicians, 'give themselves.' Here then is a clear demarcation between two visions of science.[30]

The difference between these visions is of considerable importance today, an importance that transcends academic debate because the constructivist vision has already had a considerable impact on campuses and is beginning to have one on journalism, on science administration, and on the public at large. The issue here is *not,* as some scholars have (unfortunately) written, whether science is to be funded at higher or lower levels. Much more than funding is at stake: At least since Husserl, it has been clear that science is a central feature in the character not only of Western culture but of a culture that is rapidly becoming global. If we seriously *misunderstand* this feature of our character, with all its relations to other important features, then we misunderstand ourselves in a way that could prove grievous.

Certainly, the constructivists deserve credit for their detailed investigations (when these are carried out with due care), for highlighting the interpretive aspect of the natural sciences, for showing hermeneutic circles in these sciences even if under some other name, for engaging in a much-needed criticism of historical insensitivity in textbooks and popularizations; and also for attempting something very difficult, which could yet turn out better than now seems likely—a more general critique of science from the outside. Yet, while acknowledging all this, and indeed learning from it, we should point out those features in the constructivist literature that are fundamentally different from the hermeneutic and are a *misrepresentation* of science. We should point out as well that these two failings have something to do with each other.

If, in other words, we make greater efforts to show that a genuinely hermeneutic approach leads to better understanding of the *same* kinds of historical examples constructivists treat in an engaging but often misleading way, then this approach may yet gain wider appeal as a corrective to present trends—one that retains the critical attitude, but is essentially serious.

[4]

Language and the Double Hermeneutic in Natural Science

ABSTRACT: The argument of the "double hermeneutic," advanced by Anthony Giddens and Jürgen Habermas, among others, is designed to separate off the natural from the human sciences by claiming that only in the latter is *language* part of the object of study; in the natural sciences, presumably, the mediating scientific languages belong to the equipment of the scientist, the subject. This puts the spotlight on the role of language in natural science. But if "language" is understood in the broad sense—including concepts, classifications, theories—and if the actual *practice* of scientists and students is examined, then a good case can be made that in natural science, too, language is, at least sometimes, part of the object. More precisely, in regard to language, the boundary between object and subjects *shifts* hermeneutically as inquiry proceeds.

I use the work of Patrick Heelan to show the relevance of this shift to the critique of the argument of the double hermeneutic, and then extend the discussion to introduce another metaphor—'the two books of science'—to exhibit this same boundary shift from another side. In this context, examples from the history of physics are discussed, and Richard Feynman's reinterpretation of quantum mechanics proves to be especially illuminating. The gist of the argument is that *interpretation,* and therefore hermeneutics, takes place in natural science in the manner of a 'cascade,' from the highest stages of competence and creativity all the way to the normal process of 'mere study.' [—M.E.]

Introduction

Perhaps nothing about hermeneutics is better known than its role, or putative role, as a demarcation criterion between the natural and the human sci-

This paper was presented at a conference in Veszprém, Hungary, organized by the Hungarian Academy of Science in 1993, at which the International Society for Hermeneutics and Science was founded. The conference proceedings were published as *Hermeneutics and Science*, M. Feher, O. Kiss, and L. Ropolyi, eds., (Boston Studies in the Philosophy of Science 206, Dordrecht: Kluwer, 1999), 265–280. The stories about Hertz and Feynman on pp. 81–84 are very similar to passages in Paper 2, but they are retained here because the propositions which they illustrate are somewhat different in the two papers.

ences. Faced with the danger of creeping scientism, social philosophers have repeatedly grasped at the idea that if hermeneutics can be shown to be both essential and unique to the human sciences, foreign or secondary to the natural sciences, then an important freedom will have been won to reject in the social realm that misplaced, alienating stance of the 'neutral' investigator, and to regard human beings as human once again.

It is not surprising, therefore, that after the recent work of Mary Hesse and others[1]—denying the existence of such a hard boundary, suggesting instead that hermeneutics is relevant in varying degrees across a continuum of sciences—new, refined arguments should have been put forth, attempting to show that in so far as hermeneutics is at all relevant to the natural sciences, it touches only on technical problems, not on those vital questions of meaning that in the social sciences lead to the true hermeneutical circle. Because arguments of this sort from Charles Taylor, Hubert Dreyfus, and Jürgen Habermas, among others, have a common structure, the whole genre has acquired a family name—the "argument of the double hermeneutic."[2]

In what follows, it is not my aim to deny that important differences exist between the two types of sciences, nor do I take lightly the distorting effects of scientism. Yet in confronting scientism, care must be taken to avoid distortions *of the opposite kind*—distortions of our understanding of *natural* science. Therefore, in turning now to this argument of the double hermeneutic, to criticize it, I do so not on behalf of some program of the 'unity of science,' but on behalf of a better understanding of natural science itself, and, especially, a better understanding of the role of hermeneutics in it.

The Double Hermeneutic in Natural Science

In outline, the argument is as follows: Of course, the anthropologist trying to understand the behavior of some remote culture and the physicist trying to understand the remote aspects of nature are to this extent in the same position; both will have to *interpret* what they observe, and construct theories. But that is only one stage of what goes on. Actually, the anthropologist's situation is different because, unlike the physicist, he finds a pre-interpreted world and a *language,* or symbol system, already in being when he comes on the scene. And this is the language to which he must gain access—hermeneutically—*before* any interpretation can begin regarding the object of study, the world inhabited by the people who speak this language. And so, Habermas tells us, in such a situation, there is already a problem of understanding below the threshold of theory construction, namely in *obtaining* data and not first in *theoretically describing* them. . . . If the paradigm-dependent theoretical description of data calls for stage 1 of interpretation that confronts all sciences with structurally similar tasks, then we can demonstrate for the social sciences an unavoidable stage 0 of interpretation, at which there arises a *further* problem. . . . *Prior* to choosing any theory-dependency, the social-scientific 'observer,' as a participant in the process of reaching understanding through which alone he can gain access to his data, has to make use of the language encountered in the *object* domain. (Habermas 1984, 110, emphasis in original)

In other words, wherever a language in being, a part of the "object domain," must be understood, interpretation mediates *first* between the language of the investigator and this native language of the people studied—just to "obtain data" (stage 0). It is needed a *second time,* to mediate between the life-forms of these people—now made accessible—and the investigator's newly enlarged language, including his theories (stage 1). In this case, the object of study has two distinguishable though not necessarily separable components—the language and the phenomena. In contrast, the natural sciences have only the phenomena as object, and therefore need only stage 1—since the atoms, bio-organisms, and galaxies do not possess a language of their own. Although stage 1 seems to be universal, stage 0 is not. By directing attention to whether or not a *language* is included in the *object* of study, this argument by Giddens and Habermas again cuts off the natural sciences from the human, and at first it seems convincing.

Nevertheless, the argument is shallow. Precisely the focus on language highlights a relevant omission, and a surprisingly positivist view of natural science despite three decades of in-depth criticism of that view, and despite the specific objections of Mary Hesse and others on this very point.[3] It is implied in all such arguments that whenever a natural scientist comes on the scene to work on a new project, he finds no pre-interpreted world, *no language* already in being, which, initially at least, must be regarded as object. Surely this is not so. The natural scientist finds there the language of the particular science within which the new project belongs—a language not necessarily his own.

When specialists in relativistic cosmology converse with one another, not all physicists can follow. If I wish to gain access to such a field, for example, in which I am not at home—just to talk meaningfully with those who work in it—must I not first study the various models presently employed, together with their characteristic concepts? Would I not have to go through some kind of a hermeneutic process to *understand* this language in terms of 'local' practices, on the one hand (astronomical measurements, computer experiments), and on the other hand, in terms of wider concepts of physics? Is there really no parallel between this situation and that of the anthropologist? Initially, on entering a new field, the physicist too is a student of symbols created by others, as is everyone else—the novice, the scholar from another discipline, the layman, anyone who for some professional or personal reason wishes to understand a particular domain of nature—since there is no way, in any such situation, for the phenomena themselves to be understood unless or until the relevant language has been interpreted. Don't we have then in natural science *also* a "double hermeneutic"?

I give this apparently naïve response, at the very outset, not because in itself it could be convincing, but in order to convey the kind of intuitive recoil from Habermas's description that should be expected of anyone who has spent his life in the world of physical science, and reflected upon it. It remains now to unpack this first-order response.

My own, preliminary view is that what we have in physics, for example, is not merely a "double hermeneutic" but a *triple hermeneutic*—three distinct phases in which interpretation is needed and practiced: just in 'obtaining data' experimentally, Habermas's "stage 0," which he denies in this realm; next, in *constructing* both intermediate and high-level theories that fit the data (acknowledged by Habermas as "stage 1"); and finally, in interpreting *established* theories in the effort to understand them. This phase—let me call it "stage 2" (to continue the terminology of Giddens and Habermas)—is rarely discussed, and so another goal of this paper is to call attention to it.

Consider now the first obvious objection to the response I have called the "intuitive" recoil: Since few people today accept Galileo's view that the "book of nature" has a language of its own, which it is our task to decipher, the languages of science can in no way be compared to the natural language of a people. Scientific languages do not *inhere* in the objects of study. When a scientist works creatively on the *construction* of a theory language appropriate to a phenomenon, that is stage 1 hermeneutics; when he or she is *merely learning* the accepted language, that is the realm of education, of preparation, not of science itself (an inquiry aiming at some new physical effect or prediction or explanation). The being of a scientist should be distinguished from the being of a student. To conflate the study of a language of science with a study of the natural phenomena themselves is to conflate the *formation of the subject* with an *outreach toward the object*.

Thus, in effect, this objection raises in yet another form the well-known issue of the subject/object dichotomy, and it is this issue, at stage 0 *and* at stage 2, that I would now like to discuss. Of course, this is not the only issue raised by the argument of the "double hermeneutic." The important matter of self-understanding must also be dealt with at some point, and later I intend to turn to it briefly. But for the time being, this is the issue on which I will concentrate: whether or not language is part of the object in the natural sciences, whether or not *interpreting* language is part of these sciences.

It is in a sense an ontological issue. What *is* science? What actions or circumstances constitute its *being,* as distinct from other modes of being? and, therefore, *inter alia,* what is the role of language, and of study, in this mode of being? Habermas appears to be taking for granted a *certain kind of role*: Language, for him, is part of the "expert culture" or "cultural action system" of the scientific group, from *within* which the scientist approaches the phenomenon.[4] But, as a scientist, he or she has *already entered* this culture, so its language is taken for granted as part of the approach, and in this sense as part of the subject. The scientific object is what confronts this subject.

Such a view, it seems to me, misses the following key feature: Since the world of science is not the original life-world, we, as scientists, do not find ourselves *already inside* the paradigm, enveloped in a scientific fore-structure of meaning; we must first *get into* that fore-structure. What Habermas neglects is the *event* of entering. That this is by no means a routine or unproblematic event was made clear in Michael Polanyi's treatment of "indwelling" and further in Patrick Heelan's and Don Ihde's philosophy of "embodiment."[5] The crucial moment is a certain kind of 'shift' in attention, or awareness on the part of the subject, from what is generally called "focal" to "subsidiary": So long as the percepts themselves are in focus as the object of attention, their *meaning* for the subject remains obscure—as the meaning of a sentence remains obscure for a reader scrutinizing each word individually. It is only when the percepts, as such, fade from awareness in a process of integration that their meaning emerges as the result of that integration. When this happens, then, in the language of Polanyi (and Heidegger), the subject begins to 'dwell' in the subsidiaries as in his own body, and 'sees' or 'feels' *through* them to perceive the meaning. But all writers who take a phenomenological or hermeneutical approach to science have pointed out that such "indwelling" is *achieved,* takes time, and can be reversed.

In one of Polanyi's most effective examples, in which a medical student tries to learn pulmonary pathology by listening to experienced doctors as they examine x-ray pictures, the student is at first completely in the dark. "The experts seem to be romancing about figments of their imagination." Only after *weeks* of this, if the student is diligent, will the meaningless shadows on the x-rays and the technical terms he hears recede to a subsidiary position, and "a rich panorama will be revealed to him . . . of physiological variations and pathological changes, of scars, chronic infections, and signs of acute disease. He has entered a new world" (Polanyi 1958, 101).

Consider now this initially bewildered student, listening to the senior doctors, trying to fathom their language by connecting words with the motions of fingers on the picture; and compare him with that anthropologist discussed earlier, trying to fathom the language of the tribe in order to access his data—the typical social scientist at stage 0. Is there no useful parallel here? It is true, of course, that the doctors also speak the language of the novice, the ordinary language of their country, while the natives studied by the anthropologist share with him no language at first. But this makes little difference: The scientific language the novice wishes to learn cannot be translated into the natural one—it must be correlated with the experiences and actions of this other world, the world of pulmonary medicine; and this is much the same sort of thing the anthropologist does.

To shed more light on what is going on here, consider also a real, live anthropologist in a science laboratory. There have been quite a few lately, such as Bruno Latour, studying 'science in action' precisely as one studies a remote culture.[6] At the very beginning, the position of such an outsider is more or less the same as that of the student; but soon afterward an important difference emerges, and leads to radically different outcomes. The student eventually becomes a doctor; Latour does not become a biochemist. Why? Not because of a difference in time or effort devoted, but a difference of intention. The medical novice wishes to enter the world of the doctors, he wants badly to *see what they see.* By contrast, certain anthropologists and sociologists studying science have made it a professional principle *not* to see what scientists see, *not* to enter that world but to describe it *from the outside* (cognitively speaking) in their own, sociological, language.

Latour and Steve Woolgar say explicitly: "We have tried to avoid descriptions . . . which take as unproblematic relationships between signs and things signified. Despite the fact that our scientists held the belief that the descriptions could be representations or indicators of some entity with an independent existence 'out there,' we have argued that such entities were constituted solely through the use of these inscriptions. . . ."[7] In other words, Latour and Woolgar make sure that they never allow the actions of the scientists, and the "inscriptions" on instruments, to recede to a subsidiary position, so that the *meanings* of these actions, *as seen by the scientists themselves,* might come into focus.

I mention this to point out the difference, as well as the similarities, between such anthropologists of science and students of science in the ordinary sense—to contrast the student's *intention to enter* (and the effort involved) with Latour's *effort not to enter.* In this perspective, the science student is like an anthropologist at first, but an anthropologist who wants to 'go native.'

That this 'going native' is the *event* of embodiment, in which the subject partially *enters the object,* has been emphasized, for example, by Patrick Heelan, and described in terms of the *linguistic* metaphor—the favorite example of the human sciences. The shadows on the x-ray, the sort of 'inscriptions' made by any instrument, are like the marks on an ancient tablet that an archaeologist might try to decipher. In Heelan's view, "These marks are at first studied and compared as *objects*: the subject/object cut places them on the object side of the cut" (Heelan 1977, 12). But after the students of this language have become sufficiently familiar with it, and begin to read the signs 'for their meaning,' the signs cease to be

> objects vis-à-vis the activity of reading, since the object is now the meaning meant by the signs. Moreover, the signs themselves are not [now] part of the object, since it is immaterial to the reader what signs are used . . . provided they convey the meaning meant. . . . The signs then have come to be displaced to the subject side of the cut. . . . After deciphering the language, and to the extent that the language is being used as a text to be read or spoken, the signs are *part of the being* of the subject . . . [whose] noetic intention operates intentionally through them in order to objectivate a horizon of meaning. (Heelan 1977, 12)

The point here is that this shift in the subject/object cut occurs as described whether we are talking about physical instruments (hardware) or cognitive ones like language and theories—and that in both cases it is an intentional act. But now two further observations must be added: The transparency of the instrument is never perfect, and the intention in this respect *varies*—sometimes shifting the boundary back toward the subject so as to place the instrument itself again at the focus. This happens, of course, whenever instruments are checked or adjusted or redesigned; and therefore, at such times at least, language and inscriptions also move to the focus and join there the objects of study (Ihde 1979, Chs. 1–3).

So much, then, for "stage 0"—*obtaining* data—which, according to Giddens and Habermas is where hermeneutics plays no role. This is just the stage so vividly described by recent sociological studies of the laboratory, and so convincingly tied to interpretation at nearly every step.[8] But this observational or data-gathering or laboratory aspect of inquiry is not the whole of science by any means (as sometimes implied)—only its most empirical stage. Then comes "stage 1," the mutual adjustment of data to higher level theory, which is not in question. And then? Then *more* interpretation, an even higher level of integration, the one I earlier called "stage 2." Let me turn now to this further issue—which, sometimes, is the earliest of all.

The Two Books of Science

To do so, let me also switch metaphors: from the relatively new one of science as a perceptual system, an extended 'body,' to one that is much older—science as a reading of the "book of nature." The body metaphor is powerful because it is so comprehensive, taking into account physical tools as well as cognitive ones, and even hybrid systems consisting of hardware-software combinations executing schemata. For some purposes, however, just because it is so comprehensive, it becomes unwieldy: Specifically linguistic and textual features of the situation cannot easily be portrayed in terms of a body, yet it is these features that come to the fore in the stage we wish to examine now.

For if once again we focus on that medical student—or any science student—we know that he does not walk into such an advanced laboratory to read x-ray pictures without also studying other pictures in other books, accompanied by much prose—real, literal books, this time, in which the diseases shown on the x-ray are related in specific ways to physical processes, to microbes, and to each other. He has studied, in other words—or is studying—theories of pathology, products of "stage 1" interpretations, that explain what the x-ray shows by means of what it does not show. But in so doing, the student faces a problem whose structure is, *once again,* the same as already described—he must decipher and find the meaning of a language that is new to him. Just as, at stage 0, the meanings of the signs produced by the instrument had to be constituted or constructed, so also must the meanings of theories encoded in texts be constituted, since these meanings are neither evident nor unique—as I will try to show.

In science as a whole, then, taking a high-altitude view of it, we notice people engaged in several *different kinds of reading:* The doctors, using x-rays and other imaging devices in a subsidiary way 'see' through these devices 'into' the human body, and there 'read' the "book of nature," occasionally adding new words as needed. But simultaneously, in another room down the hall, students and other doctors surrounded by piles of real, printed books, are trying to *read* an already existing *theoretical* language, and to achieve understanding at *this* stage.

It seems, then, that *two books* are being read—one figurative and one real. There is the primary book that Galileo had in mind, which, today, we might identify (following Heelan, Ihde, and even Latour) as the inscriptions on instruments. Then there is the secondary book, the book of interpretations started, more or less, in Galileo's day—to which we have been adding ever since at an increasingly furious pace. We could call the first book, the "book of nature," and the second, "the book of science." This second book is, of course, a sprawling, rambling, collective work, with its own gradations as to depth and importance. In Karl Popper's language, it is "world 3"—the sum total of recorded theories, reports of experiments, problems, solutions, and so on.[9]

The same question now arises in regard to the second book as we have already considered: If reading the first book, the book of nature, is admittedly interpretive, is reading the *second* book also interpretive? Is there another tier of interpretation at this stage? The most common answer, usually implied rather than spoken, is "no": Understanding scientific results *after* they have become established, or verified, or constructed is a matter of teaching and learning, which is the realm of education or popularization. Certainly interpreting observations and experiments is part of science, but this belongs to the reading of the *first* book, or the *writing* of the second. It does not belong to the *reading* of the second book—except in the common pedagogical sense in which the content of textbooks, monographs, reports, must be properly interpreted by the student; and this, according to the prevailing view, is not constructive interpretation in the serious sense. All real construction, the kind that 'makes a difference,' takes place in the 'reading of,' or interpreting, the original book of nature.

We have come upon yet another form of that familiar separation, which, in natural science, allows for only a single hermeneutic: The line here is between 'science itself,' as the *construction* of particular interpretations of nature (of phenomena) and the *study* of science as the *reception* of these interpretations. On one side is the writing—or, rather, the composing—of the second book, the book of science, to which attaches all the creativity associated with the writing of original books; on the other, is a *reading* of this book, regarded as a matter of 'getting it right' but not essentially interpretive, and therefore uncreative.

The idea that true interpretation inheres only in the composing of the book of science, not in its reading, derives from the common belief that although raw data does need interpreting, by the time results reach the textbook all interpretation must surely be over. Famous 'exceptions' serve only to reinforce this belief. Today, even outsiders know that a major debate is still going on about the meaning or interpretation of quantum theory, long after that theory has become established. But invariably, this is attributed to the uniqueness of quantum theory itself. Except for historians of science, nearly everyone tends to forget all the previous interpretation debates, such as those about Newton's laws, about the statistical interpretation of the second law of thermodynamics, or about Maxwell's equations—all taking place *after* the laws themselves had entered the book of science.[10]

Should these cases be thought of rather as continuing research (a prolonged stage 1), and therefore clearly not in the realm of the 'study' or 'reception' or 'understanding the results'? Perhaps not. In the end they often appear as research, but, on closer examination, many instances cannot easily be placed within the normal paradigm, in which acknowledged problems are attacked with experiment or calculation, and some new result is obtained. Frequently the impulse to interpret arises from a *vague unease,* a lack of *understanding.* Moreover, the unease tends to occur precisely at the moment of the *reading* of the book of science—sometimes literally a textbook.

A famous historical example is that of the physicist Heinrich Hertz, in the 1890s, who undertook a fundamental re-interpretation of mechanics despite the fact that at that time no crisis, not even a serious anomaly, troubled this basic science. He did so, as he says,

> solely in order to rid myself of the oppressive feeling that to me its elements were not free from things obscure and unintelligible. What I have sought is not the only image of mechanics, nor yet the best image; I have only sought to find an *intelligible* image. (Hertz 1956, 33, emphasis mine)

It is extremely interesting, even now, to read the long introduction to this book, where Hertz explains why Newton's *physical interpretation* of his own equation of motion should seem absurd to any "unprejudiced person." Hertz then proceeds to offer an alternative interpretation of this same law—a mechanics without forces. It predicted nothing new, caused no revolution, and is now only of historical-philosophical interest; but it did demonstrate convincingly the large degree of interpretive freedom at the higher theoretical levels; and it had an acknowledged effect on Einstein in regard to the general theory of relativity.

Feynman's Hermeneutics

That is what came of one man's attempt to understand one part of the book of science; it is not a unique experience. My favorite example is contemporary, and offers even more revealing insights: Richard Feynman's re-interpretation of non-relativistic quantum mechanics—the well known "path-integral" method. Feynman too was puzzled by the standard formulation of the physics of his day. To the consternation of colleagues, even near the end of his life he insisted that "no one really understands" quantum theory.

Yet in the 1940s, while still a graduate student at Princeton University, he *was* struggling profoundly to understand it, using, among other things, the famous textbook by Paul Dirac. In these readings, he came upon something so unclear and at the same time suggestive, so frustrating, that he decided there had to be a better way of expressing it. The problem was this: Dirac had said that a certain quantum transition probability was "analogous" to a certain exponential function of the classical action.[11] But, Feynman asked, "What does he mean they are analogous; what does that mean, *analogous?*"

Having put this question to a visiting physicist who tried to help, and having received no satisfactory answer, Feynman started to interpret for himself—that is, he began a kind of dialogue with Dirac's text that we can recognize as a *hermeneutic circle* on a small scale: Assuming first that "analogous" stood for simple equality, he checked the consequences against other parts of the text—standard theory, the Schrödinger Equation—and found a discrepancy. Next, he changed the assumption slightly to a proportionality and went through the same calculation. On this second try he found that indeed the consequences were consistent. So then, he says,

> I turned to Professor Jehle, not really understanding, and said, "Well, you see. Professor Dirac meant they were proportional." (Feynman 1966, 37)

This was the beginning of a new way of thinking in quantum theory—its *re-interpretation* on the basis of entire histories of particles (paths) rather than just their states at a particular time.[12] Let me try to characterize, from a large-scale point of view, what was actually happening here. In mechanics generally, we have two major formulations—in effect, two languages—the Hamiltonian and the Lagrangian. The difference between them lies in what we take to be the independent variables of the world. In other words, the Hamiltonian language cuts the world up one way, the Lagrangian, another way. When physics had its famous revolution in the early part of the century, in which its basic laws of motion were transformed into their quantum versions (in part, by Dirac), the language used in this transition was the Hamiltonian; and that is why the simplest quantum mechanics, found in most textbooks, is presented in the Hamiltonian language.

But Feynman, with his thesis advisor, had been studying some prequantum theories in a very deep way, and he had used the Lagrangian language. Now he wanted to see whether these ideas could work in quantum theory also. Unfortunately, the quantum aspects of nature appeared not to lend themselves to a description in the Lagrangian language—except in this one obscure section of Dirac's work. There, Dirac picks out a basic statement of the Lagrangian formulation—the "action principle"—and tries to show what its analogy would be in quantum theory.

This is what Feynman could not understand. Why just this hazy analogy? Why not a full translation? But this could not be done without developing also the Lagrangian language in quantum theory—or, rather, quantum theory in the Lagrangian language. This is precisely what Feynman proceeded to do, and this is what we call the path-integral formulation. Here we see the hermeneutic circle on a larger scale: To even begin to understand the full meaning of what Dirac had written, Feynman had to *assume* that the Lagrangian language *could* be used in the quantum world—by actually starting to use it in a brash, partial way that looked 'wrong' to many physicists. In the process, as he put the parts together, he developed that language to a point where it could not only be used to formulate the original statement—the action principle—but much of quantum mechanics (and later, radiation theory). One could say that Feynman found quantum mechanics presented to him in a Hamiltonian context; but he wanted to see its profile in a Lagrangian context.

By traditional, positivistic criteria, this work was not in itself an 'advance' in physics (which explains why the *Physical Review* refused to publish it). It did not solve any puzzles, did not explain any physical phenomena; it did not prove any law or principle that had not already been accepted. In short, it did not predict or control anything. It was an alternative interpretation. Precisely because, in natural science, such work rarely wins prizes, the feeling has grown up that hermeneutics plays here only a marginal role.

An interesting question comes up now regarding that crucial dialogue with the text—Feynman's attempt to construct some sort of meaning for Dirac's use of the word "analogous": Was Feynman's construction a *correct* interpretation of Dirac, or was it a *creative* re-interpretation? In other words, did Dirac *know* (and was he actually trying to say) that the two mathematical expressions in question were proportional, but chose, for some reason, not to develop this fact as Feynman did; or, was he (and everyone) unaware of it until Feynman came along? Feynman himself said,

> I thought I was finding out what Dirac meant but, as a matter of fact, [I] had made the *discovery* that what Dirac thought was analogous was . . . equal. (Feynman 1966, 37 [emphasis mine])[13]

On the other hand, Julian Schwinger, fellow Nobelist with Feynman, thinks "Dirac surely knew" (Schwinger 1989, 45). Clearly, the question was of some importance to Feynman because on one occasion he tried to find out directly. At a Princeton conference, in 1946, Feynman came up to Dirac and asked, point blank, whether the great man had known all along that the two quantities were proportional:

> "Are they?" Dirac asked.
> Feynman said "Yes, they were."

After a silence, Dirac is reported to have walked away, and the question remains undecided (Gleick 1992, 226). I have raised it here only to illustrate the point often made by constructivists and hermeneutic philosophers that the line between receptive and creative interpretation—between *finding* the meaning and *constructing* meaning—is not easy to draw.

At present, Feynman's re-interpretation of quantum mechanics is far better known than Hertz's of classical mechanics because it affords an interesting new way of conceptualizing phenomena, and the thinking it arouses has contributed not only to Feynman's subsequent, more famous reformulation (of quantum *electrodynamics*) but, as a technique, to a number of other areas in physics as well. Even in the strictest sense of achieving something really new, no one can say with finality that the Feynman path-integral formulation is a 'mere' alternative; recently, the possibility has arisen that in cosmology it may play a unique role (Gell-Mann 1989).

One of my aims, in both these examples—with Feynman as with Hertz—was to show that interpretation is something that is done when facing the *book of science,* not just the "book of nature." It is something that occurs at the highest levels of competence, yet it involves (perhaps more frequently than we think) problems of understanding not different from those faced by the average student at various stages of education. The examples show also how such problems can lead one to *re*-interpret radically. For contemporary, philosophical hermeneutics, this close tie between understanding and interpretation is a major feature, of course,[14] and it was indeed a feature in the scientific life of Richard Feynman—a feature of which he was fully conscious. In a letter to a friend, written in the same year in which he was finishing the path-integral method, Feynman expressed his view in these words:

> I find physics is a wonderful subject. We know so very much and then we subsume it into so very few equations. . . . Then we think we have *the* physical picture with which to interpret the equations. But there are so very few equations that I have found that many physical pictures can give the same equations. So I am spending my time in study—in seeing how many new viewpoints I can take of what is known. . . . I dislike this talk [of there] not being a picture possible but we only need know how to go about calculating. . . . The power of mathematics is terrifying—and too many physicists finding they have correct equations without understanding them have been so terrified they give up trying to understand them. (Feynman quoted in Schweber 1986, 472)

Notice, please, that this is not what is usually meant by "under-determination of theory by data"—which is actually a problem at stage 1 of interpretation. In such cases, alternative theories make different predictions, and additional data can (at least in principle) decide between the alternatives. Here, in contrast, Feynman is talking about alternative 'pictures' that *in principle* yield exactly the same predictions. These are alternative ways of 'seeing' or 'reading' *the same* "laws of nature."

Do such attitudes as Feynman's represent anything more than interesting sidelights on the thinking of outstanding individuals, or do they represent hermeneutical moments and impulses to re-interpret that permeate science throughout? In a longer paper which develops this point further, I argue more extensively that the process of science involves, in effect, "a cascade of interpretations," from the highest levels of the kind just mentioned, all the way to teacher and student at the various stages of education (Eger 1993b, this volume, 29–51). Here I began from the Giddens-Habermas argument that natural science requires hermeneutics at only one of its many stages, and I have shown, I hope, two other such stages—so we have not just a single hermeneutic in natural science, but a triple hermeneutic *at least.*

No doubt, the reluctance of many philosophers to accept this kind of evidence as support for a genuine hermeneutical element in areas such as physics stems from the traditional desire to maintain intact a clear boundary between these 'objectivizing' sciences and the 'hermeneutical' human sciences. Any weakening of this boundary appears as a threat to the latter— the old threat of scientism, of 'colonization' by an unsuitable, dehumanizing methodology, smuggled in from the other side of the boundary. So great and deep-seated is this fear that even when the 'smuggling' takes place in the opposite direction—from the human to the natural sciences—the habit of defending the boundary re-asserts itself.

Karl-Otto Apel is one who clearly sees the hermeneutic underpinning of the physical sciences. He even offers as evidence Einstein's re-interpretation of space and time, and emphasizes that without such intersubjective agreement on meanings the most objective of the sciences cannot exist. Yet, on the very same page, Apel flatly denies that in these sciences language can belong to the "objects of knowledge"—any more than can sense organs or the technological instruments used by scientists. All these, he insists—language, sense organs and instruments—are *"presupposed* in order that objects of knowledge may be constituted" (Apel 1988, 322–23). They are "the condition of the possibility" of meaning intentions, and as such *cannot be a part* of "objective" science, but are instead "complementary" to it in the sense of Niels Bohr, in the sense of mutual exclusivity (Apel 1972, 28).

From the point of view of this paper, it is easy to agree with everything that Apel says here except the assumption that the complementarity of which he speaks exists between *different sciences.* It is easy also to notice the congruence between this complementarity and the focal-subsidiary relation discussed earlier in regard to both instruments and language (and patterned, of course, on sense organs, as in Polanyi's work). Yet Polanyi, as well as Bohr, took the two sides of the complementary relation to be occasioned by two *types of questions*—incommensurable and mutually exclusive at any *one time* but certainly contained within the same science, and possible even within the same research project. As physical scientists, both men knew from their own experience that one often shifts back and forth precisely between these two modes—now being "embodied" in equipment and taking it for granted, now re-focusing to examine the equipment itself, the meaning of its inscriptions, and the proper language to deal with it all. Apel's radical disjunction has the disadvantage, among others, that it would lead us to *consign to 'metasciences'* such hermeneutical re-interpretations as those of Einstein and Feynman, rather than include them in our view of physics itself as do most physicists, historians, and philosophers.

Hermeneutics and Self-Understanding

Finally, I would like to return to an aspect of this issue that has been left out thus far—the question of self-understanding. I have concentrated on showing that one reason hermeneutics is required at several stages is that language itself is part of the object of study of all sciences; and that in this respect the human and the natural sciences are not that different. But I have avoided the fact that natural languages embody the *self-understanding* of a people or a culture (the very object of the human sciences) while, of course, the language of stars and atoms and blood cells says nothing about the self-understanding of *these* objects.

It is also for this reason that certain philosophers have distinguished between a "theoretical" circle and a true "hermeneutic" circle (in the sense of Heidegger or Gadamer or Apel) while I have conflated the two and blurred the difference. The "theoretical" circle is circular regarding the relation between natural phenomena and representations of these phenomena constructed by the investigator; but the *being* of the investigator himself, the subject, is not within *this* circle. True "hermeneutic" circularity, on the other hand, includes more than representations; it includes practices and life-forms. In effect, it includes the investigator too because the understanding imbedded in the practices that language expresses is a self-understanding of humans like himself, which can both affect and be affected by the self-understanding of the investigator. The issue here is not whether the hermeneutic in natural science is single, double, or triple but *what sort* of a hermeneutic it is.

This surely is a serious challenge. Yet it is a challenge that can be met, I think; and in meeting it, in pursuing even the question of self-understanding in physics, chemistry, biology, a whole world of missing contexts might emerge for many of the sociological investigations recently carried out. Human self-understanding is not altogether absent from natural science. It is involved in different ways from self-understanding in anthropology and sociology, but not in ways as radically different as is often assumed. To analyze this and exhibit it convincingly is of course where the hard work lies; and this job has not been undertaken by many people.[15] In closing, let me just indicate how the subject of this paper might bear also on this further question.

The whole challenge rests essentially on the assumption that the language of science is about such things as stars and atoms exclusively, rather than about the interaction of stars and atoms with humans or their relation to humans. That is what some hermeneutic philosophers are really pointing to when they say—as they do so often—that the objects of science are "meaningless." But if, as I believe, the languages of science do include such relations, implicitly and explicitly, then the study of scientific languages—

not just by 'anthropologists' of science, but by scientists themselves—is more than a study of stars and atoms in their invariance under changing points of view. It also is a study of the *understanding* we have of our relation as *human* beings to these and other kinds of beings, which, *inter alia,* is a study of the self-understanding of science as a mode of being in the world. This can be seen, I am suggesting, in those activities (carried on by all scientists, not just 'students' properly so called) where the various languages of science are still the *objects* of study—as with the young Feynman—before they are fully 'entered into' and inhabited.

Today, however, after all the work by the various writers I have mentioned, the time is surely past to be still asking whether or not there is such a thing as hermeneutics in natural science; or whether it is the same full-blooded thing as in social science. The task now is to *extend* the work already done, to undertake more detailed historical, philosophical and sociological studies, and thus to deepen our understanding of this whole realm.

[5]

Natural Science and Self-Understanding

ABSTRACT: Heidegger and other hermeneutically grounded philosophers often claim that the objects of the natural sciences are 'meaningless' for human existence, in contrast to the objects of the social sciences. This strong dichotomy has been challenged by Mary Hesse, who postulates a continuum of degrees of involvement of hermeneutic self-understanding in scientific disciplines. Her thesis is elaborated in the present paper by recognizing the interplay of science and self-understanding not only in theoretical matters—e.g., evolutionary explanation vs. creationism, and steady-state cosmologies vs. big-bang cosmologies—but also in practical matters like science education. Furthermore, not only are scientific theories claimed to have implications for self-understanding—e.g., the attempts at grounding ethics upon sociobiology—but there are sociological explanations of the genesis and receptivity of deep physical theories as a consequence of human concerns. A notable example of the latter type of analysis is Forman's thesis that the indeterminism of the new quantum mechanics was preferred and accepted because of the chaotic condition of Germany after the first World War. Several diagrams are given which represent different types of interaction between scientific conceptions and self-understanding. [—A.S.]

The Question

To what extent does natural science involve self-understanding in the sense in which self-understanding is involved in hermeneutics? That is the question I would like to consider today.

It is usually taken for granted, in hermeneutically grounded philosophies, that a science like physics requires decontextualization (Heidegger), or detachment of the subject from the object so that the former is not affected by the latter. As a consequence, the objects of natural science are

---

This paper was originally a talk presented in 1994 at the European Association for the Study of Science and Technology (EASST)—Hermeneutics Section, in Budapest, Hungary.

often said to be 'meaningless' for human existence. That is to say, a theory of molecules or stars has no meaning for us in the life-world not only because it speaks in a language disjoint from that of ordinary experience, but because such a theory in no way interacts with our conception of ourselves as humans imbedded in language and community. The interaction of natural science with the human realm is all at the level of technology. Detachment serves well the interests of prediction and control.

In hermeneutical thinking, on the other hand, we have the famous expression *"sich in der Sache verstehen"*—to understand oneself in the matter, or issue (Gadamer 1975). Aside from the obvious fact that by studying history we get to understand our present situation better, by studying remote cultures we come to see our own in a new light, and so on, this expression reminds us also that in such studies we always start with a pre-understanding of ourselves and of the world, which necessarily must change in the dialogical process of seeking understanding of the object. To understand oneself-in-the-issue, therefore, is to understand one's previous limitations in regard to that issue; to have 'come to terms' with oneself regarding that issue; and this means to understand one's mode of being in the world differently by having dealt with the issue.

The Stakes

Now before going on to discuss whether such reflexive relations between objects and subjects occur also in natural science—despite the prevailing view to the contrary—let us be clear about what is at stake in this question. One major issue is the putative demarcation between the human and the natural sciences. If self-understanding is *not* involved in the natural sciences, if the objects and claims of natural science are (in the sense I have already indicated) 'meaningless' to us, then the "circle of understanding," insofar as it does occur in such sciences, is merely a 'theoretical circle'—the circular relation between theory and experiment, for example. In such a circle, since the Heideggerian *Vorhabe* is *not* involved, interpretation does not touch the being of the scientist; and therefore, at least the more recent type of hermeneutics—philosophical, post-Heideggerian hermeneutics—is neither involved nor required (Gadamer 1975, 235ff).

By contrast, to the extent that human self-understanding is in some way involved in natural science, to that extent hermeneutics becomes relevant, whether or not actually practiced. In this way, the question of self-understanding is related to the question of whether natural science is hermeneutical at all; and this, in turn, is a major point in the Diltheyan and more recent assertions of a fundamental, in-principle, difference between the natural and the social sciences (Rouse 1987, Ch. 6).

That is one thing that is at stake, but that is a theoretical issue. In addition, practical, social issues are also affected—for example, in certain educational controversies now taking place in various parts of the world.

In California, in the 1970s, the issue was not so much the teaching of creationism (about which everyone has now heard), but the way in which evolution is to be taught and discussed—whether as 'fact' in itself, to be learned and applied, *or* as an *interpretation* of facts, having a bearing on our self-understanding as humans, and subject therefore to hermeneutical discussion (Eger 1988b)[1]. Very similar problems are also occurring in a different context in New Zealand—in teaching evolution to Maori and Polynesian children. During a stay at the University of Auckland, in 1992, I found the situation there even more sharply defined because Christianity is not necessarily involved. One Maori high-school boy put it this way: "If you believe you are descended from apes, you cannot be a Maori."

But classic Darwinian evolution is not unique as a problem of this kind: Today we have controversies over sociobiology's meaning for human altruism and for relations between the sexes[2]; arguments over the meaning of artificial intelligence or cognitive science for the understanding of our own intelligence and expertise[3]; and controversies over the meaning of the big-bang model and the anthropic cosmological principle for the understanding of our place in the universe (which is a kind of 20th-century replay of the 19th-century controversy over 'our place in nature').[4] Just to list such issues is to affirm that, to a degree perhaps underestimated by many people, natural science does indeed enter into our self-understanding.

However, even more is involved: Self-understanding can affect theories of science, as well as the other way around. Mary Hesse, who takes account of *both* directions of influence, puts the matter as follows: On the one hand, "it is impossible in studying theories of evolution, ecology, or genetics, to separate a mode of knowledge relating to technical control from a mode relating to self-understanding of man. . . . the very categories of these theories, such as functionality, selection, survival, are infected by man's view of himself." On the other hand, she says "theories have always been . . . part of the internal communications system of society. Society interprets itself to itself partly by means of its view of nature." Therefore, she concludes, "this is a sense in which nature does indeed take part in the dialogue of man with man, and can itself be said to be informed by human meanings, and subject in its theoretical aspects to hermeneutic methodology" (Hesse 1980, 186).

Although Hesse discusses these interactions in only a general way, others have been more specific: Paul Forman (1971), Evelyn Fox Keller (1985), Joseph Rouse (1987), and, most recently, James Cushing (1994), have all tried to specify in some detail this reverse influence—*from* self-understanding *to* natural science. Some of these efforts are more successful than others, and later I will say something about their relative significance; but now let me just indicate briefly what these arguments are.[5]

Paul Forman (1971), "Weimar Culture, Causality and Quantum Theory, 1918–1927," <u>Historical Studies in the Physical Sciences</u>, v. <u>3</u>, pp. 1–115.

James Cushing (1994), <u>Quantum Mechanics, Historical Contingency & Copenhagen Hegemony</u>

Evelyn F. Keller (1985), <u>Reflections on Gender and Science</u>

Joseph Rouse (1987), <u>Knowledge and Power</u>

Georg H. von Wright (1974), <u>Causality and Determinism.</u>

Fred Hoyle, (1982), "The World According to Hoyle," <u>The Sciences</u> v. <u>22</u>, pp. 9–13.

Overhead 1

Examples

Forman's thesis has to do with why the non-causal or non-deterministic interpretation of quantum theory was so quickly and easily accepted by physicists in the 1920s (Forman 1971). His first point is that this development was overwhelmingly a German phenomenon, a phenomenon significantly related to the Weimar culture of the 1920s. His second point is that, in the aftermath of Germany's World War I defeat, because of the association of science with that war, the German public reversed its previously favorable view of science; and it did so *at a time* when the academic classes of that country were under the sway of Spenglerian, Bergsonian, and similar *Weltanschaungen* that identified contemporary science with a form of thinking dominated by exaggerated objectivity and 'rigid causality.' The result was, according to Forman, that German physicists and mathematicians felt themselves under pressure to depict their science as not that rigidly causal and not that objectivistic after all. Therefore, when radically non-causal interpretations of the new physics emerged along with others that still retained causality in some form, German scientists were relieved to be able to choose the former with little serious consideration of the alternatives.

Forman supports his thesis by numerous examples drawn from speeches, letters, documents and events of the times; and concludes that our present quantum theory, with its underlying non-causal picture of the physical universe, owes its dominant position, in part, to the self-understanding of Germany's academic classes at a particular historical period.

James Cushing—physicist, historian of physics, and philosopher of science—offers new support for at least part of this thesis (Cushing 1994). Cushing centers his historical-analytical study around the figure of David Bohm, and the attempt by this physicist in the 1950s to reverse the 1920s decision, and return to a more deterministic interpretation. Bohm's theory also received little consideration, despite the fact that it is admitted to be technically flawless, and that recent experimental results make its philosophical disadvantages perhaps more acceptable.

Cushing argues that nature, or experiment, so greatly underdetermine physical theory that the causal and non-causal versions of quantum mechanics stand at present on more or less equal footing if considered from purely logical and empirical points of view. And so, in Cushing's words "we can reasonably ask whether the 'laws' of nature tell us more about ourselves than about nature" (Cushing 1995). Or, rephrasing this in the manner of hermeneutics, the question might well be: To what extent do such theories represent a stage of self-understanding of the scientific community, a stage of physics "understanding itself in quantum phenomena" (in the sense of *sich in der Sache verstehen*).

Turning now in a different direction, there is the group of feminist writers in the United States who have argued in various ways that in specific theories, as well as in its general features, science incorporates a masculine or male self-understanding. "Objectivism" is again mentioned, this time as a gender-based rather than a human pre-conception; and primate ethology is offered as an example of a field particularly susceptible to biases regarding relations between males and females of the species (Haraway 1989). Going a step further, Evelyn Fox Keller argues that even physical theory is influenced by a peculiarly human, perhaps masculine, view of causality as embodied in the so-called "master molecule theories." These are theories like those of DNA and Crick's "central dogma of molecular biology," and certain organic theories in which effects originate in a single molecule or species of molecule, and are transmitted hierarchically to lower levels of organization. These theories, exhibiting a one-way causal relationship, an authority immune to influence from below, are contrasted with more collective modes of molecular or cellular control in which no single type is distinguished as more dominant (Keller 1985, 154, referring to the work of David Nanney).

Joseph Rouse, who has discussed the issue of self-understanding in natural science at some length, accepts Keller's ideas on the significance of such "master molecule" theories, and cites also, in the same vein, G.H. von Wright's theory of the pre-conceptions underlying *any* idea of causal connection between events (Rouse 1987, 189–190). According to von Wright, our concept of causality, and our whole framework of causal explanation, is conditioned on our ability to alter the sequence of natural events by deliberate intervention. By causing outcomes different from what they would otherwise be, we acquire the notion of causal necessity. In this way, it is suggested, our self-understanding as agents enters into the very character of practically all our theories of nature.

Turning in still another direction, toward cosmology, the epic battle between the big-bang and the steady-state models of the universe suggests itself immediately. It is well known that the recent flowering of a detailed theory of emergence of the universe from an initial fiery flash bears such a striking resemblance to Judeo-Christian beliefs that a whole new dialogue between physicists, philosophers, and theologians has commenced—on such questions as 'design' in the universe, and man's role in it.[6] This in itself clearly reflects on human self-understanding, but there is more. Behind the competitor model—the steady-state—we find preconceptions of precisely the opposite sort. In this model, nothing ever changes on a large scale; the universe is infinite and has neither beginning nor end. A major argument in its favor and, indeed, a factor motivating the original construction of the steady-state theory by Fred Hoyle, Hermann Bondi, and Thomas Gold, was that by eliminating a 'beginning,' science is spared the need to explain it.[7]

In other words, a self-understanding of science as an all-inclusive and 'demystifying' enterprise, an enterprise embarrassed by unexplained loose ends, seems to have played its role in the development of this major theory, dominant during the two post-war decades. When, in 1965, the astrophysical community dropped the steady-state in favor of the big-bang, Sir Fred Hoyle was aghast, and could not refrain from saying publicly that the reason for such a hasty decision (before conclusive evidence was in hand) had to lie in the deeply ingrained Biblical understanding of our existence, from which—according to Hoyle—not even sober physicists are immune.

Discussion

Other such examples could easily be added, but let me end this list now, and make some preliminary points. First, it is just not possible to say that human self-understanding is involved *only* in the human sciences, not in the natural sciences. That argument, if there ever was such an argument, must have been some sort of miscommunication. Of course, there still remain questions as to what sort of self-understanding is involved, and *how* it is involved in natural science—which brings me to my second point: Even the small number of examples I mentioned show considerable variety in this regard; but it is important to be aware of this variety because, frequently, the term 'self-understanding' is used without any specification, as though it were all of one kind.

But there is, of course, the important difference between self-understanding *affecting* scientific results, and *being* affected by these results. In addition, there are different kinds of self-understanding—self-understanding as an individual, as member of a group, as scientist, or as a human being. And beyond this, one might also wish to ask whether it is the *theories* of science that interact with self-understanding or the *practice* of science, and make other distinctions of this sort. Not to introduce new classification schemes, but just to convey a picture of the different sorts of things we are talking about, I have tried to schematize these differences on a chart.

Directions of Influence

$$\boxed{\text{Natural Science}} \xrightarrow{\text{R}} \xleftarrow{\text{L}} \boxed{\begin{array}{c}\text{Self-}\\\text{Understanding}\end{array}}$$

Kinds of Self-Understanding:
1) As a human.
2) As a scientist.
3) As member of other groups.
4) As an individual.

CLASSIFICATION:

Hesse's general observations	R1, L1
Forman's thesis	L3
Bohm's interpretation of q.m.	L2/L4
Cushing's View	L
The evolution controversies	R4
Keller on gender & "master molecules"	L3
Rouse & von Wright on causality	L1
AI & brain research	R1, R4
Steady-state cosmology	L2
Big-Bang Cosmology	R1, R4, (L1?)

Overhead 2

Not all the interactions between self-understanding and science are equally powerful or important. To me, it seems that the classic effects of which we have long been aware are still the most significant. These are the R-type interactions like the effects of the 17th century revolution, the Darwinian revolution, and of present-day developments in brain research, artificial intelligence, and so on.

Next in significance are the L2-type interactions: the possible influence of self-understanding on alternative interpretations of major theories like quantum mechanics and cosmology. This is an area where more historical research, such that by Cushing, could be very interesting.

The L3-type interactions seem to be least significant, and certainly the least documented: These are things like the alleged bias for "master molecule" theories. In fact, I am surprised that such flimsy arguments based on so little research are taken seriously by anyone. However, gender influence may well be important in restricted border-sciences like ethology.

Let us return now to some of the issues at stake, which I mentioned at the beginning of this talk. Why is it that despite the array of examples one sees here (Overhead 2), the dominant view, still quite unshaken, is that the objects and products of natural science are 'meaningless' in the life-world?

Everyone is aware of at least the Rl-type interactions, of course. But these are influences from very high-level theories that are believed to have generally long-term effects, over decades if not centuries; they are contrasted with the more immediate, short-term, interactions that take place, say, between an anthropologist and the community he studies. In such situations, each concept under consideration—kinship, bargaining, etc.—can, *by itself,* alter the scientist's own pre-understanding of human relations, and therefore of himself.

That theories of molecules and stars cannot have such an immediate effect largely accounts for the prevailing view about natural science. This view is correct on one important point, incorrect on another. It is true that in natural science, the strong interactions with self-understanding involve high-level theories almost exclusively; and since most science work is low-level, in the sense that it focuses on small segments of theories, this kind of interaction cannot be felt by the scientist himself, or by a student, on a day-to-day basis. It is only when molecules and stars are seen as part of biological and cosmic evolution, for example, that our self-understanding becomes involved.

On the other hand, it is not true that these Rl-type interactions take place outside the life-world altogether, in some remote, abstract realm of cultural history. They take place here and now, with powerful effect on real individuals in real-world situations. The examples of the evolution controversies are not the only ones that can be given. Similar repercussions have been documented for artificial intelligence and cosmological theory (Hoyle 1982; Turkle 1984; Chaisson 1988).

It is simply an error to dismiss this kind of effect as not of the same order or significance, or not on the same time-scale as the effects of historical, anthropological, or sociological research. That the effect of these high-level theories on self-understanding is often forgotten or dismissed is, I believe, one of the factors exacerbating the kind of education controversies taking place in the US, Canada, New Zealand, and elsewhere. Often we teach such theories *as though* they had nothing to do, could not possibly have anything to do, with the students' sense of their own humanity. And then, when problems arise, we again expect the students, their parents, and everyone else, to discuss the problems *as though* only matters of science were involved, as though these interactions I am talking about were *in fact* absent.

However, this whole discussion does imply that not all sciences are equally involved with self-understanding and for those that are, distinctions still have to be made regarding the level at which the involvement occurs. What is suggested therefore is a kind of continuous, or nearly continuous, spectrum of all sciences in which on one side are those where the interaction with self-understanding is strongest, on the other side those where it is weakest.

The point of Overhead 3 is to show that although the human sciences are still more or less on one side, and the natural ones on the other, no

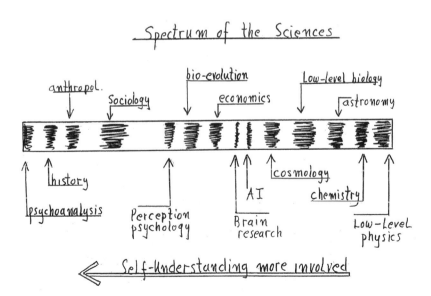

Overhead 3

sharp dividing line really exists. It would not be absurd to argue, for example, that natural sciences like biological evolution or mathematical sciences like artificial intelligence have more to do with human self-understanding than a "social science" like the psychology of perception.

In 1980, Mary Hesse suggested a similar continuous array for the sciences on the basis of their involvement with, or susceptibility to, hermeneutics (Hesse 1980, 225). My chart could represent her idea as well if the left-pointing arrow were labeled "increasingly hermeneutical." This correspondence is not surprising. A continuum regarding self-understanding implies a nearly identical continuum regarding hermeneutics because whenever self-understanding is involved in any science, in any of the ways I have discussed, a hermeneutic approach is called for at some stage of that science. Influences of the R-type involve interpretation of human life in the context of the science; influences of the L-type involve interpretations of the science in the context of human life.

However, the idea of this chart, and of this whole discussion, should not be confused with any program of methodological unity of the sciences, certainly not the positivistic one. Rather, my interest is to point out that in our fear of the illegitimate encroachment of the natural sciences on the social sciences, we should not distort the character of the former by ascribing to them far more objectivistic, far fewer hermeneutic features than they in fact possess.

Final Note

Many years ago, a most interesting and important idea was introduced into the discussion of hermeneutics when Apel suggested his version of the complementarity principle: This posits a complementary relation between the objectivistic and the hermeneutic approaches in the same sense as Niels Bohr's complementarity in physics; it means that the two approaches are supplementary but mutually exclusive at any one time (Apel 1988 [1968]). Apel has discussed at length the way in which this complementarity shows up in the sciences on the left side of our continuum, the human sciences; regarding the right side, he says that the hermeneutic aspect pertains only to the pre-condition, or condition of possibility, of these sciences.

Characterizing it as a condition of possibility implies that in natural science, hermeneutics is at most a background sort of thing, realized some time before the actual science begins, and absent actually when the science itself is practiced. But I would now suggest—without argument—that if it can be shown that even in these sciences hermeneutics is more than just a condition of possibility; that although complementary to objectivistic work, just as Apel claims, it nevertheless enters at various stages of actual scientific practice, then the continuum of the sciences in my chart, and Hesse's idea, could also be thought of as a continuum based on the relative weights accorded to the two components of Apel's complementarity principle.

[6]

Meaning and Contexts in Physics:
A Case Study

ABSTRACT: The mathematical astronomer George Darwin was deeply concerned not only with questions of lunar and planetary evolution, but also with larger questions of general evolutionary mechanisms arising from instabilities in dynamical systems. Although his thinking can be seen as part of a widespread cosmological interest in the latter decades of the 19th century, involving such figures as Herbert Spencer, John Fiske, and Ernst Haeckel, his approach stands out because of its fusion of carefully worked out physics with sweeping cosmological ideas. This is especially true when viewed in the context of present-day cosmology, chaos theory, and the all-embracing evolutionary outlook that points up links and similarities between physical and biological evolution.

In our contemporary context, therefore, in contrast to how it appeared 30 to 40 years ago, George Darwin's work seems surprisingly modern in spirit. This paper focuses particularly on the relation of his writings to biological evolution, to bifurcation theory, and to the philosophical cosmology of his day. [—A.S.]

Introduction

The subject of my talk lies somewhere on the hazy boundary between physics, the history and philosophy of science, and the world of education. Although I am a member of a physics faculty, and still teach physics for the most part, I have, in recent years, become interested in the other areas—and that accounts for the way I am going to treat my subject today: That subject is a part of physics, but my point of view, and the questions that motivate that point of view, arise *outside* of science altogether—as science

This paper was originally presented as talks at the University of Auckland and the University of Waikato, New Zealand, in October, 1992. Fig. 1 is from J.A. Ripley and R.C. Whitten, *The Elements and Structure of Physical Science*, 2nd ed., New York: Wiley, 203. Fig. 2 is from D.U. Wise, "An origin of the moon by rotational fission during formation of the earth's core," *Journal of Geophysical Research* 68, 1547–554.

is now understood. So, before taking up the subject itself, I would like to say something about these questions and this motivation.

At the most general level, the questions are directed at the *meaning*, or significance, or cognitive value of particular scientific results. Now every physicist is aware of the historic problem of the interpretation of quantum theory, and how difficult it is to find in that theory something beyond instrumental meaning, something that would give an *understanding* of phenomena—not just to the professional physicist, but to a broader spectrum of scientifically competent or interested people. *However*, the problem of meaning precedes quantum theory; it arises out of the abstract nature of modern science, and the separation of science from the life-world of even highly educated people—*of scientists even*.

This is why certain humanist writers have, for over a hundred years, tried to distance themselves from the natural scientists by emphasizing what they perceive as the crucial difference between the two kinds of study. Briefly, this difference can be put in the following way: The goal of the natural sciences is *explanation* by means of *law*, while the hermeneutical or cultural sciences seek *understanding of meanings*. Law supposedly answers the interest of prediction and control; and this, at bottom, is what hard science is about. In contrast, the search for meanings expresses the human need for orientation in the world, for self-understanding, and for mutual understanding in the light of tradition. These disciplines pursue understanding through interpretation rather than fixation or determination; and they communicate through narratives rather than timeless propositions.

It is not hard to see how the increasingly abstract character of science, coupled with this kind of philosophical thinking, leads to a view of scientific knowledge as *commodity*, something of great value but only as a professional asset for those whose career plans make it relevant, while for everyone else it remains *irrelevant*. The French philosopher Jean-Francois Lyotard pushes this point of view to its limit: "Knowledge," he says, especially scientific knowledge, "ceases to be an end in itself . . . it is and will be produced to be sold." And he continues: "The question now asked by the professionalist student, the state, or institution of higher education is no longer 'Is it true?' but 'What use is it?'"[1]

The result of such a change of attitude toward knowledge is a change in the role of the educational system. We are on the threshold of a new era, he believes, in which traditional teachers are replaced by intelligent terminals connected to memory banks, and the only thing left to teach is techniques of questioning the information system.

One doesn't have to accept all of Lyotard's extrapolations to notice that much of what he describes is indeed happening. Some people go further; they connect what is happening in this regard with the decline of interest in the study of physical science among students in certain industrialized countries. Others, going further still, try to relate this entire process—that is, of knowledge turning into commodity—to the way in which science is *presented* in Western society—in other words, to the way it is understood by most teachers, depicted in most books, and taught to most students.

That teaching, at all levels—from elementary school to graduate school—is too fragmented, according to these critics. Scientific activity, like all other human activity, takes place in a context; and the full meaning of its results arises out of this context. Wrenched out of context (or "decontextualized," as people now say), the results lose their larger meaning, appear as mere instruments, and invite being handled as so many salable commodities.

Bring back these contexts, and science also becomes a narrative; consider its history and then you have more than explanation through law—you gain *additional* meanings by recovering the *questions* that motivated the *search* for the laws in the first place. Usually, scientific contexts *connect* to more general human contexts; and through such connections science loses its 'alienness.'

It is in this way that some of us have come to think of the history and philosophy of science as the major missing contexts.

And so, today, I would like to pick out one piece of scientific knowledge—though not one that looks especially exciting—take it as it stands in the textbooks and try to *re-contextualize* it—to see what it might look like, what new meanings might come into view, if some of the missing contexts were re-inserted. Even if no great insight comes from this, it's an enjoyable thing to do. And if we don't look at specifics, the kind of generalities I've indulged in so far are also meaningless.

The Phenomenon and Its Contexts

A. IN THE TEXTBOOKS

The topic I chose is from the middle levels of the curriculum—a *minor* topic that might be taken up in the early years of college or upper years of secondary school: It is the recession of the moon, and the resulting lengthening of the day, due to tidal coupling of the earth-moon systems. What the textbooks give, aside from a basic description, varies. Most often, the relation to angular momentum conservation or torque laws is mentioned, as in figure 1.

Sometimes, only numerical values are given:

Recession of the moon = 3.7 cm/yr, +/− 0.2

Lengthening of the day = 1.4 ms/century^2

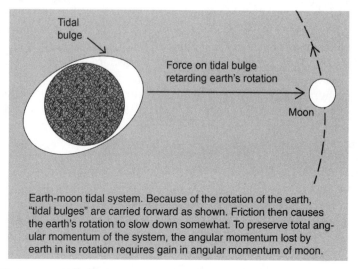

Figure 1. Earth-moon tidal system. Adapted from Ripley, J.A., and Whitten, R.C. (1969), *The Elements and Structure of Physical Science*, 2nd ed. Wiley, NY.

In physics textbooks, this is usually the end of the matter. The only context offered, for the phenomenon and for our interest in it, is the context of universal laws. For the student, then, the *meaning* of the phenomenon can be no more than this: that the moon does recede is a feat in itself, *and* a good example, and instance, of the conservation laws. Whether that is particularly meaningful, in the sense of *significant* or interesting, depends no doubt on the individual. But *what else* could it possibly mean?

B. The Context of Discovery: The Fission Idea

Well, suppose we look at the context of discovery during the second half of the 19th century. Qualitatively, the effect was known long ago (Kant, Helmholtz), but in terms of serious calculations, the subject was 'opened up' only in the 1870s when all sorts of people began to be interested in the *rate* of the slow-down of the earth's spin. It was then that the question was first asked: Since the day and the month have both lengthened, but at different rates, was there a time when they were equal? And where would the moon have been at that point? The man who made his reputation grappling with this problem became a leading astronomer at the turn of the century, a world authority on tides, an admired Cambridge don, and a statesman of science. For the time being—because I want to treat the question of personality in another context—let me just call him *Sir George* (or *George*).

In first approximation, part of George's scientific question can be answered right away, and makes a good student problem. If we leave out solar effects and the obliquity of the orbits, earlier states of the earth's spin and the moon's orbital angular momentum can of course be found from the conservation law alone, without any knowledge of the coupling mechanism. Assuming the plane of the lunar orbit in the ecliptic and the earth's spin vector perpendicular to that plane, one easily finds that when the *day and month* were 5 hours and 36 minutes, the moon was about 16,000 km. away—center to center.

Now this is highly suggestive because in this state the earth and the moon are behaving as one rigid body, and, as Sir George put it, "this small distance seems to me to point to a break-up of the earth-moon mass into two bodies."[3]

We have here the beginning of the fission model of the origin of the moon. In *this* context, the present slow lunar recession acquires a *new* meaning: it could be a *fossil*, a relic of a cataclysmic event in the history of both the earth and the moon; it could be a hint also about the origin of other satellites, perhaps of planets—of double stars, even. Sir George thought just that.

C. THE BIG QUESTION: BIOLOGICAL EVOLUTION

But let's continue our search for contexts. Why all this interest in looking backward, and in the length of the day? It is true that young scientists often work on problems just to work on problems, because problem-solving is their business; but not infrequently, among mature scientists, one finds larger questions in the background. Sir George—at that time, *young* George—inherited his problem personally from Lord Kelvin. A paper of George's caught Kelvin's attention; the two met, got along well, and the younger man became informally a kind of protégé of the famous physicist.

But Kelvin was at that time the most prestigious opponent of Charles Darwin on the question of natural selection in biological evolutionary theory. Kelvin's belief in design in nature, coupled with his interest in thermodynamics, propelled him on a course of calculations and arguments to limit the age of the earth and show that the time required for natural selection was just not there.

Within a decade of the publication of *The Origin of the Species*, Kelvin had developed no fewer than three independent arguments, all based on various forms of energy dissipation: The first was simply that, given its rate of radiation, the *sun* could not have lasted more that 20–100 million years[4]; the second, the famous "rate of cooling" calculation using the temperature gradient near the surface of the earth, gave a similar figure for the age of the earth.[5] Finally, in the third argument, he focused on the frictional dissipation of energy due to terrestrial tides.

This is an ingenious argument: the shape of the hot molten earth, before solidification, would depend on its spin angular velocity; its present shape must correspond to this velocity at the time the crust solidified. If we knew the rate of decrease of this velocity due to tidal dissipation, we should be able to say how long ago the crust solidified and how much time was available to biological evolution. Hence, it is understandable why it became important for Kelvin to investigate the rate of change in the length of the day. And this is the problem he urged upon young George, after himself laying out the framework of the calculation.

This, of course, is the difficult part of the problem. The evolution of the earth-moon system *in real time* depends on details of the coupling mechanism, on the actual tides, and on this young George spent an enormous amount of labor in the course of the next several years. In his 1879 paper, the one where the fission theory was first proposed, he finally came out with a lower limit of 57 million years from the fission event to the present, although that was just an estimate based on body tides, not ocean tides, and maximal assumptions for the earth's viscosity and other parameters.

Young George had gone far beyond Kelvin's argument: In George's hands, the treatment proceeds to a major deformation of the earth, not just to the 'freezing in' of a small change in shape. Along the way, the Laplacian nebular theory was being amended: The rotating earth "gave birth"—as George put it—to a moon all in one piece; she did not merely blow off an equatorial ring to condense later, as the nebular hypothesis demanded. And yet, the number itself—the 57 million years—was well in line with Kelvin's ideas: Biological evolution would surely not have begun before the fission event.

Here we have another meaning for the barely perceptible recession of the moon: In this expanded context, it points to a possible earliest time when life on earth could have begun; it unites astronomy with biology; and the problem of calculating the recession takes its place in the major 19th century debate over natural selection—the debate that featured Charles Darwin and other naturalists on one side; Lord Kelvin, supported by several physical scientists, on the other.

D. The Human Context

And so, we are brought to the human context. *Who was young George?* It is no secret, of course, that he was the fifth child, and second son, of *Charles Darwin*—which makes the story even more interesting.

What we have here is a situation where the son of a famous scientist obtains a result that seems to contradict his father's theory, a theory of world importance—scientifically, philosophically, socially—a theory which is under intense debate. From our viewpoint, the son appears to be in the camp of his father's adversaries. Yet, nowhere in the literature have I found any hint that the father resented it or that the son thought there was anything wrong with his position. To the contrary: The father was proud of the son's friendship with the great Kelvin; the son was proud of, and deeply impressed by, the father's theory. Father and son seem to have had the greatest affection and respect for each other, as attested to by Sir Francis Darwin, George's younger brother.

Apparently, both Darwins assumed that science is *independent* of personal relations, that a true scientist follows the argument wherever it leads, and that a good theory will be able to absorb counterevidence—old fashioned ideas and *clichés* like that. How this contrasts with some of today's sociology of science is a point I need not belabor. In this context, George Darwin's early work offers a possibility for studying the role of philosophical bias or worldviews in theory formation—because there is no doubt, as we shall see in a minute, that George's sympathies, and his scientific worldview, from the very beginning, were close to his father's.

In this light, his early support of a young earth appears as an instance of *self-criticism* of the Popperian sort, since Kelvin's program, and his intention, certainly *was* to use thermodynamics to poke holes in the Darwinian theory of natural selection.

E. THE CONTEXT OF SCIENTIFIC WORLDVIEWS

As it turned out, George Darwin did not stay long in the camp of the enemy. Conflict with some of Kelvin's assumptions, and his own growing awareness of the sensitivity of the results to the specifics of the model, led him throughout the 1880s to question publicly the confidence that was being placed on Kelvin's numerical results. His own experience with many tedious calculations of the type that we would now call 'what-if' questions—but without computer—showed him, for example, that the time to the fission event could be as long as several hundred million years, *if* ocean tides instead of body tides were given greater weight in the model.[6] Later on, with the discovery of radioactivity, Kelvin's thermodynamic arguments crumbled, of course; and in all this, George took part on the side of the ever-increasing antiquity of the earth, ever-increasing availability of time for natural selection to work—which certainly was more in tune with his own evolutionary worldview.

I have already mentioned that evolutionary thinking came early in George's life, as might be expected. But in order to really understand the context in which he, and others at the time, had come to regard the evolution of the earth-moon system, we must take into account the prevailing view of evolution in those days. Surprisingly, perhaps, and in contrast to what might have been said in the 1950s, that view now appears quite modern. In the late 19th century, the word 'evolution' had stronger links to astronomy than to biology. It was not then thought of as a process having much to do with chance, but with an orderly, lawful succession of physical states, as epitomized by classical dynamics. The paradigm of 'evolutionary' thinking was the Kant-Laplace nebular hypothesis, rather than Darwinian natural selection; and that paradigm embraced 'stellar evolution,' for example, more naturally than evolution of biological organisms.

The feature of this 19th century evolutionary view that now makes it seem modern is its *cosmic scope*, and its wide reach across all the sciences. For example, in the voluminous and influential writings of Herbert Spencer, what he called the "laws of evolution" were *physical*, or at least modeled on physics, but they were so *general* that they seemed to apply everywhere: the law of the instability of homogeneity, the law of increasing heterogeneity, the law of the "dissipation of motion." These were *tendencies* that could almost be *seen* at work on the large scale, as in stellar evolution, possibly galactic evolution, certainly in planet formation as pictured by Kant-Laplace; they could be demonstrated in many human-scale situations of chemical and dynamic instability; and they could also be seen in human structures such as states, economies, and languages.

But Spencer, in England, was not the only one thinking in this way: In the United States, philosopher John Fiske elaborated the same theme at great length; in Germany, biologist Ernst Haeckel had his own version. It is in this context that biological evolution took its place, and, as George Darwin put it, gave to cosmic evolution a "great impulse."[7]

Some historians of science (Steven Brush, for example) make a case for the reverse influence—astronomical evolution on biology. No doubt the reinforcement was mutual. What I am pointing to, however, is the relation between worldviews now and then: If one had one's education sometime in the 1940s, 50s, or even 60s, one was not likely to hear much about evolution in physics; and biologists, likewise, thought it prudent not to assert anything beyond biology. It was only in the 1970s and 80s that books began to appear in quantity emphasizing *universal* evolution—not just 'amoeba to man' but 'big bang to man': atomic species, stellar species, and living species all at once. Today, all this *seems* ultramodern, but *this*, more or less, was the spirit of the 19th century—of which George Darwin was a prime example.

In 1905, seven years before his death, Darwin made a speech—his presidential address to the British Association—entitled "Cosmical Evolution." It makes interesting reading for at least half a dozen reasons. Among other things, it shows that this *physicist* preceded Stephen Gould in suggesting "punctuated equilibria" as the mode of biological evolution (but that's an aside). To us, the most interesting thing is how, at the dawn of the 20th century, George—as a *representative* scientist, not a maverick at all—came to see the various processes of *physical* evolution.

To him, steeped as he was in 19th century thought, and looking back on his own work, the crux of all evolution was the question of stability—whether the subject was a satellite orbit, an atomic structure, or a biological species. About stability in all these cases, he said that it "is a property of relationship to surrounding conditions; it denotes adaptation to environment." Radioactivity as disintegration of "atomic species" was new then, and it had its philosophical effect. Darwin compared the long-term instability of apparently stable atoms to the long-term instability of apparently stable planetary orbits, and both of those to the instability of living species.

To drive the point home, he asked his audience to visualize the following situation: a sun like our own, circled by a single large planet, like Jupiter. To this, add a swarm of hundreds of what he called "meteoric stones" and planetoids of various sizes—these are the species. "We know," he said, "that under the combined attraction of the sun and Jove, the meteoric stone will in general describe an orbit of extraordinary complexity"; countless times it will graze, and just escape being swallowed up by one or the other of the two giants. But, because of long-term instability, sooner or later, the great majority of them will indeed meet their end in this way. So, he explained, the large bodies "sweep up the dust and rubbish of the system," and in the long run only a few small ones remain, in the most stable orbits, arranged "according to some definite law." Of all *possible* orbits (and I quote): "The unstable ones are those which succumb in the struggle for life, and the stable ones are the species adapted to their environment."[8]

If this language sounds forced, the stuff of popularization, let me point out that just about this time, when he was speaking to the British Association, he was also engaged in the last major research project of his life—the question of the stability of rotating spheroids. As is well known, that problem had been worked on by Maclaurin, Jacobi and Poincaré.[9] Together, these investigations had shown that, as the spin velocity increases, spheroids of increasing oblateness will go through bifurcation points (see figure 2).

First, they become unstable to the formation of Jacobi ellipsoids (three different axes); then, at the higher velocities, they become unstable to the formation of Poincaré pear-shaped bodies. Finally, Poincaré himself hypothesized that the pear-shaped sequence might lead slowly to the formation of two bodies of unequal size.[10]

Darwin called the different geometrical types "species," explicitly comparing them to biological species.[11] And we see immediately how his fission theory of the moon's origin was tied at both ends: From the present recession, he could calculate backward to the five-hour day, and find there what looked very much like the offspring of the final bifurcation from the Poincaré "species." In the context of our present awareness of nonlinear dynamics, of chaos, bifurcation points, and all the rest, George Darwin's

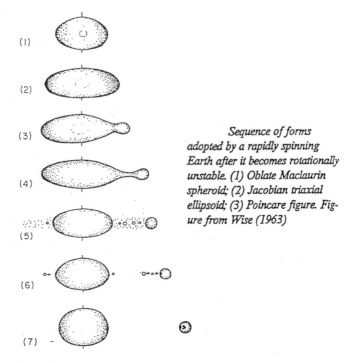

Sequence of forms adopted by a rapidly spinning Earth after it becomes rotationally unstable. (1) Oblate Maclaurin spheroid; (2) Jacobian triaxial ellipsoid; (3) Poincare figure. Figure from Wise (1963)

Figure 2. Sequence of Forms. From Wise, D.U. (1963), "An origin of the moon by rotational fission during formation of the earth's core," *Journal of Geophysical Research* 68, 1547–554

language no longer seems forced or far-fetched. What he was looking for was a general theory of transformations of structural types or species— good for physics, astronomy and biology. In our own time, this has been made familiar by another physicist, Ilya Prigogine, who also points to a connection between new types of order, biology and just such bifurcations in physical systems.

In this context, we see the moon's recession and the earth's slowing spin as an instance of *cosmic evolution*. Perturbations of all sorts form an environment that is at once hostile and transition-inducing, an environment in which a structure, or species, must prove its ability to survive, or evolve into another species that does survive. Seen this way, the moon's present recession is a kind of *microevolution*, of a species that Stephen Gould would consider in long-term *stasis*. And its *significance* now is to remind us that even these stable states do evolve, and eventually pass away. Ultimate stability is nowhere to be found in this universe.

F. THE CONTEXT OF THE PHILOSOPHY OF SCIENCE

There's still one other context that should not be left out—the context of the philosophy of science. We can ask, for example: What made George so confident about his fission theory—and he was confident—despite the various approximations, the paucity of data, and so on? Was it 'good science'? To discuss this, I have to at least mention some aspects of the problem that I've omitted. Of course Darwin's retrodiction had to keep track of *all* the relevant variables, including the inclination of the earth's and moon's orbit to the ecliptic, and the moon's eccentricity. And it turned out that the way these changed was also interesting. The result as a whole became more impressive because it explained several things in one shot.

Shortly after fission, it seems, the plane of the earth's orbit was much closer to the ecliptic (11°), and the moon's orbit was less eccentric—both of which are to be expected on a general nebular hypothesis of planetary formation, and fission model of the moon's origin. In other words, starting from a more 'aligned' but unstable original state, Darwin 'explained' the present length of day, the obliquity of the ecliptic, the distance to the moon, the eccentricity and inclination of the lunar orbit—and he explained all that with one mechanism, tides, a process admittedly taking place. It is on this account that, even in 1905, he could still say that the theory has "strong claims to acceptance."[12]

We have here an instance of what was called in the 19th century, a "consilience of inductions," to which the closest contemporary concept is "explanatory power." The tides, as a coupling and dissipative mechanism appear to explain a lot; and, as I mentioned earlier, George Darwin believed they could explain a lot more still, beyond the solar system. Philosophically, the structure of this justification is the same as that used by his father in arguing for natural selection; and even today, that argument is still invoked by defenders of Darwinian evolution, sometimes under its old, 19th century name.[13]

George Darwin's theory of the evolution of the earth-moon system is an interesting case for the study of explanatory power as a principle of theory construction. And in *this* context, the recession of the moon—our original bit of scientific knowledge—invites us to ask how much of George's theory still stands today—how *good a guide* was *explanatory power* in this case?

It goes without saying that the earth-moon system has been studied thoroughly with the use of a whole array of models. And it is no small tribute to the younger Darwin that even after the Apollo missions and well into the age of computer simulations, his theory (or a variant of it) still appeared as a contender at a major conference on the origin of the moon in 1984 (the Kona conference).[14] It was not, however, a leading contender, and it is fading fast under the blows of modern technique. In simulation studies, fission of the earth does not give rise to a single new body, as Darwin thought, but to a stream of *debris*, resembling more the original Laplacian idea. And even this process faces serious dynamical counter-evidence.

So, we may ask, what does survive of all those correlated effects whose "consilience" was so impressive a century ago? That's hard to say, actually: the general framework of the discussion, key parameters, limiting factors, and above all, the very idea that such variables as distance to the moon, length of day, inclination of the earth's and moon's orbit *are* correlated.

But because the system is so complex—oceans forming, internal structure of the earth perhaps changing, and so on—it is hard, even today, to say definitely that any one of the variables just mentioned was increasing or decreasing in the remote past. It all depends on the specifics of the mechanism—the viscosity of the earth, for example, and other assumptions in the model. Because the calculation was lengthy and tedious, few people realized the significance of these assumptions; and the public, as usual, was thrilled by the idea of the earth 'giving birth' to a whole moon—the pictures (as you recall) show a surprising resemblance to biological fission.

The moon still recedes a few centimeters every year, the earth is still slowing, its inclination still changing more or less as Darwin explained; but now these things no longer point to what he thought was a fundamental mechanism of cosmic evolution.

To our philosophical question, the answer seems to be that *explanatory power* is a wonderful thing, but, as Cardinal Bellarmino once said to Galileo, God might have done it another way. Today (if I'm abreast of things) the latest promising idea is that the moon was formed as a result of a collision between the earth and some sort of planetesimal.

That *a theory* can be largely sound, can explain precisely how a specific mechanism brings about an observed state of affairs, can lay the foundation for a hundred years of further development, *and yet*, in its most significant conclusion (the fission of the earth, the thing that got everyone excited) be *essentially wrong*—is one thing the history of *George* Darwin's evolution theory shows. And that is one more way, not mentioned by textbooks, in which lunar recession is meaningful.

The Variation of Profiles

Our recontextualization is finished now. As a final comment, let me stand back for a moment and try to *place* this kind of exercise in the total picture of things. What we have done is to take one, simple, fairly well-defined piece of physics—which we can call an 'object' of the study of science—and we have inserted, in *back* of this object, several different contexts, one at a time. A picture of the procedure might look like this (see figure 3):

Contexts and Profiles of Lunar Recession as "object"

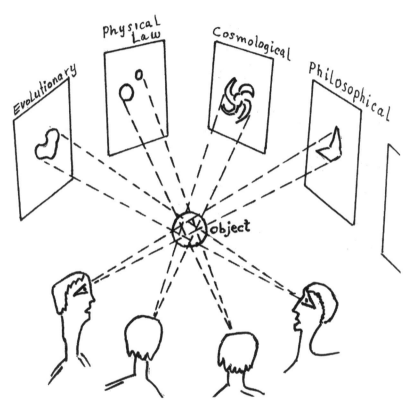

Figure 3. Contexts and Profiles of Lunar Recession as "object." Diagram by Martin Eger.

Viewing it pictorially makes it easy to see that changing contexts forces us, literally, to change our *point of view*, and *project*, or *interpret* the object against a different background. In this sense, each projection exposes a different *profile* of the object. This whole idea is well known in philosophical phenomenology and the hermeneutics associated with it. It is called the *variation of profiles*. Its purpose is precisely to reveal new or hidden or unexpected meanings in an otherwise familiar object. And one could say that, in a rough way, I have merely adapted this basic procedure of phenomenology to objects in the world of science.

In our example, taking as our object the recession of the moon, what I have tried to show is that, in a sense it 'looks different' or 'means something else' when viewed against the background of Charles Darwin and Lord Kelvin from against the background of the Kona conference and the latest computer simulations. And both of these profiles are different still from what one sees against a background of conservation laws only. But these different meanings are not to be *set against* each other; on the contrary, they tend to amplify and enhance each other—within a *total* meaning which is personal appropriation of a particular weighted sum of all these profiles.

Some people might object that these contexts are not *within* physics; they are really *extensions* of the physics: Whatever the moon's origin, its present recession and the relation of that recession to angular momentum are the facts of physics we can teach *simply, unambiguously, and with certainty*.

To this I would say: *yes, of course*. But I would add one thought. Simplicity, lack of ambiguity and certainty are great virtues—with significant trade-offs. The simpler, the less ambiguous, the more certain we make it, the less meaningful it becomes.

[7]

Hermeneutics in Science and Literature

ABSTRACT: Hermeneutics, which was originally an art of interpreting texts, was generalized and radicalized by philosophers Heidegger and Gadamer to become constructing rather than reconstructing; the process of interpreting was ontologized to become an activity of "human beings in the world." Recent philosophers of science, notably Feyerabend and Kuhn, similarly transformed scientific methodology by ontologizing "being in a scientific paradigm." Although this radicalization is excessive when applied to scientific research, it is illuminating when applied to scientific education, where the student is indeed confronted with a new language. [—A.S.]

I would like to address the question of constructivism from the point of view of the *debate on hermeneutics*. My purpose is to see what we can find out by comparing the way hermeneutics has been understood and applied in such disciplines as literature and history, on the one hand—and on the other hand, the way it is understood in relation to the natural sciences.

Let's recall that hermeneutics began in a modest way as the attempt of historians, theologians, and literary critics to judge the work of interpretation they had to do. When faced with a text that's ambiguous, fragmentary, contradictory—because its sources are remote—the *methods* and criteria used in arriving at its meaning are not obvious. You have to reflect. As natural science calls for philosophy of science, textual interpretation calls for hermeneutics. Now it used to be, long ago, that finding the meaning intended by the author was the goal of any interpretation, and therefore also the main criterion in judging interpretations. Since each text had behind it its intended, its true meaning, each interpretation had before it its true goal—the very same meaning.

This talk was given at the 1991 Conference on Cybernetics: Its Evolution and Praxis, in Amherst, Massachusetts, where Martin Eger also took part in a plenary session on "Constructing Realities."

In our century, however, this honest craft attracted the attention of certain philosophers—the philosophers of meaning we could call them, of whom the pioneer was Martin Heidegger,[1] and the second, in our line of interest, Hans-Georg Gadamer.[2] They, in effect, reinterpreted interpretation. With them, hermeneutics was universalized *and*, at the same time, took the constructivist turn. It was no longer just a matter of puzzling over old texts; now it was any sort of communication or activity—from a spouse's incomprehensible complaints to the dance-ritual of an African tribe, to a pile of metal at a museum of modern art confronting a class of school-kids and their lecturing guide. And in all such cases, the conscious intentions of speaker, actor or artist ceased to be the source of meaning; instead, the interpretation itself took on that role, and came to be viewed as a *con*struction, not a *re*construction.

The reason for this turn to constructivism had to do with the so-called "hermeneutic circle of understanding"—which is just the well-known and very general problem of the *whole and its parts*: The meaning of a word may depend on the whole paragraph, but to understand the paragraph, of course, you have to know the words. Where and how does one start? The answer, according to hermeneutics, old and new, lies in "projection"; we rush ahead of ourselves, so to speak. We throw out a kind of "pre-understanding," to break into the circle. In Gadamer's words,

> A person who is trying to understand a text projects before himself a meaning for the text as a whole as soon as some initial meaning emerges. . . . The working out of this fore-project, which is constantly revised as he penetrates into the meaning . . . [this] is understanding. . . . This constant process of new projection is the movement of understanding and interpretation.[3]

We see here exactly how the radicalization came about. What Heidegger and Gadamer did was to 'ontologize' the act of interpretation: In their view, the meaning-seeking activity belongs to *human being in the world*; it is not optional. For the 19th century, the hermeneutic circle was a *methodological device*. It was thought that as the interpretation proceeds, the circle gets smaller; the back-and-forth motion between pre-understanding and text subsides; and at the end, you should find yourself standing still, feet planted firmly on the ground of truth. But for Gadamer, being in the circle *is itself* the act of understanding.[4] The interpreting that is done is like the interpreting of musical or dramatic scripts; it is a *performance*, not merely a recovery, and every performance is different. So the circle does not vanish, and the circling does not stop. Some 'facts' may, of course, be established more or less permanently, as history tries to do, but *meanings*, which are context-dependent, cannot be fixed this way; and that is why alternative interpretations coexist—they are not really 'decidable.'

One aspect of the radicalization is that prejudice, pre-understanding, and 'the tradition' now play a double role. They still are something to watch out for, to bring to awareness so as not to be misled; yes, but on the other hand, they must also be viewed positively, since they are part of *the event of human understanding*. One can come to understanding only *within* a tradition, however much that tradition may prejudice the understanding. To avoid distortion of view, one does not eliminate the pre-understanding. Rather, in the course of the circling and the back-and-forth switching of perspectives, mutual adjustment takes place until the horizon of the text and the horizon of the pre-understanding fuse. Pre-understanding, or tradition, is not discarded but assimilated.

Now if we look at what happened during the past three decades in the philosophy of science, the parallel could not be more obvious. Here too, the role of hypotheses, theories, and paradigms was known before the 1960s. What Kuhn, Feyerabend, and the others did was to ontologize *being in the paradigm*, in much the same way as Heidegger and Gadamer, in the 1920s, ontologized *being in the hermeneutic circle*. In what we now call the "new philosophy of science," the paradigm is not *a means* to truth, but the *way of being of science, the way of scientifically dealing with nature*. The paradigm—or the tradition, or research program—here plays the same dual role with which we are already familiar: On the one hand, its narrow vision hinders transition to the next paradigm, but on the other hand, it and it alone enables the student and the practitioner to see *what is there*. This is why, according to Kuhn, "normalcy" and indoctrination must dominate the schools.

I describe all this as an *ontologization* to emphasize the primacy and necessity of *process* in the new outlook. Paradigm is the *being* of the scientist qua scientist. And the reason for the turn *in this case* is, once again, a radicalization in the way we view the circle of understanding. In the past, the back-and-forth movement between theory and experiment was assumed to *converge*. Just as in old hermeneutics, at each turn of the circle, with each correction of the pre-understanding, the differences were supposed to get smaller as truth was approached. In both cases, the "circle of understanding" was a kind of cybernetic mechanism—an *effective* procedure. Which hypothesis one started with could determine *how fast* one got to the truth, but it could not affect the truth itself. The Kuhnian turn radicalized this situation by declaring the self-correcting capacity of the interplay between theory and observation *inoperative* on the large scale. Echoing Gadamer's approach to history, Kuhn described self-correction as affecting only the *quantitative* but not the conceptual results, only numbers but not meanings. This is "conceptual non-convergence," which leads to the famous "incommensurability of paradigms"—which is, of course, the analogue of "undecidability" in textual translation.

These kinds of resemblances have convinced a number of philosophers that interpretive understanding is something that applies to the natural sciences as well as the humanities. Foremost among them is Mary Hesse, who sees the sciences not as divided into contrasting types, but as a continuum, with hermeneutics playing a strong role on one end, among the humanities, and a weaker role on the other end, where physics might be located.[5] But it's interesting that even an experimental scientist can adopt a hermeneutic perspective. When Gunther Stent, for example (the well-known biologist), tackled the nervous system of a leech, to investigate how it generates the swim pattern of its body, he found himself criticized by colleagues for the simplifying assumptions he had to make in arriving at the final theory. *You have gotten out of your model what you projected into it*, they said. Let me quote his reply:

> Since the topology of the circuitry of any nervous system . . . is so highly interconnected that there is usually some neural . . . pathway leading from any point to any other point, it is simply not possible to start any functional analysis without having some prior theoretical insight into the problem. Hence . . . neurobiology takes on some of the characteristics of *hermeneutics*: the student of a complex neural network must bring considerable *pre-understanding* to the system as a whole before attempting to interpret the function of any of its parts. Accordingly, the explanations that are advanced about complex neural systems may remain beyond the reach of objective validation.[6]

However, Mary Hesse and Gunther Stent are in the minority. For the most part, the social and natural sciences, and their philosophers—each group for its own reasons—agree that hermeneutics does not apply to things like physics. Natural scientists want to emphasize that *undecidability* is not characteristic of their fields, while social scientists are anxious to erect a well-articulated demarcation criterion between themselves and the natural sciences—to protect their *humanness* from the imperialism of an ever-hungry, ever-expanding natural science. When Gadamer universalized hermeneutics, he made one important exception—natural science.

What *arguments* do these people offer, in view of what looks like a convincing picture that no *qualitative* difference exists between the sciences? At this point, we come to a concept that I consider especially provocative; it is called "the double hermeneutic." It comes from European sociologists like Anthony Giddens and Jürgen Habermas. Here is what it says: Of course there is also a kind of circle of understanding in natural science, between theory and experiment. But the human sciences are different on account of *what* they study. They need an additional, a deeper level of interpretation because—as in anthropology, for example—*they encounter a language in being*, a symbol-system made by or for humans, whose meanings have to be ascertained *before* anyone can even begin theorizing about the actions of the people who use that language. As Habermas puts it,

> If the paradigm-dependent theoretical description of data calls for *stage 1* of interpretation that confronts all sciences . . . then we can demonstrate for the social sciences an unavoidable *stage 0* of interpretation.[7]

If we take Gunther Stent's research as the counterexample, Habermas is saying that while there was a hermeneutic problem in the relation between Stent's final theory and the particular data on neurons used to construct that theory, Stent had no *prior* problem with a new language that would have to be used to get the data on the neurons in the first place.

Now why do I say this is provocative? Because, in effect, it invites us to look more carefully at natural science, to see if we can find analogues to this double hermeneutic, analogues to this 0-level of linguistic interpretation—which perhaps might be enlightening, useful. Well, such analogues are not hard to find.

Think of any science, not as research, but in its educative mode—as we all faced it at the beginning, in classrooms, books and laboratories. Was there no existing language, no tradition, whose meanings had to be appropriated before we could hope to understand any fairly complicated experiment? The language we had to *learn how to interpret* was in fact the language of the science studied. Habermas fails to notice this '0-level' of interpretation in the natural sciences, because his picture of these sciences is still too objectivistic: He imagines a scientist as subject, already in possession of the relevant paradigm *as his working language*, turning to nature as object— and finding there no existing, in-use, symbol-system to interpret. The minute you picture a student, an outsider, or a newcomer to the field—who might even be a scientist—then you realize that, of course, the already-existing language of that science has to be mastered first.

I think there are many consequences of this. Time allows me to mention one—in education.

If the idea of science as a language is taken seriously, then the problem of interpreting this language also becomes serious, and the whole theory of hermeneutics is applicable to the study and teaching of the natural sciences. Actually, many researchers in science education have already begun to investigate in detail the interpretive difficulties that science students have, the problems caused by their pre-understanding, and so on; entire conferences on this subject have been held.[8] Philosophical hermeneutics might well provide insight here about the way to deal with these pre-conceptions once they have been brought to light—which, at present, is not clear.

Or, regarding the role of the teacher in general, one sees immediately what the implications might be: A teacher should be an interpreter—interpreting between the language of science and the language of the life-world.

Such a view precludes, in one stroke, the two extremes that are well known in pedagogical thinking: On one side, there is the image of teacher as 'transmitter' of knowledge, to which are attached all the old stereotypes about lecturing, stuffing heads like buckets, etc. On the other side, we have teacher as 'facilitator' or 'resource person'; and the new stereotypes are non-interference, letting the student discover, etc. From the point of view of hermeneutics, however, not only do we see these images as inappropriate, but as actually missing the point—the teacher's *positive* role: In order to interpret, one must himself (or herself) constantly be *trying to understand*, in and out of the classroom, and well beyond the confines of the textbook. We get the image of the *teacher as student*—engaged with the *subject,* not just with the job of teaching—a colleague, a fellow interpreter, not just an *underling* of the professor and the researcher.

This point shows up an important and general feature of the hermeneutic perspective: It rejects both extremes regarding construction. That it opposes objectivism is clear enough; but, equally, it opposes what Gadamer calls "hermeneutic nihilism."[9] This emerges from the existential and ontological nature of philosophical hermeneutics, which I've tried to emphasize. If the human being is *interpreting*, then there must be something to interpret. The word 'construct' does not carry that kind of connotation; it does not encompass all that is happening either in natural language or in science. Interpretation implies unambiguously that beside construction, there is something more—something to which the interpreter is *listening*. The use of this image, therefore, supports and lends seriousness to the various debates about what the 'more' may or may not include—whether these debates be of the moral, social, or purely epistemological kind. A human being *interpreting* is not a human being trapped in a closed language game; nor is he just exercising his whim or his will. He is *constrained* by what he interprets, and by his commitment to *listen*. The key idea is that between the listening and the constructing, there is the moment of *trying to understand*.

[8]

Rationality and Objectivity in a Historical Approach: A Response to Harvey Siegel

ABSTRACT: A critique is offered of certain conceptions of "critical thinking" in science and science education. Harvey Siegel's just-published paper (*Synthese*, 80), developing this view, is discussed. My approach is to use epistemological reflection on episodes in *contemporary science history*; and my main point is to recommend such use for education. In this way, I try to show why recent scholarship in the history and philosophy of science need not be resisted by those who wish to retain the traditional ideal of *objective reason*. To support this ideal, I propose a "strong version" of the concept of objectivity. [—M.E.]

A widespread movement among educators today attempts to promote, as a major goal, the study of rationality and what is often called "critical thinking." Because science is increasingly viewed as the exemplary rational activity, it follows that in any teaching embodying such a purpose science ought to be a central component. One example of an approach based on this traditional theme is a paper by Harvey Siegel (1989) to which I would like to devote some attention.

My purpose is twofold: on the one hand, to show why a treatment of science that bypasses the distinctive insights of the past thirty years of historical and philosophical scholarship is inadvisable; on the other hand, to support, possibly to strengthen, the traditional belief in *objective reason*—as it relates to *morality*, to character, or, in Siegel's words, to "being a certain sort of person." Along the way, I would like also to demonstrate the extraordinary value of examples of *contemporary science history*—viewed in philosophical and social perspective—not merely by advocating this resource for the classroom but by utilizing it here, in both the critical and the constructive aspects of this paper.

This was a plenary talk presented at the First International Conference on History and Philosophy of Science in Science Teaching, in Tallahassee, Florida, in 1989. It was published in the conference proceedings, *The History and Philosophy of Science in Science Teaching*, Don E. Herget, ed., Tallahassee, FL, 1989: 143–153.

121

The Problem with Reasons

Starting with a definition of the rational as that which is "guided by reasons," and of rationality in science as "commitment to evidence" (1989, 14–15), Siegel builds the following chain of connections: When reasons (evidence) are seriously pursued, then *criticism* and testability of hypotheses are obvious desiderata. But criticism, which is the assessment of reasons, can only be based on *principles* broader than the evidence in any particular debate. In turn, this implies consistency, objectivity and impartiality; for otherwise the principles would be meaningless.

The next step is to draw the image of "a certain sort of person" to be cultivated by any education imbued with the critical spirit (Siegel 1989, 27). Again, a short, simple definition is given: The "critical thinker" is "one who is *appropriately moved by reasons*" (ibid., 21). Rationality "demands not only an ability to seek reasons. . .not only an ability to judge impartially, but a willingness and desire to do so even when impartial judgment runs counter to self-interest" (ibid., 25–26). Consequently, he recommends a healthy dose of the philosophy of science, comparison with pseudosciences like creationism, and a pluralist approach to theories to lower "the likelihood that one will be blinded or biased by one's theoretical commitments . . ." (ibid., 33).

The focus on rationality, and on *reasons* as its adequate characterization, is common to a number of educational theorists (Green 1971; Scheffler 1973; Strike 1982a&b). When no more is intended than a correction to rote memorization, one could hardly object. The trouble is that although we are all aware of the inadequacy of *mere* reasons as a *working* definition, this formulation continues to be advocated—usually in general terms, without facing the implications or contradictions or dangers to which it leads.

For example, it is of prime importance to understand and communicate the *difference* between reasoning in science, and such practices as rationalization, propaganda, and pseudoscience. At the same time, Siegel and other people wish to *associate* scientific thought with respected practical affairs because their final aim is a "commitment to the idea of rationality as a guide to life" (Siegel 1989, 34, cf., ibid., 15; see also Brush 1974, 1165). However, on account of its *generality*, reason-giving alone cannot achieve both goals. The result is that an approach of this type becomes ambiguous and misleading, if not inconsistent, despite—as in Siegel's paper—commendable efforts to overcome the inherent difficulty of the position.

Thus, in his original formulation, Siegel states bluntly, "What insures [science's] rationality is the commitment to evidence . . ." (1985, 529). The word "insures" clearly implies a sufficient condition. But, says one critic, this means that "any belief can be defended as rational so long, that is, as it is tied in with a commitment to evidential support. . . . [T]his is the nub of the creationist defense of the immutability of species, and the heart of Popper's critique of Marxism" (Baigrie 1988, 439). In reply, Siegel offers the following clarification: Commitment to evidence (answer to his Q1) "tells us in what the rationality of science consists . . ." which, in effect, only describes its "character" (1988, 444–45).[1] It is true that this "will not suffice to distinguish rational from irrational (or science from pseudoscience) belief clusters" (Siegel 1988, 447). Only "*substantive* principles of evidential support . . . entitle beliefs that meet such principles to the honorific 'rational' . . ." (answer to Q2) (ibid., 444–45, emphasis added). Apparently he is conceding here that commitment to evidence is a broad, *necessary* condition, while the substantive principles provide the *sufficient* condition. Yet, in the more recent paper, we are back again to the original wording intact: "What *insures* . . . [science's] rationality is the commitment to evidence . . ." (Siegel 1989, 14, emphasis added).

Another type of ambiguity mars the notion that to think rationally is to be "appropriately moved" by reasons. What exactly are we to understand by this in a real but problematic situation?[2] Consider, for example, the interesting case of Sir Fred Hoyle. Long after most fellow astronomers had rejected the steady-state theory of the universe in favor of the big bang, Sir Fred refused to bow to the formidable evidence piling up against him. Instead, he criticized the *warrant* of that evidence, and continued producing ever more ingenious ideas to account for the new data within the steady-state framework. Was this a failure to be "appropriately" moved? Was he so "blinded" by his own "theoretical commitments" that his former rationality was now impaired? Was his co-authorship of the steady-state idea, and therefore self-interest, preventing him from judging the evidence objectively? That, precisely, is what some scientists thought. But, clearly, Sir Fred does not think so; and neither do some other astronomers and collaborators.[3]

Criticism and the evaluation of evidence proceeds on the basis of *principles*, we are told (Siegel 1989, 22–25), and principles must be applied *consistently* by a rational thinker. What is left out in such a general prescription is the not inconsequential matter of *interpretation*. Let us see how that works. One of the most basic substantive principles in science is conservation of energy. It was natural, therefore, that proponents of the rival big bang cosmology repeatedly pointed at the violation of this principle by the "continuous creation" of matter required in the steady-state theory. Not to worry, said Hoyle. It all depends on how you *interpret* the principle: If, in the expanding universe, we take the reference volume (for measuring energy) as bounded by particular galaxies, then energy will not be conserved. But if instead we take the entire observable universe as that volume, then energy *will* be conserved—the newly created matter being offset by the galaxies leaving this volume.[4]

Consider this quandary of real scientific decision-making in light of the theory of reasons. Siegel admits that it is a "difficult task" of the philosophy of science to develop a "general account of the warrant of scientific reasons that would allow us to determine whether or not, in any given case, the reasons offered for an hypothesis actually afford warrant to that hypothesis" (1989, 20). The example of Hoyle shows how central the problem of warrant is, and how it afflicts even the outstanding scientist. Nevertheless, Siegel (1989, 24) "requires that the student be able to assess reasons and their warranting force properly." How, without general criteria for such assessment, is he or she to do it? Michael Polanyi and Thomas Kuhn have given a famous answer to this question. By *imitation*, or modeling oneself on exemplars, often *tacitly* (Polanyi 1958; 1967), one's interpretive style is formed; and since even the simplest "laws" are only "law-sketches," interpretation is needed at all levels, not just at the frontiers (Kuhn 1970, 188; see also Cartwright 1983).

Postpositivist work on science emphasizes that this act of interpreting is no rare departure from the scientific method. Because theory is *underdetermined* by data, because laws and principles are not recipes, interpretation must take place, and must include other factors (Arbib and Hesse 1986, 181). In part, Hoyle interpreted as he did because *on philosophical and religious grounds*, he rejects the big bang idea: It is *"irrational,"* he said (Hoyle 1960, 108). It brings with it the question, What came before the bang?— a question that seems to lead outside of science, and to support a Judeo-Christian theology (which he opposes). His allies, Hermann Bondi and Thomas Gold, also based their preference on philosophical (or aesthetic) principles: They sought *perfect symmetry*, an unchanging picture from infinity to infinity, in time as well as in space (Bondi 1960). How then shall we regard all this? Is it *bias*, and therefore not strictly in accord with impartial rationality, or are *these* reasons also legitimate in science?[5]

Another insight now commonly accepted is that, under some circumstances at least, viewing the conventional sort of evidence in a strictly impartial way, or being "appropriately moved" by it, is the *wrong* thing to do. When a new theory (like the steady-state) has just been introduced, the evidence is often overwhelmingly against it: So it was for Copernican astronomy and so also for quantum theory. Just what Siegel classes with pseudoscience, a period when beliefs are "systematically protected from contrary evidence" (1989, 15), is now widely seen as necessary if a radical new theory is to have any chance to show its power. From the traditional point of view, this is the most foreign and disquieting element—*faith, commitment, partiality,* as described by Lakatos's "negative heuristic" (1970, 133) or by Brush (1974). For these writers, however, a *struggle against the evidence*, like Hoyle's, is not merely understandable and rational, but in fact quite normal.[6]

Objectivity and Commitment: Two Stories

We have reached now the central problem of this paper, the apparent dichotomy between objectivity and commitment. In the traditional view, as expounded by Siegel and others, to be scientifically rational is to be objective and impartial. By contrast, it seems, a *prior* commitment to particular theories, philosophies, or worldviews is typical not of science but of politics, religion, or everyday life. For objectivity certainly requires a *detachment* or distancing from the issue at hand, while commitment is a passionate *attachment* to one side of that issue.[7] However, if scientists too have their philosophical, aesthetic, even religious attachments, sometimes to the benefit of their work, then what becomes of the commitment to evidence, objective assessment—*as a value for the individual*, not just for the profession?

In the context of education, notice, this problem carries a notorious sting: Either scientists really are objective, detached, and therefore "cold," humanly speaking (the famous stereotype), *or*, as many now rejoice to point out, they are not "cold" after all but partial and passionate like everyone else—hence, *less rational but more attractive!* It is to Siegel's considerable credit that he refuses to sacrifice for the sake of popularity what he takes to be the heart, the *moral* component, of science. Such recoil, however, should not throw us to the other extreme: It is not true that to preserve basic values we must hold fast to essentially pre-1960s views, dissociate epistemology from history, and depreciate or sidestep those insights of more recent work that appear to threaten rationality in general by questioning objectivity in science. This is not a viable response. For reality must somehow be "come to grips with," lest our students, when sooner or later they "find out," begin to resent their education, and look upon it as a meaningless collection of textbook maxims.

Yet history is not altogether on the side of Kuhn or of subjectivism. It provides other kinds of stories as well, and suggests better solutions to the epistemological problem. The basic dichotomy is false: In actual scientific life, objectivity and commitment, detachment and attachment, need not be in conflict at all. In fact, often they *coexist simultaneously in the same individual*. How is this possible? Before we theorize, it were better to exhibit such individuals in flesh and blood. Let me return, therefore, to the intrepid Fred Hoyle, whose story touches our problem at several points, and in ways that are strikingly related. From here on, then, the argument will be mainly constructive; and my effort will be to sharpen somewhat the concept of objectivity so that the important distinctions can after all be made.[8]

A key problem for any cosmology is the capacity to explain nucleosynthesis—that is, how the various chemical elements came into being. Because hydrogen and helium are so plentiful in the universe, it was natural to assume some kind of build-up in which, step by step, at high temperatures and pressures, lighter elements fuse to make up the heavier ones. Since in the steady-state theory, no unique or original "hottest place" could exist, it was crucial for Hoyle to show that a build-up of elements happens all the time in the only possibly hot-enough places known—the interiors of stars. That hydrogen atoms do fuse into helium had already been proved in the 1930s; but right after this first step, a serious difficulty arose, a kind of "gap." The crucial stepping stones, ^9Li and ^8Be, are unstable; these isotopes, in other words, would not last long enough for the build-up to continue to carbon, oxygen, and further up. Here the idea of nucleosynthesis in stars came to an impasse.

But Hoyle did not accept this adverse evidence as decisive. Being convinced that somehow the thing *had to happen* in stars, he began to look for unusual or fortuitous ways in which it *could* happen.[9] So we encounter one of the most impressive theoretical predictions of all time. Hoyle calculated that in the carbon nucleus an excited state *must* exist with an energy of precisely 7.68 Mev. If true, this would allow a "bridging of the gap" by a process in which, first, two helium nuclei fuse to form ^8Be, and then a third helium combines with the beryllium to make ^{12}C. The beryllium is still very short-lived, but the existence of the excited state (at just that energy) would make possible a *resonance* in the second stage of the fusion, and speed it up so much that a good quantity of carbon would be produced before the ^8Be disappeared.

In 1953 Hoyle visited the California Institute of Technology where experiments on nuclear structure were being carried out, and convinced the experimenters to look for the state he was sure had to exist. In the paper that reported the results, we read that the state was found exactly where Hoyle had predicted (Dunbar et al. 1953, 650). In other words, on the basis of astronomical observation, a man had calculated what the structure of the carbon nucleus had to be in order that the synthesis of elements in stars could proceed as required by the cosmology to which he was committed. Interestingly, the same work also showed that such a process *could not* take place in the rival big bang cosmology (Fowler 1984, 923). After this triumph, the need for an original, uniquely hot place decreased; the steady-state theory earned wide credibility in the 1950s and began to dominate. This episode, then, forms the faith or *attachment* part of the story, and shows what such attachment can achieve. Now for the *de*tachment.

Biographers may well see in this famous prediction the apex of Hoyle's remarkable career. But for us what happened next is even more significant. No one doubted any more that helium was synthesized in stars, yet Hoyle's cosmological perspective made him especially sensitive to the fact that so much of it was around—*too much*. A simple calculation showed that even if all the energy sent out by all the galaxies since their beginning came from hydrogen fusing into helium, the amount of helium now in existence should be ten times lower than actually observed. The conclusion was dismaying: Only a small fraction of helium could have been "cooked up" in stars. For the most part, the cooking must have happened in a hotter place, before the birth of the galaxies—in something like a big bang.

The publication of these results in 1964 (Hoyle and Tayler), together with the discovery, just a little later, of the famous 3-degree cosmic radiation, turned the situation around decisively, with the outcome that today the big bang rather than the steady-state is our "standard model" of the universe.[10] In light of this history now, let us consider again the general problem of commitment versus objectivity.

Hoyle's attachment to the steady-state theory began in 1948, when he coauthored it, and has lasted into the 1980s as we have seen. Yet, in 1964, despite all this, Hoyle himself provided major evidence against the object of his own attachment. The same man who, at one time, seems irrationally partial to his own theory, is also, on another occasion, an outstanding example of that idealized critical thinker whose objectivity does not waver even when it "runs counter to self-interest" and to his "most deeply held convictions" (Siegel 1989, 26).

The positivist approach to science paid little attention to such personal stories because only the *context of justification* seemed to throw light on the scientific method, not the *context of discovery*. It was said, as Siegel says now, that "science is a communal affair; individual commitments and passions are controlled by community assessment" (ibid., 16). Perhaps, although this is debatable (Brush 1974, 1168 and note 38). Nevertheless, even if accepted, the point loses its relevance for an education designed to nurture "a certain sort of person." A *community* cannot function as a role model for a *person*; that only another *person* can do. Further, individual histories involve more than accident and psychological eccentricity; often they suggest or illustrate general cognitive problems. In Hoyle's case, we may ask, for example: How is it possible to say "yes" and "no" to the same idea in the same breath, so to speak?

The contradiction is only apparent. The "yes" and the "no" do not occur on the same plane. Usually we are committed to *high level* theories or worldviews—evolution, the quantum picture, big bang cosmology—while arguments over particular evidence, on a lower level of generality, are more limited and localized. Therefore, even a strong commitment may be temporarily *suspended* when particular evidence is examined. If then the evidence runs counter to commitment, this by no means requires that the commitment be annulled. Because such commitments or convictions lie at the higher plane, they rest on many particulars, and on other considerations besides conventional evidence. On this plane, major turns require the most intense, comprehensive, responsible judgment—a kind of soul-searching—and do not easily occur in any case. On the lower plane, however, objectivity, or justice to particulars, *always* demands that their import be developed and acknowledged. This may indeed present problems but no logical contradiction. Since different levels are involved, both requirements can be met.

In believing that commitment and objectivity are mutually exclusive, we tacitly assume that a high level commitment must necessarily damage our objectivity at lower levels, must somehow cause us to evade or distort unwelcome evidence. The merit of history is to show dramatically that since for some people this is clearly not true, it need not be true for the rest of us. What is required, the ability to genuinely (not just apparently) *suspend* commitment, is neither easy nor common since it is a form of *self-limitation*. Yet in science, this ideal is still alive, recognized and respected wherever it is found. The special character of Hoyle's 1964 work has not been overlooked; Steven Weinberg, for example, pays tribute to it in his widely read *The First Three Minutes* (1979, 121).

The key feature is self-limitation. Its impulse derives from a commitment that is higher even than theories or worldviews (in the ordinary sense) and often centers about such concepts as truth, justice, or "how things are" (Siegel 1989, 26). Objectivity, therefore, as self-limitation in service of the highest ideal, is to science what justice is to social action. It is this moral aspect of the "critical spirit" that Popper (1968; 1972), Scheffler (1967; 1973), and now Siegel try so strenuously to defend against the apparent onslaught of the "new philosophy" emanating from the Kuhn-Feyerabend school of thought. And this indeed is worthy of defense.

Cynics might counter that no special virtue was required for Hoyle to publish his 1964 paper. Having noticed the flaw in his own theory, self-interest required that he reveal it himself and receive the credit, rather than wait for others to do so. I doubt that the details of history support this. Rather than argue the point, however, let me proceed to another example in which, because no credit could possibly accrue to the author, the element of self-limitation emerges even more forcefully.

Some years ago, the physicist Frank Tipler, well known as a specialist in general relativity, wrote a paper that accomplished the following feat: He showed that despite the mountains of evidence for an age of the earth of billions of years, despite all the data supporting biological, geological, and stellar evolution, it is possible to construct a theory of origins in which the earth is only some few *thousand* years old and "created" in a rather small number of days! He did this by means of a relativistic model of the universe, employing a judicious distribution of exploding mini-black holes (Tipler 1985). It is characteristic of this type of theory that *all* the familiar data, implying a far greater age, are accounted for, and do not contradict the *reality* of a much later beginning. What made the idea intriguing, however, is that all this was not mere ad hoc speculation; it was related to other cosmological work (Tipler 1983), and is *testable*. Certain astronomical observations specified by the author, but not carried out at the time of writing, could give evidence for or against the theory.

Tipler, who describes himself as an atheist, believes in the standard evolutionary scenario as much as do most other scientists. It is of interest therefore that when, in 1982, he lectured on his wayward model to the astronomy department at Berkeley, the following question was put to him: Is it not *unethical* to "publicly discuss such a theory, inasmuch as it could only give ammunition to the creationists, who were at that time trying to push through 'equal time' laws," especially in California? Tipler answered that "it is the duty of a scientist to tell the truth as he sees it when this truth is relevant to important public issues" (1985, 893). His model was relevant, he felt, because it was not then generally known that "a falsifiable (but not yet falsified) young universe theory can in fact be constructed" (ibid., 893). Its value lies in a shocking demonstration of the *sheer power of theory* in science, a power that conventional views underrate—even after the fall of Newtonian physics—by implying that a great deal of evidential support makes doubt "perverse" (Gould 1981, 255). In addition, the paper probes deeply the ground of our most secure beliefs: Why, despite the fact that it is testable and that no evidence was known to contradict it, did we all (Tipler included) dismiss this theory out of hand?[11]

A Strong Version of Objectivity

The Hoyle and Tipler stories, different as they are, appear to show that the driving force in developing low level evidence against one's higher level commitments is the *perception of something relevant but as yet undisclosed*. The highest objective then becomes to *reveal* what still lies hidden. This suggests a *strong version* of the concept of objectivity: It is not enough just to be "open-minded" or "willing to listen to opposing views," for how, after all, can we tell that such "listening" is not mere courtesy? It is not enough to give good reasons for the view one is defending and to amend that view when faced with powerful counterarguments, for even politicians do that much. One is fully objective only when the *wish* that opposing evidence be strengthened to the utmost *demands participation* in such strengthening, if at all possible. This ideal may sound strange to us late 20th-century folk, but it comes straight out of Mill's *On Liberty* (1951, 129, 138), is very much in the spirit of Popper, *and*, as we see, still finds its embodiment in otherwise quite normal, attractive and passionate scientists.[12]

A strong version of objectivity does seem to provide the much desired demarcation from pseudoscience, rationalization, and propaganda. For invariably the sort of reason-giving and treatment of evidence in these practices excludes precisely the element of *self-limiting* inquiry in opposition to the cherished commitment. Yet it does *not* solve the problem posed earlier: This conception of objective reason is not broad enough to *associate* the scientific attitude with other respectable activities—and perhaps it should not be. Here again reality differs from what we might wish: that the ideals of science, when exported to other areas dealing with evidentiary criteria, will flourish there, ennobling, and generally improving things. True, *techniques* of science are applied far and wide: Competence in mathematical model-building, hypothesizing, testing, and so on, are useful in business, for example. But if within the *spirit* of science we include the strong version of objectivity, then, of course, *the opposite is the case*. Nothing is more firmly tied to evidence than the practice of law, for example; but objective or unbiased it certainly is not. Rather, the system as a whole is, we hope, to some extent objective; yet the attitude and behavior of those individuals trained to seek and criticize evidence—lawyers—are, by general agreement, *required* to be partisan. The same holds of course for business, politics, and much else in modern life.

What is demanded of most professionals is that the reasoning be "solid"—that is, buttressed with facts or data if possible, logically flawless, and to the point. In short, reason-giving of a selective nature, *instrumental* rationality, is *institutionalized*. Routinely we train individuals to serve as components of a complex system which may deliver a modicum of justice and objectivity in the long run and on the whole; but this means that the components, the individuals themselves, are trained not for objectivity but for its opposite—for *adversarial rationality*. Everywhere, the person who can make a good case for *his or her side* is appreciated.[13] The spirit of science, then, in its ideal traditional sense, far from having great value in practical affairs, stands somewhat in opposition to society. As is the case with morals, whether religiously based or not, self-limitation, the distinctive mark, points to that transcendent element which, in both realms, makes for something quasi-religious: a corrective, even a prophetic note, in our modern culture of self-expanding institutions and self-assertive individuals.

To foster this spirit—the aim of a number of outstanding philosophers of science—may make better human beings, better science, and a slightly better world. But no one should be deluded that it will help also to get on in life. Here is the tragic note in our educative use of science's highest moments, which clashes with the more familiar triumphalism. But the triumphalism, we must not forget, refers to technique and practical success, not to the self-limiting spirit of objectivity. The question to Tipler (about giving ammunition to creationists) reveals, in a nutshell, the sobering truth: In comparison with *social and political advantage*, what this scientist viewed as quintessentially ethical was, in the eyes of the questioner, not merely imprudent but actually unethical.

I will not attempt in this short paper to draw far-reaching conclusions about all these serious questions. Summarizing briefly, my main points are these: The "reasons approach" to rationality is too inclusive to make some crucial distinctions that, in teaching, we do want to make. But the value, difficulty, and moral aspect of *reasoning objectively* (self-critically) do deserve special attention. Although this ideal is not in harmony with the temper of modern professionalism, it is still alive in science, and should be taught. It need not be regarded as conflicting with the newer philosophies that focus on history and on bias in theory formulation; rather, and especially in education, the two streams of thought can be combined, with epistemological reflection on history as the method of choice.

[9]

Alternative Interpretations, History, and Experiment: Reply to Cushing, Crease, Bevilacqua, and Giannetto

ABSTRACT: Cushing's contribution on Bohm's deterministic interpretation of quantum mechanics raises the question of whether self-understanding is at issue in physical science as it is in social science. This question is pursued further in the present paper. Bevilacqua's and Giannetto's belief in the importance of history and the life-world of the researcher is one I share; but their prescription of the historical context as the only context of interest leads to an extended discussion of the several types of philosophical hermeneutics now practiced, and of other existing links between science and the life-world. Historical studies, I argue, should be used to interpret current texts, current concerns, even current textbooks, not to *replace* any of these. While I do not share Robert Crease's belief in the *primacy* of experiment, I agree that it is fundamental in natural science, and must not be neglected in favor of theory. However, the pervasive nature of theoretical languages, their role in understanding experiment and in communicating results, still justify a special hermeneutical concern for this aspect of science, particularly in education. [—M.E.]

Introduction

James Cushing's essay is a welcome expansion of the theme of my (1993b) paper, emphasizing the large scope of interpretive freedom even within theoretically stable domains of science. Fabio Bevilacqua and Enrico Giannetto complain that my treatment is not sufficiently historical, not properly related to the life-world, and too semantic. Robert Crease shares in my appeal to art, agrees with the need to account for both the constructed and the invariant aspects of natural science, but faults me for neglecting experiment, for favoring language over action, and thereby sup-

This paper appeared in *Science and Education: Contributions from History, Philosophy and Sociology of Science and Mathematics*, vol. 4, No. 2 (1995), 173–188.

porting an old philosophical prejudice. I thank all these scholars for their comments and criticisms, and will try below to address the questions they raise.

In Regard to James Cushing

My example of Richard Feynman's 1940s path-integral formulation of quantum mechanics (Eger 1993b, this volume, 29–51; see also Eger 1999, this volume, 73–88) was intended to show how difficulties in understanding established texts (standard presentations of theories) can lead to major reinterpretations. Now Cushing presses the point further by raising the issue of David Bohm's 1950s reinterpretation, which itself is an improvement on Louis de Broglie's 1920s interpretation. Although Feynman did add paths to quantum theory, these paths are not classical trajectories and do not therefore reintroduce a concept of the classical particle, as Bohm's theory does. Thus, Cushing's example widens considerably the interpretive freedom under discussion here.

In the context of education, and of hermeneutics, his essay brings to the fore three points not often discussed. First, that in basic quantum mechanics we have an example of a science always presented with the same interpretation, one almost universally taken as unavoidable, despite the fact that profoundly different ones are possible; and that such famous 'discoveries' as microscopic indeterminism—dramatically described in numerous popularizations as well as textbooks—are part of this interpretation.[1] To be aware of at least this much is an example of what a hermeneutic approach to science would demand, especially of teachers. Going one step further, it would demand also that some knowledge of the alternative interpretations be acquired and that the object of study—quantum mechanics, in this case—be examined by means of the different interpretations. Pursuing this point further still, it seems clear that since Bohr, de Broglie, Feynman, and Bohm (among many others) belong to the history of quantum mechanics, the study of the history of the subject is implicit in the study of alternative interpretations—a point raised by Bevilacqua and Giannetto, but more of that later.

That one rarely has time to ponder all available alternatives, especially in a conventional educational setting, brings us to what I take to be Cushing's second major point: Some interpretations are more *understandable* than others. In his recent book, in a number of papers, and now in *Science & Education* (Cushing 1994a), Cushing has pursued the question of understanding in physics—suggesting that when we choose between alternative interpretations, intelligibility ought to be among the criteria.[2] Such a concern implies not only that the originator's 'interest' should count in theory choice, but the appropriator's as well—not merely the researcher's point of view but the student's. To me it seems that this draws into the very center of scientific activity that mode of relating oneself to science which I have called the "student" mode. It is the mode in which trying to understand some established text, rather than to use or test or to extend it, is the primary concern.

However, given this much freedom of interpretation, we may wonder what one actually comes to understand in the basic theories of science, aside from their uses. By asking explicitly "whether the 'laws' of nature tell us more about ourselves than about nature," Cushing brings to physics a major hermeneutical question usually raised in relation to the human sciences—the question of *self-understanding*. This is the third of his points that I would like to highlight here.

When it is said that in a genuine dialogue, or in a hermeneutic confrontation with a text, understanding can be achieved only as understanding of oneself-in-the-issue (*sich in der Sache verstehen*), what is being affirmed is that one's own entire pre-understanding and predisposition toward that issue become involved in a dialogue with some new view of the issue, so that whatever understanding emerges sheds light not only upon the new view but on the predispositions as well. Can we now go so far as to say that *in physics also* to understand a theory is, in part, to understand oneself?

In light of Cushing's discussions, what is being suggested in the case of quantum mechanics, for example, is something of the following sort: If one studies this theory with the aid of standard courses and textbooks, one learns the concepts and techniques, along with the Copenhagen interpretation, and *believes* that one has grasped what *science reveals* in the atomic realm. But if one adds to this first look a study of Feynman's interpretation, and then of Bohm's, and perhaps others, one learns not only more about the theory and about history (as would be expected) *but also* about human predispositions to such things as determinism, causality, atomism, which are on display here, possibly reflecting preferred philosophies; *and further*, that one's own individual preferences are now engaged.

Thus, Bohm's understanding of 'himself-in-quantum-mechanics' brought forth a picture of the atomic realm in which the nonclassical features are minimized at the expense of such things as symmetries and elegance, *and* of himself, Bohm, as preferring this picture, as living in a fundamentally deterministic world. However, there is more still—a bonus of incalculable value—because examining the object from several viewpoints allows that object to reveal its own outline. The insights are not all of the subjective sort; the 'space' of *possible* interpretations is not unbounded (Eger 1993a, this volume, 15ff). From the sample of those that 'work,' taken as a group, emerge most interesting 'facts about nature'—invariants under all these interpretive transformations—the Heisenberg uncertainty, the much-discussed long range quantum correlations, and so on.[3] These cannot be interpreted away.

Lest it be thought that only in the case of advanced, complex theories are alternative views of the laws of nature enlightening or available at all, I would once again give Feynman as counterexample. In one of his little popular books (and in his famous textbook), he shows how one can give three radically different types of explanation of the most elementary optical phenomena: first, the laws of reflection and refraction, as taught in high school; second, Fermat's law of least time; and, finally, quantum electrodynamics in the Feynman language.[4] The last two mentioned are related to the Lagrangian "action" or "least action" principle, which, as the reader may recall, is a favorite of Feynman's and a key element in his major innovations (Eger 1994a, this volume, 89–99; see also Eger 1993b, this volume, 29–51, and Eger 1999, this volume, 73–88). Any physics teacher or student not charmed by the beauty and light-shedding power of these explanations taken together is probably in the wrong field, but this is only the beginning of the story. My point is how Feynman got started on this path of minima principles, alternative explanations, and so on. It all came from his high school teacher:

> When I was in high school, my physics teacher—whose name was Mr. Bader—called me down one day after physics class and said, "You look bored; I want to tell you something interesting." Then he told me something I found absolutely fascinating, and have, since then, always found fascinating. Every time the subject comes up, I work on it. (Feynman et al. 1963, Vol. 2, 191)

What Mr. Bader showed Feynman was the action principle as an alternative to doing elementary physics with Newton's laws.

In Regard to Fabio Bevilacqua and Enrico Giannetto

Of the various reservations and criticisms brought forward by Bevilacqua and Giannetto (hereafter denoted as the 'Pavia response'), some go to the very center of the question with which we are dealing; these cannot be adequately answered in a short reply, though I am glad to have the opportunity to offer here at least preliminary clarification. Other objections seem to arise from misunderstandings or miscommunications, while still others do perhaps reflect differences in judgment on substantive matters.

Let me begin with the words "hermeneutics" and "hermeneutic circle." Disputes over their proper use arise frequently, as they did at the Veszprem Conference (mentioned in the Pavia response) when K.-O. Apel asserted—against Dagfinn Føllesdal, myself, and others—that interpretations of formalism or reinterpretation of theories in science cannot be the real hermeneutics, or at any rate not the hermeneutics that ought to occupy us. In a similar vein, and apparently in agreement with the Pavia response, some writers dismiss what they call the "theoretical circle" or "theoretical hermeneutics" because this touches only *certain* kinds of presuppositions— the hypotheses in theory construction ("fore-conception"), with possible effects on language ("fore-sight," Pavia's "semantics"). But typically, it says nothing about the "fore-having" (*Vorhabe*), which is cultural and belongs to the life-world.[5] Only interpretation relating itself to this cultural context is, in their view, 'hermeneutics.'

Let us note first that philosophers simply do not agree on such a demarcation. H.-G. Gadamer (1975) employs the term with its more inclusive meaning, as I do, embracing the 'theoretical' and 'practical' types of interpretation. Thomas Kuhn (1977, xiii) and Mary Hesse (Arbib and Hesse 1986) use the older, essentially epistemological meaning. Patrick Heelan (1983) supplements the classic and Heideggerian usages with his own extensions to "texts" written by nature or "readable technologies," and so on. My feeling is that, in science, hermeneutics applies at *several* levels and in *several* ways; that it will take many people, working along various lines, to explore all of this; and that the best policy for now is not to fore-close any of these lines by proclaiming that only this or that type of hermeneutics should be pursued.

In this issue of *Science & Education*, Dimitri Ginev also draws a distinction between two groups of writers, based on whether they practice the theoretical or cultural type of hermeneutics. But then he goes on to introduce a continuum of modes of being-in-the-world, according to varying degrees of objectivation practiced, to which varying degrees of theoretical (or cultural) hermeneutics correspond. Essentially, then, if I understand him correctly, we agree on the more inclusive conception of hermeneutics, and on the need for different versions of it for different aspects of science.

However, this by no means answers the concern underlying all these disputes. Beneath the argument about definitions lies a profound question about the roots of meaning: *Where to look?* What sorts of contexts to evoke for our hermeneutics of science? On this score, I think that when all the differences are sorted out the position of Bevilacqua and Giannetto will be far closer to my own than it now appears.

Consider their fear that history might be neglected in favor of what they describe as the "semantical connotation to the context of interpretation." Except possibly for questions of emphasis, we really have no argument on this point: So seriously, in fact, do I take the historical context that to illustrate its value, I thought separate publications were required.[6] In the papers under discussion here, my focus on the language of science had a definite aim, though its cost was to allow the kinds of misunderstanding that have now surfaced between the Pavia authors and myself. That hermeneutics is relevant to the history of science is by now so well known that I thought it less important to press that point; nor is it surprising, given the fact that history has long been the 'home' of hermeneutics. But that hermeneutics is relevant to the understanding of science as such, history aside, is not so well known, or even accepted, and it is this point, in its most general aspects, that I wanted to make first.

Hermeneutics is relevant to the whole corpus of science, the 'book of science,' *not only* because that book has a history but because the *underdetermination of meaning* (Cushing!) in many of its parts obliges the reader (and teacher) to interpret in a significant way. This is a circumstance of enormous importance, especially in education, and should not be totally subsumed under the historical interest.

I had hoped that my discussion of Feynman's early work might allay at least the worst suspicions that the whole treatment was to be ahistorical. That with Bevilacqua and Giannetto I did not succeed is due perhaps not so much to the small role of historical contexts in what I have published thus far, as to the *type* of history these contexts involved: The "interpretation processes suggested by Eger seem not related to 'ontological understanding' but to an hypothetical deductive and experimental kind of explanation. . .[which] cannot lead us to an understanding of the life-world but to the construction of the science-world."

I know enough about the work of Bevilacqua and Giannetto to put aside one implication easily drawn from this criticism—that even in the study of science as such, the problems of the 'science-world' are less meaningful than those of the researcher's life-world; and therefore, in varying degrees, the latter should be *substituted* for the former. I mention this because, unfortunately, today we do find such tendencies. However, I take the Pavia authors to regard history and sociology essentially in the same way as I do: not as a *substitute for* but as *part of* a substantive scientific understanding.

Under this assumption, the concern of Bevilacqua and Giannetto would be that my examples thus far, combined with my use of the word "text," could lead readers to believe that *only interpretations within the languages of science* are involved in the kind of hermeneutics I am discussing; and that by *this*, the *being* of science, or of the scientist, or of humans more generally, cannot be illuminated. Such a concern I do share, and for that reason, in the historical paper mentioned above, I took one small segment of physics as an illustration, and examined it in contexts that never appear in any textbook—the personal, the social, the philosophical, the interdisciplinary—using debates, original sources, biographies, and so on, just as the Pavia authors prescribe. The purpose was to show what a 'fusion of horizons' of the language of science and the language of the life-world looks like in a particular case.[7] This, however, cannot be done in regard to every piece of science; and it should never be done for its own sake alone, as an imposition. Only in those cases when life-world contexts *illuminate* the meaning of the science itself have history and philosophy and sociology a legitimate place in science education as such.

The danger of imposing, and thus distorting, should not be underestimated in our enthusiasm for history as the answer to all problems. According to the Pavia response, the incoherence often encountered in a subject like electromagnetism should be overcome not "through a deeper analysis of a textbook" but by a "clarification of the specific individual interpretations of the authors . . . of the reasons they had for shifting from one meaning to the other, *reasons embedded in their own life-world*" (emphasis mine). Let us see how that might be construed in the case of Feynman, for example. I am not suggesting that Bevilacqua or Giannetto would so construe it, but others certainly might.

The little story I told about the impact of Feynman's teacher was a life-world episode with all the drama one could desire: A precocious youngster, future Nobel laureate, bored by the slow pace of his merely human classmates, is shown the power of the Lagrangian by a perceptive teacher; he clings to it like a terrier, works it in every possible way, and years later turns it into a major scientific breakthrough. But does this lead to an understanding of the reason for Feynman's 'shift,' in his graduate work, from a Hamiltonian to a Lagrangian formulation (Eger 1993b, this volume, 29–51; Schweber 1986)? No. The main reason, in the sense of *telos*, lies largely in the texts of science itself—in the divergence problem of quantum mechanics which he was trying to solve, in Dirac's texts which had first to be *understood* in order to be of use, and so on. The deepest understanding is achieved by *adding* to the strictly scientific texts the historical texts that give us a glimpse of Feynman's life-world, not by marginalizing one context in favor of the other.

Thus we come to the question of texts and textbooks. On this, the Pavia response seems to be making two distinct points: First, that my 'processes of interpretation' are limited to the 'semantic' sort, in the sense already indicated above; and, second, that I include as objects worthy of interpretation even introductory textbooks on the undergraduate and high school levels. Regarding the first point, I would add here only that in the paper cited in note 6, in order to show explicitly the various kinds of interpretation possible, I begin with a treatment of a commonplace subject, as presented in an introductory astronomy textbook and occupying there no more than one page. What follows is the addition of contexts one by one, each expanding the meaning of the text. At the end, the problem under study emerges as much larger than it seemed at first, a problem illuminating and illuminated by some of the major questions of the nineteenth century and of our own time.

Of course interpretation is not just internal analysis of one and the same text. The whole point of talking about hermeneutics in regard to teaching and study is to emphasize that all texts, be they elementary or advanced, *need* to be interpreted (not consumed) in order to be genuinely understood; that the required interpretations are of several kinds, occurring on several levels, including, though not limited to, first order semantic understanding of new terms; and that contextual interpretations (using external materials) are, *of course*, required as well, though not all contexts are historical.

Regarding the second objection of the Pavia authors, I must plead guilty. I do think introductory textbooks ought to be taken seriously, far more so than is now being done by their authors or users; and I am at a loss to understand the open disdain shown by the Pavia response in this regard. Certainly, as a matter of 'professional honor,' all of us who teach science complain about the low quality of most textbooks. Certainly we are aware of the longing today for some radically new way of teaching that would transcend 'traditional' lectures and books. I, for one, have real hope for the kind of hypertext development carried out by Bevilacqua and his group at Pavia, and I can understand that after such an experience he might well be more pessimistic about conventional books than the rest of us. But even hypertexts are texts that have to be interpreted, not just viewed or played.

Does all this mean that we who theorize about education can afford to write off as hopeless the greater part of teaching and learning activity today? Would it not be better to add to existing teacher education programs some training in what it means to genuinely interpret a scientific text, training in how to *relate* histories, debates, etc., to textbooks? As long as students have to use textbooks anyway, would it not be better for them to be *shown* that often these books are mere residues of a kind of dehydration process, designed to be 'reconstituted' by teachers and students?

At their best, however, textbooks are not necessarily inferior to original works. Original papers frequently do things in a roundabout or one-sided way. Special relativity, for example, can be (and has been) presented more simply, graphically, and meaningfully than in Einstein's 1905 paper. Similarly, Maxwell's 1860 derivation of the "Maxwell velocity distribution" is not the best, and is rarely used, for it seems to pull the result out of a hat, giving no insight into the physical process. A textbook author who is a mature scientist with a firm grasp of his field will take into account alternative treatments, previous criticisms, subsequent improvements, and come up with what he thinks is the best *presentation*—which is to say, his own interpretation of the field. That is why some of these presentations have become classics, influencing generations of students, and why some of the best scientists take time to write such things. Feynman is known to have remarked once that, in the long run, he thought he would be most remembered for his three-volume introductory textbook, *The Feynman Lectures on Physics*.

I am particularly baffled by Bevilacqua's and Giannetto's charge that "textbooks are mainly related to technical purposes." My own under-graduate texts in modern physics were Max Born's *Atomic Physics* (1950) and David Bohm's *Quantum Theory* (1951).[8] Both of these, but especially the latter, are designed mainly to convey ideas rather than techniques— insights into the complexities and mysteries of nature, not its glib 'demys-tification.' Were that not so, we would never have had sufficient interest or drive to spend long hours in the night discussing and arguing over such books. Nor do I know anything now that changes my former view of the matter; though, undoubtedly, larger doses of historical-philosophical inter-pretation (some of which Cushing now supplies in his 1994) would have made the experience better.

A good textbook, used not as something to be ingested but as the 'script' for an interpretive performance in class, laboratory and home— something to be *made* to come alive—has invaluable organizing and refer-ence uses for which no ready substitutes exist. My reason for mentioning textbooks was to emphasize that *these* texts, above all, ought *not* to be taken as they stand; that only through interpretation can they acquire a meaning beyond the purely technical.

Yet, behind the whole complex of objections raised by Bevilacqua and Giannetto I sense a deeper reservation, more pervasive than any one of these issues. It surfaces most visibly in their insistence that if hermeneutics leads *only* to an understanding of the 'science-world,' not of the life-world, then it has failed in its main goal, and it shows itself again in their sugges-tion that the 'bridge' between the science-world and the life-world in gen-eral is the life-world of the scientist. The underlying concern, it seems to me, is that the *priority of the life-world*, as posited by Husserl and Heidegger, not be given up; that ultimately the meaning of science depends not upon this or that interpretation of a theory, taken in itself, or even on the ensem-ble of all existing interpretations, but in a relation between those interpre-tations and the *being* of humans in the world.

If this is a fair reading of the Pavia response, then, again, I must admit to sharing the concern, at least partially. But it is a concern that can be relat-ed mainly to high-level theories, or to segments of such theories through the high-level theories in their entirety, not to daily work or the detailed study of science (Eger 1993c, this volume, 261–279). It brings up once more the question of self-understanding, to which I alluded earlier; and leads me to suggest to Bevilacqua and Giannetto that while I am willing to listen to their warnings about overlooking things, I would now like to convey to them a similar warning about something marginalized, if not over-looked, in their own approach: *The life-world of the researcher is not the only life-world in which the sought-for relation to science may be found.* In other words, there seems to be something amiss with the notion that the meanings of the 'book of science' are to be found only in the lives of the *authors* of that book—*their* biographies, debates, etc.—*not in the lives of its readers.*

The point of the Feynman example was to talk not about Feynman the Nobel laureate but Feynman the student, *the reader* of Dirac's texts, and to show how Feynman's context entered into that reading. For the same reason, but more generally, the point of view of the 'student' (the appropriator, not the creator) is taken to be ontologically or existentially primary in my approach: The student—one who formally studies such things as Newton's laws and quantum mechanics, or, generically, anyone who tries to understand science in order to orient himself in the world—this person also has a life-world, which is to be taken seriously. In *this* life-world, the book of science can become a 'partner' in dialogue that may or may not stir up the student's own self-understanding.

How the theories of Newton and Darwin change the life-world *for* us cannot be understood by focusing on the life-worlds of Newton and Darwin only. To understand that question, the meanings of Newtonian and Darwinian texts have to be reinterpreted in the life-world contexts of whoever is reading these texts. This is a major point for Gadamer, one that links science with art, and projects the works of science beyond their historical and technical contexts.

It is not just a theoretical point. In various parts of the world, from North America to New Zealand, it is a point of ongoing social controversy. When parents in California and New York protest the way evolution is taught in schools, they do so not for theological reasons but because they object to having their children's self-conception changed by Darwinian texts (Eger 1988b). When a Maori boy near Auckland tells his friends about his evolution lessons they are incredulous—not because they are Christians or because they think evolution is unlikely, but because "if you believe you are descended from an ape you cannot be a Maori." Herbert Simon tells us that,

> Perhaps the greatest significance of the computer lies in its impact on man's view of himself. No longer accepting the geocentric view of the universe, he now begins to learn that mind too is a phenomenon of nature, explainable in terms of simple mechanisms. Thus, the computer aids him to obey, for the first time, the ancient injunction, 'Know Thyself.'[9]

That Simon is not exaggerating here is shown by sociologists who try actually to gauge this impact of the 'science-world' on the life-world: When Ned, an MIT pre-medical student, tells us "Some of the agents [processors in the brain] are a little smarter than the others, the way an op-amp is smarter than a transistor, but that still makes them a long way off from having consciousness," we see what Simon is talking about (Turkle, 290). And when a well-known American professor, lecturing to his colleagues, assures them that "molecular biology has proven there is no free will," we see once more the pervasive effect of science on 'self-knowledge.'

As with art, there is more meaning in a work of science than can be found in the life-worlds of its creators—because, as in art, meanings are fully actualized only in the interpretations, whether these be formal (as with Feynman), formal-philosophical (as with Bohr and Bohm), or involving more directly the life-world of the interpreter (examples above).[10] Therefore, the first picture drawn by Bevilacqua and Giannetto of a clear divide between 'life-world' and 'science-world,' bridged only in the lives of the scientist-creators, cannot be quite right—as they realize. To compensate for this, they briefly refer to a second picture, a highly condensed version of Buchdal's interpretation of Kant, in which the interpretive freedom of the reader is restored in the "realization" part of Buchdal's "reduction-realization" process (Buchdal 1992, 53–104).

However, on the one hand, I fail to find any significant difference between "realization," as described in the Pavia response, and "interpretation" as I intend it. On the other hand, if we take Buchdal's analysis in all its complexity, then its practical value for education is doubtful. "Realization," in the Pavia response, is roughly what I describe as the relation between the core of a work of science and its potential meanings, one or another of which is 'brought into being' in the interpretation (Eger 1993a, this volume, 3–27; 1993b, this volume, 29–51). This can occur on a small scale, as with a single idea, or in regard to high-level theories, as in the examples of Feynman and Bohm.[11]

There is no disagreement, therefore, between the Pavia authors and myself about the need for interpretation on the receiving end of scientific achievement as well as the producing end; and the receiving end is largely contemporary, not historical. All of us, I think, share the desire to understand better the relation between science and the life-world. The main difference is between Bevilacqua's and Giannetto's belief that to do so one should still maintain a sharp conceptual and methodological distinction between the two 'worlds,' and my attempt to show (and encourage!) fusions at several levels.[12]

In Regard to Robert Crease

There is a sense in which Crease's main criticism resembles that of the Pavia response: Again it is pointed out that I have neglected or deemphasized an important part of the story, and again, what has been neglected involves the life-world of the scientist—instead of biographies, original papers and debates, this time it is the laboratory, the world of deeds, not just of language.

Let me say, before I answer, that Crease has good credentials for making this criticism. He is a historian of Brookhaven, one of the venerable American accelerator laboratories, and has just published a book 'doing' what he preaches—developing, that is, in considerable depth, a hermeneutic approach to experiment. But that is not all. It has been a source of wonder to both of us, ever since we discovered the fact, that independently each of us has found in performing art a most valuable analogy, which, hermeneutically understood, leads to a view of science that takes into account both its objective and its subjective aspects. Since art and science are more often contrasted than assimilated, I take this convergence of thought to be something of a confirmation of the intrinsic merit of the idea.

The difference between Crease's use of that analogy and mine is that he takes experiment as primary, "performance" as exclusively performance of experiment, and theory as essentially the "script" for such a performance. Only in an experimental performance, according to Crease, is a natural phenomenon actually *presented* to the scientific community and the world. By contrast, in my treatment, "presentations" can be purely theoretical, as for example Dirac's and Feynman's presentations of quantum mechanics; and a "script" is any concise description of a work—experimental or theoretical—especially when schematized or couched in formalism. He criticizes me, therefore, on two counts: first for putting "experiment and theory on a par," *in principle* that is; and then, in practice, for devoting most of the attention to theory, thereby "overtextualizing" science. The relation of this to Bevilacqua's and Giannetto's charge of "semantic hermeneutics" is apparent.

Taking the second issue first, it cannot be denied that thus far I have given experiment less attention than it deserves. Except for the discussion of instrumental "embodiment" and the references to McClintock, my main examples were taken from theory. However, as with the concern about history, the omission is due mainly to the fact that hermeneutics of experiment is a large subject in its own right, requiring separate and extensive treatment, which could not be tacked on to the papers under discussion. These papers being only a preliminary sketch, all that was attempted in regard to experiment was to make for it a 'space,' so to speak. Now that the space has been so ably filled by Crease's book (1993), there is less need in this regard, although it still remains to be shown what all this suggests for the role of experiment in education. To me it seems that hermeneutics points in the direction of laboratory as an *experience*; as taking part in *presentations* of phenomena; as *interpretation* of theory rather than its proof or independent discovery; and as *history*—since, in any case, much of high school and undergraduate laboratory time is spent repeating classic experiments.[13]

Regarding the *primacy* of experiment, however—that science is more of a doing than a speaking, that it deals more with instruments than texts, more with physical performance than with language—I have serious reservations. On the whole, it is true that experiment has been neglected in historical-philosophical discussions of science; and Crease's book—now joining those by Hacking (1983), Galison (1987), Franklin (1986), and others—helps to right the imbalance. But in trying to make any correction, there is always danger of *overcorrection*. Is it really true that "the language of science . . . for the most part arises from and is addressed to experimental acts"? Can one say this of such major theories as biological evolution and general relativity?

There is a school in science studies today whose overreaction in this regard has gone so far as to be, in my view, a misrepresentation—a shrinking of science to something surprisingly local, parochial, and of limited import. To say, for example, that "science is first and foremost knowing one's way about in the laboratory" (Rouse 1987, 72) is simply to misunderstand. Such a description fails to capture the essence of even *experimental* work, much less of science as a whole. It was common knowledge, at Columbia University, that the graduate students and postdoctoral fellows working under I.I. Rabi (Nobel laureate in experimental physics) "didn't let him touch apparatus," and that he for his part didn't particularly want to. According to one member of the group, Norman Ramsey, Rabi's job was "the concepts of the experiment and the concepts of the apparatus, and the interpretation of the results" (Rigden 1987, 116–17).

The main work of any good scientist is *thinking*—regardless of whether that thinking is connected with physical or mathematical or computational apparatus; and thinking is usually done in terms of the conceptual language of the science. When I listen to a colloquium paper by an experimentalist, I do not hear a different language; I hear the same theoretical language of physics everyone else uses. The reason is that the object of interest, even for the experimentalist, is, in the end, not the equipment but what is 'seen' through the equipment—the physical system; and the language of the physical system is theoretical. Since the linguistic component of the enterprise is pervasive, it is not necessarily prejudice that, in the past, led to the focus on theory.

Although Crease is not swayed by the overreactions mentioned above, and is well aware of the universal aspects of science, nevertheless, his emphasis on 'doing' leads him to question the priority I still give to theoretical language. One additional reason I do so is because of the particular standpoint I have chosen, consonant with the interest of this journal—the standpoint of the 'student.' When I say that any 'student' (i.e., outsider) begins science study through the 'book of science,' not the 'book of nature' (Eger 1993b, this volume, 29–51), and therefore *immediately* faces problems of linguistic interpretation, I am merely reporting the common experience.

Rightly or wrongly, texts have a greater effect on our understanding of nature than experiments, but it is an interesting and hermeneutically relevant question whether the student interprets experiments by means of theory or theory by means of experiments, or both. Crease and I are inclined to say that experiment is an 'interpretive performance' of theory, but there is a sense (in connection with Husserl's phenomenology) in which this also includes interpreting the results of experiments by means of theory. The subject cannot be pursued here any further; those interested should look into chapters IV and V of Crease's book.

I completely agree that performed experiment is an important dimension, or level, involving hermeneutics in a different way from texts; it is the dimension previously emphasized by Michael Polanyi (1958) and Patrick Heelan (1983 and 1995), as distinct from classic, linguistic, hermeneutics. Heidegger and Gadamer bridge the gap. In discussing these authors, I intended to give the performative or 'practical' side of hermeneutics some attention, though of course there is a lot more to be said on the subject. Yet, on the whole, what Crease is saying supports my contention that hermeneutics applies to natural science at *several levels*, in several ways.

It could be objected that so general and inclusive an idea cannot be worth much; it is like saying "life is love." But I would answer, first, that even "life is love" has worth if taken as a norm rather than a tool of analysis (a point relevant to the role of hermeneutics in education); then I would insist, Karl Popper notwithstanding, that a wide scope does not necessarily make an idea useless—e.g., 'evolution.' It all depends on how the concept is applied, and what fruit it yields.

To conclude, I would like to add this thought: When Gadamer wished to distinguish his approach to language from that of Ernst Cassirer, he pointed out that the latter treats language as *one of several* symbolic forms (including myth, religion, science); in contrast, Gadamer claims that language *pervades* the other forms, and is in this sense universal. In like manner, but not to overwork the similarity, I would claim, in reply to the Pavia authors as well as to Crease, that the theoretical languages of science *pervade* both its experimental and historical side. Experiment usually confronts us (scientists included) as a report, a text, and is understood in terms of theory. Histories of experiments and of debates also come down as texts, and also employ these same languages of science.

That performed experiments and actual lives of scientists have an additional 'practical' or 'life-world' context—at times connected with politics and social concerns—is something we should certainly get to know better.

However, this context, itself largely reconstructed from texts, should not be made to overshadow the contexts of the 'book of science,' but to *fuse* with them meaningfully.

[10]

Positivists v. Positivism or the Screwtape Effect

ABSTRACT: An article by Roger Garaudy praises African culture for its cultivation of the aesthetic component in life, as a corrective to the over-emphasis of the rational component in Western culture. Although approbation is expressed for Garaudy's appreciation of the aesthetic component, his denigration of the rational component is subjected to a humanistic and historical criticism. There is no way of devaluating thought, and regarding it simply as an instrument, without devaluating man and regarding humanity instrumentally. Thought is capable of being corrected and expanded without abandoning it totally. Historically, the attack on reason characteristic of much of the avant-garde philosophizing in the early twentieth century glorified 'action' and 'intuition' and engendered fascism, which violently subverted humanism. Garaudy's polemic is an instance of the "Screwtape effect" recommended by a devil: "We direct the fashionable outcry of each generation against those vices of which it is least in danger and fix its approval on the virtue nearest to that vice which we are trying to make endemic." [—A.S.]

What Garaudy Said

The following is an excerpt from Roger Garaudy's article "We Can Learn from Africa," which elicited the response from Martin Eger.

Against the major rationalism of Socrates, Leibniz, or Hegel, and against the minor positivistic rationalism of Comte (for whom nothing has meaning, or even real existence, outside of what can be referred to reason, concept, and discourse), a dialogue of civilizations with non-Western cultures, and, above all

This essay appeared in the July/August 1973 issue (Vol. VI, Number 4) of *The Center Magazine*, a publication of the Center for the Study of Democratic Institutions, whose Chairman was Robert M. Hutchins. It was a response to an article by Roger Garaudy which had appeared in *The Center Magazine*'s March/April 1973 issue.

with African culture, can help us to become conscious of and appreciate the aesthetic component of life which is at least equal in importance to its logical component. Our Western tradition looks upon the aesthetic component as residual or supplemental. No room is left for that which Taoism, for example, calls "non-knowing," that is, an essentially non-mediate knowledge, an act of participation through which we coincide with the movement of being.

If, since Socrates, we have gotten into the habit, as Nietzsche has shown, of underestimating the importance of whatever escapes our purely intellectual procedures, hypotheses, deductions, verifications, concepts, and language, then aesthetic experience can help us see the value of these other realities. When I have analyzed a picture, I cannot demonstrate scientifically that it is beautiful and ought to move us. At best what I can do is bring you to the point where you, and you alone, can test all that I have not been able to express. This is even more evident in music and in dance. In the ancient Greek tragedy, the chorus, through singing and dancing, expressed what words or mime alone could not express. The anguish of death or love; the faith that makes the believer, as well as the militant revolutionary, joyfully confront sacrifice; the emotion evoked by the beauty of a landscape or another human being—all these are irreducible to the notion of "concept."

But this irreducibility does not mean that concepts are of no value. Utilitarian and technical activities, and the objects they can make, are aptly expressed in concepts and language, as, for example, are the movements of the stars or the atoms. But to express a vital experience or an act that is specifically human, and that therefore transcends everyday knowledge or practice, one must also transcend purely conceptual language. . . .

If we really want to overcome our dualisms and to harmonize the aesthetic with the logical, shouldn't we welcome the prophetic promise of Nietzsche, and seek, through the dialogue of civilizations, to recover the dimensions of being as well as of knowing?. . . All of us must look beyond the language of myth to the future that it symbolizes and that we must not so much discover as create. African culture—like the high cultures of Asia—can help us to escape positivism, which, by confining thought within the limits of the given, also confines action within the limits of order.

African culture has this vital ferment. It teaches us that thought is not a copy of the "real," but rather both the negation and the transcending of the real. It teaches us, therefore, that the real is also the possible. African religion, like African art, expresses a marvelous faith, what Kierkegaard called the "passion for the possible." In this way man's relation with his future can no longer be conceived of as an extrapolation of the past or the present. Rather, it is the invention of the future.

What we see, therefore, is that African arts are not merely an expression of a particular culture but reflect a culture which contributes to universal civilization. African culture is a way of living the world. Everything in its poetry, music, dance, sculpture, "masks," houses, and tapestries, renders the invisible visible and reveals the meaning of the confrontation between life and death, and between liberty and necessity.

Martin Eger's Comment

In "We Can Learn From Africa" [March/April, 1973], Roger Garaudy offers an argument for the long-advocated idea that a sympathetic appreciation of non-Western cultures can help us surmount some typically Western limitations of perception—limitations that are believed by some to be responsible for such modern problems as the degradation of the environment. In place of Tao, Buddhism, and Hinduism, which have often served as model alternatives to "Western positivism," the focus this time is on "African culture," and the thrust of the discussion is directed at the identification of those points of contact between Africa and the West that could be used as channels for intercultural dialogue. Thus far, Mr. Garaudy's objective surely deserves the support not only of the educational community but of every well-intentioned person. Yet the shared elements actually chosen, and the arguments presented to recommend them, are such as to suggest it would be well—before talking further of Africa—to bring to bear upon his article that power of analytical reason which not so long ago was still regarded as one of the more wholesome fruits of our own civilization.

Mr. Garaudy has a point, a point more and more frequently made: that in the Western world, but especially in America and Britain, life would certainly be better if aesthetic values and "harmony with nature" played a greater role in our daily lives and in the fabric of our institutions. In this respect, it might well be that Africa has something to teach us. Yet when it comes to reforms on a continental scale, few would deny that a thorough understanding of the problem and its history should first be seriously undertaken. It is, therefore, all the more regrettable that, in looking for the causes of this sad situation, Mr. Garaudy has been able to find nothing other than the same philosophically based argument that has in the past few years become the conventional wisdom of the magazine world; an argument which so completely fails to make the crucial distinctions in modern intellectual currents that, far from holding out any promise of help, it may—if hammered in long enough, and taken seriously—lead us once again into those disastrous quicksands from which the West has only recently climbed out.

The general attack on the primacy of reason is now about one hundred years old in our society, and shows no sign of abating. On the contrary, it seems to be gaining new groups of followers and adding new elements to the style of its polemic. The original contribution of Mr. Garaudy is the suggestion that in its art, in its view of nature, and in its communal inclinations, African culture has surpassed the values embodied in the Western social process, and is comparable only to the latest insights of our most advanced sciences, which emphasize a field rather than a particle concept of matter. "This ultrascientific way of thinking about nature coincides with the African experience of the arts, of life, and of culture. In such a perspective, the human being is continuous with nature and zealously concerned about his harmony with the environment." Only in certain of the most recent forms of Western thought have these insights played a central role, according to Mr. Garaudy, and it is through these forms, he hopes, that a dialogue with Africa may be most beneficial to us; for they are the means "through which contemporary epistemology is trying to react to rationalism and positivism." Thus, in African culture there is an ally to be found for all those who struggle "against the major rationalism of Socrates, Leibniz, or Hegel and against the minor positivistic rationalism of Comte . . . ."

In this context, as possible bridges between traditional Africa and the avant-garde of Western thought, Mr. Garaudy lists these movements: structuralism; views aiming at "the restoration of the *aesthetic* and immediate components of knowledge to the same plane as the mediate and conceptual components"; and philosophies of knowledge emphasizing the *active* role of the subject, such as operationalism. In spite of the fact that the connections made by Mr. Garaudy between Africa and certain popular criticism of our own life are easily recognizable, and that the ecumenical spirit and desire for intercultural respect that pervade the article can hardly be faulted, there are disturbing features in this argument; and those inclined to accept his analysis on account of these attractive qualities would do well to ponder also the less visible but nonetheless important implications it contains. This requires an attempt to see his overall view in a perspective of philosophical and historical relations to experiences that, as a culture, we have already had.

To begin with, one should ask why Mr. Garaudy believes that operationalism can really "help us to escape positivism which by confining thought within the limits of the given, also confines action within the limits of order." Of all the characteristically twentieth-century philosophies, operationalism is now recognized as one of the most narrowly thought-confining. By insisting (in its extreme form) that only concepts operationally defined be allowed to enter into analysis, or even conversation, it overtly (or in its less extreme form, covertly) delineates "reality" as precisely these operational "observables." Nor is it news to anyone that it is just this modern "physicist's approach" that provides much of the philosophical underpinning of behaviorism in the social sciences—a direct result of the efforts of such outstanding positivists as Rudolf Carnap.

Behaviorism, operationalism, and more generally instrumentalism, are in fact positivist philosophies—positivist, that is, not in the nineteenth-century sense of Comte but in the sense of the specifically twentieth-century movement known as logical positivism, which was as highly influenced by the revolution in physics as by logic. The vital connecting link between all these isms is precisely that factor which so appeals to Mr. Garaudy: the primacy of act over thought in the process of coming to know, and the identification of the act with the "real," and of thought with instrumentality. Thus, Mr. Garaudy invokes positivism in order to free the West from positivism.

On second thought, though, this is perhaps not so strange. John Dewey did it, and did it quite explicitly and persistently over a period of many years. As long ago as the turn of the century, influenced by the rapidly changing scene in physics and its philosophy (the introduction of field theory, statistics, and the works of Duhem, Poincaré, Mach, etc.), this American pragmatist began to evolve a type of outlook for which he aptly chose the name instrumentalism. But it was not until 1927 that two important events in the philosophy of science convinced him he had been on the right track all the time: the advent of Heisenberg's and Bohr's interpretation of quantum mechanics, and the publication of P. W. Bridgman's *The Logic of Modern Physics*, a fundamental treatise on operationalism. These departures from the classical scientific ethos had, it seems, the same effect on Dewey as on Mr. Garaudy forty-five years later: They appeared to him as the means—and the mandate from "ultrascience" itself—for the *coup de grace* to the rationalist heritage of the Enlightenment. Does not Heisenberg himself say that prior to these developments "it had not been possible to see what could be wrong with the fundamental concepts like matter, space, time, and causality," though many had for a long time felt them to be inadequate? But now "the dissolution of this rigid frame" could be seen as the most important change brought about by modern physics. Therefore, in his *Quest for Certainty* (1929), which is saturated with the language of operationalism, John Dewey makes his clearest and boldest formulation of the new theory of knowledge: "the true object of knowledge resides in the consequences of action," and thus "all reflective knowledge as such is instrumental." As does Garaudy, Dewey also took the latest science as his model: He never made the mistake, so common today, of speaking as if the rationalism of the Enlightenment was still in the philosophy of modern physics. To him it was perfectly clear that in twentieth-century science, theory is actually nothing more than "formulae for the prediction of the probability of an observable occurrence."

How, we may ask, does such a philosophy arise? What was Dewey's motivation in all this? Exactly the same as Mr. Garaudy's, it seems: to raise experience and aesthetic values to a higher level by demoting rational concepts from their position as being-in-itself to that of instrument. "There has been repeated occasion to note," says Dewey, "that the claim of physical objects, the objects in which physical science terminates, to constitute the real nature of the world, places the objects of value with which our affections and choices are concerned at an invidious disadvantage. . . . From this derogation of things we experience by way of love, desire, hope, fear, purpose, and the traits characteristic of human individuality, *we are saved by the realization of the purposely instrumental* and abstract character of objects of reflective knowledge" (emphasis mine). Thus two of Mr. Garaudy's forms of modern thought, in which he sees hope for reform in the West, have been highly interrelated and repeatedly advocated, by Dewey among others, since the beginning of the century.

We have not undertaken this recapitulation of intellectual history to quibble over priorities, nor merely to point out that, except for the use of Africa as a model, Mr. Garaudy's critique of the philosophical basis of Western culture has been reformulated now every few years for nearly a century, and sometimes brilliantly so, as in the case of F.S.C. Northrop's *The Meeting of East and West.* Rather, the point is that when it comes to old, well-known problems, there must be an important difference in the way we react to an old suggestion for a cure, from the way we might receive a truly new proposal. The crucial question that has to be asked in the former case is this: If the suggestion has been made in the past, and made again, and no effort has been spared to make it clear and widely understood, why, then, do we need to make it once more? Why, in other words, did it not already do its work? One suspects that the answer to this may be connected with a previous question: How can positivism be fought with positivism?

Why such an idea should arise in the first place, and why it should retain its appeal in the course of generations, is not hard to see. There is a most significant difference between the older positivism of the nineteenth century and its modern version. While classical positivists might indeed have accepted the end-point objects of physics as the basis of an independent reality, their twentieth-century successors, realizing this to be untenable, shifted the notion of ultimacy from rational constructs to experience. Now the operation or verification or function associated with a concept becomes its sole anchor in reality. The purpose and hope is that by laying down precise rules restricting the manner in which reason is to be used, and reducing the ontological status of its products, science and philosophy will never again make the mistake of taking their own inventions for those of God or nature. It is, therefore, not surprising that the idea should occur to people like Dewey or Garaudy that in some form of the new positivism there are weapons that could be directed against the older positivism. The positivists themselves inaugurated this battle; and the new version does indeed reduce the role of "purely intellectual procedures."

Yet positivism remains positivism. New or old, it still imposes some kind of "rigid frame"; it still chooses, out of the whole range of possibilities, one type of object—the experience of operations and consequences—as "positive," as having the only warrant on credibility. This means that, willy-nilly, sooner or later, it takes on the role of being-in-itself. And here we return to the crucial point: There are dangers in fighting fire with fire. Nor do we need to look far afield—to something like behaviorism—to see the kind of pitfalls that lie along the path of the "active" philosophies of knowledge. It is already there in Dewey's original formulation. Starting from the policy of demoting rational concepts and "mere" intellectual activity in favor of more specifically "human" concerns, Dewey—quite understandably—ends up by demoting man himself! Incredible though it may seem to some, the path from the view of mind as an instrument to that of man as an instrument appears to be short indeed. For if experience is the ultimate object of all thought, then man himself becomes a mere instrument for the bringing about of experience.

As Dewey puts it, if his theory of knowledge were carried into the field of ethics "men would think of themselves *as agents, not as ends*; ends would be found in experienced enjoyment of the fruits of a transforming activity" (emphasis mine). It is this kind of belatedly discovered consequence that disillusioned many of Dewey's young enthusiastic followers such as Lewis Mumford and Randolph Bourne. Thus, a man of unquestionable humanistic sentiments, of laudable intentions, having put prodigious efforts into his attempted "reconstruction in philosophy," becomes at last a symbol for the most dehumanizing aspects of the contemporary outlook; and the method of applying instrumentalist thought to social science is now under heavy attack in every area from epistemology to politics, quite aside from the fact that his naïve view of the philosophy of science has long been superseded.

We have said that all this is quite understandable, and so it is—to anyone who still takes seriously the idea of *Homo sapiens*. There simply is no way of devaluing thought without devaluing man. There are ways of correcting thought, of replacing bad or fruitless thought with better thought, of augmenting thought, without this kind of disastrous result; but there is no way of attacking thought itself while leaving man unscathed.

The answer to our question as to why it should be necessary once again to restate the old suggestion that reason must be devalued in favor of aesthetic experience, is not that the previous efforts somehow failed to have an effect; on the contrary, they have all had quite noticeable effects. There is hardly a student of the social sciences who does not know the effect of operationalism and other forms of modern positivism on his field; while in Europe, philosophies emphasizing the "act" and the "immediate" or "intuitive" component have bred some strange and gruesome offspring indeed. One could conclude, not unreasonably, that many of the previous demotions of "pure reason" have succeeded only too well. But partly as a result of this very success our conception of what constitutes reason and what its role should be is now so blurred, so atrophied, that often that which has been substituted for reason is then in turn attacked as reason: and with each succeeding attack reason itself sinks a little lower. With each "revolution," the new reign is quickly seen to be so dismal that another revolution is required; but in the meantime, the revolutionaries have either forgotten or they do not care to acknowledge that the old king is no longer on the throne.

Those who a generation or two ago gained ascendancy promising to curb "mechanistic thought" and "abstract reason," and who have honestly labored to carry out their mandate, are now assailed by the same slogans which they themselves once used so effectively. In sociology, psychology, political science, the young and old revolutionaries are now calling for an end of the ill-conceived borrowing of techniques from physical science: They are calling for a more "humanistic" approach, and they object to the idea that "theoretical models should be tested primarily by the accuracy of their predictions rather than the reality of their assumptions." But this rejection of what is patently instrumentalist methodology is usually accompanied by a rhetoric more appropriate to an attack on eighteenth-century rationalism. It is assumed by nearly everyone that physics is the epitome of reason; therefore any fault or limitation found in its method is a fault of reason itself. It seems to be equally natural to see technology as reason, and thus also the entire modern industrial state and all its institutions. The school or university becomes a representative of "pure reason" in this way, and so, too, the military with its "think tanks." Few bother to point out that the philosophy of physics is no longer that of Galileo; that Washington DC is not today the embodiment of Jefferson, Franklin, or of Tom Paine; and that Dewey himself has had quite a hand in making the American schools what they are.

It appears that the intellectual critics have simply lost their way. Something is very wrong, not just in our culture, but also in the manner in which that culture is criticized; and this opens up far larger questions than are specifically raised in the Garaudy article. The suspicion arises that one symptom of the malady of our time may in fact be the strange role which the intellectual has now taken upon himself. Is it not strange indeed that those who have chosen the life of the mind as their calling should so loudly inveigh against the value of that very activity; and that those who for centuries formed a unique group that stood for the possibility of transcending our limited, partial, bleak, human condition by means of the one gift that allows us to see where eyes cannot reach and to create a common ground of meaning and hope that might, at least in part, offset our ultimate separateness—that these should now be heard consigning concept to "utilitarian and technical activities," but not to "vital experience or an act that is specifically human" [*sic*]?

Julien Benda called this the "betrayal of the intellectuals." In a widely read book published in 1929, he made the following important distinction: There is nothing especially wrong when someone preaches the "act," or "experience," or "intuition" at the expense of reason. There is nothing wrong either when some call upon the thinkers and the teachers to come down from their "ivory towers," from their preoccupation with "timeless questions," and immerse themselves in the dirt and sweat and bustle of "real-life" concerns. Things more or less in this vein have always been heard. What is wrong is when the people saying this most loudly are the thinkers and teachers themselves—the "clerks," as Benda called them. For it is one thing if the man in the street proclaims "there is no God" and quite another when the priests themselves start saying so. The specifically human value of "concept" and "reason" was never really accepted by most people. Only a small minority throughout history has attempted to keep this idea alive, to rejuvenate it whenever it seemed paralyzed, and to cleanse it when the accumulated remains of too much makeup, too many accommodations to fashionable tastes, threatened to blot out its meaning altogether. This was the unique function of the "clerks," which no other group had ever taken upon itself, and which they have now apparently abandoned.

Certainly, Benda's situation was different from ours in many respects. What alarmed him were the large numbers of students, philosophers, writers, theologians, and critics who in the France of the nineteen-twenties had apparently gone over to the side of the antirational "action philosophies" or "life philosophies" that soon turned into outright fascism or served as fascism's ally. Thus, it may seem that the connection we are making is far-fetched. How, after all, can one compare the humane, self-effacing, thoroughly civilized Dewey with the tough-talking chauvinist ideologues of pre-fascist Europe? And yet, is it not worth noting that in those days Benda himself saw a connection, for he explicitly linked the pragmatists with the "clerks" whom he was describing. From the viewpoint of this French Jew—calling upon his Christian countrymen not to forsake their Christian tradition, nor to trample on the heritage of Descartes—pragmatism, instrumentalism, operationalism, Bergsonianism, and the philosophy of Nietzsche all had one thing in common: "they are full of scorn for the mind which seeks to grasp general states of being," and teach instead that "intellectual activity is worthy of esteem to the extent that it is practical and to that extent *only*" (Benda's emphasis). Therefore, "these new clerks declare that they do not know what is meant by justice, truth, and other metaphysical fogs . . ." for it is only in the *act* that such "abstract" intellectual concepts can acquire meaning; a meaning that must finally be determined by "experiment," by what *works*. Benda calls this kind of stance in politics "the Romanticism of Positivism."

The work of people like Benda or John Dewey provides the kind of potent reminder necessary for ridding ourselves of the notion that it is somehow given to our generation to "see through reason." Philosophies tending to demote the mind may then be viewed in their historical continuity, making it easier to perceive in what directions certain types of criticism lead.

It should be emphasized, however, that there is no objection in anything said here to the positive aspect of Mr. Garaudy's proposals, only to the philosophical justification upon which he bases these proposals, and specifically to the polemic against the rational component of life which he continues along the line laid down by Dewey and other instrumentalists. Indeed, Mr. Garaudy violates his own injunctions (against overconceptualizing) when, in order to recommend such reasonable and generally desirable goals as a dialogue with Africa and the enhancement of our lives through greater aesthetic awareness, he feels obligated to erect a truly Byzantine structure, linking these worthy aims with "ultrascience," with the contrasting nature of field theories and particle theories of matter, and with the derogation of reason in "specifically human" affairs. Such curious theorizing ignores experiences like that of Japan, where, in spite of a strong traditional involvement with ancestors and with the aesthetic dimension in all aspects of life, there had nevertheless come into being an even greater environmental problem than in Europe and America. Mr. Garaudy's argument proceeds by stringing together vague resemblances, not by evaluating the consequences of man's activity.

It remains, as a last thought, to point out the truly ironic situation in which we find ourselves, when from all sides assurances are given that "the Enlightenment is dead," while at the same time an ever-rising chorus of voices cries out against the dangers of too much reason. One stands back, gaping, as at some tragic-comedy wherein the hero's every move is cleverly contrived to weaken still another support beneath the very stage upon which he acts; and one wishes there were a way of understanding such a fiendish note in the historical process. Then there comes to mind that wonderful series written by C.S. Lewis for the newspapers during the Second World War, and published as a book under the title of *The Screwtape Letters*.

In these letters, an old experienced devil named Screwtape sends advice to his younger nephew, an apprentice devil serving in the field, and explains to him the nature of the work of their organization. "The use of fashions in thought," says Screwtape, "is to distract the attention of men from their real dangers. We direct the fashionable outcry of each generation against those vices of which it is least in danger and fix its approval on the virtue nearest to that vice which we are trying to make endemic. The game is to have them all running about with fire extinguishers whenever there is a flood, and all crowding to the side of the boat which is already nearly gunwale under. . . . Cruel ages are put on their guard against sentimentality, feckless and idle ones against respectability, lecherous ones against Puritanism; and whenever all men are really hastening to be slaves or tyrants we make liberalism the prime bogey. But the greatest triumph of all is to elevate this horror of the Same Old Thing into a philosophy so that nonsense in the intellect may reinforce corruption in the will."

The social phenomenon here described—strange, though surely not unfamiliar—has in our family discussions acquired the name of the "Screwtape effect." Thus, when "the Enlightenment is dead," when astrology is endorsed by educators and taught at universities, when in "therapy groups" all over the country people daily practice the art of suppressing "conceptualization" to encourage "feeling," and reason is everywhere on the defensive—it is a "Screwtape effect" if at such a time distinguished philosophers can find no greater danger of which to warn us than an excess of mind. And when in the same issue of *The Center Magazine* that carried the article by Mr. Garaudy—a journal representing the one institution in America officially dedicated to the preservation of the rational Socratic dialogue—there appears a statement by a man who, having listened to these dialogues less than a day, declares to the world that the Center is too rational (because "our culture frowns upon emotion")—that also is the "Screwtape effect."

[11]

Academic Dialogues and the
Postmodern Mood

To:	S. Stearns, L. Nachman, M. Affron, L. Quart, A. Levine, J. Hartman, R. Chiles, D. Kramer, G. Rozos, R. Carey
From:	M. Eger
Subject:	Academic dialogues and the postmodern mood (a comment on our discussion of April 3 and its aftermath)
Date:	April 23, 1995

At first I was puzzled as to why a long memo from Steve about Larry's views was addressed to me. But, after reading it, I am glad of the chance to send to colleagues some thoughts on what I think is wrong with today's academic dialogues—an issue that is on many people's minds. It certainly was not "distress at the logical untidiness of our discussion" that moved me to make the remarks that, apparently, brought forth Steve's memo. Lots more than logical tidiness was missing from this discussion.

The presuppositions of dialogue, and therefore the very nature of much contemporary debate in the humanities and social sciences, are quite different now from what they were when we (people over fifty) were students. I see effects of this at every non-technical academic conference I attend, and on every campus I visit (most recently, Cornell)—effects that extend even to the study of the natural sciences (history, philosophy, sociology of science). Hardly ever does one attend a discussion now of which one can honestly say that it 'worked,' that it was more or less what it should have been.

This is a memo written to a group of Martin Eger's long-time colleagues at the City University of New York's College of Staten Island. Most of them had been part of the original founding faculty of the college almost thirty years before.

To characterize this failure by saying that scholarly exchange has become too polarized, too ideologized, and too adversarial is easy, but this misses the more interesting question: What are the cognitive roots of the new situation, and how, in that situation, should one comport oneself? Liberals and conservatives we have always had with us—certainly so in the 50s, when communism and Cuba and sit-ins were in the news. But then, the intellectual confrontations were carried out under different presuppositions.

Then we assumed that however committed we might be to a general philosophical or political orientation, we still were obligated—as intellectuals and academics—to 'enter into the question' and to pursue it, as much as possible, on the basis of objective analysis, reasoned argument, and factual information. Though our commitments might cause us to enter on opposite sides, these values would be the common ground on which we could meet, even if temporarily, and from this meeting profit. With such an understanding it was possible for two people to be philosophical opponents but intellectual comrades. A professor was not expected to engage in 'total war.'

Where the situation is different today, as most of us know, is precisely in regard to these classic values—objectivity, rationality, facts. New concepts have emerged recently, in apparent competition with the older set—'situatedness,' 'standpoint,' 'contextual rationality,' 'local' or 'personal' knowledge, and hermeneutics. Within the new perspective, the very idea of objectivity is not only naive but possibly a cover for maintenance of the socio-cognitive status quo. Rationality, far from being our 'common ground,' is the contingent, historically constructed, mode of thinking of a dominant class. And facts are no more reliable than the interpretive debates and negotiations that led to their being accepted as such. Interestingly enough, support for all this is believed to lie in the newly rediscovered power of language (logos!)—a power now seen as in opposition to the persuasive power of facts and reason.

Even if few of us accept all these assessments, especially when phrased as baldly as above (to make the point), they still have enormous effect on campuses. Indirectly, tacitly, as presuppositions, they are 'in the air.' They set the mood of human interaction, they structure the pragmatics of debate, and they arouse mutual expectations among debaters. Moreover, they are not without warrant. In many areas, as research penetrates deeper, we do encounter new grounds for questioning the received understanding of objectivity, reason, and fact. But as this questioning goes on, many people assume that the outcome can only be the demise of the received understanding; that its further development, reformulation, and deepening is not a possible outcome. In this way, some theorists have been led to assert that, at bottom, language is not the way to communication and discovery, but rather to agonistics (Lyotard): In no sense can we consider ourselves on the 'same side' of a specific argument—together pursuing implications and insights—even when we are on different sides in our larger commitments. Socrates is for children. The purpose of argument is, in Lyotard's words, to "take tricks." (It was Merilla, I believe, who thought she spotted this very thing in our discussion with Larry and objected.) Thus, what used to be a lapse from accepted ideals is now a theoretically supported strategy.

But in such an environment there is an additional factor at play, contributing to a vicious circle: If one already knows that one's partner in dialogue is aiming to "take tricks," then, to take his arguments seriously would be to make a fool of oneself—the most dreaded thing of all. For there surely is a performative factor in the pragmatics of language: We must, to some extent, trust the intention of a speaker in order to listen seriously, even when we 'already know his agenda.' And today, this trust is thoroughly undermined by just those philosophies—expounded sometimes with great erudition—that make of dialogue either a "kibitzer's" game (Rorty) or a power game (Lyotard and many sociologists). It is this kind of mood that I sensed in our discussion of April 3.

What to do? To believe that everything would be fine if only this or that participant were more reasonable, more open-minded or more centric in his approach, is to ignore the enormous pressures on ourselves, on all of us, in this regard. The very notions tending to make genuine dialogue increasingly difficult are now taught, formally and informally, in colleges throughout the country. The only thing I can suggest by way of antidote is increased consciousness of precisely this situation. It helps, in any weather, to know which way the wind is blowing. For if we are aware of the pressures, then those of us so inclined can make the needed corrections: Liberals who think that most conservative arguments are 'garbage' might try to find the best of the conservative thinkers (e.g., Sowell, Gadamer) to keep themselves from believing that all true insights lie on the left; conservatives might likewise seek out those on the left whom they can respect (Charles Taylor, Richard Bernstein, etc.). This will seldom change our commitments, but it can substitute a genuine hermeneutic consciousness for the bogus kind of which we hear so much.

I am sorry to be so simple-minded as to give the kind of advice one might find in some high school primer of a bygone age. It only reflects the seriousness of our situation. Of course, there are more 'scholarly' approaches. One can join the ongoing debates on the basic points and perhaps discover that one has been credulous after all—where one least expected to be. Are we as skeptical of the latest 'insights' as of the 'cherished ideas' we all love to question? Is it really true that serious argument necessarily leads to a cognitive black hole—that it's always 'interpretation all the way down'? Is it true that a large number of differing but valid interpretations preclude clear-cut rational judgment (and so, only rhetoric remains)? Is it really true that prior commitments make of objectivity a vain dream? Is it true that a socially constructed cognitive fact cannot have validity in a universal sense?

Perhaps it is not true that postmodernity sheds all credulity; perhaps it just redirects it.

PART TWO

Philosophy and Education

[12]

Physics and Philosophy:
A Problem for Education Today

ABSTRACT: The teaching of the process of science in physics courses is discussed with reference to recent intellectual trends. Arguments and examples indicate that 'process courses' must necessarily involve philosophical questions and questions of social criticism currently raised. A description is given of an interdisciplinary upper-division course for science and nonscience majors that incorporates some of the features advocated. [—M.E.]

Introduction

Much concern and activity has been directed recently toward the development of new physics courses for the nonscience major, an effort that is now being described by some as "one of the most urgent tasks of the physics community." Among the pivotal ideas about which a good deal of the discussion revolves, none has been more crucial than the problem of purpose and the concept of relevance. Lists of goals have been compiled; novel means of relating physics to current social questions have been attempted; and a variety of new courses have been developed to implement these goals and to present physics in a new light.[1]

One of the welcome trends within this reform movement has been the increased attention to the *process* of science, as distinguished from content. Updated subject matter, rearrangement of the syllabus to begin with the most modern topics, and similar changes, do not seem to be enough to give a satisfying answer to those who continue to raise the age-old question of relevance:

> Whether or not a poet or writer or classicist knows the second law of thermodynamics or the "overthrow of parity" is his personal business: It certain-

This paper appeared in the *American Journal of Physics* 40 (1972), 404–415.

ly would not do him any harm. It may also be entirely irrelevant for what he has to say. For the "natural order" which the quantifying sciences define and master is not *the* natural order. . . .[2]

The turn to process with its emphasis on method, on asking questions, on formulating rather than solving problems, on conceptualizing experience, may be seen as a partial return to an earlier tradition, the tradition of the 17th and 18th centuries, when such questions were central in physics itself, and central in the dissemination of science to outsiders—the tradition of 'natural philosophy,' which prevailed before the modern separation between the strictly 'scientific' and the philosophical aspects of the study of phenomena. It is hardly surprising that physicists then became involved not only in the formulation and application of basic laws, but also in a common concern for the effects that these new formulations would have on mental attitudes in general: on beliefs, and on those conceptual schemes in other fields of thought that were part of the existing culture.

In time, of course, "normal science" takes over; the critical mood, the wide-ranging concern, recede and fade into the background. As long as most people retain their faith in the essential validity of the prevailing paradigm[3] and are more interested in finding out *what* it is than in understanding or questioning its foundations, *learning* the paradigm remains the sum total of science education, and no other need is ever openly acknowledged. However, as we intend to show, such an attitude cannot be assumed to exist among students today.

Along with a concern for the way science is applied, or misapplied, there now exists also a deeper uneasiness about some of the most fundamental aspects of science itself. For years, men like Gerald Holton, C.-F. von Weizsäcker, and René Dubos have labored to convince laymen that "science shapes culture not necessarily through its technological aspects, but rather by providing new points of view and facilitating new attitudes."[4] Today, it seems, the ideas stimulated by such claims, combined with recent events, have indeed brought about a wider acceptance of the proposition that "science shapes culture" in this way; but the conclusions drawn have not always been of the kind originally hoped for: To many nonscientists—students especially—such 'shaping of culture' has been a mixed blessing, to say the least.

The recent interest in 'process courses' reflects, perhaps unconsciously, the reality of the new atmosphere and opens up possibilities for the reintroduction of philosophical thinking into physics, not as a separate though related activity nor as a sweetener to entice students into 'hard' science, but as a serious and integral part of science itself. In the 20th century, in almost all the sciences, it has become a point of pride to keep the philosophy separate from the 'real work.' We need not deny the obvious advantages of this approach in day-to-day research, and in the training *for* research or for applied work. But in *science as education*, it may well be that an entirely different attitude is called for.

The problem has such important roots in the recent rapid changes of attitude at the universities, and within the intellectual community, that a more detailed discussion of those attitudes appears to be necessary before we delve into the particulars of curricular innovation. Specifically, we wish to consider the nature of the relation among the three basic factors involved—physics, philosophy, the student—in light of events and currents of thought of the past decade.

Physics, Philosophy, and the Student

Customarily, the nonscience student's interest has been sought either through his desire to understand something of the wonders that physics has accomplished, or through the novelty of modern theories, or through the *game* aspect of problem solving and the idea that "physics can be fun." *The process of coming to know*, the role of concept, the implications, and other philosophical aspects have not been prominent in any of this. The main argument of this paper is that the essential connection between physics and the interests of many concerned laymen is not now what it used to be, that philosophical discussions are no longer remote for many students, and that this is in fact one of the visible intellectual changes which have recently occurred. One might even go further and claim that questions involving philosophy, which always exist, are usually suppressed by modern educational practice, but that the present critical attitudes toward science, toward the University, and toward education in general, have served to bring them to the surface. In any case, in contrast with the recent past, there are today certain objective reasons for believing that the philosophical dimension of science is once more prominent in people's minds.

Though many signs point in this direction, we can start with the kind of first-hand evidence that comes from the personal experience of teaching. Several instances come to mind, of students more concerned, more insistent, about the *formulation* of theory than they have ever been before. During a discussion of the concept of energy, for example, a psychology major was very skeptical about *value* of the whole enterprise: "Your definition of work is arbitrary," he said, "it's an abstraction; and kinetic energy's another definition which *you* have chosen to make. Then you show all kinds of relations between these abstractions—but what has all this got to do with the real world?" The question is, of course, a good one and not unusual. What made me remember the occasion is the difference in *attitude* and *expectation*—which one could hardly help noting—from many similar questions asked by the same kind of person 10 years ago. Then, the student usually assumed that he had not yet gotten the point and that only a little clarification was needed to dispel the mystery; invariably, a few minutes effort on my part would bring about assent. But in recent discussions of this type, it has been obvious that the student *does not expect* an acceptable explanation; he has the confident manner of one who has put his finger on "what is wrong with science." The first attempts at an answer produce more questions and an even greater skepticism. Finally, the predictive power of a conservation principle is grudgingly admitted as useful, but the suspicion that the overall picture might still be arbitrary, and even harmful, remains. Only a subtle difference in emphasis? Perhaps, though examples of this nature seem to be too numerous to ignore.

What is probably more significant is that these attitudes cannot be taken as mere random samples of an intensified antiscience; they show a consistency and a well-defined common viewpoint, corresponding remarkably to the dissatisfied accusatory stance toward science found in the writings of many social critics. They are part of a reaction that expresses itself with varying degrees of articulateness and sophistication and is heard among well-known scientists as well as college freshmen. It is not surprising that on those occasions when intellectual leadership is explicitly acknowledged, one of the names frequently mentioned is that of Herbert Marcuse. For Marcuse has gone further than most others in expounding his conviction that the dehumanizing aspects of modern society are "inherent in pure science." The traditionally emphasized neutrality and objectivity of science are brushed aside by Marcuse, Lewis Mumford, and others as deceptive, because the basic concepts with which science operates predetermine the nature of the conclusions; the approach structures the result; or, paraphrasing McLuhan, one might say that in this view *the method is the message*:

> Pure and applied operationalism, theoretical and practical reason, the scientific and the business enterprise execute the reduction of secondary to primary qualities, quantification and abstraction from "particular sorts of entities." . . . While science freed nature from inherent ends and stripped matter of all but quantifiable qualities, society freed men from the "natural" hierarchy of personal dependence and related them to each other in accordance with quantifiable qualities—namely, as units of abstract labor power, calculable in units of time. "By virtue of the rationalization of the modes of labor, the elimination of qualities is transferred from the universe of science to that of daily experience."[5]

In and out of class, the suspicion of abstraction, quantification, and precision is almost always central to the discussions that occur; and the connection between these features of science and the mechanistic, bureaucratic aspects of modern life, is invariably made. The fear that 'concrete experience' is somehow lost or falsified in the scientific outlook finds its expression in the exploration of such alternatives as Eastern religions and personal encounter groups, as well as drugs. The literature that has developed around these needs can be found prominently displayed at most college bookstores and is revealing for the contrast it deliberately draws between the concepts of 'western science' and the suggested alternatives.[6] But one need not be a devotee of yoga, sensitivity groups, or other means of turning the mind away from 'abstract Western rationality' in order to find ample and respectable confirmation of doubts concerning the scientific method. A serious student of the social sciences could hardly help but be aware of the present reaction against the behaviorist approach in his field and of the connections with physical science that are constantly being made. In the December 1969 issue of the *American Political Science Review*, for example, the two lead articles, including one by the president of the association, were devoted to questions of method, the current redirection of interest, and the illusion of "value-free science." No time is wasted in pinpointing the problem:

> The essence of the post-behavioral revolution is not hard to identify. It consists of a deep dissatisfaction with political research and teaching, especially of the kind that is striving to convert the study of politics into a more rigorously scientific discipline modelled on the methodology of the natural sciences.[7]

The analyses that are then presented in these two articles are replete with references to such names as Descartes, Bacon, Kepler, N.R. Hanson, Karl Popper, Michael Polanyi, and above all, Thomas Kuhn, whose *Structure of Scientific Revolutions*[8] serves as a standard reference on the conceptual development of the physical sciences. The arguments against behaviorism are, of course, ostensibly confined to political science, but some of the statements have been given such wide and deliberate scope that their import for other sciences could hardly be missed:

> The impoverishment of education by the demands of methodism poses a threat not only to so-called normative or traditional political theory, but to the scientific imagination as well. It threatens the meditative culture which nourishes all creativity. . . . An impoverished mind, no matter how resolutely empirical in spirit, sees an impoverished world. Such a mind is not disabled in theorizing, but it is tempted into remote abstractions which, when applied to the factual world, end by torturing it. Think what must be

ignored in, or done to, the factual world before an assertion like the following can be made: "Theoretical models should be tested primarily by the
accuracy of their predictions rather than the reality of their assumptions."[9]

It must be remembered that the kind of student who reads these articles, or similar ones in psychology, economics, etc., sits among the 'nonscience' majors who enroll in the courses we are considering. Others
absorb the gist of it from their teachers and colleagues. For the past few
years, these students have been coming into our classes with just such
philosophical and methodological questions already on their minds.
Thus, a preoccupation with content, even if somewhat diluted with doses
of historical and cultural background, may be profoundly disappointing
to the most concerned students of the present college generation:
because the results themselves do involve the methods by which they are
discovered and validated, and the student whose suspicion of formulations has already been aroused will be left hanging while those questions
which are most relevant to him are hurriedly passed over by the very people who could indeed shed some light. Worse yet, the absence of an honest and critical discussion of approach may even confirm his most exaggerated fears that method does indeed determine the result and that a
course in science is a form of intellectual indoctrination into the dominant world view of a mechanistic industrial state.

It is not the purpose of this paper to analyze even a fraction of the
possible approaches to the problem posed by this situation; nevertheless
there are some dangers when physicists venture onto philosophical
ground, which should perhaps be discussed. One of them is the strong
temptation to play the role of defender and even propagandist for some
version of the positivistic and reductionist outlook which has been characteristic of physics in modern times. The tendency in this direction
(from which the author is not exempt) is due to at least two factors.
When the method of physical science is emphasized to laymen, and its
many triumphs exhibited, there is built up an unavoidable implication
that the same method is, at least to some extent, valid in other fields also.
Why otherwise should it be expounded to those who will never become
scientists? Secondly, there is the awareness that present-day youth culture
has already shown itself at least skeptical—when not openly hostile—to
just those values and conceptual formulations which are embodied in the
practice of physics. Viewing this mood as just another eruption of antiscience, many physicists feel the need to stem the tide:

> What is of concern to the scientific community, and what we must come to
> grips with, are those aspects of the new wave that distrust and reject the very
> processes of reasoning and rational discourse. This type of antiscience is the

air we breathe today . . . in the unrestricted hedonism, the emotions conveyed by pop records and the drug subculture. . . .[10]

In meeting the new wave, it is only natural to stress the thought processes of 'normal' physics—the very antithesis of the deplored emotionalism and irrationalism. Thus, the restrictive principles of logical positivism and operationalism occupy the center of the stage, and appear as the quintessence of science. Once the main direction of allegiance is established toward the 'normal' models—authoritative, and tightly deductive—no amount of hedging asides about the 'other' elements of creative research can be expected to change the picture much. No anecdote about the subjective sources of Einstein's insights or Kepler's mystical drives will then erase the overwhelming impression that the physicist's real love (and real faith!) is more or less where antiscience has always said it was: with the straightjacketing of language and imagination, and the confinement of the mind to obligatory patterns, for the purpose of imposing a fixed, calculable order on all of life and nature.

Teachers who now redouble their efforts in plugging operationalism and similar aids to rationality are not unaware of how this may be viewed by students with opposite inclinations. It must be assumed that they consider an honest, outright defense to be the best available option, and a vigorous attack to be the best defense. For example, the author of the preceding quotation, Adolph Baker, takes his stand squarely on the wide applicability of the operational definition; he recommends it as an improved method of thinking not only for relativity and quantum mechanics, but also for such problems as the nature of consciousness:

> Questions like "Does this drug expand my mind?" are simply not useful. They could more profitably be replaced by operational questions such as the following: Does this drug increase or decrease the rate of heartbeat? Does it stimulate the flow of adrenalin?. . . Do people who use it tend to be better physicists?. . . Do people who use it tend to be better musicians?[11]

While I found Baker's book unusually urbane and stimulating—and have used it with students—it is regrettable that the philosophical stance taken in sections such as the above does not transcend the latest version of the clash between the two cultures; rather, it has the effect of intensifying the conflict. Operationalism, that viewpoint which appears to be the epitome of behaviorism in social science, is defended not just within physics, but as a universal cure for perennial controversies over 'meaningless' questions. Such cheerful advocacy—as if operationalism were something new to the behaviorist in psychology or political science—runs the risk of being considered old hat, and linked with modes of thought that are becoming increasingly discredited in the student's own field, for it takes no account of the fact that the climate of criticism today exists *in spite of*—and possibly *because of*—just that application of the objective, operational outlook to social science that is here proposed.

Yet Professor Baker's obvious concern with the need for going beyond subject matter, and the need for explicit treatments of methodology, makes his book a timely contribution. In a vivid and highly successful chapter on *ad hoc* theories, he begins with the following reflection,

> What does it mean for the human race to have reached the stage where it is capable of this kind of reasoning? It is appropriate to interrupt our story in order to address ourselves for a while not to the physical world, but to the thought processes which enable us to learn about it. This, after all, is what is most important.

Such an orientation, if taken seriously, is a step toward the revival of the mood of natural philosophy: It rekindles the sense of wonder and personal involvement among students and echoes the hopeful attitudes associated with the "new philosophy" of the 17th century: In the words of Descartes, science exists not for "resolving this or that difficulty of scholastic type, but in order that man's understanding may light his will to its proper choice in all the contingencies of life."

Others in the field of science education, equally inclined to teach the scientific method, have deliberately avoided direct discussion of philosophical questions. Instead, they try to get the student to participate in problem solving or in the research process itself, and thus absorb 'by doing' what might appear remote and abstract, if taught overtly. Dr. J. M. Fowler, of the Commission on College Physics, has experimented with an approach wherein the aim of transmitting a more empirical outlook is not only consciously adopted but even objectively evaluated:

> I made them etch mica, decide how to recognize fission tracks, count the tracks, decide whether they were uniformly distributed, work out the age of the mica. . . . They learned very little content. I hoped that they would learn some process and I tried to go beyond hope. At the beginning of the course, I gave them an old article by Teller discussing the dangers of fallout . . . and asked them to discuss and criticize the argument. I then gave them the same assignment at the end of the course and looked for changes and (hopefully) improvements in the criticisms.

He and a colleague identified several types of criticisms which were of the sort they wanted to encourage and several of the kind they wanted to discourage, and measured the frequency of occurrence of the good and bad types of criticism at the beginning and end of the course:

> "Good" criticisms included those in which the students asked questions about the statistics, the size of data samples, and the lack of controls, or

offered suggestions for more definitive experiments. "Bad" criticisms were, for example, the expression of pure opinion, or criticisms of Teller himself.[12]

There can be little argument with the advantages of this kind of student involvement as a way of coming face to face with typical science problems. But from the point of view of the critique of modern habits of conceptualizing and thinking, would it not appear to many that the effect of such teaching—regardless of intentions—might be to shift the focus of attention away from the larger framework of aims and concerns, toward the purely technical? That by practicing methodological procedure without an explicit and critical discussion of the method itself, in science and in the search for knowledge, the student is willy nilly pushed into the acceptance of a specific technique as the most vital element of criticism, and scientific criticism as the most valid approach to *any* problem? In other words, can such 'learning by doing' be successfully defended against the charge that 'the method is the message' and that ultimately the message is not just 'how scientists reason' but what *good thinking in general* has to look like?

The questions raised here should not be misunderstood. The teaching of empiricism, operationalism, or logical positivism, as part of the scientific philosophy of our time, need not in itself be harmful, provided it is not done in such a way as to suggest that all misgivings which may still remain in the mind of the student must be due to his own misunderstanding, rather than any genuine limitation of the method. This however is bound to happen when teaching is accompanied by an overt or covert implication that the method being expounded is *the* method, while all previous and competing methods are mistaken, that *this* is the heart of science and the latest wisdom, and that its application to other areas of thought is likely to bring the kinds of results to which the world has become accustomed through the successes of physicists and biochemists. The modern example for this type of confidence was set by P. W. Bridgman, whose earlier work suggests that the most valuable contribution that science makes to thought is the operational principle:

> To adopt the operational point of view involves much more than a mere restriction of the sense in which we understand "concept," but means a far reaching change in all our habits of thought. . . . Not only will operational thinking reform the art of conversation, but all our social relations will be liable to reform. Let anyone examine in operational terms any popular present day discussion of religious or moral questions to realize the magnitude of the reformation awaiting us.[13]

This was written in 1927. Now, one is tempted to reply, the reformation has been tried, and the counter-reformation is upon us. In light of that, would it not be better to stand back and re-examine the reformation with an open mind rather than persist with it in the face of mounting opposition and increasing evidence that something is being missed somewhere? There are indications that physicists and other scientists are now beginning this type of reappraisal. In discussing Lewis Mumford's claim that the logical-positivist view of science continues to foster a mechanistic world picture which in turn determines the shape of the technoeconomic system, the editor of *Physics Today* makes this comment:

> The idea that this philosophical viewpoint could have played a fundamental part in shaping our cultural values may not seem so completely farfetched when we consider the inhibiting influence this philosophy has had on the "softer" sciences.[14]

The Course

We conclude that a 'process course' *must* involve philosophy in a fundamental way, and the only remaining question is how this philosophical component is to be handled. Of necessity, the course to be described explores only one of the conceivable solutions to this problem; and admittedly, that solution is based partially on the interests and inclinations of those who design the course.

The guiding principles that we have chosen are, first, to avoid the defensive attitude and the battle with antiscience, but not to ignore the real and very important suspicions, objections, and fears which exist. These can, in fact, become a powerful motive for serious examination of what science really is, and what its effects on society have actually been. Second, we do not try to bring out relevance through discussion of such topical applications as environmental pollution, the problem of resources, the population explosion, etc. Rather, we restrict our attention to those controversies that have to do with the *way* in which problems are formulated, the inherent restrictions and imperatives of method, and the nature of the side effects. Of course, a completely general, theoretical discussion of such matters would be unbearably arid and futile. It is only in the study of selected discoveries and theories, in the fullness of their historical situation, and together with their ramifications and effects in distant fields, that the clarification of philosophical questions is to be sought.

We therefore turn to process, as do others; but perhaps more consciously than others, we seek out those important developments in physics which have had the most clear and far reaching philosophical and methodological effects outside of physics itself; it is interesting that these also turn out to be intimately connected with some of the misgivings that people have concerning science and the so-called Western rational styles of thought.

In order to insure that equal weight is attached to the philosophical aspects of the subjects, the course is listed as "interdisciplinary" and is taught jointly by a member of the faculty in philosophy and a physicist: Mr. David Hemmendinger and myself.

We focus on four major developments:

(1) Galileo, Copernicanism, and the controversy over 'truth' and absolute knowledge;

(2) Newton, the method of reductive analysis, and the change from internal cause to external cause, from organic to mechanical models of nature;

(3) the quantification of qualities: motion, heat, energy, color, order, symmetry, and the attempts to quantify social variables;

(4) Einstein, relativity, and the philosophy of operationalism.

Subject matter of physics is certainly taught: Newton's law and relativistic equations are discussed and applied. However, we differ from other courses not only in replacing much of the problem solving practice with more attention to questions of validity, approximation, assumption, etc., but also in taking up seriously the problem of relevant variables (primary and secondary qualities), abstraction from 'reality,' and the role of reductive analysis in other fields of thought and in society. It is at this point that modes of thought associated with 'youth culture' come to the surface. A student raises her hand: She has understood everything; the value of dividing the problem into simpler elements is clear; the way it is all put back together at the end leaves her with no objections on grounds of *correctness* or *utility*. But in spite of all that *she does not like it*. For, in this chopping up of the 'real' situation, in this 'compartmentalization,' habits are built up which take attention off the 'whole.' What's more, these habits are extended beyond the problems of the laboratory, and their effects can be seen in the way larger social situations—indeed, *life* itself—are handled in modern society.

Again, students sense the problems and raise the questions, sometimes in an almost incoherent form; but it is the older intellectuals who elaborate and disseminate the theories. In his latest book, Lewis Mumford devotes a great deal of energy to the proposition that the evils of today's world are directly traceable to the overemphasis on reductive analysis introduced at the very beginning of the scientific revolution by such men as Galileo, Descartes, Newton, etc. In his view, the scientific community appears as a group of specialists who, in order to understand a clock, must first take it apart: "The glass, the face, the hands, the springs, the wheels and ratchets, and so forth, until the clock is completely dismantled."

> Then each part is accurately measured, photographed and analyzed by qualified physicists, chemists, metallurgists, mechanics, each working in his private laboratory. When their reports are assembled, every part currently open to scientific investigation will be accurately known in "objective" reductionist terms. . . . But meanwhile the clock itself has disappeared. With this disappearance, the design that held the parts together has vanished, along with any visible clue to the function each part performs, how the assembled mechanism interlocks, and for what purpose the clock once existed.[15]

While all this may seem like the chewing of old bones, the new elements in the present debate cannot be ignored: In the first place, as already mentioned, the positivist program for the social sciences is no longer something to be envisioned as entirely in the future. For the past several decades, we have been watching its growing and spreading influence; thus, the present reappraisal can base itself on a critique of tangible results and effects, rather than mere speculation. Secondly, the ecological and social crises of today can easily be depicted as the application of reductive, mechanistic thinking habits to socioeconomic life. If Lake Erie is dead, is this not because every 'expert' is using its waters for his own *limited* aim, and no one is seriously bothering about the region as a *whole*? Is not the ultimate purpose of the economy, the *quality* of life, forgotten in the concentration on infinitesimal *components* of that economy? Is not the neglect of side effects and interactions precisely the kind of thing that laymen usually find in introductory science courses? Where are the courses that emphasize the importance of *looking for* interactions and collective phenomena, rather than the need to 'isolate the system'? In advanced study, yes; but does the nonscientist ever see that? Certainly, it is not difficult to come up with a fitting answer to arguments such as those of Lewis Mumford. But should we really be content to point out that caricatures of science serve only to reveal their authors' own misunderstandings? If so, what then would we reply to the more careful statements of others, such as this from the scientist-critic René Dubos:

> In the course of reductionist analysis, the scientist tends to become so much involved intellectually and emotionally in the elementary fragments of the system, and in the analytical process itself, that he loses interest in the organ-

ism or the phenomenon which had been his first concern. For example, the student of man who starts from a question singled out because of its relevance to human life is likely to progress seriatim to the organ or function involved, then to the single cell . . . then to the molecular groupings. . . . But in practically all cases the phenomenon itself is lost on the way. . . .[16]

It is not our approach to dismiss views of this type with such unimpeachable statements as, "The organismic philosophy has been tried for over one thousand years before Copernicus, and has led nowhere" or "This attitude has been rejected by the overwhelming majority of practicing scientists today." In the first place, scientists should be the last to claim that the majority is always right; as for the past failures of competing philosophies, well, it could he pointed out by an alert student that the atomic theory had also been around for over a millennium, in its vague and fruitless form, and it too was called 'unscientific' by a majority of scientists, not so long ago. Surely there must be more to it than that. Why should we expect the undergraduate to be satisfied with answers that a scholar in the philosophy of science would never accept?

For this reason, we partially share the critical attitude of many students, and ask with them *why* it is that the method of reductive analysis has been so widely adopted and so strongly held by those whose job it is to grapple with problems in a wide variety of fields. What generally follows is a discussion of the applications of analysis to a number of different types of problems inside and outside of physical science, and a critique of the results. At this point, student initiative is sought in suggesting areas to be examined. The object is not only to distinguish the successful from the unsuccessful applications, but to search for possible side effects in the largely successful cases. Conversion of the student to a more 'pro-science' point of view is not sought. If his initial skeptical attitude is strengthened as a result of this inquiry, but is now founded on a more secure understanding of how and when analysis may lead to an approach that misses the forest for the trees, then we take it that one of our main goals is achieved.

In the discussion of Galileo and the Copernican revolution, again the content of physics quickly leads to major philosophical questions: The well-known controversy with the Church is, of course, usually oversimplified. Freedom of research and publication is not the only important question involved. A look at the arguments of all parties quickly shows that the Church was not at all interested in which method of calculation Galileo wished to recommend—geocentric or heliocentric—but only in the claims made for the 'truth' of one of the competing 'world-systems.' Thus, we are led directly to modern debates on the question of whether the 'pictures' put together by science do actually make some contact with 'reality' or whether they are mere 'instruments for calculation.' The majority position of present day physicists—that 'truth' is not a scientific concept, not at all something they wish to argue over—meets with great surprise on the part of most students, and adds obvious complications to the question of relevance.

In tracing the continuing process of quantification of aspects of experience that had previously seemed unquantifiable, we come up against one of the truly problematic and characteristic features of modern science. Of course, any study of physics is bound to include much of the material treated under this heading. The difference in a philosophically motivated approach is that we group together such disparate phenomena as work, order, color, heat, information, and learning, with the aim of investigating the so called method of 'abstraction.' This mostly classical subject affords ample opportunity for the study of 'choice' and the problem of 'relevance' within science itself. Finally, in relativity we have an unusually fertile field since the resulting reappraisal of concept formulation is just the type of relevance this course is aiming for. No other subject elicits as much interest and willingness to work on physics itself, for the philosophical implications are only too visible, and provide the incentive to master some basic technique. The unexpected properties of time and space do, of course, engender a good deal of discussion in themselves; but the most important feature for us here is the direct and clearly traceable connection from a purely physical problem to a well-defined philosophical position, and even further, to an outlook, a posture, a general attitude to problems adopted and preached by physicists, philosophers, and others. As has already been suggested, operationalism represents an especially important subject for students today, because they connect it with behaviorism, and with a style of *human interaction* that goes well beyond science itself. A reading of Bridgman is in order here, and again we take up areas outside of physics wherever it is believed that the operational viewpoint has been applied. Behaviorist psychology is most frequently mentioned. At such times, a representative of this school from the psychology faculty is invited to speak and answer questions.

Summary and Conclusion

We have tried to call attention to the existing relations between ideas emanating from physics, methodologies in social science, and current problems, and the philosophical concerns of students. In the course taught at Richmond College for the past two years—open to science and nonscience majors alike—a conscious attempt is made to view selected subjects in physics in the context of these relations. It is too early yet for a conclusive evaluation, although anonymous written questionnaires have been encouraging. More significant is the fact that there are always some students who discover a special interest in one or another of the topics mentioned, and continue to pursue it with readings on their own, well beyond the scope or time limit of the course. As one of our aims, we hope to provide a general framework to be used by students of the social as well as the physical sciences in examining their own disciplines more critically and seeing them as part of a larger phenomenon. In this, we have had modest success. A number of students have claimed that their entire view of science and scientists has changed.

We return now to the question of the relevance of physics. In research, we know that the choice of variables and effects that are to be considered relevant is a most crucial and difficult one. Surely it is no less crucial in the field of education. Yet, with regard to the teaching of physics as a whole, we often overlook the strongest and most far-reaching effect: its impact on the problem of knowledge. If science is, above everything else, a search for the most secure understanding of phenomena and their interrelation, then the methods of science have to be examined and understood by every serious thinker, whatever his field may be. It is a matter of history that this has indeed been true of many of those involved with social and philosophical ideas—from Voltaire and Goethe to the present—in spite of the frequent assertions concerning the separation between the 'is' and the 'ought,' and other theories about disjoint realms of discourse.

When Skinner cites historical examples from physics and medicine, and then goes on to say, "The methods of science have been enormously successful wherever they have been applied. Let us then apply them to human affairs,"[17] we have a concrete example of how the relevance of physical science is actually understood by others, and how its ethos is diffused throughout the culture. When we see that social thinkers like Mumford, Marcuse, and Hannah Arendt,[18] to name only a few contemporaries—none of whom are especially interested in science for its own sake—feel the need of putting into their books at least one chapter on the historical or conceptual development of the physical sciences, then the long-range relevance of physics, and the manner in which it might be studied, become clearer. It was Marcuse who was quoted in the introduction as saying that a poet or writer has no real need to know about thermodynamics or the overthrow of parity; but in his own analysis of modern thought and society he makes repeated references to Heisenberg, Philipp Frank, von Weizsäcker, Galileo, etc.[19] Yet there is no contradiction here, for in one place he is clearly referring to specific results, while in the other it is the concepts and methods that force themselves upon him.

Relevance may sometimes appear as too superficial and opportunist a goal to be applied to the study of science. The idea of "art for art's sake" has put an end to many a good dialogue in modern times. And yet, is this phrase not simply a way of pointing vaguely towards purposes, hopes, and insights that are too complex and too much on the dim border of knowledge itself to be easily communicable? The present ferment in education presents new opportunities for further inquiry into these hopes and insights; from which something may also be learned as to the proper role of science on the campuses.

ACKNOWLEDGMENTS

I wish to thank Mr. David Hemmendinger for an ongoing discussion of many of the problems treated here and for suggestions on the improvement of this paper. I am indebted also to Mr. Joseph Kelley, a student, for calling my attention to the political science papers by D. Easton and S. Wolin.

Appendix: Partial Bibliography Used in a One Year Course

Many of the excerpts below are distributed to students with extensive annotation, while the technical aspects of scientific subject matter are based almost entirely on specially prepared notes. This bibliography does not reflect the order in which topics are introduced in the course, and is given only to suggest the type of materials we use.

A. GALILEO, NEWTON, AND THE SPIRIT OF CLASSICAL PHYSICS

Drake, S. (ed.) (1957). *Discoveries and Opinions of Galileo.* Garden City, NY: Doubleday. Contains the shorter original works. Especially worthwhile are "Letters on Sunspots" and "The Star Messenger," as superb examples of semiquantitative reasoning; also, "The Assayer" for a statement of the emerging modern philosophy of science and its relation to religion.

Brophy, J. and Paolucci, H. (eds.) (1962). *The Achievement of Galileo.* New Haven, CT: College & University Press. Chapter II, "The Copernican Controversy," contains a long excerpt from *Dialogue on the Great World Systems* and also the Letter from Cardinal Bellarmino to Foscarini stating the Church's position on the role of science in relation to 'truth.'

Koestler, A. (1963 [1959]). *The Sleepwalkers*. New York: Grosset and Dunlap, Chs. II, IV, XIII and IX. Highly readable discussions of Kepler, Tycho, and Galileo.

Mumford, L. (1970). *The Myth of the Machine*, Vol. II. New York: Harcourt, Brace, Jovanovich, Chs. 2 and 3. A very critical view of Galileo related to modern social problems.

Holton, G. (1970). "The Pentagon of Power." *The New York Times Book Review*, 13 Dec. A critical review of Mumford's book.

Butterfield, H. (1952 [1949]). *The Origins of Modern Science*. New York: Macmillan. For pre-Galilean background and post-Newtonian *Philosophe* movement.

Thayer, H. S. (1953). *Newton's Philosophy of Nature: Selections from His Writings*. New York: Hafner Publishing. Introduction to *Principia*. Newton's Letters.

Feynman, R., Leighton, R., and Sands, M. (1963). *The Feynman Lectures on Physics*. Reading, MA: Addison-Wesley. The chapter on gravity in Vol. I. Supplemented by instructor's notes on Kepler's laws, the law of gravity, and $f = ma$.

Bernoulli, D. (1965). "On the Properties and Motions of Elastic Fluids Especially Air." Reprinted in S. G. Brush (ed.), *Kinetic Theory*, Vol. I. (Oxford, England: Pergamon, 1965). Annotated and supplemented by instructor's notes on kinetic theory, heat, and energy.

Gillispie, C. C. (1960). *The Edge of Objectivity*. Princeton, NJ: Princeton University Press. On early energetics. Annotated excerpts.

Films by R. Feynman, in the series "The Character of Physical Law," obtainable from the Educational Development Center, Newton, MA. See especially the films "Seeking New Laws" and "The Great Conservation Principles."

B. Relativity and Related Philosophy

Einstein, A. (1949). "Autobiographical Notes." Contained in P. A. Schilpp (ed.), *Albert Einstein Philosopher-Scientist* (New York:

Harper and Row, 1949). Technical sections deleted. Instructor's notes on technical content of the theory of special relativity. Examples, problems, paradoxes.

Baker, A. (1970). *Modern Physics and Antiphysics*. Reading, MA: Addison-Wesley, Chs. 4–10.

Einstein, A. (1934). *Essays in Science*. New York: Philosophical Library. The essays entitled "Principles of Research," "On Scientific Truth," and "On the Method of Theoretical Physics."

Mach, E. (1968). "The Economy of Physical Theory" and "Explanation in Physics." Contained in B. Brody and N. Capaldi (eds.), *Science: Men, Methods, Goals* (New York: W. A. Benjamin, 1968).

Bridgman, P. W. (1928). *The Logic of Modern Physics*. New York: Macmillan, 1–25. A well-known introduction to the idea of operationalism.

Carnap, R. (1955). "Logical Positivism." Contained in M. White (ed.), *Age of Analysis* (Boston, MA: Houghton Mifflin, 1955).

Popper, K. (1968). *Conjectures and Refutations: The Growth of Scientific Knowledge*. New York: Harper and Row, Ch. III. Good discussion of essentialism and instrumentalism.

C. RELATION OF PHYSICS TO OTHER FIELDS

Descartes, R. (1954). *Discourse on Method* (complete) and selections from F. Bacon, *Novum Organum*. Contained in S. Commins and R. N. Linscott (eds.), *Man and the Universe* (New York: Washington Square, 1954).

Burtt, E. A. (1932). *Metaphysical Foundations of Modern Science*. New York: Anchor. Selections.

Comte, A. (1954). "The Positive Philosophy." Contained in Commins and Linscott (1954).

Mill, J. S. (1961). *Auguste Comte and Positivism*. Ann Arbor, MI: University of Michigan.

Skinner, B. F. (1965). *The Science of Human Behavior*. New York: Free Press, Chs. I–IV.

Brown, R. (1961). "The Congruity Model." Contained in R. Brown, E. Galanter, E. Hess, and G. Mandler, *New Directions in Psychology* (New York: Holt, Reinhart, Winston, 1961). An example of the application of physical reasoning to social science.

Dubos, R. (1965). "Science and Man's Nature." Contained in G. Holton (ed.), *Science and Culture* (Boston, MA: Beacon Press, 1965).

Marcuse, H. (1965). "Remarks on a Redefinition of Culture." Contained in Holton (1965).

Bronowski, J. (1956). *Science and Human Values*. New York: Harper.

Brain, W. (1965). "Science and Antiscience." *Science* 148, 192.

Whitehead, A. N. (1925). "Religion and Science." Contained in A. N. Whitehead, *Science and The Modern World* (London: Macmillan, 1925)

[13]

Philosophy in Physics and Physics in Philosophy

ABSTRACT: A single course exhibiting the relations between philosophy and physics is desirable, especially for nonscience majors, for several reasons: (1) specific high-level physics theories have philosophical suppositions and implications; (2) epistemological and ontological questions arise in the process of validating scientific theories generally; (3) paradigms in physics are often illuminating for disciplines like the social sciences, which seem at first to be remote from physics. A rich classical and modern literature on these relations is accessible to students and can be taught either by one person with interdisciplinary interests or by a team. [—A.S.]

The purpose of this paper is not to belabor the well-known historical relationship between philosophy and physics, but to justify, on grounds of educational values, as well as proximate pedagogical and institutional requirements, the proposition that the two subjects can and ought to be studied within the framework of a single course.

The simplest argument for the conjunction was given by Philipp Frank: "Everyone who is to get a satisfactory understanding of twentieth-century science will just have to absorb a good deal of philosophical thought." The connection of physics to philosophy is thus qualitatively different from its relation to literature or socio-political history. Not just cultural broadening is involved here, but a *necessary element* in the study of physics itself. A course in philosophy and physics restores to the latter that vital component of which it has been deprived during the past century of specialization and departmentalization of knowledge.

This paper was written in 1976 for a publication planned by the American Physical Society Forum Committee on Courses in Physics and Society, a project never completed. It will appear in a forthcoming volume edited by Calvin S. Kalman.

Broadly speaking, there are three avenues through which the intercourse between physics and philosophy takes place: 1) The philosophical import and presuppositions of specific high-level theories such as quantum mechanics, relativity, and statistical thermodynamics; 2) the more general epistemological questions involved in the validation of theories, and the ontological problems raised by the meaning of the results; and 3) the fact that established paradigms do, at times, have a long range influence on philosophies apparently unconnected with natural science, e.g., social thought of various types. The teaching of physical principles and techniques that largely ignores or slights the first two aspects does not do justice either to the subject or to the students; while omission of the third aspect severs an important *intellectual* link between science and society, thereby strengthening the trend toward what Jürgen Habermas and his school have called the "orientation to instrumental action." The problem is how to incorporate these aspects in a single course, and also do justice to the technical content. Neither facts and formalisms without reflection, nor mere reflection without facts and formalism.

The subject itself points the way to the solution: A focus on a few appropriately chosen major theories, their treatment in depth—with due attention to all three of the enumerated aspects—should be the *modus operandi*. Space limitations on this brief summary preclude discussion of anything more than one example, which must serve as the paradigm case: special relativity. It has the merit of the highest 'philosophical cost-effectiveness factor,' i.e., the ratio of philosophically interesting questions raised and conclusions reached to the mathematical work required.

Inevitably, the relativity of time order of point-events and other startling features of the subject have an effect like an awakening from 'dogmatic slumber.' When this happens, it is advisable to backtrack over the basic derivations, and show—perhaps by an alternative route—the logical connection between the conclusions and the original, seemingly innocuous and simple-minded idea of *operational definition*. The reiteration of this concept, in the study of the usual problems, paradoxes, and experiments, reinforces the connection between the *epistemological* innovation and the strictly *physical* predictions and verifications. Thus, the first aspect of the relation of physics to philosophy is dealt with at the beginning, center, and end point of an important and well established theory.

Now the formulation and its meaning must be analyzed more generally: How justified was the adoption of Einstein's approach over Lorentz's? What is the role of operational definitions in physics and science as a whole? Does the development of relativity fit the hypothetico-deductive method, or perhaps Lakatos's more complex "logic of discovery"? And what is a "theory" anyway? What relation to reality does it bear? At this point, even the most philosophically sensitive textbooks must be supplemented with additional reading: above all, the relevant essays by Einstein himself; then the physicist-philosophers like Mach, Bridgman, and Frank. The last of the questions mentioned above is of special significance, forming a continuous thread from the Galileo-Bellarmino clash about truth and the limits of science, to what today is called the instrumentalist-realist controversy. Here, Karl Popper's well known essay, "Three Views Concerning Human Knowledge," (in *Conjectures and Refutations: The Growth of Scientific Knowledge*, 1968) is most enlightening.

In regard to the third avenue linking physics and philosophy, historical evidence is quite explicit and good reading easy to provide: William James and John Dewey make numerous references to specific developments in the physics of their time, and use the results to support their pragmatist philosophies. For James, the cardinal fact was the final breakdown of belief in exact, universal laws of nature, which turned him into a staunch supporter of the instrumentalist theory of knowledge (*Pragmatism*, 1907). This was even more forcefully expounded, and applied across the board from education to morals, by Dewey—who made extensive use of the emergent quantum mechanics of the late 1920s, in which the operational viewpoint, modeled on Einstein's, proved its power again (*The Quest for Certainty*, 1929). In addition to pragmatism, positivism, and kindred twentieth century thought, there is the equally interesting relation to certain theories of psychology: Historical essays such as Sygmund Koch's fascinating sketch of behaviorism can be combined with primary sources like B.F. Skinner's paper, "The Operational Analysis of Psychological Terms."

The study of these influences should not, however, imply that they are necessarily appropriate or beneficial. Instead, this part of the discussion can make contact with the students' awareness of recent dissatisfaction with many such adaptations in the social sciences; the question is then raised as to how conceptual and methodological transference is to be usefully carried out and rationally evaluated.

Can such a course be taught successfully under the conditions of present day institutional reality? If the teacher possesses sufficient experience in both areas, there is no reason why one person could not do the job alone. If it is team-taught, however, problems of faculty credit apportionment may arise: When the financial situation was good at CUNY, both instructors received full credit; now, this is unimaginable.

Enrollment has not, in general, been a problem, though the figures for my courses varied between thirty and six. One does have to make an effort to inform and encourage those who could truly benefit. But this effort is the heart of the matter, because many students cannot by themselves imagine how relevant these questions can be to the problems in their own fields. When the course is designed primarily for nonscientific people, the technical content is elementary, and therefore, as at Richmond College, natural science majors may have to be denied credit. At Boston University, however, an interesting and successful senior course is taught by Prof. Abner Shimony, presupposing a fair amount of physics and mathematics; thus, a relatively small number of physics and philosophy majors enroll. But for these groups, such a course is unusually valuable.

[14]

Review of V. Weisskopf's
Knowledge and Wonder

It may well be asked: What advantage is there if an elementary, qualitative science book for the general reader happens to have as its author not merely a competent scientist and a good writer, but one of the outstanding physicists of our time—a participant in the 'heroic age'? What follows is largely an answer to this question, though its main purpose is to point out three specific features of this book—not often encountered—that potential users will probably wish to take into account.

First, the overall framework exemplifies what may be called the *cosmological interest* (in contrast with two other main types—the *disciplinary* and the *methodological*). Initially, the cosmological outlook appears almost naïve; for its goal is nothing less than to *tell what the whole world is like*— from the "very large" to the "very small" to the "very complicated"—and to see, if possible, "the *one* in the many." This is, of course, the most ancient interest of all, Lucretian in a sense, only recently revived and raised to new levels. And Weisskopf's book has the distinction of being one of the first of this now growing genre.

It begins with stars and galaxies ("our place in space"), then quickly narrows the view to the planets, the Earth, geology, and the "clocks" of radioactivity ("our place in time"). Then the author pauses to make some typically philosophical remarks in typically simple language: We have "set the stage," he says, for the main drama—which is the *forms* of things and their *regularities*. There follows a discussion of basic forces, waves, electricity, light, and quantum theory—all leading up to *chemistry*, where the book is especially effective in explaining and depicting the different types of bonds. Finally, the culmination: life, DNA, "the master plan," and *evolution*. It is in this second half of the new edition that the main revisions and expansions appear.

This book review appeared in the *American Journal of Physics* 49 (1981), 605.

The very last topic is particularly important because it is implicit throughout the "drama," and is a major element of the "new cosmology": a universal evolutionary perspective, supported by the spectacular developments in astrophysics and microbiology.

Naturally, the distinction drawn here between the cosmological and methodological approaches is one of emphasis only; it would be a poor writer on science who failed to address to some extent the question of methodology, of 'how we know'; and Weisskopf goes further. For although he does adopt the evolutionary framework, his caution in the matter may be recommended to others: "This," he says, is "how we believe today it may have happened."

The second feature to be noted has to do with intellectual style. The book lives up to its title more than one might expect; for "wonder" is indeed a dominant mode, deliberately excited. "How could we imagine," asks Weisskopf, "that electrons will find their way into exactly the same orbits when the atoms are evaporated" (after being condensed to a solid), so that the frequencies emitted are the same as though solidification never took place? "It is as if the planet Venus, after being knocked out of its orbit in some collision with another star, should suddenly glide back into its previous orbit when the star has gone." The issue is simply the identity of matter: Why is a piece of gold the same regardless of how we obtain it?—a rather basic question, but not one likely to occur to the average reader. Weisskopf, however, insists that *wonder* is an appropriate response to everyday reality, common objects—not only black holes.

Finally, there is one aspect of a master's treatment that sometimes surprises us: Not only might he say some things better, he might actually be *more correct,* even on such an elementary level. As a nontrivial example, consider one very common description of nuclear *fission.* This, or an equivalent version, may be found in comparable volumes by experienced textbook writers: The stored energy is *nuclear* energy; the force that holds nuclei together, the *nuclear force,* is much stronger than the electrical forces that hold molecules together; therefore the nuclear rearrangements release much larger amounts of energy than chemical rearrangements—such as oil burning or the explosion of TNT. Now, this from Weisskopf, "If one could split a nucleus in such a way that the halves were separated by an amount which, though small, would be greater than the range of nuclear forces, the two parts would no longer stick together but fly apart, driven by the *electrical repulsion*" (emphasis mine).

Aside from the vividness and precision of the second version, it has the merit of avoiding an obvious puzzle: How does this strong force that "holds nuclei together" suddenly change character and blow them apart? But the advantage goes beyond pedagogy. The increase in energy per nucleon in the very heavy species—and thus their instability—is, after all, due mainly to the Coulomb term (in the semiempirical mass formula). The large energy involved here results from the small nuclear *distances*. Weisskopf does cut through much detail, but in the end his 'laymen's explanation' is more in line with quantitative theory, and does not place *fission* and *fusion* in the same category.

The comparison is not intended to be invidious; strictly speaking, of course, the other version is not 'wrong.' However, this book shows once again that when it comes to knowledge, especially of 'simple things,' no scientist is 'overqualified.'

[15]

Philosophy of Science in Teacher Education

ABSTRACT: Among teachers, preconceptions about *what science is like* involve questions that are metaphysical, methodological, and historical. Because many preconceptions derive from older philosophies of science and are affected by the recent historical philosophies, a problem is posed for teacher education: how to communicate the insights of the new understanding to science teachers; how to challenge preconceptions, despite continuing disagreement among scholars on basic issues. As an attempt to address this problem, a nonsystematic, historical treatment (currently used at CUNY) will be outlined. It starts from a simple core-structure description of the growth of science, and proceeds by successive approximations (attempting to fit history) to illuminate significant points in philosophy of science. Learning takes place dialectically—in discussion of *alternative* conceptual schemes, and therefore alternative interpretations of science. [—M.E.]

The philosophical preconceptions held by science teachers have aroused greater interest recently. I can certainly confirm that such preconceptions do exist, and that epistemologically they are, in general, of an empiricist nature. But there is more. Although my purpose here is not to offer statistical data, it may be useful to preface the main subject of this paper by indicating briefly the general character of these preconceptions as they appear to me in the course of teaching philosophy of science to groups of in-service high school science teachers in New York City.

Epistemological empiricism surfaces often in such remarks as this: "I believed that anything stated in a science text was a fact unless otherwise identified. . . . Theories had been taught to me as fact throughout high

This paper was presented in 1987 at the Second International Seminar on Misconceptions and Educational Strategies in Science and Mathematics, held in Ithaca, New York. It was published in the conference proceedings, Joseph D. Novak, ed., Ithaca, NY: 1987, 163–176.

195

school, college, and graduate school." However, many teachers also evince a thoroughgoing ontological realism of two different kinds—in regard to concepts and in regard to laws. That is, such constructs as inertial mass are assumed to be pre-existing and *discovered*; while in regard to laws, one often finds that these also are taken as pre-existing, discovered, *and* absolute. Thus, another teacher was very skeptical that 'continuous creation of matter' (as it appears in 'steady-state' cosmology) could possibly have been proposed by legitimate scientists, even if the amount 'created' was below the threshold of detection. "It violates conservation of energy," she insisted. When it was pointed out that big bang theories involve a similar, perhaps more drastic notion, this teacher confessed that, indeed, she had always had trouble understanding the big bang for the same reason—which just shows that conservation may be even more difficult to *un*learn than it is to learn.

Of course, among philosophers of science, realism (of a certain kind) is today a most praiseworthy stance; and I do not mention it here as a fault. What I wish to highlight is the unlimited and unreflective way in which it is often held by teachers—certainly without awareness of alternatives or of the problems to which it may lead. Moreover, the viewpoints I have just described suggest that such preconceptions are not the result of any sort of consistent, though perhaps outdated, philosophy. What I see instead is an odd assortment of fragments: snatches of empiricism, isolated *metaphysical* principles, and generalizations representing varying degrees of 'construction'—all taken as *fact*, and held as core-concepts that exert influence on newly encountered knowledge. More systematic studies of all this would, I think, be very welcome at this point.

My object in this paper is two-fold: first to indicate what I believe is an important *continuing* reason for this problem of incoherent preconceptions; and then to outline one way, the way I have used, of confronting the problem through formal course-work with teachers.

Concerning the first point, and putting the conclusion before the evidence, I suggest that a major source of the teachers' and the public's 'misunderstanding' of science is *ourselves*; that is, scientists, textbook writers, popularizers, teachers of students, and teachers of teachers—in short, most of the academic community. This thought will not be shocking to those working in the area of misconceptions, and so I will not belabor it. But to make sure we have before us a vivid image of what actually takes place today in the realm of philosophical-scientific education, let me give one especially worthwhile example. It is taken from writings that are deliberate attempts to explain to the educated public what science is like, among the best received such writings in our time, and the work of a well-known and acclaimed scientist—Stephen J. Gould.

I quote first from a widely reprinted essay called "Evolution as Fact and Theory":

> Facts are the world's data. Theories are structures of ideas that explain and interpret facts. Facts do not go away while scientists debate rival theories for explaining them. Einstein's theory of gravitation replaced Newton's but apples did not suspend themselves in mid-air pending the outcome. . . . In science "fact" can only mean "confirmed to such a degree that it would be perverse to withhold provisional consent."[1]

Notice how thoroughly empiricist this is (as is the rest of this essay). Facts are first of all observational data, like falling apples. Later Gould includes also what he calls "confirmed" inferences from direct observation; but neither the observations nor the inferences are dependent on theory in any way. And so, we are told further on, the inferences are "no less secure" than direct observation.

But now let me take an excerpt from another essay of Gould's,[2] which actually appeared earlier; it is titled "Validation of Continental Drift":

> I remember the a priori derision of my distinguished stratigraphy professor toward a visiting Australian drifter. He nearly orchestrated a chorus of Bronx cheers from a sycophantic crowd of loyal students. . . . Today, just ten years later, my own students would dismiss with even more derision anyone who denied the evident truth of continental drift. . . . During the period of nearly universal rejection, direct evidence for continental drift—that is, the data gathered from rocks exposed on our continents—was every bit as good as it is today. It was *dismissed* because no one had devised a mechanism that would permit continents to plow through an apparently solid oceanic floor. In the absence of a plausible mechanism, the idea of continental drift was rejected as absurd. The data that seemed to support it could always be *explained away* (emphasis added).

Continuing the story, Gould tells us that with some new data and a heavy dose of "creative imagination," we have now fashioned a new theory of planetary dynamics:

> Under this theory of plate tectonics, continental drift is an inescapable consequence. *The old data from continental rocks, once soundly rejected, have been exhumed and exalted as conclusive proof of drift*. In short, we now accept continental drift because it is the expectation of a new orthodoxy (emphasis added).

The rest of the article describes some of these data that were previously rejected but later exhumed and reinterpreted. Finally, toward the end, Gould spells out the lesson: "The new orthodoxy *colors our vision of all data; there are no 'pure facts' in our complex world*" (emphasis added).

In viewing these two essays side by side, we ought to note the element of *consistency* as well as the obvious divergence of meaning. In both cases the 'facts' did not 'go away.' But, it turns out, there are different senses of the phrase 'to go away.' In the first essay, Gould-the-positivist drives home the major point about the *givenness* of facts, their rootedness in the nature of things, by pointing triumphantly to the unimpeachable assertion that no one has yet seen an apple fall up from a tree. In the second essay, Gould-the-Kuhnian shows by means of some dramatic recent history precisely *how*, in a different sense, facts can indeed 'go away'—not by nature changing its ways, not necessarily by the discovery of error in the process of observation, but by being 'explained away,' or ignored, or simply 'dismissed' as not sufficiently significant.

When writing about evolution, Gould is what Hilary Putnam calls an "externalist"—facts are external to theory. But in regard to continental drift, he is an "internalist"—the facts, if not completely constituted by theory are, in a way, *coaxed* by the theory, and given their meaning and significance by theory. My point here is not that Gould has to be wrong somewhere (though I will return to this matter) but that, like many other scientists, on philosophical questions he is just plain *careless*; and carelessness of this sort, even when practiced with a most engaging style, can only leave confusion in its wake—*incoherence*. This example, I think, when taken with much other evidence of the same sort, tends to place in a different light the oft-repeated complaint about the public's 'misunderstanding' of science. At least in regard to such things as 'fact' and 'theory,' or comprehension of structures of ideas as wholes, or the capacity to gauge meaning, it is probably as much a problem of *misteaching* as misunderstanding.

But *why* the misteaching? Well, as we know, scientists normally feel that the ground they stand on is their professional, technical achievement, not their more general, philosophical comments. What we see here is the divergence of two different interests: The interest of education, even in science, is not entirely the same as the interest of the associated professional community or the discipline. The interest of the discipline may at times countenance not just philosophical carelessness but even a degree of philosophical *opportunism*. On the other hand, the interest of education includes (as we often say) conveying a coherent picture of what science is like. If this is accepted, then serious attention to philosophy of science becomes, for all levels of teaching, an obvious *desideratum*.

Assuming this much as a goal, I now shift abruptly to the second task of this paper—the description of an implementation designed for in-service teachers that bases itself on the primacy of the historical viewpoint over the analytical. I take the scientific-philosophical education of the teacher *as an end in itself*, without regard to how that education may be used in the teacher's own work. In defense of this, I offer two considerations: 1) that the teacher corps in an open society is, from an intellectual point of view, a significant sector of that society, whose opinions on science and culture are important *as such*, and not merely as means toward more successful teaching of subjects; 2) that even when our aim is to use philosophy of science to improve the teaching of science itself, it is still undesirable, if not inconceivable, that teachers employ and transmit the insights of scholarship without themselves consciously absorbing these very insights. Therefore, in what follows, I make no suggestion that the method described, or any part of it, can be directly applied by teachers in their own classrooms.

The approach is based on the following features:

A graphical scheme representing the structure of scientific fields—an adaptation of William Whewell's "induction tables"—applicable (within limits) regardless of specific philosophy.

Discussion of particular scientific theories, including their technical aspects.

Application to scientific or social or educational *controversies*.

While the last two features are by no means secondary, the remaining discussion is devoted mainly to the first—because much of what I want to say about the other two can be said in that context, along the way. And toward the end, I will return to the more general implications.

We start with a very unsophisticated 'three-tier' diagram of a scientific field or sub-field (Fig. 1). Although we know that the distinction between empirical and theoretical laws is not sharp or universally recognized, and neither is the separation of 'observables' from 'non-observables,' nevertheless, in the educational context I regard such objections as of second order. For those who think this so oversimplified that it has little value, let us recall first that the basic terminology of the scheme is widely used in science textbooks on the college level, and that this specific hierarchy has been implicit in the writings of philosophers from William Whewell to Ernest Nagel.[3] So it is something relatively familiar. Second, precisely because it is crude, it is not *this* picture that occasions major differences between the modern philosophies; we can use it in discussing positions that range from extreme empiricism to at least a moderate constructivism. Third, and most important, we only *begin* with this diagram. Later, other features are added, and it does, of course, get more complex.

I have found that even this crude scheme, when thoroughly discussed and illustrated, already introduces an important change into people's views. For many, it alters the landscape of science from one of flatness—all real science is fact—or from a hierarchy based only on degree of confirmation—fact, theory, hypothesis, speculation—to one where there is some depth (in several dimensions).

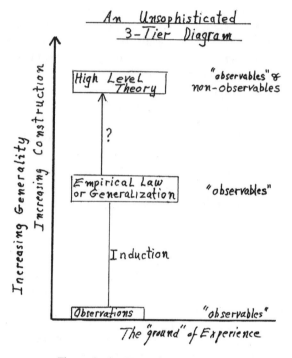

Figure 1. An Unsophisticated 3-Tier Diagram

However, the only way to really teach this is to move quickly to a number of well-known exemplars: In physical chemistry, with Boyle's law at the middle level, the kinetic theory is high level. In cosmology, with Hubble's law in the middle, the big bang theory is at the top (Fig. 2). In mechanics, if Kepler's ellipses and Galileo's law of falling bodies are at the intermediate level, Newtonian theory at the apex unifies these two disciplines, as the textbooks say (Fig. 3).

Many features of science can now be discussed by reference to these diagrams, with suitable illustrations from the exemplar cases. The *lower* upward arrow often, but not always, stands for relatively unproblematic induction—like extrapolation and interpolation. The movement of historical development is generally (but not always) upward. The direction of explanation, and deduction, is typically downward. And a major difference between the higher levels and the realm of merely empirical laws is that the former characteristically introduce concepts and processes that are not directly observable, sometimes even in principle *un*observable—though it should be noted that Nagel, for example, declined to make this the demarcation criterion between experimental and theoretical laws.

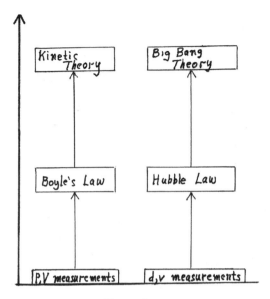

Figure 2

Obviously it is possible to draw such diagrams for segments of optics, geology, electromagnetism, and many other fields. And in doing so one is led to deal with scientific disciplines *wholesale*, and in a comparative way—often a new and dizzy experience for those narrowly trained. This, of course, cannot be done without assuming, or imparting to the students, a certain amount of knowledge, including historical knowledge, of some of

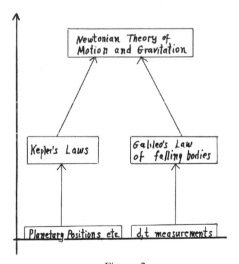

Figure 3

these fields. It is at this point that my second feature—dealing with science itself—comes in, and does indeed take up at least 50% of the time.

One benefit is that students of certain sciences, like biology, where the distinction between empirical and theoretical law is not often so clear, have a chance to refocus on a field like physics, where it is far more prominent. Hugh Helm has pointed out that many beginning students do not recognize the difference between definitional or tautological laws and laws of nature.[4] To this I can add that many high school science teachers do not recognize even obvious differences between experimental and theoretical laws, or between induction and deduction, and have trouble with other distinctions of the more abstract kind. This simply tells us that misconception or preconception, whose origin may sometimes be profound, are mixed also with plain ignorance, especially in regard to general concepts concerning science as a structure of ideas.

But now the point: After some work with the three-tier diagrams, most students begin to see that the situation cannot be as simple as that. Should Boyle's Law and Keplerian orbits really be at the same level? The former seems to fit well the Baconian prescription—collect data, and discover patterns. But it is surely debatable whether 'seeing' the elliptical orbit in the data of planetary positions is at all in the same category. In fact, this was the subject of the famous nineteenth century debate between William Whewell and John Stuart Mill, with Whewell arguing for what we now call the more constructivist position.

Historically speaking, there were real choices in moving from the ground of observation to what, from our present vantage point, is the intermediate level; and for this, a separate three-tier diagram can be drawn. With diurnal motions, planetary positions, etc., at the bottom level, there were three different candidates at the highest level: the Ptolemaic system, the heliocentric, and Tycho Brahe's compromise system. Between observations and the major 'world systems,' we would place the empirical generalizations accepted at the time—maximum elongations of the inner planets, the retrograde motions, etc. From this point of view, and our present hindsight, all candidate systems are imaginative constructs. Therefore, the first change in the 'unsophisticated' diagram is to allow for bands, or many levels, instead of just three tiers (Fig. 4). Elliptical orbits and Boyle's Law might still be somewhere in the middle band, but with the former higher than the latter.

The need for other kinds of corrections, or refinements, is even more glaring. The ground level of the cosmology diagram contains such 'observations' as distances and speeds of galaxies. But since these are not in fact directly observed—another point calling for scientific discussion prior to the philosophical—some structure must be introduced here as well, to account for *background theories*. And so, depending on how far one wishes

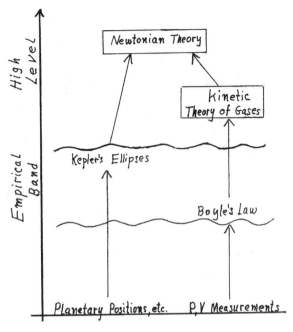

Figure 4

to probe, the diagrams can indeed become very cluttered. The heuristic point to all this is not so much in 'getting it right'—although, within bounds, that must of course be the goal—but in the discussions one is forced to go through in deciding where to place a particular element.

We have to consider what its relation is to other elements, what sort of concepts it involves, which is the direction of deduction, what its historical development was, and so on. Clearly this is an *exercise*, a *critical* exercise— an exercise in a particular kind of concept mapping, which shares therefore many of the virtues and shortcomings that are already known in regard to such exercises.[5] But in contrast to some other kinds of mapping, the basic format and principles here are clearly laid down at the start, and are taken from the historico-philosophical disciplines relating to science. It should therefore be no surprise that such diagrams are a very old thing; in a somewhat different form, William Whewell called them "induction tables." He drew up large, complicated ones, paying particular attention to history, and took them very seriously not only as a means of *understanding* science, but even as a way to—"truth."[6]

Let me turn now to philosophies as such. The diagrams certainly do not depict everything; we cannot, for example, show any difference between realist and instrumentalist viewpoints (usually discussed in such a

course), for that would require some portion of the diagram to refer to
'reality'—and locating 'reality' on this plane is not easy. But let me point out
those things the diagrams can do. It may seem at first that the very structure of this scheme—everything proceeding from the 'ground' of experience, and the resemblance to Whewell's thinking—already has built into it
a philosophical bias. Perhaps, but if so, the bias can be overcome. We do
start with a number of variants of positivism (which, by the way, is certainly not dead); but then we go on to the Popperian viewpoint, and to the
Kuhnian version of what is now called "the new philosophy of science."

The typical positivist concern, simply put, was to *verify*, or make
secure, the valid inductions represented by the upward trend in these diagrams, *and* to screen out those inductions which could not be so secured.
Whewell's way of doing that—what he called the "consilience of inductions" (and what we might call convergence) was simply that *the more*
upward arrows converging on the same high level theory, the better. So if

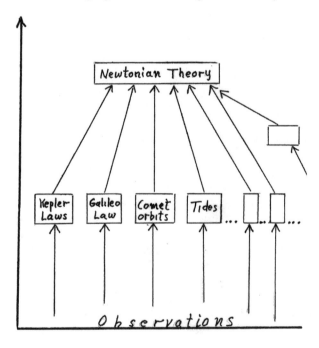

Figure 5

to the two inductions of Fig. 3 we add orbits of comets, satellite orbits
around Jupiter, tides, the oblateness of the earth, and much more, we then
have the paradigmatic case of consilience (Fig. 5). And on a diagram such
converging arrows do look impressive.[7]

An alternative method was to concentrate on quality rather than quantity. Thus, the approach of positivism in our century was to examine meticulously each upward arrow and develop, whenever possible, special procedures to make that arrow more solid, more reliable. Examples of this would be operationalism or the quest for a pure observation language. On a diagram, we might just draw a thicker arrow to indicate security.

Finally, among the positivists we had also the famous radical branch which included Ernst Mach. For them the problem was that regardless of the number of arrows, or how secure they are, high level theory usually contains *constructed concepts* not accessible to direct measurement—which might well be fictions. Their bold solution, doing away with high level theories, or at least some of them, means that the top tier is simply crossed off.

When we get to Karl Popper the discussion becomes particularly interesting. On the one hand, Popper liked to emphasize his difference from the positivism of his time, but on the other, some philosophers have continued to include him within that general designation. Let us see how, on these diagrams, both the differences and the similarities appear.

The main distinction, of which he is so proud, consists in the now-famous characterization of the upper arrow as a 'conjecture,' a guess. Inspired by Einstein, and the difficulties inherent in contemporary versions of the positivist program, Popper concluded that "induction is a myth."[8] One way of indicating this graphically is to replace the solid arrow by a broken (dashed) arrow—*conjecture* replaces *inference*.

But Popper's resemblance to positivism can be seen in his partial return to the consilience of inductions. Quality *and* quantity are both emphasized by him, couched in new terminology and a different interpretation, and presented as an improved version of the hypothetico-deductive method: The conjecture must give rise to 'interesting,' 'risky,' or 'improbable' deductions—denote this by wiggly lines to the lower levels—which, if *corroborated*, yield what in the older language would have been inductive support. The risky deduction, whenever successful, elicits a solid upward arrow, which is no longer just a conjecture (Fig. 6).

The resulting picture looks very much like the old consilience, with the following changes: First, the new analysis relies much more on corroborations *after* the initial conjecture than on the historical path—we call this the dominance of the context of justification over the context of discovery—and second, the corroborations that really count must be of a certain kind.

It is important to emphasize at this point that such a use of diagrams, in themselves, or as part of a purely analytical discussion, would only add to mystification. Although there may be other influences, the various philosophies are strongly related to events in science itself. Twentieth century positivism cannot be understood without some idea of the rise—in the late nineteenth and early twentieth century—of new sciences and concepts

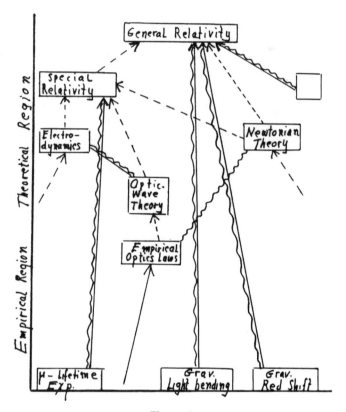

Figure 6

outside the Newtonian framework, and of the resulting collapse (as ultimate explanation) of the best-corroborated high level theory in all history.
Nor can current Popperian and Kuhnian themes really be understood
except as a response to the twentieth-century revolution in physics. For this
reason, a historical account of scientific change, utilizing the more recent
works in this area,[9] is, I believe, the best way to attempt this sort of teaching with any group of people who, for the most part, are not philosophically inclined.

As a final illustration of the use of the diagrams, consider now Kuhn's
well known reaction to both positivism and the Popperian viewpoint.
Graphically, this is indicated by a series of thick, looping arrows downward
from the higher levels (Fig. 7). These represent not only the 'theory-ladenness' of observation, but also the effect of theory on methods, on standards,
on problem choices, etc. It is important to distinguish sharply these downward arrows from any kind of deductive inference in the hypothetico-
deductive procedure. (Here a degree of graphical consistency is called for.)

Deductions lead typically to particular testable observations, which either pass or do not pass the test. To distinguish these verbally from the Kuhnian

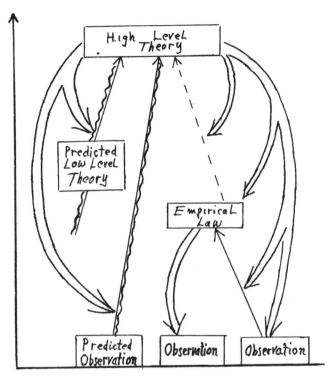

Figure 7

feedback loops, I use for the latter the term 'reverse induction' or 'downward induction.' In this way, I mean to highlight the *generalizing* nature of such feedback.

Consider Gould's example. Before plate tectonics, *no observation* could support continental drift because such drift was *known* to be impossible. That is a very general kind of downward inference concerning evidence. One result of the success of the special theory of relativity was to legitimate operationalism in quantum theory. That is a very general influence on *procedures* in research. And, of course, the hypothetico-deductive method itself acquired its articulated and acknowledged status largely as a result of the success of Newtonian theory.

The new picture—which, as a whole, can be called a Kuhnian "paradigm"—now has the kind of look that does seem to reflect our verbal descriptions. The feedback loops make it more dialectical, more self-contained, and more like what we today call constructivist. Naturally, this does not take the place of reading Einstein, Bridgman, Nagel, Popper, Kuhn, Toulmin, and others; in the end, the diagrams must become merely symbolic or mnemonic aids.

However, in regard to the value of using alternative philosophies of science, I should like to make one comment à propos of a paper by Joseph Nussbaum (Nussbaum 1983). He pointed out, in the *context of research* on students' conceptual change, that there are significant differences between the post-Kuhnian philosophies, and that these differences must be taken into account. Everything I have said so far surely supports the value of attention to philosophical differences and alternatives; but when it comes to teaching itself (as distinguished from research) I should like here to raise a flag of caution. I am sceptical that such differences as exist, for example between Kuhn and Lakatos, can provide any fruitful lessons. After much discussion, Kuhn thought that "Lakatos's position is now very close to my own."[10] And Feyerabend called Lakatos's philosophy an "anarchism in disguise."[11] Even the protagonists do not agree on many of their differences. My experience has been that it is a very demanding task just to convey clearly the significance of the distinctions between the *main* branches—that is, classic positivism, Popper, Kuhnian constructivism, and the ontological schools—even when the students are mature adults, teachers of science, diligently trying to improve their grasp of the issues. For that reason I prefer to take one or two representatives of each school, keep the diagrams as simple as possible, and concentrate on the relation between the philosophy and science itself.

But a fair question to ask at this point is what sort of results are we to expect from such studies, beyond the general feeling that the widening of horizons is good for everyone. In partial reply, let me return to Gould's writings and to the role of controversies in this approach.

In the essay, "Evolution as Fact and Theory," Gould's aim is to convince readers that evolution is *both* fact and theory, and that those who now emphasize the word 'theory' are improperly, deviously attempting to cast doubt. Gould's response, and that of many others, is to separate the *basic proposition* (Darwin's "descent with modification") from the question of specific *mechanisms—and*, speaking in his positivist mode, to pin the label 'fact' on the basic proposition. This argument is now very familiar, and it all but *indicts*—for ignorance or worse offenses—anyone who fails to treat the basic proposition as 'fact.'

But after weeks of using these diagrams in a comparative manner, and having seen family resemblances between laws and theories across disciplinary lines, it becomes possible for some to view the fact/theory controversy in alternative ways—ways which often reflect a deeper grasp of the process of science *and* at the same time a deeper insight into alternative modes of education.

As an example, consider this: William Whewell also regarded a complex, indirectly established proposition as *both* fact and theory; but he was referring to something quite different from Gould's (in this case) more positivistic *dissection*: "All attempts," says Whewell, "to frame an argument by the exclusive or emphatic appropriation of the term *fact* to particular cases, are necessarily illusory or inconclusive."[12] Why? The answer, in his work, received special emphasis:

> The distinction of *fact* and *theory* is only relative. Events and phenomena considered as particulars which may be colligated [i.e., subsumed] by induction, are *facts*; considered as generalities already obtained by colligation of other facts [i.e., induction from them], they are *theories*. The same event or phenomenon is a fact or a theory, according as it is considered as standing on one side or the other of the inductive bracket [on our diagram, the arrow].[13]

Although here we are listening to a voice from the nineteenth century, which on many other points is now outdated, the above statement was actually ahead of its time, and is not likely to raise opposition from many philosophers of science today—since modern examples of what he is saying are easy to find. When, all over the world, calculations are carried out for atomic phenomena, quantum effects are of course taken for granted—treated as facts. But when Aspect and his collaborators performed their celebrated experiments in 1982 (using Bell's theorem), quantum mechanics was in the full sense a *theory*, pitted against other possible theories (hidden variables).[14] In other words, it all depends on what the goal of the inquiry is.

But it is in education that the implications of this last point have the greatest scope; for education often *has* a number of distinct goals. If the aim is to train people in an existing scientific paradigm, then everything that is *at present* well established is 'fact.' If, on the other hand, the goal is to understand *inquiry* more generally, and the significance of the different kinds of *products* of inquiry, then it follows just as surely—whether we are looking at relativity, or at the basic proposition of evolution, or at Newton's laws, or at anything else—that inductions or conjectures (upward and downward), not just the paradigmatic deductions, ought to be at the center of attention. And then all these major achievements of science are indeed 'theories' about which one can argue.

From this perspective, and aside from the very different import of the two Gould essays I mentioned, it is possible to see that even the first essay (the positivistic one), taken alone, suffers from certain limitations. By insisting on 'the fact of evolution,' it merely describes the existing state of affairs *in biology as a research discipline*. When used to correct misconceptions or misleading statements about that discipline, this argument is perfectly in order. But in the wider context of education, it is important to realize that how much priority we give to imparting information about the state of the discipline is a *goal-dependent judgment*, a judgment of philosophy of *education*, a pedagogical judgment, sometimes a social or legal judgment, *but not a matter for authoritative scientific decision*.

I have found that the discussion of controversy, such as the current one on creation/evolution, the nineteenth century debates on the same issue, the ones on sociobiology and on the computer model of mind—all of which involve science *and* more general human concerns—have the advantage of simultaneously putting the various schematizations to use, and of *testing* them. Therefore, if this approach brings an otherwise abstract and 'academic' discipline—philosophy of science—to life, to intellectual use, this is by no means of secondary value.

American teachers, for example, receive from state education departments various guidelines on how to teach evolution. They also receive in their mailboxes literature from a number of outside organizations, questioning evolution or the manner of teaching evolution, and attempting to involve these teachers actively in the controversy.[15] Aside from other good arguments for the serious study of the philosophy of science, it does seem that teachers ought to have more adequate intellectual tools for coping with such real-life problems.

[16]

The 'Interests' of Science and the Problems of Education

ABSTRACT: Some reflections are presented concerning the widespread complaints of declining interest of both pre-college and college students in the natural sciences, on the resulting 'scientific illiteracy,' and on the serious shortage of high school teachers of science. Three 'interests' of science are identified and considered: (1) technological interest, (2) methodological interest, and (3) cosmological or more generally philosophical interest. The first of these is central in standard courses in physics and other sciences, and is served by teaching laboratory techniques and assigning problems for students to solve, but it does not appeal to students who have no intention to pursue technological professions. The second is commonly used as a justification for teaching science to a wide spectrum of students, on the ground that the scientific method is a model for rationality in general—a justification that is weakened by comparison of scientific reasoning with the actual reasoning processes in law, business, journalism, etc. The third is generally neglected by educators, in spite of evidence that it is actually or potentially widespread—evidence based upon the enormous popularity of serious books exploring in detail the implications of science for philosophy, religion, psychology, and social problems, and upon the complaints of both radical and conservative students that science as generally taught is 'irrelevant' to their fundamental concerns. It is argued that the declining interest in science will be reversed by well-informed courses addressing the cosmological interest of science. Such courses should be conducted with wide-ranging open discussions, should not be based upon pre-packaged educational materials, and should above all be taught by teachers whose technical knowledge of a science is well supplemented by immersion in its cosmological and philosophical implications. [—A.S.]

This paper was presented in 1989 at the First International Conference on History and Philosophy of Science in Science Teaching in Tallahassee, Florida. It was published in *Synthese: An International Journal for Epistemology, Methodology and Philosophy of Science* 80 (1989), a special issue containing selected papers from that conference, 81–106.

211

Introduction

Reorientations in the understanding of science, such as that of the past quarter-century, can have educational impact far beyond the level of pedagogy. If educative activity is the means for each generation to pass to the next its view of the major features of life, then a new theory of the *nature* of science could lead to a new view of science in society, and therefore also in education. It is this broader aspect of the possible influence of recent philosophy of science that forms the subject of the present paper. Because I will attempt to link a number of apparently disparate problems, however, and to make unorthodox suggestions about interests of science that are at present suppressed, the treatment here must of necessity be general and programmatic.

Our starting point will be the recent work of Mary Hesse, with some reference to Jürgen Habermas. I will assume, therefore, the reader's familiarity with earlier ideas introduced by such figures as Thomas Kuhn, N. R. Hanson, Paul Feyerabend and Imre Lakatos—ideas that have ushered in the age now often called 'postempiricist.' Briefly put, I would like to draw out the possible consequences for educational philosophy of two key innovations treated by Hesse, and then, to reformulate these innovations in a way particularly suitable to contemporary educational problems.

I refer, first, to her surprising acceptance of Jürgen Habermas's proposal that science ought not be regarded as neutral inquiry because, in fact, the meaning of its findings are constituted by its own specific interests (1980, xxi); and second, her even more provocative suggestion (going beyond Habermas) that hermeneutics is involved in all the sciences, including the most analytic and nomological (Arbib and Hesse 1986, 181).

To appreciate fully the import of this last point, it must be recalled that until recently—for many philosophers, especially in Europe—hermeneutics (interpretation of meanings) was in effect one more demarcation criterion: It was relevant to the human sciences, where the objects of investigation include socially constituted concepts and norms; but it did not apply to natural sciences, whose domain of study was susceptible of 'objective' definition and measurement. To raise the possibility of hermeneutics in physics, for example, is therefore to blur the line between the natural and the human sciences.[1] Of course we are aware that blurring this line has long been a major goal of many theorists, but only by a process in which the human sciences take on increasingly the appearance of the natural sciences; in other words, only by a kind of reduction of the latter to the former. To blur the opposite way, to suggest that natural science incorporates characteristic features of the humanities, is to press against powerful currents even today—notwithstanding the 'new philosophy of science,' which, since Kuhn's influential work (1962, 1970) has moved in precisely this direction.

However, we will by no means confine ourselves to the realm of abstract theory. With John Dewey, I assume that education is the ideal and perhaps primary ground for the application of philosophical ideas.[2] Therefore let us begin straight away with those features of the educational situation to which the philosophical ideas may be related directly.

'Interests' Socially Perceived: Two Awkward Questions

Consider first a certain comparison between two general, fundamental subject areas pursued on all educational levels: natural science and the natural language. What are the possible interests with which the latter may be approached? Reading, writing and speaking are 'basic skills,' of course, in which everyone has an interest. Next, we have certain differentiations within each of these broad areas: writing of the kind suitable for business and the professions, or for journalism, or for ambitious 'creative' aspirations; 'speaking' as presentation—of self or ideas—or representation on the communal and political levels; 'reading,' not merely news and entertainment, but literature, with all the accompanying knowledge that demands. And finally, there are the higher intellectual interests—literary history, criticism, theories of language.

Although some of these are associated with particular types of work, notice that we are not talking about job categories here. Such interests are recognized already in pre-professional education by means of institutionalized choices available to the student: creative writing courses, debating clubs, newspapers and literary magazines—all common in many American high schools.

Against this background let us now ask the analogous question regarding any one of the natural sciences, keeping in mind that, in our advanced society, the importance of science is indeed comparable to the dominant language itself. What are the corresponding interests of physics, say? Moving quickly beyond the most basic skills—which, at a much attenuated level, may possibly have been acquired in the elementary grades—in what way can the newcomer or outsider approach this subject?

The answer, alas, is well known: There may be 'a physics course,' possibly an 'advanced physics course'; and at the college level they come 'with' or 'without' calculus. Such strictly professional differences as, for example, between 'pure' or 'applied' (engineering) physics are not in question prior to the *later* years of college work. In effect, then (and simplifying only slightly), we offer a 'one and only' treatment of physics, designed or approved by the profession itself, and occasionally updated.[3] But compared with the situation in English as a field of study, we just do not have in the

natural sciences any overt recognition of a variety of interests, with courses, extra-curricular organizations, and materials to serve these interests.

We are all so accustomed to this situation, the 'given,' that no doubt
many readers will be puzzled at such a juxtaposition, and at the thought
that some serious point can be made this way. Yet is it not strange that this
should be the case? Is a natural science really so unifunctional that the question of *fundamentally different interests* regarding it is too awkward to ask
seriously, and merits no place in a discussion of education?

Let us postpone for a while the implications of this, and turn now to
another, equally awkward question. It has become commonplace in recent
years to bemoan the declining interest in the natural sciences, and the
resulting 'scientific illiteracy' (Rigden 1981): as indicated by course enrollments (Holton 1967, 31; Layman 1983, 27), as reflected by science's 'tar-
nished image' (Matson 1966; Davis 1970), as shown by the lamentable
shortage of high school teachers of these subjects (Van Hise and Nelson
1988, 47), and finally, as confirmed by impoverished requirements in even
the most prestigious universities (Westheimer 1987). Among the conjectured reasons for this decline, we hear of such things as science's failure to
'deliver' on the 'promise' of a happy, healthy, war-free world. We hear even
more ominously that, in regard to these perennial worries, science itself, far
from being the solution, may well be among the problems. Finally, we are
reminded that scientific knowledge has become so technical and complex
that the nonprofessional can follow it only superficially, while the student
contemplating a scientific career is, for the same reason, tempted to pass it
by.

Yet there is something odd about this development, if rightly portrayed—for it comes concurrently with perhaps the greatest upsurge ever
of high quality science literature, amid such acclaim and acceptance as
would have surprised us only a decade ago. I refer not only to the well-
known success of 'popular' essayists like Stephen Gould, Lewis Thomas or
Jeremy Bernstein (in the United States) but to the emergence of what
should perhaps be regarded as a wholly new kind of science writing, neither professional nor 'popular'—a 'third genre,' with profound cross-cul-
tural influence.

My examples are these: Jacques Monod's *Chance and Necessity* (1971),
Eigen and Winkler's *Laws of the Game* (1981), Prigogine and Stengers's
Order Out of Chaos (1984), Steven Weinberg's *The First Three Minutes*
(1979), Joseph Weizenbaum's *Computer Power and Human Reason* (1976),
Steven Rose's *The Conscious Brain* (1976), Douglas Hofstadter's *Gödel,
Escher, Bach* (1979), E. O. Wilson's *On Human Nature* (1978), and some
of the volumes on cosmology by Paul Davies (see his 1983). Many others
could of course be added to this list, but the type should now be well
enough defined. It consists largely of full-length books (some selling in the

hundreds of thousands), includes among the authors Nobel Laureates, and among the products Pulitzer-Prize works. This much attests to both scientific and literary seriousness. But what distinguishes the genre most sharply is its *intent*: For the books cited do not merely convey scientific knowledge in an interesting way; their main goal is to distill the deeper meaning of recent advances on fairly broad fronts, calling attention to cognitive implications that bear on social and on personal self-understanding. They are, in short, philosophically oriented: While staying for the most part within the cognitive domain of science, while teaching various aspects of science, while insisting on the value of science itself, they nevertheless deliberately direct the import of their message to classic humanistic interest. And it is as such that this literature is widely cited and used in professional work outside natural science proper.[4]

Thus, our second awkward question: *Why* do outstanding scientists choose to devote time and energy to write for a diverse public outside their own disciplines, and *why* does this public respond enthusiastically *if*, as widely reported, interest in natural science is fading, and the old esteem vanishing so alarmingly?

It will have been noticed that both questions involve the concept of *interest* in some way. Concerning the 'third genre' of literature just mentioned, 'interest' is used primarily in the sense of *interest in*—the interest on the part of nonspecialists, *in* some aspect of a natural science. The first question also includes this sense of the word; but because curricula are determined by professions and recognized practices, there is the additional meaning, *interest of*—the interest *of* science, as viewed from the inside. This latter sense of the word does not refer to what may be reported by individual practitioners, but what is implicitly expressed by the very structure, the character, and the history of the enterprise.

Both of these questions point to the concept of 'interests,' suggesting that, if we examine it, something revealing may come into view—about education, and perhaps more generally about the place of science in society. We proceed, therefore, to consider 'interests' from a theoretical viewpoint, though later we will take up again the questions just raised, and, in a new conceptual framework, propose some answers.

'Interests' Theoretically Perceived: Two Philosophers

It was Jürgen Habermas, in the 1960s, who first systematically set forth the idea that—contrary to the positivistic view of value-neutrality—*all* sciences in fact have certain inherent interests. By "interest" (in the sense *interest of*), he meant that the conceptual and methodological structure of a particular science already predisposes that science toward certain kinds of applications

(Habermas 1971). 'Interest,' in other words, is a link between a form of inquiry and a form of action. Thus, the natural sciences, being empirical, analytic, and nomological, lead inescapably to the accumulation of 'technically-exploitable knowledge.' And in this sense Habermas believes their interest is exclusively technical; because from the very beginning, the objects of possible investigation in these sciences were so constituted that the products of investigation are oriented toward control of natural processes, a form of goal-directed, instrumental action.

Of course, this is not a description of the subjective interest of individuals, but of the objective form of an activity any person must accept when he or she begins to 'do science.' Although this form is presented as free of any *telos,* and may appear so to the unreflective mind, every new generation of initiates in fact pursues the inherent *telos* of a given science by continuing, refining, expanding precisely its form of activity.

In contrast, the humanities and certain social sciences that do not fit the empirico-analytic mold, like history, sociology, anthropology—which Habermas calls "historical-hermeneutic"—incorporate an entirely different interest. Because these originate in symbolic interaction between persons, their methodology is oriented toward mutual understanding within a community rather than instrumental or strategic ordering of objectified processes. Here the relation of structure to possible action is a *communicative* ("practical") interest. We pursue history and anthropology in order to facilitate mutual and self-understanding against a background of events distant in space or time but deemed relevant. We argue about the proper interpretation of Greek antiquity or the Azande culture because we believe that these questions—and the *discussion* of these questions—help us understand the meaning of the here and now.

Thus, the natural sciences channel thought toward the manipulation of material objects; the 'human sciences,' on the other hand, dealing with persons and speech, have as their *telos* the orientation of the human being in a social environment.

At first sight, this neat typology might appear incongruous. After all, from Galileo to Newton to Einstein there surely was more to it than a technological interest. What about the long effort, lasting over millennia, to put together a credible, coherent picture of the world we inhabit, the cosmos, and our place in it? Yes, Habermas admits, in the past there was that too (and, of course, subjectively it continues), but objectively that interest has now become "apocryphal" (1971, 304). From the classical age of Greece to perhaps the nineteenth century, it was believed that theoretical natural science had a direct educative, cultural effect, through *mimesis* of a cosmic harmony. Because 'understanding the cosmos as a whole' could yield norms of individual human behavior and lead to a 'thoughtful and enlightened mode of life,' there was then indeed a *socially* orienting interest in the cultivation and study of these subjects (Habermas 1970, 54; 1971, 302).

Now, however, we are aware that all this was illusory: In its modern form, science is pursued in an institutional manner, and exerts power instrumentally, *without transforming the knower as a social being*. Thus, whatever the personal attitude of Einstein may have been, today his theories are effective only through their application to nuclear reactors, lasers, and other technologies. No one is changed as a humanly interacting person by studying relativity. Most important, according to Habermas, the present situation is unavoidable: for the very form of natural science, at least from the seventeenth century on, *always had* this deeply imbedded, technical interest—which only in our time has fully emerged, discarding all pretense.

In this way, a demarcation line is drawn between two types of scientific knowledge: that which enables the knower, as a subject, to confront objects; and that which enables him to communicate with and understand other subjects. Of the two, only the latter affects people as socially competent persons; the former, in effect, treats the knower *himself* as an instrument—a *producer* of technically exploitable knowledge, a component of the technical-economic system. The educational consequences follow easily. Such a philosophy appears to support, or be supported by, the popular-pragmatic notion that the 'cash-value' of science is indeed technology alone; this is what the public pays for. But from a scholarly, theoretical point of view, we see another interesting correspondence: In Kuhn's picture of normal science (1970, Ch. 4), the entire scientific enterprise is essentially a "puzzle-solving" instrument, attempting in its day-to-day work to apply the accepted paradigm as widely as possible. To the extent that Kuhn's "normal science" is a good description of how things work most of the time, to that extent Habermas's "technical interest" can be visualized in concrete practice. Finally, there is the tangible effect on the curriculum itself. A whole class of textbooks appeared in the 1970s reflecting just this Habermassian (or popular) view that "statements about the phenomenal domain of things and events . . . can only be translated back into orientations for goal-directed action (in technology or strategies)" (1973, 20). This approach takes, for example, technical-economic energy problems, or environmental problems, or urban problems, as themes for courses in physics (Marston 1970; Saperstein 1975; Romer 1976).

It is true that in teaching the sciences Habermas also perceived another dimension: "The translation of scientific material into educational processes of students requires the very form of reflection that once was associated with philosophical consciousness" (1970, 8). And "this dimension must not be closed off," he says. However, because Habermas does believe strongly in fundamentally different interests (or "value spheres"), he thinks reflection on the meaning of knowledge is possible today only in *philosophy,* not within science itself.

It is important at this point to understand Habermas's motive in emphasizing this kind of division of labor. Basically, it was a reaction to the *positivism* of the first half of this century—a positivism and a scientism that threatened to engulf the cultural spheres as well, and reduce increasingly greater portions of our everyday activity to the "purposive-rational" type— "colonization of the life-world" he now calls it. By "uncovering," as he thought, the true interest of the natural sciences, and showing their deep roots within a *limited* realm of life, Habermas hoped to safeguard the other realms (communicative) for a more hermeneutic approach—one where meaning of humanly created values and institutions would be understood by participatory interpretation rather than by a distancing objectification.

Despite the flaws in Habermas's analysis (to be discussed shortly), the value of this kind of typological thinking must not be minimized. Its achievement was to describe, delineate, and raise in importance the *role of science in cultural communication,* notwithstanding that (in his scheme) that role was reserved for sciences other than the natural. Habermas did for social theory what in philosophy had been done decades ago: Against Marx (and Dewey), he stresses language rather than work as the essentially human trait; against Max Weber, he insists that the meaning of 'rationality' is not exhausted by an appropriate matching of means to ends—an altogether different kind of rationality pertains to the communicative realm.

Meanwhile, however, in the philosophy, history and sociology of science, something unexpected happened—the postempiricist revolution of which we are now all aware. The thrust of this revolution could, in one sense, be called *counter-scientism:* That is, features of the historical-hermeneutic sciences were being imported into the analysis of the natural sciences, reversing the positivistic trend of the past one hundred years. The particulars of this movement are well known. They involve such ideas as the 'theory-ladenness' of data, the dependence of meanings on theoretical coherence rather than correspondence to 'facts,' and the inexact nature of technical language, which (at high level) results not in 'laws' as popularly understood, but only in 'law sketches' (Kuhn 1970, 188) or 'metaphors' (Hesse).[5] Naturally, all this made the earlier Habermassian treatments of science more deficient than ever, which he of course realized and to which he has reacted. But since natural science itself has never been Habermas's main interest, it was left to others to try to retain his original insight in some amended form. Most important among these efforts, in my view, has been the work of Mary Hesse. To this let us now turn.

While Habermas treats natural science with broad, sweeping language, implicitly using as his model low-level generalizations, Mary Hesse insists on the important distinction between relatively direct, empirical knowledge on the one hand, and on the other, *high-level theory*. The former is, so to speak, a locally anchored residue, which remains invariant when high-level theories undergo revolution. As Stephen Gould is fond of saying, apples continued to fall down from trees (at the usual rate) while the theory of gravitation changed drastically from the Newtonian to the Einsteinian paradigm. In contrast, the important thing about high-level theories is precisely that they are not so strongly anchored; for in the aftermath of Kuhn and Feyerabend, there is greater awareness today that these are always underdetermined by the empirical evidence.

Consequently, in large-scale theorizing, other considerations besides prediction and empirical testability must enter. When remote from human experience, nature is perceived as through a glass darkly; yet we succeed in making it intelligible. How? We do so as with anything foreign or obscure—we do it metaphorically. High-level theories, then, are to be regarded as metaphors of the environment, not as pictures, not even as partial, always to be improved, correspondences with some underlying but pre-existing domain of natural kinds. To say that an object falls—with time and distance related by a certain formula—is to speak fairly directly about a repeatable, testable experience. This is that 'pragmatic' part of science, technically exploitable, to which Habermas refers. But to say that the object's fall is governed by a metric tensor, even though the mathematical expression can be written explicitly, is to speak just as metaphorically as those who use the popular phrase "space-warp."

This distinction within the natural sciences themselves is the key, in Hesse's theory, to a view of 'interests' which, though similar to that of Habermas, nevertheless brings the human and natural sciences closer together. For if metaphor is indeed part of the latter, then interpretive understanding, hermeneutics, is needed here too, not just in the *Geisteswissenschaften*. "Scientific theory is a reading of the 'book of nature,' requiring circular reinterpretations between theory and observation and also theory and theory, and also requiring 'dialogue' about the meaning of theoretical language within the scientific community" (Hesse 1986, 181).

Kuhn had already described how hermeneutics entered his own work in science history (1977, xii–xv). Hesse is now extending the realm of application of this concept by claiming it holds not just for the history of the practice but for the practice itself. I take the above quotation to mean that hermeneutics enters natural science in two distinct ways, or at two separate points in its process: First, at the center of research there is the dialectical relation "between theory and observation and also theory and theory." This is the part that has been extensively discussed in recent philosophy and history of science. An outstanding example is the decades-long debate about interpretations of the quantum formalism and the experiments inspired by this debate (Herbert 1985, Ch. 12). But in addition, the need for dialogue about "the meaning of theoretical language" extends also to the periphery of research, where meanings are discussed not with a view toward experimental or theoretical development, but in order to convey the outcome of inquiry to newcomers or outsiders. This follows from the principle that the appropriation of knowledge demands cognitive performance similar to the original acquisition of that knowledge. So 'dialogue' concerning meanings must take place in the teacher-student relation, and also the student-student and teacher-teacher relations, as both groups try to cope with their task.

The picture that emerges, then, of all the sciences, is a continuum: The most empirical, the most technically exploitable, lie at one extreme, with a small hermeneutic component; the least empirical, with the largest hermeneutic needs, lie at the other end. No sharp line can be drawn between just two basic types of knowledge.

The two 'interests' discussed by Habermas are present in some degree across the whole spectrum of the sciences because high-level theories (large-scale metaphors) do not embody instrumental or technical interest exclusively; nor are they generally understood that way. What in fact such metaphors do accomplish by filling in the nonempirical spaces is to present the surprising depth and complexity of nature in terms accessible to the human mind. And this makes communication possible. It makes possible not only mutual understanding about nature as such—speech at various levels, from ordinary talk to scientific exposition—but also self-understanding. For the image of our own humanity can only be formed against a background of relations to the natural environment (outer and inner), however these relations may have been constructed. It "has been abundantly demonstrated in the history of all natural sciences," says Mary Hesse, that "theories have always been expressive of the myth or metaphysics of a society. Society interprets itself to itself partly by means of its view of nature" (1980, 186).

In short, Hesse is saying that the cosmological interest in science is far from apocryphal; and therefore—contra Habermas—it also becomes an interest *of* science. Habermas's reason for dismissing this was based on too restricted a view of how cosmology could affect life. Taking into account only direct imitation, he did not fully appreciate the far more pervasive indirect influence on world views. Today this indirect effect is so widely felt—in disputes over the teaching of evolution, in scientifically based theological theorizing, in the use of science to further environmental values, in philosophies of mind and culture—that further elaboration would surely be superfluous.[6] To exert such an effect, at an intermediate level of expertise, is of course a major motive behind that third genre of literary works previously mentioned.

I should like now to gather from these recent analyses what seems most relevant to the questions raised at the beginning, especially as it bears on the possible neglect or suppression of interests that have been explicitly recognized. The next two sections, then, will be concerned with drawing some conclusions for the contemporary educational situation.

The Three Interests of Science

Essentially, I adopt Hesse's view that there are *two* fundamental interests of science. But in the context of education, we gain some advantage in treating explicitly a third interest—the methodological—which she, like Habermas, took for granted and subsumed under the others. I propose, then, that we recognize at least three distinct interest domains, to be regarded in both the senses already discussed *(in* and *of* science)—technology, cosmology,[7] and methodology.

Methodology requires separate discussion not only because of the recent emphasis on 'process' in science teaching, but because I want to distinguish two of its aspects: First, by an interest in methodology we mean the need to develop various procedures—from specific laboratory techniques to model building, to such general forms of theorizing as the hypothetico-deductive approach and the principles of correspondence. So understood, methodology appears in its *instrumental* aspect. But in addition, there is also a *reflective-philosophical* aspect—when we question method or its newer applications in order to ascertain the validity, or significance, or meaning of the result.

This reflective aspect of methodology is both an interest *of* science, since the latter must be self-corrective, and an interest *in* science by outsiders. The outsiders, in this case, are not only philosophers. All sorts of people have a tendency to transfer 'scientific methods' to other areas of life—such as modeling in business, controlled experiments of social science, and operations research in government and the military. More subtly, however, this interest reveals itself in a form that has far greater impact: the inclination to use what is sometimes called the 'scientific method' as a model for *rationality in general.* In discussing an ideal curriculum, for example, where science and civics courses would be closely coordinated, Michael Martin explains that "the emphasis throughout would be on the importance of scientific method and knowledge for understanding social issues" (1972, 159). Contrary to some lingering popular beliefs, then, and to theories such as the early one of Habermas, science has broadened its impact (for better or worse) to a point where we must now view it as a *multi-purpose social enterprise.*

If this much is granted, it will be useful for what follows to show certain relations between the three interests, as defined here, and the two basic types of action to which these interests lead.[8]

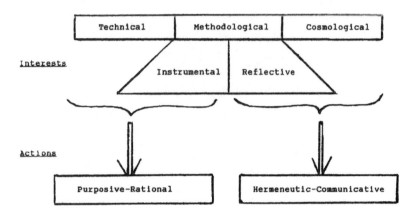

Figure 1

Borrowing Habermas's terms (because of their greater familiarity), I will call these the "purposive-rational" type of action, and "communicative-hermeneutic" action. The former is any behavior pursued rationally in terms of success while using a knowledge of the external world and of other people as means for one's own ends (Habermas 1981, 281–86). The latter is a symbolic interaction, based on consensual norms, and involving at least two acting subjects; its aim, in contrast to an objectified idea of success, is the attainment of mutual understanding or agreement (Habermas 1970, 91ff).

If, now, we decompose the methodological interest of science into its two aspects, and join instrumental methodology to the technological interest, we see that both these components are ideally oriented toward purposive-rational action. In technical exploitation of science, methodology of the instrumental kind is an indispensable ingredient, while its reflective-philosophical aspect plays almost no role. On the other hand, the cosmological interest and the reflective-philosophical aspect of methodology interact naturally whenever our intentions are toward the hermeneutic-communicative type of action: when science is to be explained, interpreted, and appropriated for the purpose of mutual or self-understanding, or for orientation within the socio-natural environment.

The relations between these interests and actions can be summarized by means of the table in Fig. 1. The resulting picture, in which the different interests of science *lead* to the different types of actions, can be formalized also in an alternative way:

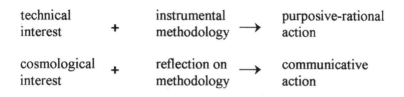

Figure 2

Once such a scheme has been set down and is to serve as the basis of discussion, then, from our educational perspective, the next question is obvious: What sort of priority ought to be given to these different interests and actions at various levels of the educational system? and what *in fact are* the priorities as things stand?

The Suppression of Interests

With the help of the diagram, my previous references to the suppression of certain interests can now be phrased concisely. What is pursued with ardor and pride in formal education are the 'problem-solving' interests oriented to purposive-rational action (research as well as application). What is suppressed is the entire bottom tier on the diagram—the cosmological-philosophical interests bearing on hermeneutic-communicative action.

By "suppression," I mean, in the direct sense, simply the absence (in general) of courses, textbooks, lecture series or informal discussion groups where these interests are seriously addressed. In the indirect sense, 'suppression' reaches further. It means, in particular, that the science teaching profession itself (the corps of high school and college instructors) is by training, individual inclination, and professional behavior oriented almost exclusively toward the interests on the top of the diagram. The result is that informal, personal, student-faculty discussions—still a tangible aspect of education—exhibit the same bias as does the less tangible element called "atmosphere" (of a campus, department, school).

Such 'suppression' is, of course, neither intentional nor usually conscious nor monolithic. A student asking a cosmological or hermeneutic type of question will certainly elicit some sort of interesting answer from most faculty members, at least at the college level. The point is, however, that both the question and the answer will be a mere aside, or interruption, in the ongoing business of the purposive, problem-solving oriented activity.

Why this should be so today, and why—sporadic efforts to the contrary notwithstanding—the trend remains in the same direction, may be easily understood on the basis of Kuhnian theory (1970, 137–143). If formal study is conducted within the bounds of a dominant paradigm, as it must be in times of "normal science," and if textbook-based instruction is dependent on the mastery of exemplar "puzzle-solutions," then the paradigmatic process itself is always oriented to instrumental, purposive-rational action. In other words, the puzzle-solving aspect of "normal science," emphasized in education, almost by definition suppresses the hermeneutic-communicative type of activity.

To verify this, lengthy documentation is unnecessary since anyone who has taught or taken such courses need only call upon his own experience. However, it is not just practice, but educational policy itself which often explicitly asserts this interest. Consider the following example:

> The essence of being a scientist is that of being a puzzle solver. Consequently science education ought to equip science students with the skills necessary for puzzle solving in specific scientific domains. (Wagner 1983, 605)

This prescription, given in a recent paper on the goals of science education, leads the author to explain why even "rote memorization and techniques for getting students to behave as we behave, are not only appropriate but unavoidably essential" (1983, 611). The justification is quite simply Kuhnian—Kuhnian "normal" practice, not that "overromanticized notion of something called revolutionary science" (1983, 608).

As a reading of Kuhn, at least, this *seems* warranted. Nor is Wagner especially narrow in his views, for he does include a "critical spirit" as part of "normal" practice. What is missing here, though—and this distorts even the purely Kuhnian message—is any thought of the student clientele and their interests *vis-à-vis* the potential of science. It is an omission that Kuhn himself does not make. For the Kuhnian description, in so far as it is normative, applies only to the novice, the initiate into scientific professions. It is not for everyone:

> Of course, it is a narrow and rigid education, probably more so than any other except perhaps in orthodox theology. But for *normal scientific work* . . . within the tradition that the textbooks define, the *scientist* is almost perfectly equipped. (Kuhn 1970, 166, emphasis added)

As do many others, Wagner addresses himself to the problems of introductory science teaching *in general*. By implication his conclusions apply, even at the high school level, to a very heterogeneous student population.

Yet the interests expressed in his approach are those of a small minority at best. There is no hint in Wagner's discussion that anyone might study science to learn what the *world* is like (not what *science* is like[9]), or that anyone might be concerned with the significance of the results of inquiry and their competing interpretations.

To what extent such interests are real for outsiders, or for the still uncommitted student, is certainly difficult to ascertain[10]; and the situation may differ from one culture to another. But if we assume that this is in fact the case to some appreciable degree, then, at once, we have an answer to both 'awkward questions' posed at the beginning of this essay.

The success of the 'third genre' of science literature cannot be accounted for in the same way as the success of truly 'popular' writing, that is, by its simplified, easily comprehensible form (compared with formal science study). For, as Steven Weinberg admits about his own book, this literature does involve "fairly complicated ideas." Indeed, more than ideas are offered: The proofs in propositional calculus and number theory contained in *Gödel, Escher, Bach,* for example, are above the level of many college courses. People looking for exciting but easy reading about science will not tackle such a book.

If, therefore, we seek an explanation for the success of this literature in some other common factor; and if, again, we utilize our schematic representation of interests, the conclusion is hard to resist that precisely what education suppresses is prominently featured in just this class of writing—the cosmological interest—often magnified to epic proportions, as by a Wilson,[11] a Prigogine, or an Eigen. Addressed here too are such classic philosophical concerns as ultimate origins, religion, consciousness, personhood, free will, determinism: all without apology or condescension, eschewing attempts to 'explain away,' trampling over disciplinary lines, combining freely the 'value spheres' of the objective, the moral, and the esthetic, extrapolating boldly but more carefully than 'popular' writing, and therefore managing to stay scientific. Apparently, such treatments satisfy a need.

In summary, then, the hypothetical answers to our two 'awkward questions' are: (1) Science, much like other socio-cognitive realms, does have more than one interest, whether we look at it from the internal or external point of view; (2) but in our educational institutions, only *some* of these interests are *expressed,* the others largely *suppressed;* (3) the result of this suppression is an *intellectual (cognitive-moral) hunger* on the part of a significant portion of the public, a hunger now beginning to be satisfied, *completely outside the educational system,* by that third genre of science literature and its 'popular' approximations.

Because only general considerations are offered here, without systematic data to back the conclusions, I do call my proposed answers "hypothetical." Nonetheless, a minimal discussion of supporting evidence can he given. To begin with, if we take the first of the 'awkward questions' as our basic problem—Does science, in its educative context, have a multiplicity of interests?—then the very fact of the success of the 'third genre' can he taken as evidence for a positive reply. In addition, however, it is worthwhile to recall certain events of recent history.

Since the 1960s, education has been confronted with a number of protest movements emanating from the political left as well as from the right. Many different factors contributed to these protests, of course; but, in regard to science, we can in retrospect identify one common theme—the feeling that somehow (though inadequately understood), an undue constriction exists precisely in the realm we call "interests." On the left, during the 1960s and 70s, this feeling was characterized by slogans such as "relevance" (the campuses having too little of it), or "factory" (the campus being too much like it), or "one dimensional man" (the unfortunate product of the "campus-factory"). On the other hand, the battles waged by American conservatives in the 1980s, which concern essentially 'values,' and 'morals,' have nevertheless been explicitly related to such scientific subjects as evolution, anthropology, ethology,[12] and human sexuality. During the decade of the 1960s, while at Berkeley, Mario Savio orated on the fragmentation of knowledge at the 'factory-university,'[13] philosophers of the stature of Herbert Marcuse tried to make the theoretical link between the operational-instrumental structure of physical science and a particular kind of dehumanizing, purposive-rational action to which our society was apparently addicted. Both the young activist and the old thinker (Marcuse 1964, 231) openly claimed that some different sort of learning was in order, even in the natural sciences as such.

Today, it might at first appear that protests from the right in effect favor restriction rather than expansion of the interests of science. But in looking closely at the most explosive of these controversies, evolution/creation, it quickly becomes apparent that against the official defense of science study from alleged 'external interests,' protestors have insisted that *indirectly*, and without acknowledgment or discussion, more than 'pure science' is in fact being taught.[14] Conservative parents have not missed the significance of that 'evolutionary epic' so forcefully described by Edward Wilson, and already having its effect in the schools. But that effect (not universally welcome) on their children's self-understanding looms larger for many than the technical competence promised and touted by the educational system. Hence these discontents on the right also seek a broadening, but it is the broadening of *recognition* that something is happening in an uncontrolled, unreflected manner. The philosophical counterpart of Marcuse's charge of 'one-dimensionality' is today the cry 'secular humanism.'[15]

It is not necessary, of course, to condone the exaggerations, the pressures, or the simple lack of scientific understanding that have surely played their part in these protests (left and right) in order to consider them—at least a priori—as possible reactions to suppression of interests that are in fact quite real. The allegations of 'one-dimensionality,' as well as of 'secular humanism,' reflect an intuition that interests of science, other than those officially defended, are in fact vitally involved but somehow either *ignored, distorted,* or *channeled* for ideological purposes.

'Interests' and Teacher Education

From the positivistic past, we are heirs to a dogma that is no longer tenable, but one that still shapes our educational policies and our stance *vis-à-vis* the public. According to this dogma, curricula in history, social studies, even literature, do affect the self-concept of the student, and must therefore be evaluated from this point of view also. But any science bearing the label 'natural' or 'physical' is, in terms of human interaction, sufficiently neutral to be exempt from such assessment.

The most general implication of what has been said here is, at the least, that this dogma must be seriously reconsidered. Large sectors of the public are questioning it today because of the rapid expansion of the school's purview from a relatively restricted realm of factual knowledge and skills, to high-level theories in biology, cosmology, anthropology, and other fields. That schooling in such matters is merely 'teaching science' no longer seems credible. In addition, however, leading scientists themselves are increasingly claiming for their fields a socio-philosophical significance. The latest book by Roger Sperry, for example, argues that "a different approach to the public support and role of science is envisioned in which science is upheld, not because it begets improved technology, but because of its unmatched potential to reveal the kind of truth on which faith, belief, and ethical principles are best founded" (1985, 107).

The convictions of this famous brain physiologist are by no means eccentric. Their candor and heightened confidence help us realize that today a wider *cognitive* influence of natural science is openly discussed and accepted by many. Some educators, in fact, have been in the forefront of this drive: "The complete manifestation of the spirit of science goes beyond the confines of what is usually called science into practical, moral, and even religious contexts. The goals of science education should also be conceived of in these contexts" (Martin 1972, 158). But such goals create, within the educational community, an obvious and troublesome contradiction. For in some situations, e.g., when urging 'science literacy' on the public, its wide, ethical influence is treated positively, as a value to be prized. In other situations, however—socio-political disputes, for example—science is again described as neutral, beyond ideology, serving only the interest of establishing 'facts.'[16] This inconsistency may be seen as another outcome of the suppression we are discussing.

Of course, to cite these difficulties, and suppression as their main cause, even if it is all accepted, is a far cry from suggesting an appropriate response. That, in any case, could not be achieved by one person. The most that I will attempt here, in these final paragraphs, is to point out one promising direction of reform: It is in teacher education, I think, that something might be achieved, with the smallest risk.

There is no question that high school teachers of science (at least in the United States) are almost completely out of touch with what I called the 'third genre' of science literature, and indeed, with much good popular science, too. Because their knowledge is regarded as a specialty, and their work itself a problem-solving task, their ability to respond to the cosmological interest is almost nonexistent. What is of genuine concern to many parents and students, as well as to leading scientists—the meaning and significance of formal results—is, for most teachers, 'outside their field.'

Yet it is not at all outside their *interest.* Presented in a way that reveals clearly the connections to the teachers' own sciences, the discussions provoked by Sperry, Wilson, Prigogine, and the others do have the expected effect of broadening the horizons of those sciences, stimulating intellectual involvement, and raising key historical and philosophical questions which (most noteworthy!) *lead back to the sciences, with increased attention to detail.* In my course, for example,[17] after some study of philosophy, and some discussion of current controversies (punctuated equilibria theory of evolution), in-service teachers returned to their college textbooks to see how their preconceptions had been formed, how these subjects had been taught to them. Several continued to read on their own about current developments,[18] and expressed the view that "not until now" had they actually understood what they were teaching.

This entire discussion supports, I think, greater attention to programs that place scientific knowledge in a context of interdisciplinary effects and socio-cognitive concerns. As others have done,[19] I advocate for teacher education a healthy dose of philosophy of science, but not necessarily the standard fare. The goal must not be more knowledge, or different knowledge, but a changed attitude: An attempt ought finally be made to involve the teaching profession in a scientific culture. That means, *inter alia,* the continuing practice (accepted as a professional obligation) of keeping up with such literature as the 'third genre' represents, and of passing on, at least to their own best students, a desire to do the same. Absent some such renaissance, pre-college (and early college) science education may well continue to lose its 'relevance' for the professionally uncommitted—a perfectly understandable trend. For if the 'interest of science' really is as Kuhn and Habermas depict it, then what possible motive could there be in systematically studying these disciplines for those not interested in 'puzzle-solving' or 'technological exploitation'?

Yet, in our time, science has been enormously enriched by interests other than the technical: Not only philosophy and history, but also the sociology of science, have gained new attention. Never in our century have there been so many, within the research world of science and its professoriat, who also take an active part in these other pursuits. The problem is that this enrichment, these interests have yet to reach the middle levels of education; and that is why I point to teachers as the key intermediate group.

Unfortunately, it is still standard American procedure essentially to bypass teachers, in the seemingly reasonable belief that our 'true clients' are after all the mass of secondary and elementary pupils. Teaching systems, or strategies, richly financed and elaborately executed, are devised at the highest levels, with children as targets to be affected in a particular way, and tested for results (Fowler 1969, 1198). Teachers, in this scheme, are merely *used* to *implement* the systems. If science does have a single interest—technology—then this approach may succeed. But when cosmological and philosophical interests play such a prominent role in a newly revived culture of science, shared by specialists of all sorts and by many laymen, then the teachers' continued nonparticipation becomes inappropriate. In this situation, where the teaching profession is, in effect, left behind, we should not be surprised that education in the sciences is, more and more, taken over by others.

In a recent statement, the editor of the *American Journal of Physics* called for a whole "new ecology" in education leading to "fundamentally different relationships" between "teacher and that which is taught, between the learner and that which is learned" (Rigden 1988, 203). This idea was provoked, no doubt, by a realization that highly focused, isolated projects have little effect.[20] The same editor wrote nearly ten years ago, in regard to courses for nonscience majors, "after all the papers . . . and after all the meetings . . . I have not yet heard definitive arguments that enable me to adopt one set of objectives and to reject the others" (Rigden 1979). Why?

Let me suggest one reason for this. The idea that *the* proper science course should emphasize process, not content (or vice versa), or that *the* course for nonscience majors should have *this* particular set of goals rather than that one, allows scientists and educators to construct such a course, complete with text, experiments, video loops, and discussion booklets—a system ready for delivery. By whom? If *that* were really a secondary matter we would now have success with all these projects. But if, in fact, a single approach or a single set of goals cannot be universally used, then the teacher is more important than the system, and that is where effort must be directed, not by training teachers to deliver specific systems, but by educating them as the bearers of scientific traditions that include all the interests.

Orienting teachers strongly toward technology and methodology, predisposing them to rational-purposive action, is, in a sense, *mis*orienting them. For, as teachers, they do not really take part in typical scientific activity of this kind. They do, on the other hand, play a major role in the hermeneutic-communicative action of science: They interpret and convey science's metaphors, well or badly, as their day-to-day task. It is they who largely impart to each generation the cosmological background which will endow experience with meaning. It is they, through their own attitudes, their response to the material, who determine, more perhaps than the researcher or curriculum reformer, how science appears to the public.

Therefore, as an ingredient of the new ecology of education, the new relation between 'the teacher and what is taught,' I propose a shift of focus: from the teacher as a means of transmitting prepackaged curricula, as a trainer of puzzle solvers for the science-system, to the teacher as carrying out primarily the hermeneutic-communicative action of science. In so doing, he or she does not neglect the other interests, but puts them in perspective. Above all, the teacher must *embody* that perspective.

[17]

Hermeneutics and Science Education: An Introduction

ABSTRACT: This paper is a programmatic sketch of a line of theoretical investigation in the philosophy of science education. The basic idea is that philosophical hermeneutics is an appropriate framework for science education in most of its aspects. A brief discussion is given of hermeneutics in general, of the version of it developed by H.-G. Gadamer, and of the reasons for its relevance to science and to the problem of meaning in science education. A key element in this approach is the suggestion that each science be viewed as a language. Arguments against the appropriateness of hermeneutics to natural science are also discussed. One application of the theory to ongoing educational research—'misconceptions'—is specifically treated. [—M.E.]

The Problem of Meaning

It is difficult to think of a more important problem for education than the problem of meaning—which, in relation to science, comes up in several different forms. The physicist Stephen Weinberg, for example, writes that after all the observation, the calculation, the modeling and theorizing that gave us the big bang scenario, he finds the result 'pointless' (Weinberg 1979, 144). In Europe, philosophers habitually differentiate between the humanities and natural science by saying that while the former seek 'understanding of meaning,' the latter is 'based on causal explanation' (Apel 1988, 325). Debates abound on whether theories are to be understood literally,

This paper appeared in *Science and Education: Contributions from History, Philosophy and Sociology of Science and Mathematics* 1, No. 4 (1992), 337–348. Martin Eger took an active part in the establishment of this journal and served on its Editorial Committee. An earlier version of the paper was presented in Kingston, Ontario, also in 1992, at the Second International Conference on the History and Philosophy of Science in Science Teaching.

metaphorically, or instrumentally; and in educational research, the issue of 'meaningful' learning is central in the work on preconceptions, conceptual construction, conceptual nets, and critical thinking.[1]

For science education, the immediate problem is, of course, that increasing numbers of students fail to see meaning in crucial scientific ideas, though they may be competent enough in first-order knowledge and technical manipulation. Further, it appears that this phenomenon is true not only of beginners; the same thing has been noticed, for example, among senior engineering students. Although the educational research mentioned above seems, in general, appropriate and well oriented, the question arises whether a more comprehensive look at the entire issue might not be in order. Is it possible to see things in such a way that the ongoing reform initiatives become pieces of a larger picture, to which still other pieces may be added once the full landscape is clear? Aside from the hints it would give about the still-missing pieces, the theoretical advantages of an overall viewpoint are well known. They include, among other things, the possibility that underlying causes, hardly detectable in the individual parts, may come into focus.

On this level, the inquiry can lead to such questions as: What *sort of thing* is science education? What sort of thing is it from the point of view of the student? of the teacher? of society? Do scientists like Weinberg, philosophers, students, and educators—each group in its own way—face the same 'problem of meaning'? But these questions point immediately to an even more basic question: In what kind of thinking or academic pursuit do we find an approach that tackles the question of meaning in general, meaning in *any kind of study*? It is in this way that one is led to consider hermeneutics—which, precisely, is the reflection on how understanding of meaning comes about. The issue, then, is science education in a larger context. The ultimate aim is a philosophical framework, embracing the reform thinking already going on, but opening up more theoretical space. That aim is not pedagogy as such, though pedagogy lies nearby, along the way, and out in front. The hermeneutic approach, we will find, has the great merit that it relates a general philosophy of human *being* in the world to rather specific issues of practice, in education and elsewhere. The idea can only be indicated in this paper: The development will appear elsewhere.[2] Let me start by sketching in some aspects of hermeneutics that clearly show its relevance to science.

Interpretation

Hermeneutic thinking comes to us from the 19th century as a byproduct of repeated efforts to improve interpretations of ancient texts—an unlikely source, if ever there was one, for an exciting philosophical trend. Nevertheless, this musty, bookish activity quickly led to a most fundamental epistemological problem: In a fragmentary text, written in a strange language, and belonging to a people whose concerns are radically different from one's own, the meaning of individual words usually depends on the meaning of the text as a whole; yet, the meaning of the whole depends on the individual words. How can one deal with this 'hermeneutic circle' in a nonvicious way? Clearly, some sort of a pre-judgment (prejudice!) must enter in, followed by corrections. Taking a part of the text as a starting point, we 'project'—that is, we throw out ahead—a general meaning for the text as a whole. By 'breaking into the circle' this way, with a 'pre-understanding' that may be conscious or tacit, we give to the parts their initial meaning; and this, in turn, allows us to see whether the parts add up consistently to support the whole. As more text is taken up, discrepancies lead to a second, corrected pre-judgment, and then to another, and so on. Interpretation thus takes place in a back-and-forth movement between the whole and its parts, involving mutual adjustment, and leading (one hopes) to ever smaller corrections.

It did not take long to discover that an almost identical process occurs in many other situations as well: in anthropology when trying to understand a foreign culture, in law, in art, and quite generally whenever someone is trying to tell us something unfamiliar or complex. By the time we get to H.-G. Gadamer, the principal contemporary figure in this development, hermeneutics is universalized: ". . . the whole human experience of the world. I call this experience hermeneutical, for the process we are describing is repeated continually throughout our familiar experience" (Gadamer 1976, 15). Today, the term "philosophical hermeneutics" refers to this generalized concept.

But before we apply to science any part of this framework, one key feature needs more discussion. To some extent, especially when the pre-understanding is tacit, it really is prejudice in the obstructive sense. A person with a true hermeneutic attitude is expected to be aware of this possibility, to minimize it at the outset, and in the back-and-forth movement of interpretation, to reduce it further. Originally, everyone thought that whenever the remaining prejudices could be made insignificant, the true interpretation would be approached. In the 20th century, however, largely through the influence of Martin Heidegger (1926), it was realized that in the most urgent situations, 'true' interpretations are far more elusive. When a language partitions the world in a really different way from our own, when it speaks of experiences we have not had or cannot have, then the pre-understanding includes also an unavoidable and *indispensable* part—intuitive projections that light up the unknown land, even if only for a time, and reveal pathways. For without any idea beforehand of what sorts of things to look for, no progress at all can be made in dealing with the 'circle.' Thus the 'prejudices' take on a positive role.

It follows that an irreducible contribution is made by the interpreter through the approach he takes, especially as it shapes the conceptual orientation. Interpretations are now viewed as *con*structions, not just *re*constructions, and the starting point is no longer irrelevant to the outcome. This does imply, when different interpretive attempts reach different conclusions, that the decision between them may not be possible. How large is 'interpretive freedom'? How wide may the gap be between competing interpretations? That certainly is a question to be faced when we consider different types of subject matter. For now, we need merely note that in the human sciences, of course, the gap is sometimes gigantic.

Science As Language

Recall now Gadamer's universalist claim for hermeneutics—that it inheres in art, law, history, and the "whole human experience of the world"—and in this light let us ask whether it inheres in natural science as well. Good reasons for a positive reply can easily be given. The notions of science as a *reading* of the 'book of nature,' or *interpretatio naturae*, were there already at the beginning of the modern phase—in the thought of Francis Bacon and Galileo, to take famous examples. Today, the oscillating motions called for by the hermeneutic circle are recognized immediately by all physicists as similar to the 'self-consistent' calculations used in many areas of research. Finally, if one compares recent changes in the philosophy of science with the changing view of textual hermeneutics, striking resemblances can be seen: Here too we find a newly-discovered 'tacit' component, a new focus on 'theory-ladenness' of observation, a new questioning of the putative 'approach to truth.'[3]

Having noticed all this, and much more besides, a small number of theorists have indeed begun to describe natural science too as a hermeneutic activity.[4] Nevertheless, the consensus—of philosophers, historians, and scientists themselves—is overwhelmingly the other way. Although it is not possible here to discuss adequately the reasons for this— nor is such discussion necessary just now—let me briefly mention three major points.

The most prominent factor is that, traditionally, hermeneutics has been used to *distinguish* between the human and the natural sciences (as indicated by the earlier quotation from Apel). Hermeneutics, so it is believed, through its attention to *meanings,* bestows on the human sciences their *humanness,* and marks them off from disciplines where *elimination* of the specifically human perspective has become a principle. Related to this is the issue of 'decidability'—that natural science does decide contradictions and disputes as it progresses; that it 'logs in,' so to speak, layer after layer of established knowledge—while hermeneutics remains associated with the phenomenon of 'undecidability.' But the most telling objection usually raised is that a structural difference between the subject matter of the human and the natural sciences makes the former only, not the latter, well suited for hermeneutics.

This difference consists in *language*: In the human or social sciences, the phenomena faced by the investigator are imbedded in a symbol system made for and by humans—a symbol system whose *meanings* must be understood as part of the investigation. The natural scientist, on the other hand, faces nature itself, not its symbols already in being; and the language that finally emerges for dealing with nature is the *scientist's* language, not nature's. Such an activity, therefore, cannot be construed as interpretation.

No doubt, all objections of this kind, and more, have to be dealt with sooner or later by anyone proposing a hermeneutic approach to science in any sense. For the moment, however, let me address the last point—the matter of language. What if we focus our attention not on science as research but on science *as knowledge*, as it faces us all when we *first encounter* it? Suppose we consider not the relation of humans *to nature* but their relation *to a particular science*. In that case, surely, what they encounter *is* a language already in being—the language of that science. And this language, before it is mastered, is for everyone as remote as any the anthropologists have studied, since it too partitions 'reality' in a way different from the language with which we start, the natural language of the 'life-world.'

This shift of focus, from the study of nature to the study of science, is of course a shift to the *educational* situation. By turning immediately in this direction, we avoid for the time being the difficult philosophical question of whether the position of the *researcher* in natural science is itself hermeneutical or, as is often said, 'monological.'[5] Whatever the answer to that may be, regarding the position of the *student* (scientists included!), we are on much more solid ground: Whenever a strange language is encountered, does it not need interpreting? Whenever there is interpretation, does it not entail hermeneutics? For anyone with the experience of learning and teaching science, these questions ring true because, in such situations, what the human being faces is not really the phenomena of nature themselves, but various forms of written and spoken *text*, from lectures to research reports, to textbooks proper—literally, *texts*.[6]

What is proposed here, then, is to look upon science study as the *interpretation* of the language of science, and upon the teacher as the chief *interpreter*. The resulting task is to draw out fully the implications of this point of view. Although the task stretches well beyond the scope of a single paper, I will try here at least to indicate what sorts of implications these may be.[7]

That a science is a language is to be understood not as a provocative analogy, but quite literally, as is done in philosophy; for if we look, we find quickly many of the structural components seen in natural language. The semantics of physics receives much attention because in defining the objects of study—electrons, quarks, angular momentum—it goes far beyond the experience of the life-world. But from the point of view of the outsider, the syntax is no less a problem. It too reflects an as yet uncomprehended ontology. Beginners often say, for example, that a projectile 'has force,' thereby revealing a common and historic preconception, held even by Newton (Westfall 1980, 416). That bodies 'exert' or experience force, but do not 'have' it as they 'have' momentum or energy, is not a minor point in the study of physics, not merely a matter of 'saying it right.'

What is involved is a struggle by the newcomer for a clear view of the ontological landscape, in which forces, being relational terms, are different in kind from entities used to define the state—like energies and momenta. Today, this struggle, a recapitulation of the original Newtonian struggle, is made easier by the fact that an appropriate language already exists. But, as many educators now realize, to understand this language it is not enough merely to learn the definitions of the terms, and go over a few examples of their use. That is why there is an increasing demand for 'real world' experiences in education, an end to 'lecturing,' and so on.[8]

Despite a whole series of reforms during the past decades, the feeling remains, in science and elsewhere, that for some reason the *study* of things is still remote from the things studied; that it does not *enter into* those things, but deals with them from a distance. Hermeneutics suggests here a failure of interpretation—not of *translation,* which can be passive or automatic, but of interpretation as a mode of *being in* that which is interpreted. In developing this point along lines parallel to the work of Gadamer, a crucial transition has to he made, from an epistemological to an ontological point of view. The remoteness we sense in education involves a kind of double distancing: Not only does science look upon the objects of its study from the outside, but, in addition, the student looks upon science from the outside. If the first distancing is to a great extent unavoidable (but even this has been questioned), the second perhaps is not. Addressing just this problem, *ontological* hermeneutics questions the modes in which we relate ourselves to the 'texts' of this world—the mode of the teacher, the student, the authors of the texts. This transition to an ontology of *science-as-education,* is, I believe, the most interesting and promising avenue for reconceptualization of the issues, a development that cannot be pursued here any further. We can, however, take a quick look at an area where the situation, though simple, is highly suggestive.

Misconceptions and Preconceptions

The most clear-cut evidence that hermeneutic thinking is already under way is the work of the past decade or so in the area called 'misconceptions' (Champagne et al. 1980; Novak 1987; Mestre 1991). Out of the realization that meaning is being lost in present-day science teaching, a branch of research has arisen that tries to analyze in some detail how new meaning arises within an already existing conceptual structure. In these studies, using in-depth interviews and other techniques, researchers attempt to identify the ideas students already have about natural phenomena, prior to formal schooling. It turns out that some of these, though contradicted by science, nevertheless persist after the relevant subjects have been studied. Significantly, such 'misconceptions' do not always prevent students from attaining a certain 'success' in problem solving and in their courses. While they have learned some elements of the grammar of the new language, their pre-understanding of the *events* handled by this language has remained unchanged.

However, in the literature that describes all this, after the problems are identified, discussion often becomes diffuse because the origin and character of the 'misconceptions' are poorly understood. The remedy most often prescribed is dialogue, designed to show up internal inconsistencies in students' beliefs. It is here that, even in their epistemological sense, hermeneutic categories and ideas could prove their value most explicitly.

In some discussions, for example, all preconceptions are subsumed under one label (*mis*conceptions), all equally mistaken, it is implied, all to be addressed by guiding the students, first, to see the 'error' of their ways, then to 'construct' the same concepts or ideas that science has already constructed. Consider the much-discussed case of the succession of seasons: For many people who attribute seasons to the earth's changing distance from the sun, all that is needed is a reminder that south of the equator the seasons are reversed; this quickly breaks down the preconception, and clears the ground for other possible causes. So far, so good. Here we had simply a plausible but bad theory. We supply relevant data, elicit normal reasoning, and students usually turn off the wrong path. But other preconceptions on the same list are of a different kind altogether.

When asked about a ball thrown upward, nearly everyone will say that at the top of the trajectory (where the ball is instantaneously at rest), the acceleration is zero; and here no amount of dialogue or reasoning is likely to lead to the desired result, *as the research seems to show*.[9] Why? Hermeneutic categories distinguish preconceptions due to experience in the life-world from other kinds. Such preconceptions (called "fore-having") are formed prior to scientific reflection because we 'already have,' in some natural way, the thing later treated in a scientific way. What actually happens then is not 'correction' by science of a mistaken idea, but an *extension of language* reflecting an extension of concept. We naturally associate acceleration with the experience of cars and, to some extent, with elevators or sports like skiing. In all these situations, acceleration without motion is hardly conceivable, much less can such a thing be recalled. It is not an error, therefore, to surmise that an object at rest, even instantaneously, cannot be accelerating.

Physics says otherwise for internal reasons, not through observation. To understand it, we need some idea of the limit process, at least qualitatively; we look to Newton's second law for coherence; we graph the motion and examine slopes. All this shows why science *does* in fact make this extension of the life-world concept of acceleration—an extension beyond experience, to bodies instantaneously at rest. And it shows clearly why high school students cannot be expected to do this on their own, with or without coaxing. The understanding of the new extension of 'acceleration' relies exclusively on use of the rest of the language of physics, most of which has not yet been grasped. This, precisely, is a job for the interpreter. It is not that students must then be passive, that no further construction is left for them. Rather, interpretation becomes a common task in which the teacher-interpreter, bridging the 'horizon' of physics and the 'horizon' of the life-world, and using the preconception itself as starting point, shows the available routes.[10]

We cannot discuss here what a good interpretation of this kind might look like, or what its possible ontological aspects may be, but hermeneutics suggests at least this: Since the student's use of the word "acceleration" was not really 'wrong' within his own horizon, in the sense of the life-world, the whole exercise should not be treated as a 'correction.' There was indeed a *pre*conception, but no *mis*conception. To the contrary: With great difficulty and good reasons, physicists have used the preconception—the knowledge of acceleration we 'already have'—to construct this new extension of a common concept. But to ask people to construct it anew, *ex nihilo*, during a short dialogue, reflects yet another misunderstanding. When they begin, the conceptual horizon of students does not extend far enough to make sense of any such phrase as 'the acceleration of a body at rest.' As the interpretation proceeds, as students move back and forth between the *part* (the trajectory) and the *whole*—the language of science with its limiting process, its laws of motion, and the rest—their *horizon expands*, increasingly overlapping the horizon of the 'text' (physics itself). Just this, this activity of interpretation in which horizons 'fuse,' Gadamer takes to be what understanding *is* (Gadamer 1975, 273).

Hermeneutics in Science Education

Enough has perhaps been said to indicate that a hermeneutic approach to the language of science is neither an arbitrary idea nor a spurious one. Much has been left out. One goal here was to show that a very general theory of this kind, embracing realms beyond science and education, can also have specific consequences for practice; and that was the point of the discussion of preconceptions. It remains now to take up again, however briefly, some of the other objections that arise whenever hermeneutics is suggested as appropriate to natural science. These do seem formidable at first, and explain why European philosophers like Jürgen Habermas and Gadamer himself have eschewed the step proposed here.

The whole drive to mark off sharply the human from the natural sciences represents, for the guardians of the former, a fear of *scientism,* and for the guardians of the latter, a fear of the *loss of objectivity.* On one side, the formal, operational *methods* of science are invidiously contrasted with *human* concerns; on the other, vagueness and undecidability are seen as inimical to the scientific ethos. Let us begin with the question of humanness. If it is true that hermeneutics in the social sciences promotes there a specifically human perspective, and if, in science education, problems of meaning arise (to some extent) because of the *lack* of such a perspective, then, does it not follow that in the latter realm too hermeneutics should arouse some interest? For the same reason it is valued there, could it not be of value here? The loss of meaning is a general, cultural phenomenon of our time; but too often it is taken for granted that science precisely is the source of that problem—the origin of a detached, objectivizing attitude that brings, first to its own realm, then to the rest of the world, a lifeless, alienated experience. But what if science too—especially the appropriation or study of it—were a *victim* rather than the chief purveyor, of whatever it is that increasingly spoils much of our public and private life, draining their human meaning? Is it possible for natural science itself to be afflicted by scientism?

This is a large question, of course, and it is broached here only to counter the idea that the human/objective dichotomy must necessarily be accepted as commonly understood. I am suggesting that it is not in the best interests of education, nor of science, nor of society—and certainly not in the interest of students and teachers—to 'write off,' so to speak, the *human component* involved in the appropriation of science. How to address this component remains a problem, but it is arguably the most serious long-term problem for science in relation to students and to society as a whole.[11]

Finally, there is the issue that, to the scientist at least, may be the largest stumbling block: Hermeneutics is about interpretations, and especially about interpretations that do not converge—differing interpretations. How, in the study of established science, can that possibly enter? For it has always been a goal in education to show that *here,* where there are *right answers and wrong answers,* where *secure* knowledge is sought, we speak briefly, to the point; and we avoid rhetoric. The response to this basic question cannot be given in a few paragraphs. But I would like to indicate, at least, the kind of response suggested by the framework toward which this paper is a small step.

As background, recall that even in the hermeneutic disciplines *par excellence,* in history for example, there are plenty of facts, and plenty of questions for which a unique answer exists. Any interpretation of the American Civil War has to retain the facts, for example, that Lincoln was president, that slavery was an issue, and so on—these are the 'invariants,' so to speak, under interpretive transformation. Reinterpretations of the Second World War that deny the Holocaust have been produced, but so have 'crackpot' theories about science. None of this need concern us here. The point is that neither 'facts' nor questions with unique answers are the distinguishing mark of the study of nature.

In this regard, it is especially thought-provoking that lately, out of the heart of the engineering profession, have come demands for a more 'interpretive' type of education, more work on problems where there are *no* unique answers.[12] Even engineers, it turns out, cannot merely 'consume' texts and routinely solve cut and dried problems without real context, without meaning. On the one hand, graduates of this kind of study are proving inadequate; on the other, the students themselves are increasingly dissatisfied. There are signs that this accounts for some of the loss of enthusiasm among Americans for an engineering career. It is clues like this that lend credence to the speculative question raised above: Should we begin to view science itself as territory penetrated by 'scientism'? And should we perhaps, as we look for deeper causes, look also for a better name for this disease?

Certainly, the remarks above do not address the broader question of the multiplicity of interpretations. For those who might point to Newton's laws and similar established science, implying that no such multiplicity exists there, I would, for the time being, go only one step further. We cannot take up now the philosophical discussion of the general issue of the underdetermination of theory, though this issue is not irrelevant.[13] Rather, as the final point, let me turn to a favorite analogy in the literature of hermeneutics—the performance of a piece of music or the staging of a play. In all such cases, we start with an accepted script or musical score, which may not be altered. Yet, by common assent, what the audience sees in the work depends on the *interpretation* (the performance) as much as on the script or score itself. Yes, in its bare essentials, in its skeletal form, as a 'script,' Newton's second law must be the same in all textbooks, all courses; but that is only the beginning, not the end, of the process of meaning-giving and interpretation.'[14] It is true that in *normal* physics the various interpretations will not conflict, as they do in some social sciences. This much, then, but no more, may still be left for those who would press the argument. However, to insist that differing interpretations matter only when they conflict, is to miss the whole point of philosophical hermeneutics.

The present paper is a programmatic sketch of a possible line of theoretical inquiry. If the pursuit of this inquiry sheds light on the problems of *understanding* science, one may perhaps hope that by reflection, as it were, some fraction of the light might fall even on science itself. The separation, in thought, of the activity of science and the activity of studying science is, after all, one of those dichotomies questioned by this inquiry.

[18]

Philosophy and the Physical Sciences in Undergraduate Education

ABSTRACT: Perennial problems in science study are examined in the light of recent philosophical formulations dealing with concepts of 'rationality,' 'humanness,' and 'tacit knowledge.' The ideas of Kuhn and Habermas, especially, are used to gauge the degree and significance of the instrumentalist component in the teaching of the physical sciences. Analysis as well as experience suggest that the dominance of this component results in a stunted view of human reason—the essence of which lies in freedom of choice in asking questions rather than in the rigors of logic. But since the role of dialogue in the natural sciences has been rapidly diminishing due to increasing de facto instrumentalism and algorithm-centeredness, an important task for philosophy may well consist in reintroducing the dialectical component to these disciplines, thereby enhancing both their rationality and their humanness. [—M.E.]

Problems in Science Education

It may be expected that, on the basis of experience in mediating education-ally between philosophy and the *physical sciences*, one's view of the role of philosophy in undergraduate study would be restricted to a narrow angle.

This is true; and it is the reason why all my specific examples will have to be confined to this area. *But*—perhaps naively—the intention is larger. There is a saying that if you only know a single blade of grass, but you know that blade very well, then you also know a good deal about the rest of creation.

I would like to say something concerning the role of philosophy that goes beyond what might be expected, given its well-known relation to physics, astronomy, biology, and so on. It seems to me that some funda-mental problems of education today are not so species-specific, and that if

---

This talk was presented at the annual meeting of the American Philosophical Society in Boston, Massachusetts, in 1976.

we discuss them in relation to one or two areas, their relevance elsewhere will become apparent.

However, in line with the blade-of-grass theory, in order to pursue this very goal—of saying something more general than is perhaps warranted by my credentials—I will have to at least begin right in my own back yard, and with very concrete problems.

But that is just it.

Looking in my own back yard, I do find serious problems. And now I ask you to bear with me for a time, while I describe things which, on the surface at least, pertain only to the natural sciences or, even more specifically, to physics. I believe these issues are in *themselves* appropriate for consideration by philosophy—and I will, in the end, explicitly discuss ways in which philosophy can address them; but in addition, I want to repeat that these particulars are also intended as the proverbial blade of grass.

I propose, first of all, to take up two commonly made criticisms of physical science and its education; to show that these are found as well in theoretical (including philosophical) works concerning science; to offer some of my own experience in evidence; and then to draw one or two conclusions about underlying causes. By that time, it will be obvious what kind of role I have in mind for philosophy.

The two questions I now submit have been voiced at the highest professional levels, as well as in more popular literature—and are easily recognizable:

Is physical science human?

Is science really educational?

In regard to the first point, I want to emphasize that this is not just a reference to attacks made by students and educational critics of the 1960s and early 70s. The problem is older than that, and it continues to concern outstanding leaders in the physics community. For example, just this past summer Victor Weisskopf—one of the world's best known physicists—published an article in *Physics Today* with precisely this title—"Is Physics Human?"[1] Another physicist, Stephen Brush, addressed himself to the same question in *Science* about two years ago, when he suggested that perhaps a dose of the more recent historical works, with their surprising new perspectives, may help to humanize disciplines like physics.[2] And one could cite many others.

What are the specific problems involved in these discussions? First the question of antihuman results of technical applications—intended or unintended—is *not* at issue. Most people now do see application as a separate problem; the remaining questions—those that concern us here—are confined to the actual content and process in the study of the physical sciences themselves. Now, in the article referred to, Weisskopf lists essentially three difficulties:

> The problems of physics—especially modern physics—are far removed from ordinary human experience.

> The formulation of these problems involves highly abstract concepts.

> And handling of these concepts, solution of the problems, involves mathematics.

Certainly these are familiar complaints, but to make the list more complete, we have to add still another—one that is especially prevalent—namely, that the practice (and therefore the study) of science is inhuman because it is mechanical; and it is mechanical because it requires precise rational performance of a multitude of prescribed procedures, which thus reduce human thought to something "robot-like" (quoting Brush, for example).

Many ways of dealing with these obstacles have been offered, ways of emphasizing the 'human angle of science'—among them the use of histories that penetrate the veneer, and show the scientist as he 'really' is: not quite so rational, and even not quite so honest. As Brush and other historians remind us, Newton was a master with the 'fudge-factor,' Mendel 'cooked' his data, and Maxwell was inexplicably stubborn in the face of solid adverse evidence. And then one could also mention Paul Feyerabend, who flatly states that rationality is a "boring path," and that the kind of science we need should be more like a yielding courtesan who anticipates every move of her lover, than a stern mistress.

Later, I will suggest that some of these oft-mentioned difficulties are perhaps not really so serious, and most of the professed solutions, while helpful, largely miss what I take to be the central problem. But before we reach this point, where philosophy will become involved more explicitly, I want to state briefly the second major criticism of science education— namely, that it is not at all *educational*. It will, of course, be seen that these two accusations are not unrelated.

In asking whether science is educational, I mean to include all those doubts and objections that have always been a part of the debate as to whether the study of science is of value to anyone other than a future scientific professional (and whether even to him it is anything but training). In other words, does it have educative efficacy comparable to good literature, let us say; is there a 'teaching' in it beyond specifics, beyond application—that has the power to reorient the student in the world, and affect his communicative behavior?

Habermas

One contemporary thinker whose answer is fundamentally negative is Jürgen Habermas. I am referring here to that aspect of his theory of the influence of science on society that is embodied in the metaphor of the "two channels."[3]

To put it briefly, there once were two channels through which this influence was believed to flow: in Habermas's words, "(1) through technical exploitation of information, (2) and through the process of individual education and culture during academic study." However, while the first channel has widened greatly since the last century, and become extremely important, the second has been totally shut off (if indeed it ever functioned at all).

The reason for this has to do with how we understand the nature of scientific theory. It was once thought, says Habermas, that theories in science were concerned with the immutable essence of things. Therefore, "through the soul's likening itself to the ordered motions of the cosmos, theory enters the conduct of life. What ultimately produces "scientific culture," in this view, "is not the information content of theories, but the formation among theorists themselves of a thoughtful and enlightened mode of life." The primary aspect of theory here is that which is oriented toward communication or interaction within a community of human beings.

Unfortunately, Habermas reminds us, this is not the way theories of the nomological sciences function today. Since the past century, we have known that theory can be translated directly into technical power, without exhibiting any influence at all on communication, and without therefore having any separate educative function. What is more, it is now clear that this state of affairs was inevitable—because the hidden interest of the empirical-analytical sciences is oriented not towards communication, nor towards self-understanding, but towards *instrumental* action, that is, towards the control of objectified process. Habermas takes the view that:

> Theories comprise hypothetico-deductive connections of propositions, which . . . can be interpreted as statements about the covariance of observable events; given a set of initial conditions, they make predictions possible.[4]

Thus today, the second channel—the directly educative effect of science—appears as an illusion.

Clearly, to the extent that his picture is correct, it also strengthens the first charge—that of inhumanness. A subject that has no influence on the communicative life of the student, that is in effect already to be viewed as proto-technology (but is itself not yet technology, and therefore not actually productive) is at least in this sense not particularly human.

Here is a well-articulated, attractive formulation of the *innate* and *unavoidable* lack of humanly educative value in the physical and natural sciences. And this, viewed in the context of the more specific ongoing complaints—such as enumerated by Weisskopf, Brush, and others—constitutes a formidable indictment.

But this is not all.

Kuhn

As one reflects on the bleak picture drawn by Habermas, one cannot avoid being reminded that there exists at least one more well known theory of science—and its education—that corroborates this picture at several crucial points: the ideas of Thomas Kuhn.

We recall Kuhn's description of science as an instrument for the solution of puzzles; and we recall his essentially instrumentalist view of theory.

What, after all, could be better "oriented toward the technological exploitation of knowledge" than an instrument for the solution of puzzles? But even more relevant is the fact that *indoctrination, imitation,* and *tacit knowledge* play such a fundamental role in Kuhn's account of scientific learning. And in this context, his discussion of the nature of the typical science textbook is especially apt. For pinpointing imitation as the true mark of "normality," Kuhn has been criticized insofar as it refers to research activity; but as regards education, he certainly penetrates the core of the matter: In the past we tried to imitate in our *lives*, and within ourselves, the proportions and harmonies of the cosmos; now we need imitate only the elements of a paradigm. Once we thought that in studying science we approached—however slowly and deviously—some truth concerning the world; and therefore the interpretation of formalism was a serious business; now we are allowed to imitate anything that *works*, for that is all that is of interest to science—its 'cash value.'

Moreover, this imitation is based on recognizing resemblances *tacitly*—in a way that is not consciously controllable or accessible to verbal analysis, because it is lodged in the very process of perception, the stimulus-to-sensation route. To 'know' Newton's laws tacitly is to absorb as many Newtonian problem solutions as possible into one's senses, to lodge them in the very act of seeing, so that their sole function will be to make possible the solution of other Newtonian problems in the future. It is not a matter of *understanding* these laws, or of using criteria to determine when they are applicable. It is not, in other words, a matter of reflection at all.

Now it is interesting, and symptomatic of the complexity of our cognitive situation, that appeal to tacit knowledge has also been made in the psychological literature (for example) as a cognitive mode that is *particu-*

larly human: It is associated with intuition, with a way of doing justice to the 'contradictoriness of experience,' a way of mitigating, or at least complementing, the rigid, logical, Western approach. And for this reason, some have been inclined to welcome the revelations concerning tacit learning and tacit practice in science itself, and to encourage this mode.[5] But, strangely enough, in Kuhn's expositions, tacit knowledge presents a different countenance altogether. It is, in his words, "just as fully systematic as the beating of our hearts."[6] And precisely on account of this, it is involuntary—an involuntary programming of transformations within the sensory apparatus that may even be studied by means of a computer simulation.

Undoubtedly, those who have seen in this something particularly human have more often referred to Michael Polanyi than to Kuhn. But, on checking, I find no essential difference: Polanyi explicitly recognizes the instrumental character of tacit knowledge.[7] Even for him, the uniquely human is *not* so much the *tacit*, as the imaginative function of the mind.[8]

There is however one important invariant in all these different uses of the concept of tacit knowing—that which makes it a potential antidote to what is seen as objectionable in rationality, and that which makes of it the basis of "normal" science—I mean its *unverbal*, unreflective nature.

For Kuhn, science—at least the normal phase—not only is, but has to be *unreflective*; it holds philosophy "at arm's length, and probably for good reason."[9] Why? So that the puzzle solving instrument may be more efficient. *Habermas agrees*—of course, that is its *real* interest—and consequently its significance for *man as man* can be found only at the level where its *products* have a social impact.[10]

It would be possible to spend more time discussing how particular features of the Kuhnian description could be seen as filling out the more sketchy outline drawn by Habermas, and how these theories are related to various streams in more popular literature. But my object for now is merely to show that, from many different directions, certain problems have repeatedly been pointed out in science education which now also have a place in well-elaborated and respected historical and socio-philosophical analysis.

Response to Habermas and Kuhn

Let us turn now to the matter of a response to these related criticisms.

The type of complaints that were mentioned in connection with Weisskopf's article—remoteness, abstraction, mathematics—appear to skirt a far more vital issue. My reason for this suspicion is based on the characterization 'inhuman,' which is made too often and too consistently to be accidental. I think we must take seriously the fact that to describe an activity as subjectively uninteresting, or unenjoyable, or useless, is very different from saying it is *inhuman* or *antihuman*. Geometry, for example—elementary Euclidian geometry—shares all these features: It is mathematical, abstract, and surely remote from common experience. And for that reason many students are not attracted by it. But it is significant that even today, one does not often hear that geometry is 'inhuman'; and from history, we know that at various times and places, it was considered a most important element in humanistic education.

Nevertheless, in the case of physics, we have this strange, extreme idea: not just remote and abstract, but 'antihuman'—and the same accusation is then transferred to all those other modern studies where the methods of physics have been adopted—some of the social sciences for example.

For this reason, I believe serious attention ought to be given to the alleged *mechanical* nature of the physical sciences: that is, the robot-like quality of the work they set forth. For this alone—if true—would justify and make intelligible the charge of inhumanness.

And it is precisely at this point that the theories of Habermas and Kuhn agree—roughly speaking—and appear to bear out the allegation.

If "normal" science indoctrinates its practitioners into the cognitive mode of tacit imitation, organizes them into a super-instrument for puzzle-solving, and shows no interest in affecting human communication, then this explains the sense in which it can be viewed as robot-like. And the teaching of physics, for example, is then nothing more than a necessary evil; for it emits into the intellectual atmosphere this robot-like mode of thought, which we can only regard as another toxic side-effect of the needs of modern technology.

Of course, the pictures drawn by Kuhn and Habermas are not universally accepted; and we know who has spoken out on the other side. At this point, however, my own experience tells me that, unfortunately, in spite of all objections, Kuhn and Habermas are at least partially right; although the word 'partially' is here all-important.

Insofar as they speak *descriptively* and *sociologically* concerning the points mentioned, and provided that we understand the descriptions to be *mostly* rather than wholly applicable, I have to confirm that, looking from the inside, this is the way things appear to me also. However, when Kuhn admits to a *normative* component in his work,[11] and Habermas argues philosophically about *inherent* interest, *inescapable* development, etc., there I dissent sharply, and I could not agree more with Karl Popper when he said that if the science that Kuhn calls "normal" really is so then it endangers civilization.[12]

Kuhn is surely right about the central role of imitation as a mode of learning and practice *today*. This is precisely one of those points where he has hit a responsive note: Many of those who are immersed in science can read this and say 'of course.' It simply rings true. Likewise, most physicists will readily admit that their theorizing does indeed consist almost exclusively of propositions viewed as correlations between observable events, intended for predictions from one set of conditions to another.

As Kuhn says, a teacher can easily convince himself of the role of imitation by noting how often students complain that even though they completely understand the *theory* they have no idea of how to apply it to the problems in back of the chapter. What happens then is that the teacher (or the text) demonstrates how one does it in a few exemplary cases, and says in effect, "Now go you, and do likewise." The student then begins to see the resemblance between his problem and the examples, which allows him to imitate the method of solution. Only in doing this does he really *learn* the needed correlations—i.e. the 'laws'—and he learns them *as correlations* between 'what is given' and 'what is to be found.'

No one who has not experienced this could possibly have been a teacher of physics. *But there is a side of the story that Kuhn and Habermas do not discuss*—and it is equally important.

If one has taught a sufficient variety of students and courses, one should also be able to recall experiences which are almost the exact opposite of that just described: Some students come to you and say they are perfectly capable of solving the assigned problems, of using the necessary techniques, but they do not quite understand what they are doing, *why* the techniques work, and what the theory really says. For example, one boy will say "F=MA, I know how to use it, but what does it mean?" A group of the best seniors in electrical engineering approach you timidly at the end of a course on electromagnetics and confide that although they have taken many courses in circuit theory electronics, etc., they really do not quite understand what an electric current is—and could you please shed some light on that. Then there is the visiting former student, now working hard in graduate school, who describes with hardly concealed distaste that if he *now* ventured to ask what is actually happening *inside* the system being studied, he would certainly be viewed as odd; he would be advised to pay attention to the input-output relation which will produce the desired result, and the rest is not his business. Most interesting of all is the fact that very often the student is not satisfied just to be shown that this type of problem belongs under one theory, or model, while that type belong under another: He very explicitly asks for explanation, criteria, general principles, discussion, etc. *He wants to know why* you group things in *these* similarity classes rather than others—just what Kuhn says takes place *tacitly*. And mostly *it does*. But many students at the beginning of their career habitually ask such questions as "How do we know?"—"Why couldn't it be that rather than this?"—"Are things that way in reality?"—or—"Isn't this whole picture arbitrary?" In time, of course, the frequency of such questions decreases.

In the context of Kuhn and Habermas, such examples are double-edged: On the one hand, they clearly confirm that much of our education is effectively instrumentalist; on the other, they also testify to the fact that for quite a number of people, this is far from desirable, and probably unnecessary.

You will recognize that Popper, and others, have already said this. The retort was that Popper—a philosopher—was telling scientists how *they ought to act*, and this has nothing to do with how they, *in fact want to* or *have to act*. What I am now saying is that if you look at an unspoiled bright young student—a science major, a future engineer, or even a nonscience major altogether—you find that he clearly *wants* to act more as Popper prescribes than as Kuhn believes necessary.

I am glad that I have been able to verify for myself Einstein's remark that "the ablest students whom I have met as a teacher were deeply interested in the theory of knowledge."[13] But it is important to notice that this is true *not only* of the ablest or brightest, and often without awareness that this is philosophy. On the other hand, there are those who, having received years of standard schooling, show great difficulty in comprehending that a scientific statement can possibly be anything other than a way of solving a problem.

Now if all attention is focused on this last group, or on what happens to the others as time goes on, then the idea that things cannot be otherwise gains ground: and this, in turn, strengthens the present trend.

A Philosophical Alternative

Therefore, the first part of the statement of my problem today can be put in the following form: The physical sciences are more and more regarded as a *necessary evil*, a socio-economic instrument which we cannot do without, but which has otherwise no human 'teaching' to offer.

The reason for dwelling somewhat on the writings of theorists like Kuhn and Habermas is to emphasize that old, popular objections, which are always heard—but which were usually seriously advanced only by literary people and other humanists—are now amply backed by weighty scientific analysis itself. And that one result of this analysis is to erect a high barrier between the natural sciences and the *Geisteswissenschaften*—the sciences of the human spirit.

If we can convince ourselves that somehow the physical or natural sciences are *fundamentally* different from philosophy, different from history, different from psychology and sociology, etc.—then some of us can hope that the 'instrumental' virus that has infected the physical sciences will find these other fields of inquiry inhospitable.

However, I would now like to suggest that—*as in the case of freedom—the human value of knowledge is also indivisible—it is indivisible because ultimately all knowledge rests on the same foundation.* One cannot, in other words, draw a boundary across the intellectual landscape—where territories merge and overlap imperceptibly—and say, in effect, thus far and no farther: On this side, perhaps with the humanities and social sciences, is the human realm; yonder, heathen country; and we will hold the line. This, I believe, is the unjustifiable part in the writing of Habermas. Concerned that all university education should not be oriented to instrumental action, Habermas seems willing, in effect, to write off the so-called nomological sciences, and inject hermeneutics into social sciences as the counterforce that he hopes may hold the line.[14]

But let us be realistic: The physical sciences themselves constitute a fair portion of undergraduate studies; the life sciences are now adopting physical and mathematical methods at a rapid rate. Business and management are being transformed by the computer revolution. And while in the social sciences there is some revolt against the trend of the past few decades, I would not bet on the outcome. I fear, therefore, that the whole strategy of giving up on science as an *intellectual* asset, and holding the line elsewhere, is doomed. Already history is not very popular, and in some of our colleges philosophy itself is treated as an anachronism.

When asked whether he considered the theory of evolution one of those empirical-analytical theories (that carry no educative significance), Habermas replied that, *no*, not evolution, but of course this *is* true of *components* of evolution—like genetics.[15] Thus, we try to hold on to the humanness of *parts* of *evolutionary* theory by throwing genetics overboard. And this piecemeal salami-process must of course take place in other areas also, but does this sort of rear-guard battle offer any real hope?

My suggestion, therefore, is the *opposite course.* The instrumental orientation in science must not be given exclusive sway; *all* the sciences, even physics, do in fact have an inherent educative function, and an inherent interest in communication—even in the telling of stories.

How can this be?

The classical answer consisted of pointing out two crucial elements, which were supposed to lie at the core of all the sciences, and of philosophy as well: *truth* and *reason*.

If science consciously aims at truth, and does so *explicitly* by what it considers *good reason*, then its larger teaching consists of two themes that have profound impact on the self-understanding of the individual and his orientation in the world. But in any case, the important thing to realize is that, by its very nature, science *cannot help* but deal with these two elements in *some* way—even if only implicitly—and therefore it cannot avoid a larger teaching in this realm. This is the unseverable nexus with philosophy—in practice, in the classroom, every day, regardless of what the different schools of philosophy of science might say. It is simply impossible to offer a course in any science without communicating something about what sort of thing *truth* is, and what sort of thing *reason* is, even if no such assertions are intended. The students will in any case draw their own conclusions from what is said and done. Naturally, among the possibilities is the conclusion that there is no such thing as truth, or that reason is an arbitrary game, or that reason is 'rationalization'! Any philosophy of education that ignores this takes an ostrich-like attitude.

The questions that remain are: First, what sort of ideas *are in fact* being conveyed concerning truth? What kind of reason is *actually* being taught? Second, are these ideas necessary? And what is their over-all impact? Such questioning constitutes a philosophical critique of science—not, of course, in its research aspect, but in its educational significance. It clearly involves the problem of knowledge. And I agree that this is today, more than ever, a critical *social* problem (that is, a philosophical problem with social significance).

This is especially so if one suspects a connection between the instrumentalist habit of mind and that which is attuned or susceptible to *magic*—in other words, a feeling that man has to make his way in a world of unknown forces, which he can do only by somehow acquiring a store of formulas that deal with these forces effectively, even if incomprehensibly. It is startling sometimes to observe such an attitude in a future engineer.

De Facto Instrumentalism and Its Antidote

Thus we arrive at the second stage in the analysis of the main problem. I am suggesting that there is an *important job to done*, dealing with the philosophical—primarily epistemological—aspects of science: to bring to the surface and to the focus, and to discuss explicitly, those aspects of the content and process of science which for reasons of policy, of habit, of disinclination or lack of time, are not now receiving due attention in science courses and textbooks. In this, a central aim ought to be to counteract somewhat the de facto instrumentalist atmosphere of today, by demonstrating, and revealing in science itself, a mode of reason that cannot be characterized as 'mechanical' or 'robot-like.' Such a mode must emphasize reason's *freedom*, rather than its *necessity*; and it would, of course, run counter to the widespread view of reason based on the model of deductive logic (with premises outside the scope of inquiry!). That free choice, instead of necessary transition from step to step, should turn out to be an indispensable factor of reason is not an idea often encountered in normal scientific study—because, in practice if not in theory, such study has some time ago ceased to be a *serious dialogue*. But it is this dialectical feature (in the Socratic-Platonic sense), the foundation of science from the Greeks to Galileo—what Habermas and Kuhn now rightly see as absent—that made science manifestly and profoundly human, until our time.

The crucial factor is this: In dialogue it is always up to the participant to *decide* what question to ask, and when to *stop or postpone* questioning so as to go further. This is an exceedingly important freedom, and it places a maximum of emphasis on the seriousness of talk, on the importance of its clarity, simplicity, and *comprehensibility*. It is the unfortunate de-emphasis (if not suppression) of this freedom—as a required element—that is, I believe, more than anything else responsible for the transformation of science into something instrument-like. At the same time, the concentration on *algorithm* and puzzle-solving cannot but implant the notion that *this* is what is meant by *reason*—this tacit selection of an exemplar, this physical or mental performance of prescribed manipulations. Little wonder that when it is discovered that a computer can also perform some such work, then not only science but *rationality itself* appears 'robot-like.'

The assertion that scientific education today is becoming increasingly algorithm-centered, and that much of this algorithmic knowledge is indeed tacitly absorbed, may however be questioned: After all, doesn't science involve far more than algorithm? And hasn't this been precisely the point made by writers like Kuhn and Polanyi?

Unfortunately, the answer to the second part of the question is decidedly *no*; this widely held opinion is precisely one of the misconceptions that have to be overcome. The thrust of the arguments of Kuhn and Polanyi concerning tacit knowledge is that "we know more than we can tell."[16] Such a conviction speaks against the adequacy of language, it heightens respect for imitation, and it reinforces the notion that explicit communication, i.e. *the word,* is far less trustworthy or effective than was once thought. But it says nothing at all to imply the absence of algorithms *tacitly acquired*. As has already been mentioned, the shift is not from 'rational,' verbal rules to something subtler or more mysterious, but simply from verbalized procedures to tacit procedures.

Returning to the first part of the hypothetical objection, we assent—as any good student will after a while—that science is, of course, much more than algorithm in the strict sense. But here we are not describing research, or even teaching as it *ought to be* but rather science education as it generally *is today*. In this context, I would like to relax somewhat the usual connotation of the word "algorithm" (where a solution is guaranteed) and use instead the concept of a *procedure* (or program), which may be algorithmic or heuristic, but is in any case amenable to explicit formalization. And since we are now dealing with the central contention of the argument, the point at which philosophical questions arise, it seems worthwhile to show explicitly the role of algorithm in at least one or two simple examples.

Consider the following typical subjects included in any introductory physics course: kinematics of linear motion with constant acceleration, and the geometrical optics of thin lenses. On an elementary level, such limited topics almost always involve only those problems that are encompassed by a very small number of variables and 'laws.' Thus, not only a procedure, but an 'effective' procedure (a true algorithm), can be readily formulated for the solution of the whole set of typical problems within such a subject area, and *within the usual scope of a course.* A complete flow chart for a kinematics-problem-solving algorithm is shown in Fig. 1.

Subroutine SIMPLE (for solving kinematics problems with constant acceleration)

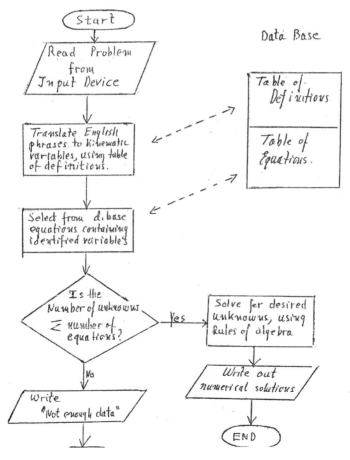

Figure 1. Flow chart for kinematics problem-solving algorithm

Since in many courses the number of equations needed for linear kinematics is no more than three or four, and the number of definitions about half a dozen, this algorithm can easily be 'hand-executed.' The important thing to note is that the data base 'definitions' need be nothing more than translations of English phrases such as "initial velocity," "acceleration," "image position," etc., to the set of symbols appearing in the equations: v_0, a, q, etc. The concepts themselves *need not be well understood*—or even understood at all—by a practical user of the algorithm. That it took Galileo, and many a serious student in every generation, much time and trouble to grasp such ideas as a nonzero acceleration with a zero velocity, or the location of images by means of "rays," does not enter the procedure, and seems irrelevant to its execution.

Has our example been too simple? Well, then, what of a higher-level calculus course? One need only check the textbooks to notice that the main difference consists of a larger data base, more complex procedures, and much more branching. Fig. 2 shows how the previous algorithm is amended to handle the wider realm of calculus-kinematics.

Alterations to Accommodate
More General Kinematics Problems

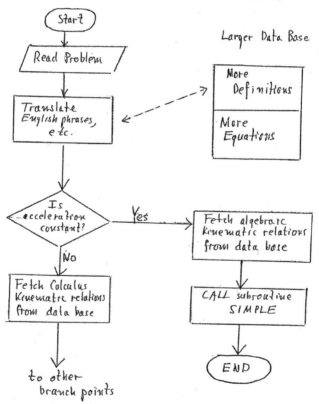

Figure 2. Flow chart for wider realm of calculus-kinematics

It is to be hoped, naturally, that few teachers, on any level, consciously confine themselves to training their students in the execution of such algorithms—which we have presented as the extreme of a continuum. In spite of that hope, however, a lot of teaching in fact comes quite close to this; and there is little doubt that most student problem-solving is actually carried out by tacitly following procedures of just this kind.

Why? Because even when a teacher does more in the way of elucidation, discussion, etc., the textbook usually does far less—while the unquestioned supremacy of *problem-solving as the ultimate goal* soon convinces many that this discussion is peripheral, ornamental, or even boring, since the puzzle-solving tasks may be largely mastered by the *procedures alone*. Habermas's "second channel" is not so much closed off, as it is beset by sclerosis—narrowing, due to lack of use.

An ambitious student will be dissatisfied, at first: The classical problem of infinitesimals is staring him in the face as it did Zeno and Galileo. Situations not compatible with the given data base immediately present themselves. Such phrases as "we may represent light by straight lines called rays" invite a host of questions. The live student will begin to ask them—and what happens from here on makes all the difference. For each answer engenders other questions; how and where is one to stop?

Here Kuhn has given us an enormously significant reply: Teachers do, of course, answer questions, but *science cannot afford to be seriously discursive*. A dialogue on the nature of motion, or of theory, or of abstraction, does *not* ensue—for there are puzzles that await solution. The data base to be covered is *vast*, the procedures are *many* and time-consuming; after a few questions are asked and answered, one must quickly remind oneself of what the real goal is.

But, from the point of view of *reason*, what then is the message?

To set the issue off as sharply as possible, I have sometimes resorted to the following 'dramatic device.' In optics, the algorithm for the solution of most problems dealing with one-lens and two-lens systems can actually be programmed on a relatively cheap hand calculator. After the students have labored long and hard on this subject, the instructor comes into class with a calculator in his hand and invites his audience to challenge the program with any problem they care to make up—magnifying lenses, cameras, projectors, telescopes, microscopes, the better part of a whole chapter, in fact. Students then provide some numbers for the object position, size, distance between lenses, etc., while the instructor punches this data in. A few seconds later the results are ready for display—position of image, size, magnification, etc. The numbers are read out, and the students are asked to check them at leisure.

After the accuracy of these answers has been verified for several such problems, the time is ripe to pose the following question: "Where now is the *reasoning* in all this? Is it in the machine, in the program, in the punching of the keys—or where?"

The demonstration is effective. The student has to face the fact that where he put in so much effort, the little calculator produced results far more quickly, more accurately, and more reliably. While he believed, all this time, that he was "using his mind," he now sees *this kind of work* as nothing more than a feeble simulation of a mere 'gadget,' held in the palm of a hand, and executing a recipe of about a hundred keystrokes.

It is in this sense that I characterize contemporary education as 'essentially algorithmic.' It exemplifies the *opposite of reason* precisely because, at bottom, it is the transfer—tacit, or otherwise—of programmable procedures. And a programmable procedure cannot possibly embody the indispensable ingredient of reason: a *choice* (and a responsibility) at every step, of what questions to ask and which path to pursue— resulting in a dialogue guided only by the *agreement of participants*, and never definitively concluded, but only more or less provisionally suspended.[17]

The task of philosophy in science education can now be stated more precisely; it is to reintroduce what is normally omitted or slighted—*reason*, and thereby also humanness. We need not shrink from this conclusion as too absurd or too much out of line with contemporary images of science and of philosophy. What is being called for is not so different from what philosophy already does in regard to other disciplines—politics, economics, psychology, etc. It is only in natural science that philosophy has taken the 'hands-off,' diffident position in the extreme. Oh, not *professionally*, not in the matter of analysis and theory construction, of course. These abound in the specialized literature of the philosophy of science. It is in *education*, precisely, that the difference, and the diffidence, are glaring.

We are speaking here simply of the possibility that philosophically-minded scientists and scientifically-minded teachers of philosophy can easily get together to design the kind of programs, courses, and materials that could in some modest measure do the job that ought to be done. Such programs need to be guided by only the following general principles:

> To teach the theories of philosophy of science is *not* the goal; but rather to reconsider, in an open-ended dialogue, the same facts, concepts, and theories that the student is already working on elsewhere, together with their implications and consequences.

> The *solving* of scientific problems is certainly not a goal—but only their understanding.

> If a goal must be named, then it would be the same as in any true philosophical discourse: to penetrate as deeply as possible whatever is discussed, to identify misconceptions, and to *experience reason*.

To specify more than this would be superfluous. Where persons exist with an inclination of this sort, they themselves will know best how it should be done.

Conclusion

It is clear that *no exciting new idea* for the role of philosophy has been proposed here. What I advocate is very old, though I do so at least partially on the basis of a rediscovery through practice.

I have tried to connect a perennial criticism of education in the natural sciences with recent *theories about science*—and to suggest that the problem is *not*, as some would like to believe, merely one of pedagogy or public relations, wherein the rational aspect must be somewhat *diluted*; or, as others say, that the rational really *does not* or *ought not* play such a prominent role. On the contrary, the rational has *already been diluted too much, and that is the problem*. The *humanness* of an activity is better demonstrated by showing *its reason* than its unreason. Therefore, the role of philosophy suggested here is not just an answer to the question, How can one spread philosophy further and wider to the undergraduate?—but a proposal for a serious effort on an important issue.

[19]

Hermeneutics and the New Epic of Science

ABSTRACT: Despite a number of important statements to contrary, such as those of Mary Hesse, the prevailing opinion about the role of Hermeneutics in contemporary natural science is still essentially negative. A major argument for this, as given recently by Gyorgy Markus, rests on a delineation of the sciences by means of key elements such as 'scientific text,' 'scientific practice,' and science's 'literary objectivations.' This paper challenges the dominant view by suggesting that the common delineation is too restricted. I call attention to a genre of scientific literature—neither disciplinary nor 'popular'—that has grown in our time and become important for communication between intellectual subcultures. The common features of this literature are a concern for the general 'evolutionary paradigm,' for unification of the sciences, and for the relation between natural science and socio-philosophical issues. I show that the authors of this literature, mostly well-known scientists, view the present scientific scene with unusual exhilaration; and I examine how their writings contribute to a coherent, cognitively powerful, influential, evolutionary vision of the world. I then argue that at least in regard to this genre, which deals in high-level scientific metaphors, hermeneutics is not only possible, it is necessary. [—M.E.]

Introduction

In a recent paper entitled "Why Is There No Hermeneutics in Natural Science?" the philosopher Gyorgy Markus accepts the widespread belief that "the natural sciences have lost their direct and general cultural significance" and that "the literary objectivations [of these sciences are] not read

This paper was published in *The Literature of Science: Perspectives on Popular Scientific Writing*, ed. M.W. McRae (Athens, GA, and London: University of Georgia Press, 1993) 186–209. An earlier version was presented at the 1987 annual meeting of the Society for Literature and Science in Worcester, Massachusetts.

today by a wider public beyond a narrow circle of professional experts."
Reasoning along this line, he concludes that hermeneutics plays no serious
role here because today science rests on an institutionalized author-text-
reader relation which is normative and impersonally imposed. In short, we
now have "cultural closure of scientific discourse upon itself" (Markus
1987, 28). However, by 'text' and literary 'objectivations' the author means
purely professional, technical reports. About popularizations he says nothing, assuming, apparently, that for the serious business of science such literature is peripheral at best.

I do not propose now to blur that well-established distinction between
'science itself' and 'popular science,' between science for peers and science
for the general reader.[1] I do intend, however, to challenge this dichotomy
as a useful classification in the 1990s, especially in the context of possible
roles for hermeneutics. For there exists today at least one other group of
scientific writings, a rather influential one, that deserves a name of its own.

Consider that literature which has come to the fore during the past
quarter century or so, most full-length books combining high-quality writing with scientific depth. Typically it deals with subjects that have philosophical or social or humane implications. It is aimed neither at the specialist nor at the casual reader. It is a body of writing in which scientific and
literary seriousness are blended and offered up consciously in an attempt to
break through professional barriers with a message deemed important
enough to be worth the effort of *demanding* reading. It comes mostly from
scientists, especially the famous; and often it involves fairly complicated
technical ideas. Of particular importance are its interdisciplinary effect and
its impact on serious thinking in nonscientific fields.

My examples are Jaques Monod's *Chance and Necessity* (1971),
Manfred Eigen and Ruthild Winkler's *Laws of the Game* (1981), Ilya
Prigogine and Isabelle Stengers's *Order Out of Chaos* (1984), Steven
Weinberg's *The First Three Minutes* (1979), Steven Rose's *The Conscious
Brain* (1976), Douglas Hofstadter's *Gödel, Escher, Bach* (1979), Joseph
Weizenbaum's *Computer Power and Human Reason* (1976), and E.O.
Wilson's *On Human Nature* (1978). Various other books, to which I will
also refer, could be included here and grouped according to how well they
exemplify a certain ideal type.[2]

Unfortunately, because the scientific/popular dichotomy *is* so potent in
our language, such books are still classed as 'popularizations.' Yet they differ qualitatively from the usual simplified expositions that do not claim to
convey a serious message; and they differ also from even the best popular
essays like those by Stephen Gould or Lewis Thomas, which, for the most
part, are easily digestible and often biographical or personal. In contrast,
the works I cite distill the deeper meaning of scientific advances on fairly
broad fronts, calling attention to cognitive implications that bear on

human self-understanding. In the 1950s a work of this type stood out conspicuously; today it is part of a genre. And so the question arises: If (contra Markus and others) this literature still has, or has *again*, the capacity to mediate between a knowledge of nature and genuine human interests, *then what interests or which modes of life does it address? And what, if anything, does this have to do with hermeneutics?*

Let me begin to answer these questions by reviewing two types of responses offered only a short time ago: the first presented in typically pithy fashion by the physicist Richard Feynman; the second, not so concise, by Jürgen Habermas. Both agree in one important respect: The *nature* of the subject matter of modern science makes it extremely difficult, if not impossible, to relate this matter to the concerns of human life.

In hindsight, we can now say that both men were wrong. And it should be enlightening to try to understand not only what they missed but why they missed it, for that may indicate where the literature of science is heading today and what role hermeneutics might play in it. With this purpose in mind, let us now consider what I will call 'Feynman's question,' 'Huxley's reply,' 'Habermas's error,' and 'Wilson's insight.' We begin with Richard Feynman.

> Poets say science takes away from the beauty of the stars—mere globs of gas atoms. Nothing is "mere." . . . Far more marvelous is the truth than any artist of the past imagined! Why do the poets not speak of it? What men are poets who can speak of Jupiter if he were like a man, but if he is an immense spinning sphere of methane and ammonia must be silent? (Feynman 1963, 3–6)

This, incidentally, from a textbook on physics. Of course we all know the conventional reply, which comes in a variety of styles. Aldous Huxley, for example, puts it as follows:

> The world with which literature deals is the world in which human beings are born and live . . . in which they love and hate. . . . [The scientist] is the inhabitant of . . . the world of quantified regularities. . . . Until some great artist comes along and tells us what to do, we shall not know how the muddled words of the tribe and the too precise words of the textbooks should be poetically purified, so as to make them capable of harmonizing our *private and unsharable experiences* with the scientific hypotheses in terms of which they are explained. (A. Huxley 1963, 8–9, 107; emphasis added)

Note that both Huxley's reply and Feynman's question contain certain assumptions about the *possible* relations of science to literature and to life. For Feynman, that link must be aesthetic; poetry being the ideal language, he does not see why the forms of expression used in classical mythology cannot also be applied to a scientific description of the world. He hungers, it seems, for a modern Lucretius. Huxley agrees about the language problem but doubts the solution is so simple, because the dissonance between *private experiences* and a symbolic, *abstract* world is too great for the kind of harmony literature tries to bring about.

It has been noticed, however, that from a historical and philosophical perspective, other relations between life and science are possible. In a series of well-known essays, Jürgen Habermas explored one perennially recognized alternative, only, alas, to reject it. According to Habermas, there are just two possible "channels" through which science can bear on "the practical consciousness of a social life-world": by means of technical exploration and by means of *mimesis* (the Greek conviction that "through the soul's likening itself to the ordered motions of the cosmos, theory enters the conduct of life"; norms of individual human behavior are ascertained first by understanding cosmic proportions). The philosophically (scientifically) educated person would then possess "practical" knowledge, or what Habermas calls "action-orientation" within a human community. And this "channel," this ancient practice of scientific study for the sake of self-formation, was still thought to be open in the nineteenth century—at least in German universities (Habermas 1970, 53).

Today, however, because scientific knowledge is overwhelmingly instrumental, because theory can be effective *without* first having to transform the person *as a person*, that channel is forever closed, says Habermas. In our time, "information provided by the strictly empirical sciences can be incorporated in the social life-world only through its technical utilization." This means that action-orienting self-understanding is now available from historical-hermeneutic study—of the humanities and certain social sciences. But "taken for itself, knowledge of atomic physics remains without consequence for the interpretation of our life-world, and to this extent the cleavage between the two cultures is inevitable" (Habermas 1970, 52). Here, then, is the more analytical answer to Feynman's question, shared by Markus (1987, 45) and widely accepted.

Yet in all this, something must have been overlooked, unless the literature to which I refer—that third category—is far less successful than we have reason to believe. To shed some light on what Huxley missed, what Habermas left out, and why Markus still ignores a phenomenon that is far from marginal, let us look at this literature more closely. Although the books I listed are well known and hardly need additional reviewing, I propose now to highlight briefly certain of their common features, to show why the mood, the attitudes, and the message of these scientist-writers is (for the most part) quite different from what one might have gathered by reading Huxley and Habermas in the past, or Gyorgy Markus now.

The first similarity is a certain overall structure typical of the most impressive works of this genre. Not that structure will be our main consideration. The argument will turn more on content than on form. But the link between these two aspects of a text is too well known for the latter to be ignored even when the former is central. Moreover, I want to call attention to the different *kinds* of resonance that exist between the works we are considering; not only the message but a certain rhythm in proclaiming that message is shared. Let me call that rhythm 'P-S-P,' for philosophy-science-philosophy. In its pure form, the book begins with a philosophical survey, sometimes only an introduction, in which certain human or social problems are posed and related to some aspect of science; that science is then expounded in a lengthy technical or quasi-technical section, and the work ends with the original philosophical problems to which solutions or responses are now offered *on the basis of the scientific content just presented*.

The initial discussion, the first 'P,' often refers to the two-cultures split explicitly, promising new ideas that might heal the wound. For example, "In the blending of biology and the social sciences . . . the two cultures of Western intellectual life will be joined at last" (Wilson 1978, 10); or: "The alienation of science from life . . . is about to be overcome by the evolution of science itself" (Yantsch 1981, 5). These philosophical introductions usually pose other kinds of dichotomies as well, such as objectivity versus animism (Monod), scientific materialism versus outdated mythologies (Wilson), timelessness versus time-consciousness (Prigogine and Stengers), chance versus necessity (Monod, Eigen), or formal logic versus human reason (Weizenbaum). In some of the works, a conceptual contrast of the latter kind may be linked directly to the two-culture split. Thus, Prigogine and Stengers say quite bluntly that "the dichotomy between the 'two cultures' is to a large extent due to the conflict between the atemporal view of (classical) natural science, and the time-oriented view that prevails in a large part of the social sciences" (1984, xxviii). As a rule, both the theme and the underlying mood of the authors are very serious, even when (as in the case of Hofstadter or Eigen and Winkler) a lighthearted mode of exposition is used. The aim is not just enlightenment but something closer to *revelation*.

The meat of the P-S-P sandwich, the scientific middle portion, ranges over all the disciplines, from flip-flops to Church's thesis, from allosteric reactions to space-time singularities, and much more. Without a doubt, resolute attempts are often made in these books to convey some scientific knowledge, considered valuable in itself and needed to develop the philosophical points. Yet the most striking feature is the flagrant excitement: the repeated assertions that only within the *recent* past have new discoveries made it possible to treat such philosophical and social subjects scientifically, and the eagerness of the authors to spread these insights beyond their own specialized community. "The selective theory of evolution did not take on its full significance, precision, and certainty until less than twenty years ago," says Monod (1971, 24). Similarly, Weinberg: "Throughout most of the history of modern physics and astronomy there simply has not existed an adequate observational and theoretical foundation on which to build a history of the early universe. Now, in just the past decade, all this has changed" (1979, 2). In general terms, of course, such developments are widely known, but these authors see similar things happening not just in molecular biology and astronomy but throughout the sciences; and the cumulative effect causes the exhilaration. Thus, Steven Rose: "Only recently have the tools become available to approach the problem of the brain in a way which makes it possible to ask meaningful questions" (1976, 46). And in *Order Out of Chaos*, based on *chemical* research, we read once again, "In the past few decades, something very dramatic has been happening in science, something as unexpected as the birth of geometry or the grand vision of the cosmos as expressed in Newton's work. . . . On all levels, from elementary particles to cosmology, randomness and irreversibility play an ever increasing role. *Science is discovering time*" (Prigogine and Stengers 1984, xxviii).

The final, philosophical 'bread' of our sandwich includes unabashed *calls for a new morality* or a new 'vision' of the world. In this regard, many of the authors view themselves as members of a like-minded group, as contributors to a historically continuous movement in which every text builds on the others, amends the others, and reflects back on common intellectual forerunners.[3] I will refer to these more committed authors as the 'core group.' Thus, Monod's version of an "ethic of knowledge"—similar to that of George Gaylord Simpson (1949, 339–349)—posits *objectivity* as the first value and foundation of moral "authenticity," from which lower-level goals may be derived. Wilson, of course, goes further, including among his values the diversity of the gene pool, universal human rights (because we are *mammals!*), and the genetic transformation of the human species to forms of higher intelligence and creativity. All this, we are to understand, follows in some serious sense from the science of the middle portion of the sandwich. Without that portion, the tone of authority in this literature would have no visible foundation; and that precisely is why the structure of the texts so often adheres to the format just outlined.

But let us proceed to the message itself, for that is what truly unites the authors of the core group, making of each individual work a contribution to something like a common project. What I shall try to show is that while the literary products of this group derive from widely different sciences, although they come in a variety of styles, and their authors differ on various points, taken together, we have here the telling of *one and the same story*.

It is the story of evolution: evolution explicated in greater detail than ever before, deepened, unified, extended far beyond biology—'universal' or 'cosmic' evolution. Although it is an old story, composed gradually with enormous labor, much of it was still not in place even within the last thirty years. Beneath an apparently adequate first-order explanation lurked a number of difficult problems, discerned occasionally even by laymen. It was possible to continue improving various segments of the theory piecemeal, ignoring the problems or shelving them for later, but the result was a partial, patchy picture, not beyond questioning by such luminaries as Niels Bohr, Wolfgang Pauli, Ludwig Wittgenstein, or Karl Popper.[4] The ultimate goal, a truly seamless, thoroughly convincing, all-inclusive science of development—if possible at all—still lay in the future.

Now, however, the excitement evident in the books I am discussing results from a feeling that *this goal is being attained*, that we are living in a time of 'the last frontier,' of the crucial unification of the scientific worldview, of the dissolution (at least in principle) of the final mysteries. And that feeling is further heightened by what we find in the corresponding literature of the nineteenth century. Even before Darwin, and quite independently among thinkers of all sorts, the idea had arisen that some sort of 'development' was going on in line with physical law, encompassing everything from stars to molecules to man. In the writings of Robert Chambers (1845), Herbert Spencer (1958 [1880]), John Fiske (1902 [1874]), Ernst Haeckel (1905), and others, we see (to our surprise, perhaps) the beginnings of what is now in full swing.[5] It is not just that 'cosmic evolution' was a concept much in vogue, but more impressively that (already then!) they 'had it right,' so to speak, about many specifics of the mature theory as well as its overall shape. That is why today when Eric Chaisson, an author on my larger list, finds kindred thoughts in the work of Robert Chambers, for example, his passion to communicate this discovery can hardly be contained (1987, 94–97). Against the historical background, twentieth-century achievements appear as so many pieces of the jigsaw puzzle falling into place, so many promises kept.

The Third Genre of Science Literature

Among the oldest of mysteries is *purposiveness*. Granted natural selection, how do mindless molecules make up those intricately coordinated, purposeful structures and processes for selection to work on? Monod's *Chance and Necessity* (which sold more than 200,000 copies in French alone) addresses this question by showing that purposiveness (teleonomy) is now unambiguously traced to 'microscopic cybernetics,' biological feedback mechanisms utilizing certain stereospecific chemical reactions. So crucial is this point for philosophical conclusions, and so interesting, that we find it demonstrated anew, at different levels of organization, by nearly everyone. Douglas Hofstadter, for example, is concerned mostly with other things, but here he is, in *Gödel, Escher, Bach*, making the same argument—for the kingdom of ants!

Achilles: Oh wait. Either the behavior is purposeful or it is NOT. I don't see how you can have it both ways.

Anteater: Let me explain. . . . Let's say a signal [team of specialized ants] is moving along. As it goes, the ants which compose it interact . . . with ants of the local neighborhoods. . . . The signal will remain glued together as long as the local needs are different from what it can supply; but if it CAN contribute, it disintegrates, spilling a fresh team of usable ants onto the scene. . . .

Achilles: . . . I'm beginning to see. . . . From an ant's-eye point of view, a signal has NO purpose. The typical ant in the signal is just meandering around the colony, in search of nothing in particular, until it finds that it feels like stopping. . . . No planning is required, no looking ahead. . . . But from the COLONY's point of view, the team has just responded to a message. (1979, 320–21)

With teleonomy reduced to the material processes of efficient causation, with intermolecular forces as the ultimate *anima*, the appearance of purpose in organisms does not imply the separate existence of some motive power, nor any extraordinary law for living matter. The long history of 'animisms,' crude or subtle, religious or Marxist, must finally come to an end, says Monod. Only a sober, austere materialism can today do justice to what we have learned. Thus an accomplished biochemist provides scientific grounding for a familiar existential position: In our cold, uncaring universe, man's "destiny is nowhere spelled out, nor are his duties. . . . It is for him to choose" (Monod 1971, 180). Accordingly, what we now ought to choose, he says, is an "ethic of knowledge," for this at least will finally overcome the disastrous split between knowledge and values.

Inescapably, it seems, the strictly biological problem of teleonomy is linked to a larger question of purpose: the meaning, or lack of it, in human life itself. If, to take one example, chance mutations are the ultimate source of the rise of consciousness, then is not this whole story—our being here, our 'ascent'—a mere quirk, as Monod thinks, improbable and devoid of purpose? And is not that, in turn, reflected in the fragmentary, jerky quality of individual existence, admittedly part of the nature of things but exacerbated by the character of our civilization and our man-made sciences? The search for some sort of meaningful *unity*, as a major theme in philosophy and culture, is hardly new. What surprises is the powerful reemergence now of this same theme in the *literature of science*.

In 1949 Nobel Prize-winner Herman Hesse published his monumental novel, *Das Glasperlenspiel* (*The Glass Bead Game*). Although central to the novel, Hesse described this game only vaguely while hinting that its function was to unify different cultural spheres at a deep level. About thirty years later, in a most unexpected way, Manfred Eigen (another Nobel laureate) and Ruthild Winkler gave precise form to their conception of just such a game (Eigen and Winkler 1981). But this construct, involving the same two elements that interest Monod, chance and necessity, is far from fictional. The authors' goal is nothing less than to demonstrate *necessity* in the working of evolution: to remove that existentially disquieting element of contingency, *and*, at the same time, to show that at different levels of organization (including the prebiotic, the linguistic, and the economic) historical processes all have essentially the same deep structure. Beneath apparent chaos, these authors claim, there is a form of unity after all, and of beauty.

The game is one of chance, employing dice, but with the important proviso that its *rules* are chosen for a purpose. In some of the most striking versions, called appropriately 'Selection' and 'Survival,' the rules for placing and removing beads from the board are designed to embody the characteristics of birth, death, mutation, competition, and Darwinian selection. The resulting 'play,' the increase or decrease of different-colored beads (representing species), becomes then a simulation of the evolutionary process. What gives the idea added glamour is that such games have, of course, been programmed on computers; and the results of many 'plays'—suggestive, intriguing patterns—are repeatedly and lavishly displayed, in full color, alongside the ubiquitous Escher.

The scientific message here is subtle: The game concept applies to just about everything. Random events taking place under particular constraints (rules) yield particular types of order, in evolution and in other natural and human processes. "It is still a matter of chance which mutations occur, and in what sequence. But it is a matter of predictable necessity that . . . selection will occur . . . defining a gradient of 'value' . . . for improvement over time" (Eigen and Winkler 1981, 59).

These bead games embody an extraordinary unification. One and the same activity offers satisfaction to scientific curiosity, to the aesthetic sense, and to philosophical interests—a resounding variation on the theme of Herman Hesse. Thus, in its wide embrace, its multi-dimensional approach, *Laws of the Game* represents well a major feature of the genre itself. For despite some disagreement on the role of chance as against necessity, despite differing degrees of enthusiasm for the significance of artificial intelligence (AI), and despite important differences of interpretation (about which more later), the impressive fact is the *convergence* of these authors on a well-defined set of interrelated themes: *extension* of the 'evolutionary paradigm' as far as possible, *unification* of the sciences, and *reconciliation* of science with a specifically human reality.

In *Order Out of Chaos* we see *all* these themes, developed even more energetically and in full awareness of the goal. First, by tracing out the 'tragedy' of how our world progressively split in two—quality versus quantity, sensible things against abstractions—Prigogine and Stengers set the stage in historical-philosophical terms. Next, like directors of a mystery play, they make sure the audience understands what the trouble is and who is hurting. The trouble is the prevailing scientific myth—still the *timeless*, deterministic, Newtonian machine-world. And the pain, which affects all of us, is caused by our inability to find an acceptable human role within the context of such a mechanical worldview. The analysis is familiar, but the recommended strategy is quite new. In his own work and that of others on the same trail, Prigogine sees a *scientific* solution to the *sociomoral problem*. An entirely new paradigm, by restoring *time* to its proper place, will help to heal the old cultural split and erase (again!) that pernicious distinction between scientific and ethical values (Prigogine and Stengers 1984, 312). How does science accomplish such a thing? By attending to the other two themes mentioned above: evolution and unification of our view of nature.

It was the achievement of Prigogine and his collaborators in chemical kinetics to show that even in the macroscopic realm (that is, aside from quantum mechanics), the behavior of a system sometimes reaches what is now known as a 'bifurcation point,' where a 'choice' is made between quite different but equally possible future states. Yet this choice is governed not by law but by small, unpredictable fluctuations. Especially significant for prebiotic evolution is the fact that sometimes these choices lead from disorder to highly organized structures: 'order out of chaos,' triggered, it seems, for no particular reason that science can discern. Here is that element of chance treated formally by Eigen and Winkler, punctuating the periods of predictable causality. The universe, it turns out, is not a 'machine' because it is only *piecewise* determined. As a result, the backward direction of time is not equivalent to the forward direction, even in physics, and the 'evolutionary paradigm' (organic, not mechanistic) applies to all the sciences without exception.

In another realm still, that thesis is powerfully supported by Weinberg's much-quoted little book, *The First Three Minutes*. Again we hear of a dramatic confrontation between 'two chief world systems,' one eternal, stationary, and timeless (steady-state cosmology), the other evolutionary (the big bang). And once more, crucial evidence piles up rapidly in favor of the evolutionary story, now extending back to the first one-hundredth of a second of the universe's existence. As the unification of biology with chemistry gave us in detail the structure of DNA, so now high-energy physics combined with relativistic cosmology yields 'the new genesis,' not just as ingenious speculation but as a highly worked out scenario. At last, as Einstein used to say, we are getting closer to the "secrets of the Old One." Neither the stars above nor the moral law within are beyond our reach. Once these antipodes symbolized the mysteries of being and pointed to a power beyond. Today, as science accelerates, the assault on both fronts is going rather well.

That this success owes much to the momentum and ingenuity of the computer sciences is widely known. Not so well appreciated, perhaps, is the influence of just the *concepts* and the *metaphors* involved, as distinguished from raw hardware power. It is the merit of books like *Gödel, Escher, Bach* that they convey those concepts in an exceptionally vivid manner and make that influence more understandable. What Kurt Gödel, M.C. Escher, and J.S. Bach had in common, according to Hofstadter, is that each in his own field noticed, appreciated, and utilized brilliantly the concept of self-reference, of a thing turning back on itself, or supporting itself, or reflecting itself, like the scenes in Escher's lithographs.

Again, as with Eigen and Winkler, the aim here is at once scientific, aesthetic, and philosophical: not only to show by means of pictures, stories, riddles, and musical notation how many different and surprising forms this self-reference can take, but to drive home the point that herein lies the deepest of all deep truths ever encountered by the human mind. The 'strange loop' is offered to us by Hofstadter as quite possibly the answer to the problem of consciousness.

How does *any* information-processing system answer questions about what it is doing? It sets up within itself a model of itself and then examines that model, examines itself. As Hofstadter puts it, a "*self*-subsystem . . . can play the role of 'soul' . . . in communicating constantly with the rest of the subsystems and symbols in the brain . . . [in] the monitoring of brain activity" (1979, 387–88). Machines do it crudely; animals do it better; in humans, just the complexity of neural hardware is enough for all sorts of unexpected effects.

> Emergent phenomena in our brains—for instance, ideas, hopes, images, analogies, and finally consciousness and free will—are based on . . . an interaction between levels in which the top level reaches back down towards the bottom level and influences it, while at the same time being itself determined by the bottom level. In other words, a self-reinforcing "resonance" between different levels. . . . The self comes into being at the moment it has the power to reflect itself. (1979, 709)

Hofstadter does not quote Kierkegaard here, though he could have done so: "Man is spirit. But what is spirit? Spirit is the self. But what is the self? The self is a relation which relates itself to its own self, or it is that in the relation [which accounts for it] that the self relates itself to its own self; the self is not the relation but consists in the fact that the relation relates itself to its own self" (Kierkegaard 1954, 146).

Now, however, as the philosophical question acquires a scientific answer, computer science and biology converge. Insofar as consciousness is a biological problem, to be accounted for by evolution, the *natural emergence* of such self-modeling 'strange loops' out of the growing complexity of organisms looks like a truly saving idea, just what was needed. At last the fog is lifting. The entire panorama is coming into sight, and the "secrets of the Old One"—from big bang to consciousness—seem within our grasp. So rapidly is the picture changing that most of us have not yet realized what is going on.

The New Epic

How the various contributions fit in as pieces of the big picture is now clear. From Darwin's original theory, the lines of extension radiate downward to prebiotic (chemical) evolution as expounded by Prigogine and Eigen; to cosmic evolution as described by Weinberg, Paul Davies, and the astrophysicists; to human culture as Wilson explains in his theories of sociobiology; and finally, through the work of brain physiologists and AI researchers, to consciousness itself. It is, as Monod would have it, an 'objective' story, free of odious dualisms, incorporating 'scientific materialism' at its foundation.

Nevertheless, and quite surprisingly, this story too is called a 'myth' by the foremost author on my list, Edward O. Wilson. It is Wilson who, seeing an 'epic' in the emerging scientific tale, gave explicit articulation to what was sometimes only elliptically said; and it is he who drew conclusions in their most challenging form. This epic, in his view, plays the same role and has much else in common with those very myths it is now displacing. Wilson's vision of the extended, fully worked out, 'seamless' theory of cosmic evolution is not normal science—it is the vision of a narrative of unprecedented scope and persuasive power, "far more awesome than the first chapter of Genesis or the Ninevite Epic of Gilgamesh" (Wilson 1978, 209). The evolutionary epic is simply the 'best' of myths, according to Wilson, and the one 'destined' to prevail because, unlike its predecessors, it does *rest* on science, even if it is not itself science.[6]

In this perspective, the literature I am discussing takes on an obvious and rather serious role. So vast is this new epic, and so detailed, that no one book can encompass it. What we have instead is a large number of major and minor works, on various levels of 'scientific literacy,' each telling some part of the story, or commenting on it, or interpreting it. Yet taken together, all these constitute the epic itself, which in turn rests on the more solid bedrock of the technical 'objectivations' discussed by Markus.

Books that have had a strong impact and have followed closely the ideal type outlined above, both in form and in the orthodox telling of the tale itself, I have called the 'core group.' But to complete the picture we must keep in mind that numerous others are part of the genre. Eric Chaisson's *Cosmic Dawn* (1981) and *The Life Era* (1987) belong here if taken together, though the scientific and philosophical components are offered more or less separately, each in its own book. Yet Chaisson, an astronomer and leading enthusiast, finds in the new evolutionary epic precisely what the old 'animist myths' proffered: "Only when complex organisms arrive at the dawn of the Life Era does *the Universe* acquire self-awareness. . . . 'Matter has reached the point of beginning to know itself.' . . . This, for me, is life's purpose and meaning, its *raison d'être*—to act as an animated conduit for the Universe's self-reflection" (1987, 229).

Roger Sperry, the famous brain physiologist, has also contributed to the genre. His *Science and Moral Priority* is so concentrated on evolutionary ethics that he nearly skips the technical part altogether (relying, no doubt, on widespread familiarity with his split-brain discoveries). For Sperry, as for Monod, since science now offers a cosmic view of life that "renders most others simplistic by comparison," a new ethics is desperately needed, one that "would lead to the designation . . . of what is good, right, or to be valued morally, as that which is in harmony with, sustains, or enhances the orderly design of evolving nature" (1985, 50). And so on. I skip over many other deserving works. The consensus is overwhelming, perhaps somewhat boring.

However, lest this be the final impression, and the whole genre seem trite, let me end the survey with a dissenting voice: Joe Weizenbaum of MIT, a prophetic figure capable of using the words *God* and *grace* without awkwardness, has made for himself a special place in this literature. The attention received by *Computer Power and Human Reason* (1976), though remarkable, is nonetheless quite understandable in view of its content. Here is a professor of computer science, a member of the leading AI research community, challenging the basic faith of that community head-on—technically, ethically, philosophically. He simply rejects the idea that human reason can be properly *understood* by means of formal logic or computation alone. It is not just that the achievement of current AI programs is *mis*understood and therefore overrated. The more basic problem lies in our failure to distinguish between the computational aspects of thinking and judgment, on the one hand, and, on the other, those aspects which are inseparable from humanness as such. What is human about our reason, says Weizenbaum, is our individual and species *experience*. No other information-processing system—however well it *simulates*—can know, for example, what it feels like to be humiliated as a human being, because no other system incorporates within itself the experiences of a human being. Yet judgment, a part of human reason, employs, among other things, just such knowledge. The computer metaphor of mind is misleading, he insists, because it points to the mind's instrumental uses only and reflects an instrumental view of being human in the world; and this in turn leads to precisely the kinds of social and communications problems that Habermas has addressed.

It is interesting that another recent addition to our genre, *The Emperor's New Mind* (1989), by Roger Penrose, takes up the same objection from a more mathematical point of view. But while Penrose writes as a scientist skeptical about the finality of our latest theories, Weizenbaum takes a more existential stance. He is concerned not just with the validity of certain claims but what we do with those claims. He worries about the uses of the computer metaphor, but even more about its reflexive effects on those who adopt the metaphor. Unlike most scientists, Weizenbaum frequently emphasizes his role as *teacher*. We are all teachers, he often says; our example counts. And it appears he means not just academics but all human beings as such.

Then why do I include him in my list? In what way does his sort of book contribute to the telling of the grand evolutionary tale? This point is important.

The significant thing is that even Weizenbaum does not question the evolutionary epic in its overall shape. He speaks, in effect, from within that epic. Indeed, his view that human reason is linked to feelings derived from experience harmonizes well with sociobiology, for example. He grants that machines with sufficiently complex internal models of themselves could acquire "a kind of self-consciousness"; he merely wants it understood that *that kind* of consciousness would be nonhuman, the self-consciousness of an organism from another planet, say (Weizenbaum 1976, 210). Penrose too accepts the main plot of the evolutionary epic. Nearly all critics within the scientific community take issue with only some *interpretations* of *pieces* of that plot; hence, in general their objections add weight to the story rather than detract from it, while arguments like Weizenbaum's penetrate so deeply that by force of relevance they more or less *join* the literature that tells that story. In other words, the serious reader who is attracted to the other books on my list but misses Weizenbaum's (or something similar) has not read *broadly* in this literature.

By including partial dissent, I wish also to call attention to the fact that some aspects of the evolutionary epic are still open, still under discussion. Although many members of the core group speak without qualification, often dogmatically, the genre as a whole must be regarded as somewhat fluid at the periphery. There, significant questions of interpretation, though not raised often enough, are still possible. Yet, this possibility (to which I will return) makes the literature of the epic more serious and interesting than would otherwise be the case. It is this, as much as anything, that saves it from being merely an intellectual fad, a passing ideology, the hobby-horse of philosophically untutored scientists.

How widely such books are read and how great is their influence can be gauged to some extent by looking at the following example. In a major work, *Does God Exist?* (1980), the theologian Hans Küng offers what he takes to be a scientifically cognizant contribution to Christian apologetics. When examining the place of God in natural history, Küng freely accepts the conceptual categories given in the works of Monod, Eigen, Weinberg, and the others. In a section dealing with the possibility of God's direct role in the creation of life, this Catholic theologian feels constrained to make the negative conclusion (1980, 642–47), largely on the authority of *Laws of the Game*.[7]

In view of all that has been said here, the answer to the question I posed at the beginning seems embarrassingly simple. Of course: From its earliest moments, science has had for us a profound *cosmological* significance, quite aside from any other interests or uses or appeals; *cosmological*, used here in the Greek sense of 'all-encompassing,' includes the large, the small, the inner, and the outer. We know this was so in ancient times and during the first scientific revolution; it was evident again in the late nineteenth century; and today once more that significance is increasingly felt, technical difficulties notwithstanding. Yet, during the middle decades of our own century, many commentators tended to overlook this deeply ingrained *cognitive* interest, an interest in the nature of 'the whole,' as distinct from all technical and aesthetic interests.

Why? The positivist period from which we are now emerging was unique in one respect. By and large, both scientists and philosophers were absorbed by details of fairly limited problems and subdisciplines, emphasizing mechanism and methodology. Quite appropriately, many have called this the "age of analysis." But today, as we witness the new drive for theoretical 'unification,' we are reminded that an age of analysis is perhaps not typical at all. To the outsider, the larger significance of science could indeed appear marginal during a time when scientists themselves insisted their role was limited to such analyses, and when historians, seconded by philosophers, supported this impression. That is why Aldous Huxley's point about abstraction and the remoteness of science from human life did have an *interim* and partial validity.

Now, however, it is becoming clearer that when analysis is pursued to such great depths, the cognitive significance of science cannot be seen in the fruits of each analytical achievement, expounded singly and in isolation. The spinning ball of methane and ammonia, *in itself*, moves few of us. Whatever images one may derive from it will not impinge on the larger picture of life and of ourselves, regardless of the language used. But if we connect that ball of gas to the evolution of stars, then to the origin of the elements of which our bodies are made, and then to a hidden language of molecular letters that spells our nature, to teleonomy and consciousness— well, that is *another kind* of story.

Contrary to Feynman and Aldous Huxley, the present interest in science's cosmological meaning, and scientists' improved ability to communicate their message, were not spearheaded by the appearance of any great poet or artist. Once again it is the writers' intense *cognitive* involvement with 'the nature of things' that spurs on their imagination. It is the cosmological interest that now leads to new, ingenious forms of expression like those of Hofstadter or Eigen and Winkler.

About the scientific view of the world, Aldous Huxley said, "Although it is a determinant of human nature and human behavior, this reality is nonhuman, essentially undramatic, completely lacking in the obvious attributes of the picturesque" (1963, 107). That the new evolutionary epic is 'undramatic' or 'unpicturesque,' none of our authors or their readers would now grant. But from the cognitive point of view, Huxley's omission looms even larger. What he meant by a 'determinant' of human behavior, was, for example, atoms, genes, or glandular physiology: the physical stuff itself, technically described. What he clearly did *not* mean was the controlling power of scientific stories *at higher levels of integration*, the power that Wilson, Prigogine, and many others today speak of, the same power that Aldous Huxley's brother, Julian, did notice, well before the present eruption of literature of the third genre.[8]

The Common Error and the Place of Hermeneutics

Apparently, Habermas's oversight is of the same kind: too restrictive a view of the *possible* effects of science on the life-world. Of course, technology has today come into its own as never before; and *mimesis* of a cosmic harmony, though not entirely out of fashion,[9] is, in our manmade environment, extremely difficult. No one understands better than Habermas the role of action-orienting thought structures, dialogically and hermeneutically acquired. Yet the traditional desire to maintain logical separation of the *Naturwissenschaften* from the *Geisteswissenschaften* (or the autonomy of 'differentiated value spheres,' as Max Weber puts it)[10] leads Habermas to assume that natural science cannot seriously "take over the function of meaning-creation" for the "communicative practice" of life (1982, 276). And so, when he reaffirmed that *mimesis* is not an option for modern man, he could only conclude that the *one* channel between cosmos and human self-understanding was closed for good (1971, 302–304).

But now, near the end of the twentieth century, although we do understand that a knowledge of stellar structure cannot affect our self-concept *directly*, that nonequilibrium thermodynamics will not give purpose to life, we are beginning to realize that this is not the end of the matter. For it is not *in themselves* that these bits of knowledge exert their effect; today they do so through the *metaphors* they support, especially the high-level metaphors—extrapolations and integrative appropriations of many disciplines. Through *this* channel the sciences do indeed have action-orienting power. As Mary Hesse puts it, "Society interprets itself to itself, partly by means of its view of nature" (1980, 186). And in a more provocative, insistent way, that is also what E.O. Wilson is saying. That is the basis of his grand view of the evolutionary epic.

In acknowledging Wilson's realistic grasp of what is actually happening between science, literature, and life, we need not accept his *interpretation* of the epic. Here Habermas's instinct to keep this grand vision at arm's length acquires a certain intellectual and moral stature. Habermas is afraid that if it is accepted in the objectivist manner, and if it does in time gain the full power prophesied by its advocates, then the ground of human self-understanding will have been even more drastically misplaced. For then the ways of seeing *from within* the paradigm of this epic, *its* modes of comprehension, will be still harder to confine to the native expert cultures. Arising primarily from instrumental interests, these modes favor an objectivizing approach, which eventually leads to something like 'colonization' of the life-world (Habermas 1985b, 209; Bernstein 1985, 23). Therefore, to keep communication 'unconstrained' by scientistic language, Habermas and many like-minded thinkers wish to preserve domains within the culture where truth is tied to consensus rather than to method.

But if Habermas's basic assumption is wrong—if the recent separation of science from the moral and aesthetic spheres is not final, as he believes—then much depends on *how* these grand myths of science are appropriated, how they fit in with the whole cognitive ecology. And this is just where reflection on meanings, dialogical clarification, and communicative rationality become especially important. By constructing high-level metaphors and extrapolating natural science to the proportions of a 'grand myth,' the new story-tellers enter the realm of hermeneutics. By impinging on the life-world, not just instrumentally but as socially orienting theory, natural science does indeed blur that boundary between 'value spheres' (of which Weber and Habermas speak), and does open itself to modes of discussion usually reserved for the human sciences.

Such a mode, *hermeneutics*, as the term is used here, is the conscious attempt to interpret difficult linguistic or symbolic messages coming from remote or unfamiliar sources, whatever they may be. It requires a circular, bootstrapping type of thought—the 'hermeneutic circle'—in which the parts are needed to explain the whole, but also the whole to understand the parts (see Gadamer 1984). Recently, philosophers of science have pretty much come to agree that such 'circular' routes to knowledge lie at the very foundation of what we call science. Data are not 'raw' and pure; phenomena are not free of theory. Rather, on the basis of phenomena *interpreted* with the aid of older theories, we construct *new* theories that *re*interpret the phenomena. If one grants even as a remote possibility that language is not isomorphic with any natural ontology (Arbib and Hesse 1986, 181) and that physical theories may therefore progress in the sense of pragmatic problem solving without necessarily converging to a 'final picture,' or without converging in a way we can monitor, then natural science itself, with all its successive models and theories, represents just such a hermeneutic effort: the effort to 'read the book of nature.'

In the philosophy of science, positions for and against this view are well known.[11] For my purpose, though, the negative arguments (such as those of Markus) are irrelevant because the 'objectivations' here, typical of that third literary genre, embody 'mythical' extensions and metaphors of high generality. At this level, interpretation is at least implicit; if it is not made explicit, the danger of naïve objectivism looms larger.

When, for example, Eigen and Winkler describe biological or pre-biological evolution as a 'game,' complete with 'strategies' (whose?) as well as rules—and they do so not just for the sake of popular exposition but for scientific understanding—they surely are saying something that nature 'in herself' is not saying unequivocally to everyone, not even to most biologists. What they do, among other things, is to reinforce a powerful metaphor, the 'strategy' of the genes, which itself becomes an object of further scientific theorizing. To treat the world this way is an act of interpretation. A further, compounded interpretation follows, as in the work of the sociobiologists, if one proceeds to build on this metaphor to draw conclusions about human nature. Thus, interpreting evolution in terms of 'strategies' leads to an 'understanding' of certain differences in human behavior that turn out to be extremely controversial.[12] Such disputes puzzle people unaccustomed to dealing explicitly with the interpretive component of science. But the preceding discussion shows why controversies of this sort should be expected, and why they do not differ much from those in the social realm. A hermeneutic approach here is useful, first, because it calls attention to the interpretive component and to possible alternative interpretations. Second, at a higher level still, it demands a historic understanding of *science as a whole* before particular theories, even the latest and the best, can be truly understood outside the context of immediate technical application.

Again, in regard to the sciences of intelligence and consciousness, the same hermeneutic need arises, only more so. Here, from the very beginning, knowledge and self are inextricably related. The problem is sufficiently difficult that interpretation often comes in layers, metaphors nested in still more general metaphors. A large, complex AI program is pictured as a 'bureaucracy' of competing jurisdictions, of courts, or of 'little societies' (of subroutines), each with its own character. Not mere professional jargon this; for in time, the same metaphors and sub-metaphors, willingly adopted, refer to the professional himself—and to us all (Turkle 1984, 287).

It is the merit of the third genre of science literature that, on the whole, it pays attention to the historical perspective, exhibits this tendency of science to generate metaphors at every level, and stimulates examination of these metaphors. When we try to understand, for example, consciousness as self-reference (strange loops), we face squarely the twin insights that science is *metaphor* and science is *self-knowledge*. How are such metaphors to be understood *if* their reflexive, self-characterizing function is to be appropriated? Note how crucial to this issue is the *purpose* of the appropriation. In the context of scientific work on computers, the question—How are such metaphors to be understood?—either has no place at all or it has an entirely different meaning with a relatively straightforward answer. Only in an educative context of enlightened cultural concern, the context assumed by the authors of that third genre of science literature, does interpretation of metaphors become a central task. In an informal way, this precisely was Weizenbaum's task in regard to the computer metaphor of mind. By critically examining the personal and social implications of this metaphor, as well as the science behind it, its development, and its effects, he clarified substantially its *meaning for us*.

It is not true that hermeneutics applies only to the 'human sciences.' It applies to any interpretive science. No doubt, natural science in its most formal modes is least interpretive; but wherever it *is* interpretive, there hermeneutics is relevant too. There we are once again dealing with humanly constructed texts, their metaphors and high-level theories, which come to us with a context and a history.

[20]

The New Epic of Science and the Problem of Communication

ABSTRACT: One of the popularizers of science on a high level, E.O. Wilson, has characterized the evolutionary viewpoint as an "epic." Although this name is appropriate, it will be argued that this new myth poses a threat to communication among individuals and subcultures, because it is a silent epic, without voices of any kind. The basic vision of the evolutionary epic is materialistic, and its dynamics is a combination of determinism and chance, without an ultimate purpose even though local purposiveness can—and with high probability does—result in an epiphenomenon from the materialistic dynamics. Hence, unlike the traditional social myths, the evolutionary myth provides no models or heroes or guides. The harmful emotional consequences of the evolutionary myth can be mitigated by changing the mode of narration and recognizing the human element in the formation of the myth—that this formation had a history, that it emerged from dialogue among the scientists and philosophers who contributed to it, and that the formulation was not just a disclosure but a construction. The constructivist narration would endow the evolutionary epic with a voice and would link the scientific and the existential instead of splitting them apart. [—A.S.]

The Old Epics and the New

In a previous paper,[1] I discussed a recent genre of science literature that has had remarkable success and influence. It includes such books as Prigogine's and Stengers's *Order Out of Chaos* (1984), Jacques Monod's *Chance and Necessity* (1971), Eigen's and Winkler's *Laws of the Game* (1981), Steven Rose's *The Conscious Brain* (1976), Douglas Hofstadter's *Gödel, Escher, Bach* (1979), Steven Weinberg's *The First Three Minutes* (1977), and E. O. Wilson's *On Human Nature* (1978). I argued that, aside from structural

---

This talk was given in Albany, New York, in 1988, at a conference of the Society for Literature and Science.

similarities, members of this genre have in common the aim of elaborating on what Wilson calls the grand 'myth,' the scientific "evolutionary epic . . . more awesome than the first chapter of Genesis or the Ninevite epic of Gilgamesh" (Wilson 1978, 209). In the present paper, I consider the threat posed by this 'myth' to the prospects of communication among individuals and subcultures; for despite the enthusiasm of the authors, the threat is a real one. However, it will be my aim to show that perhaps within the recent literature itself, some variants already suggest how, by means of a certain uncommon type of narration, that threat may yet be mitigated.

The close relation between the books mentioned is indicated superficially by the amount of mutual quoting and referencing. More important, however, are the shared philosophical and thematic commitments. All these works stemming from different disciplines, written by different kinds of people who usually expound different subjects, interlock at a more general level of meaning: Each contributes some important idea or plot to a highly elaborate story, which, taken as a whole, is what we call the grand evolutionary epic. There is, moreover, a remarkable agreement among many of our authors that, with the proper interpretation, this grand epic can indeed be the answer to the well-known problem of intellectual fragmentation: the split between the cultures—literary, scientific, esthetic, political, Eastern and Western. "We must open new channels of communication between science and society," say Prigogine and Stengers, and it is "in this spirit" that they offer their book (see Prigogine and Stengers 1984, 22).

How the various contributions reinforce and complement each other can be seen by noting briefly the major points in each of these works. Monod, for example, deals with the fundamental problem of teleonomy in living organisms: From a physicalist or materialist point of view, how can the behavior of mindless molecules bring about that purposefulness which we see everywhere in the biological world? *Chance and Necessity* addresses this question by showing that purposefulness, or teleonomy, is now unambiguously traced to "microscopic cybernetics," biological feedback mechanisms consisting of certain chains of stereo-specific chemical reactions—complex, but nothing more than the principle of the thermostat. In *Rules of the Game*, Eigen and Winkler argue that such teleonomy *had* to arise from basic, nonliving molecules, as part of a universal "play" of chance under the constraint of environmental "rules." Going back even further, Weinberg offers a cinematic script of the now "standard model" of the evolution of matter itself—the whole universe from the initial, primordial egg. At the opposite end of the time scale, Wilson and other sociobiologists extend the reach of natural selection to human behavior, while books like Hofstadter's employ ingenious devices to show that even the last mystery—human consciousness—also fits in. With the aid of information science, what Wilson had to treat in general terms can now be explained graphically, logically and dramatically: Consciousness, seen as a "strange loop" in which a computer program monitors itself, is just another form of self-organization that takes place when the tasks of the nervous system become sufficiently demanding (see Hofstadter 1979, 387–88).

Drawing the inference from these developments, and from the famous research of his Brussels school, Prigogine and his coauthor contribute an overall principle: The "evolutionary paradigm," as they call the story, is so fundamental and pervasive that far from being subject to criticism by the traditionally most incisive science—physics—it is *physics* which must now be recast (like everything else) to conform to the evolutionary paradigm. In the future, they say, the basic concepts for understanding nature will be irreversibility, directionality of time, randomness—concepts underlying evolution—not reversibility, determinism and lawfulness, the past contributions of physics.

At last, it seems, the fog is lifting. The entire panorama is coming into view, the succeeding stages of the long saga is becoming clearer, and the "secrets of the Old One," as Einstein put it—from big bang to self-reference—are within our grasp. It is a story based on "scientific materialism," says Wilson, and it is the best story we humans can have on this subject. The books on which I have focused represent only an upper stratum of a much larger category that includes several types of writing: more limited or focused essays by equally distinguished scientists, books on a somewhat higher technical level aimed at the scientifically educated nonspecialist, and many volumes of the more popular kind. For example, where Wilson is content just to draw the bounds, to place morals on a genetic "leash," Roger Sperry actually derives positive ethical directives in his recent *Science and Moral Priority* ([1983] 1985). What is to be "valued morally," he concludes, is that which "sustains or enhances the orderly design of evolving nature . . ." (50). This exhortation from a distinguished neurophysiologist is intellectually serious but nontechnical and completely accessible. On the other side of the spectrum, *The Evolutionary Vision* (a collection of AAAS symposium papers) discusses scientifically many of the key concepts related to Prigogine's work: dissipative structures, synergetics, the hypercycle, and catastrophe theory. Nevertheless, in his introduction to even such a book, aimed primarily at scientists, the late Erich Yantsch defined clearly the larger implications that stimulate many of the authors:

> The evolutionary vision tries to understand evolution as a total phenomenon. . . . [E]volutionary dynamics at all levels is considered as related in kind, not just in a formal way. In such a sweeping view, evolution becomes manifest in all kinds of *creative* dynamics, from the processes bringing about particles and atoms as well as galaxies and stars all the way to human creativity in art and science, technology and social design. . . . The ultimate importance of the evolutionary vision lies not just in its power of unifying scientific thinking . . .

but also in the philosophy it expresses—a philosophy close to life and its creativity. *The alienation of science from life . . . is about to be overcome by the evolution of science itself.* (Yantsch 1981, 1–5, emphasis added)

Students of history will notice that this is not an altogether new view of things. True, some remarkable scientific achievements in the 1960s and 70s gave this idea not only a new impetus but several new twists; so that compared with the immediately preceding decades, it may well seem like a recent development. But if we look back about one hundred years, we see clearly among leading philosophers and scientists the forerunner movement: Herbert Spencer in Britain, Ernst Haeckel in Germany, John Fiske in the United States—all produced massive works advocating a philosophy of "cosmic evolution." Then as now, books appeared treating astronomical, geological, biological, neurological, social and moral evolution as part of the *same narrative* (see Fiske 1874). Then as now, the most recent scientific discoveries—Darwinism, biological cell theory, astronomical spectroscopy—were cited as prime evidence that, for the most part, the story is true.[2] Then as now, the mere disclosure of such unity, such an all-embracing drama, was expected to lead to profound philosophical reorientations.[3]

The difference that strikes us is that in the 19th century, evolution was a *story in outline* suggested by the vaguest, most general principles, while today it is linked at every level to sophisticated experiments and detailed theories. Spencer's "law of evolution," for example, spoke of an "integration of matter," a process "during which matter passes from an indefinite, incoherent homogeneity to a definite, coherent heterogeneity . . ." (Spencer 1958 [1867], 394). Haeckel's "law of substance," from which he drew the most far-reaching conclusions about the reality of a biogenesis, was simply the conservation of matter and energy (Haeckel 1900, 211). In contrast, today the claim of a biogenesis, or pre-biotic evolution, is related to experiments synthesizing crucial organic building blocks, to detailed studies of the Earth's early environment, and to the work of Prigogine and Eigen and many others.

However, in the expected philosophical reorientation, we detect no major difference: Now as then, there is a shared belief among the converted that by finally replacing older mythologies, the new scientific epic will provide an overarching background for human self-understanding, moral reflection, and personal and social communication. As Wilson puts it, "What I am suggesting is that the evolutionary epic is probably the best myth we will ever have. . . . And if that is the case, the mythopoeic requirements of the mind must somehow be met by scientific materialism so as to reinvest our superb energies . . ." (Wilson 1978, 209). In the more radical language of Monod, this amounts to "a thorough revision of ethical premises . . . a total break with the animist tradition, the definitive abandonment of the 'old covenant' . . . forging a new one" (Monod 1972, 170–71).

Of course the historical perspective shows that this making of the 'new covenant,' this replacement of religious and other epics by a new evolutionary story, is a process that has been going on for over one hundred years. In the United states, its progress is clearly marked by two events: Tennessee 1927, and Arkansas 1981. At the Scopes trial, the new epic tried to force a legal opening into the public schools, to acquire there a place for itself beside the old. And it lost that battle. In the Arkansas rematch, having meanwhile gained the coveted place by other means, the new epic in effect formalized its victory and its *exclusive* right to the role of a *canonical* narrative in American education (see Nelkin 1982).

Yet, despite the enthusiasm of people like Wilson, Yantsch and Prigogine for the beneficial effects of the evolutionary paradigm, we know that communication between subcultures has not been improving during this long period of transition. Whether it is C.P. Snow's two cultures, or Max Weber's three cultures, or some other reasonable distinction, cognitive differentiation and alienation has increased, not decreased, ever since that era in the nineteenth century when it could be said that a truly universal humanistic-scientific culture did exist. But that was under the aegis of the *old* canonical epic (see Young 1985, Ch. 5).

There is reason, therefore, to look closely at the new epic. Perhaps, as has been alleged many times, the decline of communication stems in part from the intrinsic features of precisely this story; if so, then the enthusiasm of a Prigogine or a Wilson is badly misplaced.

The Language of the New Epic

Traditional epics—folk, national, religious—have always acted, in a sense, like catalysts of communication. It is taken for granted sometimes that the common ground required by successful communication is already available whenever a natural language is shared. No doubt, for the simpler sort of message, and the urgent need, this is adequate. But communication that is conceptually more demanding, because it deals with the more subtle aspects of life and peculiarly human situations, can be speeded and sharpened enormously (even provoked) when there is available also a shared *stock of relevant models*—models of character, of problems, of dialogues, of behavior in particular situations. A model is used as a kind of landmark, to orient oneself in the human environment. Again like a landmark, it can be used more widely in shorthand language, to communicate to others the lay of the land; further still, a model has derivative, indirect uses as material for metaphor.

It is of course the entire corpus of traditional narratives of a culture, the shared stories, that supply these models—in some societies more obviously than in others, but to a degree in all. As Alasdair MacIntyre has emphasized, it is by listening to stories of wicked stepmothers and prodigal sons that the very young first learn what a mother is, what a home is for, and what the "cast of characters may be in the drama into which they were born" (MacIntyre 1981, 201). If, as urban adults, we have some notion of what a 'wise peasant' is, and can use it when the rare occasion arises, then, in all likelihood, it is through Tolstoy or possibly Solzhenitsyn that this concept reached us. Even in modern America, in the middle of a secular discussion of social issues, it is still possible to hear someone say "when Jesus answered Peter. . . ."

Against this background, we see that the grand epic differs radically from all the narratives it is expected to replace: Essentially, it is a *silent* epic—devoid of voices of any kind. Neither a human nor any other intelligence plays a role in those happenings whereby information accumulates in the genotype. It is a story of information without deliberation, of communication without intention. What, then, in the new epic of science corresponds to the catalytic effect on specifically human interaction that traditional narratives have? Wilson, to his credit, has noticed the lacuna and tried to fill it. "Every epic needs a hero," he says, "the mind will do" (Wilson 1978, 211). Just what does this mean?

If Wilson is talking about the species mind in general, which he takes to be "an epiphenomenon of the neural machinery of the brain" (202), then the hero's story is the story of his evolution from a ganglion to the amphibian thalamus, to the mammalian cortex, and finally to the neocortex of the human being. According to the canonical paradigm, it is a story of accident and environmental constraint in which, for the most part, the hero does not engage in any dialogue comprehensible by the modern human being. Generally, the language used to narrate the epic is in the third person: A descriptive language in which the collective voice of science reveals the significant events, and provides the concepts needed for their understanding. If, then, because of the absence of internal dialogue, it is the *language of narration* of this epic that shapes future communication, we can easily predict how that communication will be altered (for of course the process is already going on).[4] Many current concepts, categories, metaphors will be displaced by corresponding elements of science—as, indeed, some writers desire.

This means especially the replacement of concepts native to the lifeworld by others that arise in the process of scientific objectification. Wilson (1978, 165) describes human altruism by situating his object on a spectrum extending from shark (individualistic) to Portuguese man-of-war to termites (social). In so doing, he introduces the concepts 'hard-core altruism' (wired in) and 'soft-core altruism' (reciprocity). It is not at all far fetched to ask how, with such a terminology, are we to understand, for example, a Raoul Wallenberg? Tradition suggests here a relevant model, the 'good Samaritan'—no doubt inadequate for the extraordinary Wallenberg, but still conveying many essential features of the situation. The new epic also provides landmark models—the shark, the Portuguese man-of-war and the termites. It is clear, however, where such models and such terminology lead.[5]

Advocates of the new epic as a means of bridging intercultural gaps have apparently not yet faced this old problem. Overwhelmed, it seems, by the momentous changes in theoretical structure within the sciences, and by the speed with which the new epic has been fleshed out recently, the enthusiasts overlook the fact that no scientific advance, no startling new idea has diminished significantly the gulf between the *language* of the sciences and that of the life-world—not Prigogine's time-consciousness, not Hofstadter's strange loops, not sociobiology's kin selection, and not the big bang. What, then, should our response be to this new epic which is undoubtedly gaining strength? Outright rejection is one possibility, and some subcultures have taken that route, so that at present the intellectual map is spotty in this regard. But rejection is not an alternative for those who understand science and take it seriously.

Two Types of Narration

In the remainder of this paper, I should like to show that elements of one possible answer to this problem can in fact be discerned in some of the literature I discuss. The potential ill effects of an invasion of the life-world by scientistic language can be mitigated if, for the new epic, one adopts a starting point and a mode of narration different from the one now most common. The difference, interestingly, depends on a choice of philosophy of science: Compared with the conception of science as a mirror, realistically reflecting the objects around us, some variants of the newer, post-Kuhnian views alter radically the import of the story. Yet the effect is more on our *appropriation* of what is narrated than on the plot itself. Let us consider what is involved.

Either by direct narration, or by implication, the grand evolutionary epic begins at the beginning—with the big bang—and, after various stages of cosmic and terrestrial evolution, ends understandably with the advent of human culture. Let us call this, or any piece of it, the "standard narration."[6] More significant even than the chronological sequence is the characteristic stance of the narrator—that omniscient, omnipresent subject familiar from the traditional novel. He speaks authoritatively, revealing the ultimate story in its ultimate form as an essentially finished product. Of course the product is not complete in the strict sense. The omniscient voice usually points to several as yet unfilled areas, calling for further work, more investigation. But while it is admitted that ongoing science could in the future change the subplots in a few places, about the main plot itself, the core-theme of the grand epic, no one harbors any doubts. The product is handed over 'finished' in the sense that a microcomputer, sold with various empty slots and sockets, is also finished. Future development may fill those slots and change the operation of the machine in various ways; but these changes themselves are largely foreseen, bounded, and taken into account. The most enhanced microcomputer cannot become a brain, nor even a Cray.[7]

Sometimes when the story is told all the way to the end, it acquires a certain mysterious twist: At that point science comes into being on earth, and—during a period of time infinitesimal in the grand epic—puts together, or discovers, or constructs *the grand evolutionary epic*. As in an Escher drawing, the contained turns out to contain within itself its own container. The story of the discovery (or construction) of the grand epic is of course a subplot of that epic itself; it is a minor epic only by comparison, for it includes all those lonely spirits, geniuses, dabblers, and philosophers, who, by public struggle, by clever experiment, by closely argued debate on the most abstruse questions, by error and guesswork and conceptual revolution, brought into being those laboratories, societies, publications, and industrial plants that now make up our very tangible world of science.

It is noteworthy that the minor epic, which caps the grand epic, is usually left out in most narrations; or, pieces of the former are specially selected to provide the grand epic with empirical support. It is left for the history, philosophy, and sociology of science to tell that tale more fully, if desired. Yet the strange, Möbius-like twist in the larger story implies that by starting at the 'beginning' and thinking our way 'straight' through time, we nevertheless arrive back at the same point—from the big bang 'itself' to the big bang of scientific reconstruction! But which comes first? On such a strange loop is it not possible to start with the latter? Is not the minor epic, though incredibly short, the proper entrance point to the gigayears of the grand evolutionary tale?

Evidently, the topology of the Möbius strip shows only that a choice does exist and that, implicitly, the standard narrative has made that choice without explanation. But what guidelines do we have for a more considered decision?

Such guidelines are indeed available. During the past few decades, the philosophy and history of science have reconsidered and redefined for us two broad orientations relevant to the question. The understanding of science that prevailed before the twentieth century, that seems most natural and therefore still endures, is some form of 'strong realism.' According to this view, the entities posited by science 'exist' objectively and independently as tables or chairs do. Even when science consciously uses partial models, restricted in purpose, still a mirror-like *correspondence* is assumed between the features of these models and *reality* itself. Consequently, what scientific work as a whole tries to attain, and does attain in the long run, is an ever widening, ever clearer *disclosure* of an objective world—a true, or approximately true, description of it (Arbib and Hesse 1986, 10, 159; see also Rorty 1980 and Popper 1972, Ch. 8).

We recognize immediately, from our discussion of the standard narration of the grand evolutionary epic, that by its very structure, it adopts and conveys *this view*. For in describing the 'first three minutes,' in describing the emergence of birds from reptiles, of altruism from kin selection, and so on, it takes for granted that the basic categories of science correspond to the basic categories of an independent, objective reality; in other words, that science approximates the natural ontology of the world. Therefore, no dialogue, no discussion is included in the epic itself concerning the validity of this assumption. Such a discussion is felt to be unnecessary, of professional interest only, since the *product* of science, the final picture that interests outsiders, is taken to be essentially *correct*.

In contrast, consider now that group of approaches I will call *constructivism*. We must be careful, however, not to confuse this with certain radical cousins like idealism or instrumentalism, or the extreme position of some sociologists of science.[8] Constructivism, as I use the term, does not deny the reality and independence of an external world, nor its role in shaping scientific theories; nor does it regard science as a proto-technology, useless for understanding the larger picture. Its basic insight, now almost universally accepted, is that scientific evidence always *underdetermines* those high-level theories, or theoretical ontologies, out of which world pictures arise (Arbib and Hesse 1986, 6, 177). Speaking plainly, this means that no grand epic can ever be *deduced* from science alone. Empirical evidence can establish with confidence only limited, local phenomena. Beyond these, however, with theoretical models *metaphorically extended*, we do 'construct' a larger symbolic world, embodying the ontology of the models used—an ontology which may not correspond to 'nature herself' (Arbib and Hesse 1986, 181). It is for this reason that conceptual revolutions, radical changes of the scientific world-view, have in fact taken place in modern times.

By no means does this imply that high level theories are arbitrary; in general, metaphoric extension is a guided process. Nor is it possible today to return to an earlier, positivistic age, when any extension beyond the data was suspect. Yet the thesis of underdetermination does tell us that other considerations must then play a role whenever high-level theory is a part of science. Therefore, in *this* view of things, a grand epic—which necessarily includes much extension—must in a serious sense be regarded as a *human construction* rather than a discovery or disclosure. While it is not fictitious, neither is it the 'objective truth.' From which it follows that the minor epic, describing the construction itself, is important far out of proportion to its space-time extent in the larger story of the whole. For it is this *human* story which affords at least a glimpse of just where empirical data leave off and extension takes over. With the benefit of that glimpse, however fleeting, the *grand* epic that emerges can be approached and appropriated in various ways; without that glimpse, the listeners are, to an extent, like the captive audience in a movie theater—the script written long ago, the cameras rolling, all questions from the audience are definitely out of place.

To summarize this part of the argument, then, my first point is that although construction is undoubtedly an ingredient of the grand evolutionary epic, the standard narration hides this fact. Its implicit philosophy of strong realism projects before us a completed scenario of *silent* objects. Kept strictly out of range of the cameras are the language-using writers, directors and special effects people.

The nonstandard alternative, beginning with the activity of science, and showing the dialogues, conflicts, decisions (of the creators of the grand epic) necessarily reveals the process of construction and conveys, therefore, along with the grand epic itself, some idea of its *mythical* character. It is the metaphoric extension beyond limited, local predictability, that allows one to speak meaningfully of *science as myth*. Because of this, scientists like Wilson (reluctantly) do use this term, and philosophers like Mary Hesse expound it at length.

It remains for me to show that besides this virtue of revealing more, of calling attention to something important, the constructivist narration has still other things to recommend it. First of all, we notice that in this alternative form, the grand epic is no longer silent. In facing the unknown universe, or any corner of it, the situation of the scientist is very much a *human* situation, at least partially understandable by most of us. The minor epic includes innumerable adventures in which countless heroes, villains, prophets, martyrs, and just plain sleepwalkers, mix it all up—with as much noise, as much zest, and as much amazement at the end as in the whole Shakespearian production. The language used by these characters is also, for the most part, ordinary natural language. Only slowly do the various disciplines begin to develop specialized languages to deal with their special data and constructs. All along, these specialized languages themselves are the subjects of controversy; and the struggle to justify their use is part of the construction of the grand evolutionary epic. Is 'caloric' a legitimate term? Does 'action at a distance' make sense? Is 'natural selection' a tautological idea?

The alternative narration describes the extension and enrichment of the natural language by these specialized terminologies; within the minor epic itself there is no sign of *displacement* of one by the other. As our heroes and villains and bystanders argue, what we hear is mostly natural language, *augmented* at crucial points by those specialized terms whose purpose is to extend knowledge beyond the immediate reach of the senses. We are shown how, under the pressure of science, the global horizon of speech expands: If a scientific language presents one (partial) horizon, and our natural language another, then the narrator of this kind of tale can bring about for the listeners a *fusion* of these two horizons—in which the specialized terms provide the natural language with new depth, and the natural language endows the specialized terms with larger meaning.

The fusion of horizons, a concept of hermeneutics, leads us to an important problem arising only for the alternative mode of narration. The standard mode describes to us, for example, black holes as part of the scene. Numerous books and articles explain for the layman what this entity is and how it comes about. But for the constructivist narrator, whose foundation is the minor epic, semantic questions about such terms as 'caloric' or 'black hole' involve actually more radical ontological and epistemological issues: Does caloric exist? Do black holes exist? And how do we know? Thus, interpretation enters here on two levels: first, that which is common in research—how, in view of the background theories, to interpret the observations indicating the black hole? How, in other words, to 'read the book of nature'? This discussion within the minor epic we hear as part of the narrative; we can participate vicariously, though in general we do not feel competent to enter it seriously.

But once we do know how the decision was made, and can speak of black holes with confidence, we still have to interpret to ourselves, or among ourselves, the meaning *within our experience* of this entire process of construction of entities that are *in principle* beyond experience. Or, since our other experiences are also theoretically mediated, the question actually concerns the meaning of these constructions in relation to the rest of our conceptual ecology: What is the impact of such an 'evolving universe' on our self-concept, our theology, our life's course? Because dialogues arising from this kind of interpretive activity involve necessarily the ontological-epistemological debates of the minor epic, the language needed in *these* dialogues extends to the furthest possible horizons on all sides. Therefore, I suggest, in this type of communication, an enriching fusion occurs rather than a displacement of one language by another or the colonization of one culture by another.

In comparing the two approaches once more, it seems possible to say that the standard narration is a *literal account*. We are asked to take as literally true all aspects of the grand epic, including those metaphorical extensions that reach out the furthest. Everything, in other words, is served up on a single epistemological plane. When Monod discusses the older epics as 'animist,' or when Wilson explains why they are inferior, a metaphoric reading of the latter is not in question; always it is the literal reading of traditional epics that is compared with the literal reading of the new evolutionary story. Since for the traditional Western religious epic, a literal reading has long ago been discredited by epistemological, linguistic, and other critiques, the result of the comparison is hardly surprising. Yet if we take seriously the idea that science too has a metaphoric component, then here also an epistemological and linguistic critique is possible.

It is typical of a standard narration to take no account of such critiques despite the fact that in the twentieth century these have been major projects in the philosophy and history of science. Thus, in his lavish praise of the grand epic *vis-à-vis* the traditional stories—its power, its comprehensiveness—Wilson is in effect comparing old narratives as they emerge after vigorous epistemological criticism, with a new one where literal meaning is to be taken for granted. This simply shows, from another point of view, that the standard narration precludes any attempt to interpret the new epic in the context of the present cognitive environment, or of our own history, or even of the history of the epic itself.

Toward Alternative Narrations

In the making of typologies there are many pitfalls. Of course the two types of narration I discuss do not often appear in pure form, and this holds especially for the one I call constructivist or alternative. Yet even among the small number of works listed at the beginning of this paper, two can be used to indicate more clearly what such an alternative narration may involve. In very different ways, I think that *Order Out of Chaos*, and to a lesser extent, *Gödel, Escher, Bach* do exhibit important elements of the preferred approach. For the sake of brevity, I will concentrate here on Prigogine's and Stengers's work.

Let us recall that, from beginning to end, the authors are vigorous advocates of the new cosmic epic, and that the scientific middle portion of the book is a contribution to a crucial link in that epic—a possible theoretical explanation of the origin of life, a solution to the puzzle of how order may arise out of chaos. Yet it is not this solution itself that is the main theme of the book. First we must familiarize ourselves with some underlying ideas of the minor epic. The story begins with *Newton*, with the 'Golden Age of Classical Science,' and with the *human* as well as the scientific problems raised by that first modern world-view. Chief among these problems, in the authors' opinion, was this epistemological-ontological dilemma: whether to regard the newly-won insights as absolute, global truths (the 'scientific myth' of classical science) or to accept them merely as a 'pragmatic recipe' for technical intervention (54).

Immediately the stage is set for the drama to follow; not the drama of the grand epic, not how life came into being (for, in itself, that is still largely unknown), but the drama of how, one by one, the philosophical and scientific assumptions behind *classical science* had to be *challenged*. That they had to be challenged eventually became clear as soon as it was recognized that classical science and *life* were in fact theoretically irreconcilable.

But more than philosophico-scientific issues were involved from the very beginning: Classical science, so the narrative continues, created a split not only between living and nonliving natural objects; that same split divided the experience of each human being into the 'objective,' 'scientific,' 'quantitative' on one side, and the 'subjective,' 'unscientific,' 'qualitative' on the other; it bifurcated life itself. Thus arose the first parting between the two cultures; and thus arose also those first feelings of 'estrangement' from the modern world, to which the philosophies of a Pascal, of a Kierkegaard, were the reaction. In other words, the scientific problem dealt with by these authors is not, in their view, a scientific problem only. The scientific and the existential are here inexorably intertwined.

The tale continues: from Newton and Laplace, to Diderot, to Hegel, to various others, and finally, in the late 19th century, to Ludwig Boltzmann. Boltzmann is not usually cast as the most pivotal of scientists, but in this version of the story that is indeed his role: It was Boltzmann whose mathematical relation of entropy to probability became for our time the clearest statement of the antithesis between (reversible) classical science and the (irreversible) processes of life. This is the climax. Here we reach the sharpest clash of our inclinations: On the one hand, the power and beauty of a structure erected over three hundred years of magnificent labor, on the other, its futility, even noxious effect, in the realm of life.

It is at this point that the narrator himself enters the story (implicitly) as a protagonist. Prigogine in effect begins to argue with Boltzmann: Yes, Boltzmann's formula holds at thermal equilibrium; yes, it holds also a little beyond equilibrium; but, no, *far from equilibrium* it does not hold, and that is just where we can show chemical self-organization into 'dissipative structures'—how life (in all likelihood) must have begun and how it still runs. This dialogue between Prigogine and Boltzmann takes us further into the subtleties of Boltzmann's interpretation of irreversibility, of the arrow of time, of chaos theory; and finally it takes us back to the larger issues with which the book began.

Whether we agree with the authors that the new 'evolutionary paradigm' (as augmented by these recent developments) can lead to a 'reenchantment of nature' (291ff) and a healing of the cultural splits, is not the issue here. What I hope to have shown, even in such a crudely abbreviated summary of the book, is the *fundamentally* different relation of the authors to the grand epic on which they discourse. Though they advocate it most strenuously—scientifically, philosophically, and socially—their understanding of it is definitely constructivist. And this understanding is not just explicitly declared (55), but *exhibited* precisely in the character of the narration of their story. The result is, I think, as earlier suggested; though to argue this convincingly would take a much longer discussion. The epistemological and ontological questions treated provoke a form of speech in which the new terminology—bifurcation point, chaos, dissipative structure—do not appear intrusive, do not displace words of the natural language. The viewing of a living organism as a dissipative structure gives an *additional* perspective, but more than that is not claimed for it. On the other hand, the concept of a 'bifurcation point' genuinely deepens our understanding of the new role of chance and law in nature, a role just now being hammered out at the frontiers of science.

Finally, I would like to make clear that my criticism of what I have called the 'standard' narration is not intended to be absolute or global. For some purposes, in limited ways, a well-executed narration of this type has its place (and the not inconsiderable advantage of brevity). To see how *The First Three Minutes* could have achieved what it did in some other way is not easy. What I have argued against is not so much this 'standard' narration in itself, but rather the possibility that the grand evolutionary epic, *narrated in the standard form*, might become *canonical* in our culture.

PART THREE

Morality

[21]

The Price of Collaboration

It was with sadness and deepening apprehension that I read Bernard T. Feld's editorial, "Recognizing the enemy," in the September 1978 *Bulletin*.

The debate over tactics has been going on for several years; and the choice between various types of pressure and boycott can certainly lead to honest disagreement between well-intentioned people: In this area, I for one have little doubt that the carefully reasoned arguments presented by Valentin Turchin (in the same issue) and similar ones by others, are by far the more valid—and the course they advocate the more likely to succeed. But my purpose in this letter is to raise perhaps an equally urgent question: What is this scientific, cultural, and economic collaboration in its present form, under present conditions, doing to Americans—as individuals, and as a community?

Some people were shocked to learn that after F. Jay Crawford was arrested by the KGB, his American business colleagues began to shun him—following, as if by instinct, the behavior pattern obligatory for all those wishing to remain in the good graces of the rulers of the Soviet Union.[1]

"That's business for you," a friend remarked, "only the bottom line counts."

And what about scientists? Surely, among those professional and human values which we sometimes even presume to teach to others, *veracity* and a respect for logic should rank high.

What, then, are we to think when we read carefully constructed pronouncements by leaders of science, officers of prestigious journals and

This letter was printed in *The Bulletin of the Atomic Scientists: a magazine of science and public affairs* (December 1978), 55–56. A statement on the editorial page of that issue reads: "Since 1945, when Albert Einstein and others sponsored the publication of the Bulletin, it has served as the conscience of the international scientific community."

institutions, containing distortions of fact, of history, and of those rules of argument which in science itself are rarely ignored?

How, for example, should a reasonably informed reader understand the "Statement on Orlov" by the Federation of American Scientists (in the same *Bulletin* issue), whose opening sentence boldly asserts: "The Soviet Union has never had, and will not have for a long time if ever, the kind of human rights we enjoy and urge upon its government"? The arguments have not yet begun: This is only the context, the common ground, which all are expected to take for granted.

However, *it is not the case* that what we urge in the way of freedom from arrest and freedom of expression has never been practiced in the Soviet Union. It was not so long ago that a book like *One Day* . . . could be published there (today it is banned); it was not so long ago that *Novy Mir* under Aleksander Tvardovsky was known as almost a "liberal opposition"; it was not so long ago that a poet read "Babi Yar" to tens of thousands in a public square, and no mock trials were staged in Moscow—now, instead, there is "Traders of Souls" on Russian TV.

What is smothered in the incredible assertion of the FAS is the very historical fact of the "Khrushchev thaw." In their zeal to accommodate themselves to Soviet reality, and to ignore the direction of change throughout the past decade, distinguished American scientists have been led to help eclipse that memory of "the hopeful days" which even now sustains many Russians who continue to hold out.

And this is not all. The FAS statement claims further that as far as the suggestion of "wholesale breaks" is concerned, the "scientific dissidents rarely want it either."

It is of course possible to check back, to reexamine once more the many appeals by Sakharov, Levich, Mel'cuk, Polikanov, letters of Chalidze and Turchin . . . the "20 Scientists" for Kovalev in 1976 . . . (and, in passing, also the pleas of Solzhenitsyn, Bukovsky, Amalrik . . . though, of course, these are not scientists). There can be no misunderstanding: The messages are painfully clear, the recommendations often explicit. Just exactly who among the dissidents opposes wholesale boycotts the FAS fails to tell us; but no matter, the argument can go on without that.

Repeatedly, in spite of the obvious illogic, one hears the same stock phrases, the same non sequiturs. Boycotts lead to nuclear brinkmanship: They are a form of emotionalism, of giving vent to indignation. In his final argument, Bernard Feld goes so far as to assert that such boycotts will also help to "throw the Soviet-American competition back into the arena of posturing and force." In this way, a peaceful act of non-collaboration with an oppressive system is identified with a return to force—a *return* to force, denying also, via a single rhetorical device, the reality of true *military* force: thousands of Cuban soldiers transferred to Africa, thousands of tanks thrown into Syria alone, not to mention the force used on a sick and helpless scholar in a freezing cell.

As the human rights situation in the Soviet Union continues to deteriorate, it is unfortunately becoming clearer that the price paid for collaboration is a price paid by our society: an ever-growing pressure to accommodate not only our political acts, but also our fundamental values—to learn to see things from a totalitarian perspective. If a man is condemned to only three years instead of ten, that is a light sentence though he may be guilty of nothing. If the era of Joseph Stalin is recalled, then we must also mention, on equal terms, the events associated with Joseph McCarthy. This is the style not only of a *Bulletin* editorial, but also of official correspondence of the National Academy of Science.[2] If a Soviet Academician lends his signature to the attack on Sakharov—well, he had to. We stand ready nonetheless, with effusive compliments, to celebrate his birthday.[3] Should we speak out more for the persecuted? Perhaps . . . but then we must remember too that "it is painful to offend colleagues in the Soviet Union who support their government's policy."[4] Such, regrettably, are the arguments now heard among leading scientists.

Of course, it still does not follow—as some would have us believe—that termination of all scientific contact is the only alternative. Confusion of the words "wholesale," "total," and "final" is an error which scientists, at least, should be able to avoid. The advocacy of cold war isolation is certainly not my aim: I wish rather to point out that "collaboration above everything" is not only a tactically unwise course, it is also a dangerous value for a still free society.

Editor's Response: It is difficult to argue with the conclusion of the above critique, particularly as expressed in the last sentence. However, having said this, we are still faced with the problem of how American scientists should act in the face both of provocations from Soviet hard-line officials and calls from anti-Soviet zealots for a complete severance of relations. I continue to maintain that our first criterion should remain the maximum support of the great majority of Soviet scientists who remain committed to the free and open pursuit of peaceful and cooperative scientific ventures in the long-standing Russian and worldwide scientific tradition.

[22]

Rationality of Science and Social Controversy

Abstract: There seems to be a great difference in the rationality of unproblematic paradigm-bound thought, as in Kuhn's "normal science," and critical, unrestricted methodological pluralism, characteristic of the investigation of values. Habermas's contrast between "empirical-analytical" disciplines, aiming at technical control, and "historical-hermeneutical" disciplines, aiming at politico-moral orientation in a social environment, provides a justification for this difference of rationalities. But Habermas's dichotomy is excessively rigid and needs to be modified by recognizing Mary Hesse's linear continuum of disciplines, in which the empirical-analytic and the hermeneutic are extreme poles. This recognition has implications for science education, for it accommodates the teaching of a human, communicative aspect of the natural sciences and of the fact that these sciences provide themes for society's self-understanding. [—A.S.]

One pivotal question relevant to the rationality of science, and frequently asked, is whether rationality in the broad sense is continuous or discontinuous across cultures and time. If it is continuous, and invariant properties can actually be identified that define at least its core, then, for one thing, the work of sociologists would seem to be confined to what have been called 'external' factors; or, in other words, the 'strong program' would rest on weak ground. If, however, significant discontinuities are indeed found, and not explained by a deeper rational stratum, then, as Harry Collins put it, "the sociology of knowledge can proceed with unabashed vigor." Behind this thinking, the basic assumption is of course that *if at bottom* reason is universal, then any hiatus, incongruity, or contradiction between its manifestations, cannot itself be a rational phenomenon. It is understandable

This talk was presented in Pittsburgh, Pennsylvania, in 1986 at a joint meeting of the History of Science Society, the Philosophy of Science Association, the Society for the History of Technology, and the Society for the Social Studies of Science.

301

therefore that the kinds of *discontinuities* so prominent in the picture of science given by Thomas Kuhn (and the movement of which he is a part) have been as stimulating and generative of new research as they are controversial.

Yet in that picture there really are *two* basic types of discontinuity, both of which were dealt with at length in Kuhn's original formulation: first, the much discussed 'incommensurability' between paradigms themselves; and then, also, a glaring contrast between two *modes of doing science*—the 'normal' and the 'revolutionary.'[1] The former kind of gap is actually less serious than the latter because it involves primarily a difference in the meaning of concepts, significance of problems, interpretation of evidence, etc.; whereas the difference between normal and revolutionary science is far more a *difference of rationality*: It is the break between work guided or bounded by a single paradigm, and a situation in which partially contradictory patterns of thought coexist with one another. As Kuhn so well described, the 'normal' mode involves constraints and a reduction of critical thought, while the other demands not only criticism at its strongest but a wide-ranging theoretical pluralism: Contradictions become, perforce, an *accepted* element of science, and not merely as part of competition but in the complementary sense (e.g., statistical mechanics and thermodynamics in the 19th century, quantum mechanics, etc.).

It is this second type of discontinuity on which I should like to focus: the abrupt change of rationality required in making a transition from relatively unproblematic, paradigm-bound thought to highly critical, unrestricted, methodological pluralism. Well-known debates as to the merits of these two types of study have in fact taken place in the case of science, between, for example, Popper and Kuhn himself; but I do not intend either to review or to continue such debates.[2] Instead, I should like to discuss a similar kind of discontinuity underlying certain contemporary social controversies in the United States: the battles concerning evolution/creationism in public schools and the disputes over 'moral education.' Because these conflicts lie in a border region between science, education, and values central to our self-understanding, the debates they generate are not confined merely to application or practice but extend also to the theories *behind* the practice. Thus they afford an opportunity to appraise the 'cash value' (as some pragmatists used to say) of certain fundamental ideas.

It is my contention that, quite possibly, a discontinuity of the second kind (of rationality) causes, or contributes to, or at least exacerbates these controversies.[3] But I want to add very quickly that I am not speaking now of the obvious differences of outlook (cultural, religious and political) between out-groups such as Christians associated with the Fundamentalist churches or Orthodox Jews on the one hand, and on the other such in-groups as the established scientific and educational professions. Of this, much has been said already. The discontinuity in rationality discussed here does *not* coincide with any dividing line between subcultures or other well defined populations. Rather, it seems to exist *within* professions, or within a closely related group of professions—the educational-scientific community itself.

In outline, and in purely descriptive terms (if I may be allowed such a provocative phrase), the problem arises from the following developments. Since at least the beginning of the 1970s, a wide-ranging program of educational reform has been under way in the area of overt, explicit teaching of values, including moral values. This is in part a response to the traumas of Vietnam, Watergate, the drug scene, and crime in the schools themselves. On the more theoretical side, it represents also a recognition that a completely value-free education is undesirable as well as unrealizable.

But public 'moral education' in a religiously and culturally heterogeneous society, with constitutional traditions such as ours, and a strong distrust of the role of government in personal matters, is an extraordinary task to say the least. How does one satisfy at the same time that characteristically modern respect for autonomy in regard to values, and yet provide 'guidance,' foster moral responsibility, and in general oppose a creeping anarchism?

Over the years, although a number of distinct approaches and philosophies have appeared, two common factors can be identified that seem to meet the challenge just described: *methodological pluralism* and *critical reason*. Both of these ingredients have a venerable history: from Socrates to J. S. Mill to Popper and other contemporaries. Pluralism of values and world views is to be encouraged *a priori*, at least as a methodological device, for the same reasons enumerated by Mill and discussed by him at length: First, no single formulation captures the 'whole truth' on any issue[4]; and in addition, a 'received opinion' taken for granted, even if true, is thereby 'enfeebled,' or its meaning is lost altogether.[5] Today this approach has the added appeal of minimizing discrimination against individuals or social groups, *and,* at the same time, encouraging youth to participate actively in the 'valuing process.'

But the critical factor is equally important, especially in the academically more respected movements such as that of Lawrence Kohlberg.[6] Differing views must be subjected to withering analysis and criticism, modeled essentially (and consciously) on the methods of natural science. That is, from goals and values *consequences* are to be deduced, and compared with those entailed by competing values. Whenever possible, intermediate values are to be subsumed under more general, higher level principles. Descriptive and normative statements are to be carefully distinguished. The procedures used are classification, ranking, branching, importance weighting, minimaxing, etc., and the conclusions must conform to the formal cri-

teria of consistency, coherence, universality, and symmetry. At the end of
the process, unique 'right answers' are *not expected*, but at least some tacit
assumptions or commitments may be brought to light; many hidden, pernicious contradictions can be exposed; and therefore, 'moral education' of
a sort can be said to take place.[7]

What is crucial here, though, is that by explicitly situating within the
domain of theoretical reason, we expose them to the formal apparatus of
analytical science. As John Rawls made a strong impression by combining
certain techniques of scientific thinking (decision theory) with a classical
approach to the question of justice, so in the schools it is believed that similar fusions might bring an enhanced rationality to bear on moral issues of
all sorts.[8] But this modern union of methodological pluralism and quasi-
scientific analysis (which has had its own troubles) heightens the urgency
of implications that educational strategists have not yet seriously considered—implications that are becoming more widely perceived as, in the
wake of the Arkansas trial of 1981–1982, the controversy over creationism
and the teaching of evolution moves from the level of courtroom and
media coverage to that of scholarly exposition.[9]

The dominant academic view now emerging in this area of education
is that in a science classroom just the mere consideration of certain questions or propositions, i.e., creationist arguments (even if *not* on equal footing) is in itself totally *irresponsible*. Regarded as 'corrosive' of the intellect,
as 'stunting' the mind,[10] it simply has "no proper place in the classroom."[11]
I will not reproduce here the many extended statements of this kind, easily found in the literature—except for one representative sample to be quoted shortly, without which, I fear, the problem will not be sharply defined.
But the overall message is loud and clear that, in the teaching of *science*,
methodological pluralism and a critical stance must *not* play a central role.
In this realm, the model of rationality enthusiastically embraced is that of
Kuhn's *normal science*. The learning of the established paradigm supersedes
all other goals.

This, then, is the discontinuity at issue. To dramatize it somewhat we
can imagine two classrooms side by side, in a high school where teenagers
are discussing two different kinds of problems. In room 101, a debate is
taking place on the question of whether stealing might be morally defensible under certain circumstances (a typical Kohlbergian dilemma). In room
102, the topic is whether 'gaps' in the fossil record really exist, and if so
whether, or to what extent, Darwin was wrong. The point for us is *not* that
on the moral issue there is no established or widely shared position in the
America of the 1980s, while on the scientific question the consensus of the
competent community of experts certainly ought to be conveyed to the
learners. That much is easily granted. And, in fact, a consensus on the
moral question might well be reached on this particular day, in room 101;

while in the other classroom, a bright student who keeps up with the latest (or reads Hoyle's books) could give the teacher a lot of trouble. The *discontinuity* in the practice of rationality lies rather in *how* the two discussions proceed: In room 101 (morality), it is *important*—indeed, the essence of the study—to *understand all* arguments, objections, premises, assumptions, etc. In room 102—unless the teaching is quite unrepresentative—it is simply a matter of students *grasping* what the textbook and the teacher are trying to tell him. Alternative ideas, historical or other, if they are mentioned at all, play an entirely minor and peripheral role here.[12]

But why should the practice of rationality in two adjoining rooms in the same school be different—and therefore the rationality practiced by the *same* students at different times of the day? At first sight it seems hard to deny that this makes nonsense of the oft-cited ideal of the *universality* of reason.[13] And if different rationalities are deliberately employed—though much effort is now devoted to make the two look alike in *certain* procedural respects—then the reason for this difference is surely *not* a part of any curriculum, not for these students and not for their teachers.

Of course the conventional reply to such questions is well known. It rests on the presumed or perceived difference between normative and descriptive realms, between 'facts' and values, which leads to a difference in object domains: The domain of the descriptive includes inanimate matter, living bodies, and even human *behavior*. Human *action*, however, is fundamentally in the realm of the normative. One sees right away that this explanation refers to a long, complex philosophical history, with its distinction between theoretical and practical reason—from Aristotle to Kant, to late nineteenth and twentieth century philosophers who tried to protect the human realm against encroachment by methods found so successful in natural science—and to contemporary, purely scholarly debates on relativism versus objectivism, on demarcation criteria, on realism versus instrumentalism, and so on.

Thus I am brought to my first point, which is preliminary but worth noting: that these classic issues in the theory and sociology of knowledge are not scholarly issues *only*; they are reflected as well in the social situation and, it should be emphasized, in social *conflict*. For many laymen as for most of the liberal academic community, "it is common," as Mary Hesse put it, "to accept with alacrity the relativity of religions and ideologies . . . to accept with somewhat greater hesitation . . . the relativity of moral systems, but to balk utterly at the relativity of scientific knowledge."[14] Conflicts arise, and are serious, because outside the academic world, this dualism in rationality is not a matter of detached, professional debates, but visibly affects the lives of people.

The details of the local conflicts, the human dramas, the specific incidents, the forms of argumentation and misunderstanding, and the power struggles that have taken place are things I cannot take up here; to an extent, I have already described this in previous papers.[15] What is important to underscore, however, is that the dualism or discontinuity in the theory and practice of educational rationality is indeed one of the bones of contention, well perceived, and often mentioned by participants in such disputes. And it is equally glaring to those on the receiving end of academic explanations that *these explanations too* appear to conflict quite sharply.

When people object to various learning games in which, after a brief introduction to a problem, students are asked to vote on such issues as whether socialism or capitalism is better, whether lying to help a friend is justified, and so on, a typical defense—from Michael Scriven, for example—goes like this:

> We refuse to give time to the opposition, we refuse . . . to engage in true critical thinking and teaching. . . . It is the threat to our own beliefs that strike fear into our hearts when we hear about alternatives to them being seriously considered in our children's schools. . . . The shock of discovering that Karl Marx's *actual words* are being read by our own innocent children. . . . This is defensiveness, insecurity; and, on this scale, it is a fatal condition. The wisdom on the other side is overwhelming. . . . Even if one believes that the old values are the best values, the reasons for—and against—them must be rehearsed by each generation or it will rebel against them. . . .[16]

There will follow supportive aphorisms by Socrates, J.S. Mill, Benjamin Franklin, even Christian churchmen. But when someone suggests—and 75–80% of the population agrees, the polls say[17]—that creationist arguments should be heard alongside those for evolution, the reaction is dismay; and the reasons, offered quite innocently, are well summarized by this from Philip Kitcher:

> One may ask, why not let creationists submit their case? Surely the truth will win out. . . . Nor should we worry about a little wasted classroom time, *when a deeper understanding of the merits of evolutionary theory might be secured* by allowing students to think through the issues for themselves. *The argument is insidious.* . . . The previous chapters [of Kitcher's book] show that there is no genuine contest, no true comparison. What is in doubt is the possibility of a fair and complete presentation of the issues discussed above, in the context of the high school classroom. . . . There will be . . . much dredging up of misguided objections to evolutionary theory. The objections are spurious—but how is the teacher to reveal their errors to students who are at the beginning of their scientific studies? . . . Even a gifted teacher would not be able to expound enough. . . . What Creationists really propose is a situation in which people without scientific training—fourteen-year-old students, for example—are asked to decide a complex issue on partial evidence.[18]

In both of the above quotations, Mill's reasoning—which he explicitly directed at natural as well as social science[19]—is reproduced with extraordinary faithfulness. But it is striking, and surely not insignificant, that while in one sphere that reasoning is gratefully embraced, in the other it is called 'insidious'—and this despite the fact that Kitcher too believes the "study of science provides important training in reasoning" and helps students to "think more clearly."[20] I do not intend in this paper to discuss why the conventional justification for the use of two such different notions of reason— the attempt, on logical and pedagogical grounds, at a rational explanation for an apparent irrationality—has, thus far, failed in the social realm; why, in other words, not only the religiously orthodox, but a large number of mainstream Americans are not convinced. Elsewhere,[21] I have already argued that this may have to do with the fact that the conventional justification is simply *weak* in purely scholarly terms, and getting weaker with every year that the philosophers, historians, and sociologists of science continue their work.

I refer to the breakdown—on several fronts—of what was once a sacrosanct boundary between the 'is' and 'ought.' I refer of course to the mushrooming socio-philosophical investigations that have placed science in a context of human concerns, of discourse, and of community[22]; and I refer, on the other side, equally to the movements in 'moral education,' 'values education,' 'decision making,' 'critical thinking,' etc., that in our time *rationalize* (in the Weberian sense) even the most private realms of what has sometimes been called the "life-world," *in a way* that gives these realms increasing resemblance to natural science. As a result, it is becoming more clear that the old conventional justification for a dualist rationality cannot be offered as before. Although it might be possible still to defend traditional positions with traditional weapons, the prospect might be brighter, at least heuristically, if the matter were addressed using a different approach altogether.

One promising path, certainly, is that marked out by Jürgen Habermas. And I should like now to state, very briefly, why following that path a certain distance is the main task of this paper. The flaws in the conventional justification are so evident that many of the arguments offered the public are not at all acceptable to a fair number of scholars in the philosophies of science and of morals.[23] For this reason, a deeper analysis, based on typologies of knowledge beyond the is/ought disjunction, seems inviting—especially if it gives hope that firmer ground for the conventional practice may thus be found. And this hope is justified (at least from a distance) by the fact that the human realm *is* clearly marked off, related to discourse in an essential way, and placed at the center of the cognitive landscape sketched by Habermas. If, however, his approach also fails to 'save the phenomenon,' then the problem is more serious even than it appears.

This too would be worth knowing; and along the way we might attain new insights—into the controversies now taking place, into similar ones that (undoubtedly) lie ahead, and possibly into the thorny question of the *cognitive* role of science in society.

When applying Habermas's conceptual framework, I will refer mainly to the earlier formulation, especially as it appeared in *Knowledge and Human Interests* (1968b). Though he has introduced changes since then— some significant, some only terminological[24]—those aspects of his theory used here remain essentially the same; and the older phraseology has the advantage of being more familiar.

The relevance of Habermas lies in the fundamental distinction he introduced years ago (and still maintains) between two types of sciences, or fields of study—a distinction not based on whether the object domain includes human action, but on what he then described as *structurally imbedded interests*. As is well known, his 'empirical-analytical' sciences incorporate the interest of technical control, while the 'historical-hermeneutic' disciplines aim at a 'practical' (politico-moral) orientation in a purely social environment. And here again, there is by now no need to explain at length that such constitutive 'interests' are not regarded as individual or subjective, but as deriving from a difference in specifically human experiences: the technological interest from *work*, from instrumental and strategic action required by material needs; the practical interest from symbolic interaction necessary for mutual and self-understanding.

Because these interests are understood as *inherent* in the formation of concepts, in the choice of problems, and in the manner of seeking answers to questions, they necessarily *predispose* each discipline to a specific type of application. But most important for the problem at hand, they also lead to two different modes of rationality: on one side, a means-ends rationality of the kind so central in the thought of Max Weber (*Zweckrationalität*); on the other, a rationality that aims primarily at "overcoming systematically distorted communication" and the removal of barriers to intersubjective understanding.[25] In the latter case, concern is especially with forces or constraints whose origins lie beyond the horizon of discussion, and remain therefore unnoticed by the participants. More recently, Habermas has, if anything, intensified this dichotomy between the two types of sciences by his shift of attention toward Weber's notion of separate 'worlds' or 'cultural spheres' (the 'objective' and the 'moral' or 'social').[26] From such distinctions and classifications—of 'worlds,' 'interests,' and 'actions'—it seems reasonable to expect that *prima facie* a Habermassian analysis might well vindicate the still-prevailing, intuitive dichotomy of the conventional view, and show that in some sense this *is* fundamental, that any attempt at a holistic approach reflects little more than futile nostalgia.

Applied to our problem of rational discontinuity, the reasoning might go something like this: Contemporary physics, biology, etc., are certainly examples of empirical-analytical sciences whose interest is technical control. Moral values are in the category of those human studies that deal with 'practical' matters, and so are subject to a rationality appropriate to the interest of communication and mutual understanding. In the former case, the role of science paradigms is precisely to use the pioneering work of predecessors in finding quickly and efficiently the best means to the problem solutions in particular subject areas—which exhibits unmistakably the instrumental rationality of these sciences. Here proliferation of routes is not only unnecessary, but actually counterproductive. In the latter case, however, the inherent interest is obviously served better by a methodological pluralism in which consensus, if it emerges at all, emerges *at the end* of every discussion, and is not presupposed. Communication and mutual understanding, particularly when they concern norms in the social world, cannot take place through prearranged channels unless the community of speakers is already cognitively homogeneous to a very unusual degree.

In short, and speaking crudely, it all boils down to goals, to purposes structurally internalized, inseparable from the enterprise itself. Where the 'knowledge constitutive' goal is really technical control, there methodological pluralism in the learning process is usually encumbering, and certainly unnecessary in normal times. Where the 'deep goal' (to use a hybrid phrase) is mutual understanding in communicative action, there pluralism—at least methodological pluralism—is of the essence of the enterprise. Thus we face no real contradiction, our hypothetical defender of the educational status quo might conclude: A discontinuity in the two rationalities is in this case quite rational, given the different *interests* of the two types of disciplines. And the substratum that relates these interests, while explaining their unavoidable 'differentiation,' is the human situation itself—its rootedness in *work* and in *language*.

In a sense, this conclusion is reached as a point of convergence of several distinct lines of research, all of which have influenced the present Habermassian standpoint: 'Differentiations' (of various sorts) in the forms of knowledge are characteristic of the *growth* of rationality—in the individual as well as in the species. From Weber to Piaget and Kohlberg, the historical studies of such differentiations suggest emphatically that the phenomenon is not one in which we have a decisive choice. Differentiation of moral discourse, where hermeneutic thought must dominate, is especially important in view of the danger of 'objectivism,' which always threatens it. But by the same token, there exists also a danger of 'moralism' in the world of the objective (scientific)—a misguided idea that, nonetheless, is still alive in our time, not only in the positions taken by some of the creationists, but also, for example, in a thinker like Herbert Marcuse.[27] This is not to say that Habermas supports some sort of strict separation of cultures—far from it. But he distinguishes 'communication' or 'mediation'—a major desideratum—from something that would amount to an invasion of one 'expert culture' by another, and a violation 'within the boundaries' of the former, of the "inner logic of the dominant form of argumentation . . . ".[28]

We can easily appreciate the advantage of such an approach over the conventional stance: This type of investigation is *itself* a historical-hermeneutic endeavor, not a logical one; instead of relying on abstract demarcation criteria increasingly questioned in recent years, it tries to uncover the genetic, the truly formative source or process. It is, in Habermas's own terminology, a 'reconstructive science.' Nonetheless—and this is my major point—despite their appealing nature and the respected tradition invoked, arguments based on interests not only fail to provide the hoped-for defense, but, when looked at closely, put present practice in question even further by opening an entirely new avenue for basic criticism.

The first thing one notices is that Habermas's way of partitioning the cognitive domain places certain aspects of natural science not entirely within the bounds of the empirical-analytic. When asked directly in what sense Darwin's theory of evolution could be regarded as oriented only to instrumental action, he had this to say:

> I think in no sense. Since the evolution theory has a methodological status which is quite different from a normal theory in, say, physics, I think that the categorical framework in which the evolution theory has been developed since Darwin presupposes some reference to a pre-understanding of the *human* world and not only of nature. . . . So in my opinion, the evolution theory is no example of an empirical-analytic science at all. But as far as bio-chemical theories about mutations go into this evolution theory, we have, of course, a usual empirical-analytic theory.[29]

Interestingly enough, the reasoning behind this has been spelled out more clearly by Ernst Mayr, one of the leading Neo-Darwinians:

> In evolutionary biology, questions like quality, historicity, information, and selective value are of special interest, questions that are also of concern in the behavioral and social sciences but not in physics. It is not at all unreasonable, therefore, to consider evolutionary biology as something of a bridge between the physical sciences on one hand and the social sciences and humanities on the other.[30]

Habermas long ago concluded that the 'social' versus 'natural' dichotomy among the sciences was misleading. The distinction he introduced instead, between the empirical-analytic and the historical-hermeneutic, was supposed to be more appropriate to the issue of science's role in society, and to the more general question of rationality of knowledge. In this picture, *some* social science is empirical-analytic (like the usual view of natural science) while *some* natural science is indeed hermeneutic (like the common view of the humanities). In light of the above quotations, this in itself casts doubt on the conventional treatment of evolution as a purely empirical-analytic science, to be studied—in this time of relative Kuhnian normalcy—strictly within the bounds of the accepted paradigm. But the doubt goes deeper. For, now it is seen that what Habermas admitted in a limited and grudging way for *some* aspects of evolution applies more broadly to contemporary science as a whole.

True, electromagnetic theory, a branch of physics, is a prime example of empirical-analytical knowledge; its impact on human self-understanding is slight, and its own concepts are not at all derived from such understanding. Yet so is *cosmology*, today, a branch of physics; and here the same cannot be said—consider merely the 'anthropic principle.'[31] The brain physiologies of Sperry, MacLean, and Penfield pose similar questions. Surely, systems programming exemplifies in the highest degree the instrumental interest of a science; but is this still so when the 'information processing model of mind' is used in a *psychology textbook* to explain human thinking? And notice, please, I have not even mentioned the word "sociobiology."

No one understands better what this implies than Habermas himself. As he readily admits,

> The wide distribution of popular scientific literature proves that cosmological findings regarding the origin and development of the universe, biochemical discussions about the mechanisms of heredity . . .— all this still touches the *self-understanding of contemporary subjects*. These findings also change the *standards of discussion for life problems*, for which the empirical sciences themselves have no ready answers (emphasis added).[32]

In other words, a Habermassian analysis in no way denies the role played by the empirical-analytic sciences in the formulation of self-understanding and in 'discussion of life problems' (mutual understanding). But this precisely is at issue in *both* of the controversies I cite. In the case of evolutionary theory, one basic question is whether the transmission of *findings* does in fact play such a role; and if so, should this role be part of the debate concerning teaching in this area—or should such a concern be ruled beyond 'the boundaries' of the 'inner logic' of a science like biology? In part, the revulsion and indignation of many scientists at the creationist movement are prime examples of belief that the boundary between the moral (religious) and the scientific *is* wantonly violated by creationist demands for consideration of certain alternatives. But on the other side—and on the part of many onlookers—there is an equally strong feeling that, on the contrary, in the context of *education*, where self-understanding is important, an a priori exclusion of any argument violates what ought to be at least that common, universal, procedural core of reason that has so often been invoked from Mill's time to the present.

In the other controversy, over moral education, focus is not on *findings* but on *methods*, i.e., on approaches, techniques, and teaching strategies whose metaethical presuppositions *constrain the discussion* to a subset of channels.[33] In both conflicts, the issue (from the protestors' point of view) is the effect of science on self-understanding. But in one case, the protest is over the effect of science's *products*; in the other, over the effect of its *process*. Now, of course, in the past Habermas has very explicitly opposed those recent forms of rationalization that rest on a "subordination of values . . . to technological procedures"[34]—an achievement which requires that we detach "society's self-understanding from the frame of reference of communicative action and from concepts of symbolic interaction and replace it with a scientific model."[35]

Within his perspective, therefore, there is no question that (1) a whole array of models developed by the 'expert cultures' of natural science—models of *the world* (findings) and models of *thought* (procedures)—do affect self-understanding, and (2) that at least the procedural models, or some of these procedural models, affect that understanding *illegitimately* and destructively. It takes but a small step in the same direction to recognize the *possibility* that *products*, i.e., the findings of science, could also exert undesirable influence on self-understanding—since a major feature of the whole recent reconsideration in the sociology and philosophy of science is precisely the effect of process on product, and the role of this effect on subsequent interpretation (which, after all, lies at the center of the most recent eruption of the IQ controversy). At this point, Habermas's formulation seems either somewhat inconsistent, or else so hazy as to be of little use. Where does 'mediation' end and 'migration of reified models' begin? When Habermas says that biological evolution itself presupposes some understanding of the *human* world, not just nature, and Mayr calls it a bridge between physical science and the humanities, we must remind ourselves that the object of these characterizations is not some minor subdiscipline but a high-level theory, regarded as one of the most important in history, and frequently related to radical changes in Western consciousness.

In the end, we might conclude, a demarcation based on interests fares no better than previous attempts at sharp demarcations. But there is yet another aspect of Habermas's theory—especially relevant to the arena of conflict dealt with here—that shows not only why these distinctions do not accomplish what might have been expected but that actually they add weight to the other side of the argument.

The controversies we are dealing with take place in a context of *education*, not of application. They have to do specifically with institutions of learning, mostly on the local level. In both cases, concern is exclusively with what happens in classrooms and family living rooms; regarding scientific research (what goes on in laboratories) or political procedures, practically nothing is said by the out-groups raising these issues. Yet the goals in a classroom are not necessarily the same as the goals in a laboratory, even if the object domain—e.g., elements of the theory of evolution—*is* the same. And similarly, the goals of a high school civics class are not those of a political campaign or judicial action.

If *interest* is the guiding and differentiating principle, as Habermas has it, then everything would seem to depend on what sort of educational institutions or programs we are talking about. Although many people have for some time expressed belief that "the struggle against the reduction of the university to a professional school is doomed to failure,"[36] this is still not the official or consensus view in this country; and it certainly is not the dominant view in regard to lower levels of schooling. That being the case, the question at hand, if pursued under Habermassian categories, leads to the following additional ambiguity. Science in a high school or college classroom is *science outside its own institutional boundaries*: It is not at all clear in this situation whether the interest and inner logic of a particular science, or the interest of education (or some condominium of the two) should dominate.[37] That these do not necessarily overlap could not be more strongly implied than by the theory of inherent interests. As fields of *inquiry*, for the most part biology or physics are empirical-analytic, but the activity of studying and teaching (general education) must, within this scheme, be classed as hermeneutic and communicative. Here *understanding*, not necessarily predictive power, takes precedence. Evidently, the problem is one of a hermeneutic appropriation of the fruits of empirical-analytic inquiries. But this we immediately recognize as a situation like that faced by historians, philosophers and sociologists of science; and consequently all the recent post-positivist work in these areas is again relevant.

In sum, then, any attempt to use Habermas's theory to account for that dualism in educational rationality which I described runs up against two difficulties, or apparent difficulties: first, that the basic distinction between Habermassian interests seems fuzzy in precisely those areas where conflicts tend to occur; and second, that in education—and perhaps in other realms—a question arises of what the 'dominant inner logic' really is (or should be) when an empirical-analytic science is approached hermeneutically. Actually, in regard to the second difficulty, the general features of Habermas's scheme point in a definite direction. Education, after all, is a communicative activity. And his emphasis on '*unconstrained* communication' (outside the boundaries of an expert culture) does in a way seem to lead to conclusions that so far only Paul Feyerabend has stated publicly and explicitly—conclusions that, regarding the evolution controversies in

California, for example, are directly opposed to the nearly unanimous academic position.[38] But such notions do not proceed merely from general ideas of communication. On this point, Habermas has been more specific: "the translation of scientific material into the educational process of students requires the very form of reflection that once was associated with philosophical consciousness." In his view, the "demand for self-reflection that such pedagogical questions create for the natural sciences and mathematics"[39] requires something very different from discussions within an expert culture, and well beyond the 'logic' of the empirical-analytic sciences themselves.

This surprising—and to some, unwelcome—drift of thought is not in itself an objection to Habermas's theory, of course. But it does raise again the very question frequently encountered in the controversies at issue: Will not such radical self-reflection, in the process of 'translation' of science into the life-world, compromise in some way the integrity of that which is to be translated? Here, on the level of mediation, we encounter the same problem—of violation, or transgression—that was already found in the initial assignment of interests to sciences. At times it may appear that for Habermas 'mediation' refers to *findings* only, while the 'inner logic' derived from interests is to be safeguarded by a respect for borders between 'expert cultures.' But if self-reflection of a science is admitted into education, and taken seriously, then as we have seen it surely will not stop at the boundary between product and process.

These difficulties notwithstanding, I think the Habermassian analysis does illuminate the current controversies and the problem before us, but not in its original form. With the introduction of one modification—admittedly, not a minor one—it seems possible to retain the vital insights, while circumventing those objectionable features that (from different perspectives) have of course been noticed by a number of critics.[40]

Such a modification was undertaken by Mary Hesse.[41] Her idea, which I fully share, is that the attention to interests is well directed; but the dichotomous relation between particular disciplines and Habermas's two basic types of rationality does not always stand up. And, I would add, it fails especially where the interesting questions arise. Instead, Hesse posits a 'linear continuum,' of which the empirical-analytic and the hermeneutic are opposite poles: All sciences include in differing degrees, and with different emphasis, both types of rationality.[42] This also implies that interests, though real and important, are not necessarily inherent or exclusive. Indeed, Hesse too realizes that "it is impossible in studying theories of evolution, ecology, or genetics, to separate a mode of knowledge relating to technical control from a mode relating to the self-understanding of man. . . ." In the history of all natural sciences, she says, "theories have always been expressive of the myth or metaphysics of a society, and have therefore been a part of the internal communication system of that society. *Society interprets itself to itself partly by means of its view of nature.*"[43]

The last statement can be taken as mere description; but it can also be understood as revealing an *interest*. In forming our 'view of nature,' or rather, in undertaking the historic project of developing and refining views of nature systematically, *one interest* has surely been to shape society's (and the individual's) self-interpretation. This does not imply that the natural and human sciences are distinguished only by the degree of emphasis on one or another type of interest or logic. Our intuition is right: There *is* a tougher kernel there. But for Hesse the crucial difference is based simply on *predictive achievement*, not on interest or empirical content.

It is especially relevant here that this 'pragmatic criterion' (predictive success) leads back to yet another, older distinction, which, though also beset by problems,[44] allows us to see just how and where hermeneutics enters natural science. That planets move about the sun in nearly elliptic orbits is as much a prediction of general relativity as it is of Newtonian mechanics—despite the fact that the former embodies a conceptually different treatment of space and time. But space-time structure is part of the *theory* of relativity, which, like all high-level theories, goes well beyond the 'pragmatic' component, and is therefore *under*determined by the empirical evidence. It is here, in relation to theory, that hermeneutics has a place, as it does in the realm of the human sciences—because theory languages do not describe reality directly but *interpret* it.[45] And of course it is the *theories* of natural science, the grand explanatory pictures (metaphors, Hesse would say[46]) that provide also the background and themes for society's self-understanding. As in Habermas's view, hermeneutics is related to self-understanding in an essential way; but now one sees more clearly why those aspects of natural science that undoubtedly affect self-understanding cannot just be exported across the boundaries of expert cultures without reflection and interpretation, and *why* this hermeneutic activity does *no* violence to the interest of even a science like physics—for even physics embodies some hermeneutic rationality and the communicative interest.

In this way, the proposal made by Mary Hesse resolves *both* the difficulties arising from the original Habermassian framework. If we assume a communicative as well as a technical interest at the center of the sciences themselves—accounting for the previously noted fuzziness of boundaries— and relate communicative interest specifically to the theoretical component, then 'translation' into the life-world via institutions of learning requires, no doubt, a shift of emphasis from one interest to another, but *not the intrusion of anything alien or distortive*. Natural science comes to be viewed as not *just* theory plus *techne*; it becomes also, in the classical sense, 'practical knowledge.'[47]

Returning now to our two controversies, what light if any does this whole discussion shed? None on creationism or evolution. And none either on methodological pluralism in a moral education curriculum. It does, however, strengthen an existing impression that the present sharp discontinuity between the rationality conveyed through a study of nature and that encountered in socio-moral education has no explicit, clearly articulated warrant, and is not beyond serious, scholarly investigation. In particular, those individuals or groups that raise the issue of self-understanding, orientation in society, frameworks for decision in the life-world, even in regard to *natural science* (though often they do this in quite different language), cannot always be dismissed as injecting *foreign or extraneous* interests—whether their socio-political loyalties are of the conservative or of the progressive kind. And if so, then classic Millian arguments for methodological pluralism combine effectively with the Habermassian attack on constraints in communication—in natural science also—insofar as *theoretical* ideas (not just technical possibilities) impinge on the life-world. Specific demands may well be misguided or, for other reasons, impossible to meet; but if the present discontinuity in the rationality of education is to continue and be coherently defended, then some other, more potent argument for it must be found than has thus far been offered—because the new element in the present situation must somehow be taken into account: that the border between the human and natural sciences is now being crossed in *both* directions. Hermeneutics is seen as relevant not just to the history of science but to science itself[48]; and technique, of course, continues to move from the natural sciences to the human realms (bearing out Weber's pessimism).

One way of seeing the change in perspective that is here proposed is to ask what accounts for the very large divergence of judgment on these matters between the academic world and the public. In Habermas's terms (as in the present conventional view), this is a glaring failure of 'mediation' between expert cultures and the life-world. But if, as I think, Hesse's formulation of the interests of knowledge is preferable, then the error is a failure to recognize and to confront openly the less prominent but by no means less important communicative interest of historic and contemporary natural science—especially in education. For the social study of science, it is the *latter view* of these controversies and the theories to which they relate that highlight an interesting question: If the discontinuity between the rationality embodied in the study and dissemination of natural science on the one hand, and on the other that characteristic of politico-moral education, is *not* easily accounted for by reference to their respective inner logics, then how shall we understand it?[49]

[23]

The Conflict in Moral Education: An Informal Case Study

ABSTRACT: In the 1980s the Department of Education of New York State issued a guide for teaching 'values formation' in state schools. Omission of this subject was regarded as irresponsible, but the project faced several difficulties: Values as contrasted to facts do not seem to be amenable to scientific investigation, and in a pluralistic society the students come from backgrounds with diverse values and diverse rationalizations for them. Designers of the program claimed to avoid these difficulties by teaching the 'rational process' of forming values rather than systems of values themselves. Strongly negative reactions to the program were expressed by parents in the small towns of Spencer and Van Etten, not far from Ithaca, typical of reactions elsewhere, and it was striking that many of the objections of parents without access to academic literature resembled the objections of educators and psychologists who were critical of the program. There were charges that the program inculcated ethical relativism, that it caused confusion in the minds of young students, and that it effectively trivialized serious matters. In this paper six questions are posed for the purpose of assessing programs like that of New York State. (1) Is it acceptable that parents be denied information regarding curricula and teacher training? (2) What checks and balances exist to prevent the 'selling' of questionable educational practices before they are competently evaluated? (3) Is it acceptable that moral education be entrusted to teachers trained only by a one-week workshop and a handbook of strategies? (4) Is the adoption by a school of a developer's system of moral education effectively a restriction on moral dialogue? (5) Isn't the methodology of the values formation program effectively an exclusion of a large part of the moral tradition of religions and of the accumulated wisdom of ethical discourse? (6) Isn't the New York State program a mechanization of the process of moral evaluation, and therefore an inadequate model for a more moral and more humane society? [—A.S.]

This article was published in *The Public Interest* 63 (Spring 1981), 62–80; the issue carried the theme "Ethics—in Education, Business, and Politics" and included an article on "business ethics" by Peter Drucker. Martin Eger's article was reprinted in the volume *The Public Interest on Education*, ed. Nathan Glazer (Cambridge, MA: Abt Books, 1984), along with pieces by Thomas Sowell, Ed Koch, Diane Ravitch, James Coleman, and William Bennett.

317

Introduction

Some years ago, the State of New York released a publication setting forth ten 'educational goals' to guide designers of programs and courses. The fifth goal on this list reads as follows:

> Competence in the processes of developing values—particularly the formation of spiritual, ethical, religious, and moral values. . . . [The responsibility of the] school: a. knowledge of the diversity of values, b. skill in making value-based choices. . . .[1]

Three words are remarkable in the above formulation: "competence," "skill," and "process." To elucidate, even in a small way, what these terms may mean within today's culture requires a long story—and we have one; but the words are significant also in a more direct sense.

"Skill" and "competence" are associated with such matters as carpentry, language, and mathematics. They hint at a science or craft, suggesting authority, implicitly relating themselves to proven methods or facts. When social support for moral values seems to be crumbling, new sources of authority are naturally welcomed, especially science.

For the uninitiated, however, there is a serious problem here, because science deals with what *is* or *is possible*, and from this no logical operation can derive what *ought to be*. To claim scientific backing in morals is therefore suspect, and could do more harm than good if the claim proves erroneous or misleading. Apparently, then, we face a classic dilemma of secular education: That schools omit values from their concern is highly undesirable, and possibly an instance of social negligence, but the appeal to science is fraught with other dangers. Moral judgment does require authority, a ground of legitimation, but where is the *science of value* fit for such a role?

For those who *are* initiated in the teaching movements that have recently commenced, there is no problem whatsoever. 'Sciences' of 'value formation' have long been sought and are now available; the 'is/ought' disjunction is no longer an obstacle. The state's booklet takes this for granted, obviously, and the key to it all is the third word we have noted—'process.' If values themselves are not derivable from facts, some theorists believe, the *process* of valuing may well be.

The studies that underlie this view—genetic and humanistic psychologies, associated with such names as Jean Piaget, Lawrence Kohlberg, Carl Rogers, and Abraham Maslow—are widely discussed today. But in addition to psychology, there is another field—increasingly important—that also contributes: the group of 'decision sciences' based on computers, initially developed for military and managerial use. What such diverse theories have in common is a concern with the 'how' of valuing, with the structure of decisions rather than their outcome. One uses 'is' to derive not the 'ought' of morals themselves, but another 'ought': the manner of thinking about morals, and the way of evaluating that manner.

Those who avail themselves of these achievements see the educational dilemma as essentially bypassed: If we teach value science as political science is taught (without telling the student which party to belong to) then *all* values are within our purview. Logic is not violated, and the student becomes morally responsible in the highest degree, for he gains the 'competence' to 'develop' his own values, and to justify them in the light of reason. There is little doubt that the apparent advantages of this approach contribute heavily to the proliferation of 'values programs' in the schools.

Yet in spite of the new outlook, and in spite of initial public receptivity, there has emerged during the past few years a widespread resistance to the project, including stunned and disillusioned parents, as well as reasoned doubts in scholarly quarters.

The clash points again to the dilemma described, and raises an important question: Is harm already being done? If so, is it related to the 'sciences' guiding such policies as adopted by the State of New York? Or are the protests we hear of incidental phenomena, reflecting undue fear, to be expected when anything new is tried?

This is the question we examine here—by probing the connection between one type of values program, in one school district, and the social conflict to which it gave rise. A controversy over *curriculum* that shakes a town is bound to reveal something contemporary about the relation of theory to practice. Recently, such a conflict took place, distinguished by the clear and extensive record it leaves us. Looking at what actually occurred, what was said and done by different sorts of people—teachers, parents, school board members and professors—we immediately find two striking features: on the one hand, a living enactment of what some analysts are trying to tell us; on the other, odd happenings and strange inversions, as thought-provoking as they are unpredictable.

The Case of Spencer-Van Etten

In the hilly region south of Ithaca, New York, only 20 miles from Cornell University, lie the small towns of Spencer and Van Etten. All around are silos, Holstein herds, barns of every style and condition—yet farming is not the only occupation familiar here. From village homes and those adjoining the cornfields, people drive away each morning to work for Cornell, or for sophisticated industries in Ithaca, or commute south toward the IBM plants along the Susquehanna. Quite a number of families are relative newcomers, having left city and suburb to seek something else still: nature, a less treacherous environment for children, peace, and perhaps a chance for the spiritual side of existence.

For such a purpose the place seems well chosen. But in the Spring of 1979, as the local district prepared for its school board election, an uncommon battle raged through these valleys, and the village of Spencer was a center of intrigue. Neighbors met in each other's homes to make policy; professors were visiting to lecture; reporters from Ithaca, Elmira, and towns unknown were dropping by—even CBS's *60 Minutes* was there, cameras shooting a classroom scene.

The reason for all this was the school system's use of a 'technique' known as Values Clarification (VC)—to many parents, the antithesis of wholesome moral guidance.[2] The election marked the climax of a protest movement already in its second year, and involving other issues also. But gradually the values program became the main target of criticism.

To picture the mode of these protests, how logic and bewilderment roused each other, let us go back still further in time. If months earlier you happened to be nearby and stopped to pick up the local weekly, you might have read an item like this:

> The following are 8th Grade Decision Making questions given to the children of our school.
>   Q. Which would you prefer to give up if you had to? A. Economic freedom? B. Religious freedom? C. Political freedom?
>   Q. Are you in love right now?
>   Q. Do you think there are times when cheating is justified?
>   Q. Do you think people should limit the size of their families to two children?
> These are my questions to those questions: Who says one "has to" give up any of these freedoms? . . . Does an 8th grader really know if people should limit their families to two children? . . . What are they talking about? Cheating in school? Cheating in society? I feel these questions are "picking at a child's brain."

Under attack is a course of broad scope called "Decision Making" (DM), designed to help youngsters choose not only occupations, but their lifestyles too. However, before people can make such choices, so it is explained, they must first determine what their true values are; and for this purpose the school had adopted the approach of Sidney Simon and his collaborators. Exercises, questionnaires, and 'strategies' brought together by Simon's group in a teachers' handbook are used extensively in this course, as is their basic metatheory of values.[3]

Since parents had not yet discovered professional criticism of this new trend, the complaint quoted deserves notice. Its *style* reveals a level of emotion typical of many letters appearing at that time; but its *content*, just as typical, expresses in terse language several objections to which pages of scholarly prose had been devoted—questionable use of forced choice, violation of privacy, and superficiality.[4]

In Spencer, however, this was only the beginning, a kind of introduction, while really tough arguments turned on the philosophical concept of 'choosing freely.' And rightly so. For if the values-teaching movement makes any idea crucial it is this: A value must be 'chosen' by the individual himself, 'free,' as much as possible, from authority, 'conditioning,' and 'social pressures.' Only in this way does it become one's 'own' and therefore something positive rather than oppressive, something to be 'prized,' 'cherished,' and to 'be proud of.'

To many parents this naturally comes as a shock—and, incidentally, to not a few children also. (Some parents became opponents when their children complained.) These people, old and young alike, had thought until now that whether to become an engineer or a farmer was certainly their own choice—truly a 'free' and personal decision—but they never believed that whether or not to cheat, for example, was quite in the same category. And this, it seemed, is what the school was saying: "If I teach my child that cheating is wrong" wrote one mother, "and VC teaches a child that there are no right answers but rather to choose freely, it most certainly upsets the house."

Have these parents misunderstood? Have they missed the point in regard to *process*? If so, they are not alone. The charge of ethical relativism is made by scholars as well—in the philosophically grounded disciplines especially. In this case, an average group of parents sees what the academic community sees. The mechanic and farmer agree with the professor of philosophy. 'Ethical relativism'—the technical term—had to be defined in Spencer, yet its significance was known long before.

But 'relativism' is not taught here, the school retorts in frustration, and 'philosophical generalities' are not relevant to what we actually do. The technique is *professional*, and scientifically tested; it raises to consciousness the cause of our ideals. "We understand ourselves as not teaching values," explains the Director of Guidance, "*but* teaching students to identify the values their parents, friends, churches, and society have already imparted to them." This is the local version of the oft-mentioned claim that VC "accepts all viewpoints," does not "promote particular values," and should therefore be offensive to no one. Moreover, the advocates insist, to avoid inculcation is not to say all values are equally good. Simon and his associates write: "If we urge critical thinking then we value *rationality*. If we promote divergent thinking, then we value *creativity*. . . . If we uphold free choice, then we value autonomy or *freedom*."

This is the theory. And many parents have shown good will toward it, even in the face of disturbing practice. When the professional speaks, citing surveys and statistics, most people listen respectfully—especially those less versed in the conceptual apparatus of recent trends. But naturally they also listen to their children.

Take for example the afternoon of May 12, 1979. At the Spencer Grange, the seats and benches around the hall are filled. A photographer walks about popping his flashbulb at whoever seems newsworthy, a television camera bobs up and down on someone's shoulder, and in back, near the door, a young woman stands up to speak:

> My little girl came home and told me how she said lying was *wrong*—because Jesus said so—and the teacher told her, "but many people do lie." Well, I don't know . . . maybe I'm wrong, maybe I shouldn't believe what she told me— but I *do* . . . I *beeeelieeeeeve* her!

The mother repeated the key word, drawing it out and raising her voice, and one could not tell whether all its meanings were intentional.

Various people were gathered in this audience—farmers, executives, housewives, engineers, and also teachers from the local school system, their administrators, and members of the Board of Education. They had come to listen to a lecture on Values Clarification and to discuss its problems. The speaker, a philosopher at Cornell, was finished.[5] Now people from the floor were telling their stories, and when the mother of the girl who thought lying was "wrong" sat down, there was a momentary silence in the place. Then a young defender of the new methods answered from the front of the large hall, "Would you like it better if the teacher said nobody lies?"

A man turned to his neighbor with a wry expression. Scholarly debates surrounding the is/ought problem have not reached most parents of the Spencer area, but this they did sense: Not the fact of lying but another kind of 'is' was making itself felt at this meeting as it does in the school—the 'is' of language, the reality of *style*. What is new and noteworthy is not that "people lie," but what educators now *say* about people who lie —or cheat or steal.

A short distance to the south at Elmira College, an assistant professor hears of the controversy; indeed, it touches his town too. A trainer of teachers in the VC technique, he gives a talk to clear up the confusion. Suppose you tell a shoplifter that stealing is wrong—what do you accomplish? "It places the other person on the defensive, and the person making the judgment on a plane above the shoplifter." Some people may clarify their values in accordance with the best methodology, and still decide to shoplift. In that case, explains the educator, "you have to respect that decision if they have reached it intelligently . . . at least in this approach, you are respecting the person as a decision-maker."

Soon afterwards a newspaper version of the professor's views is placed in front of every chair at a meeting of the Board of Education in Spencer-

Van Etten. Some parents are puzzled, others shrug; but months earlier, an answer had been given to this now familiar stance:

> If the child "chooses" freely to steal, who do I send the officer to? The school? I tell my children *stealing is wrong*. DM states "choose freely"—does that mean DM will be responsible for the child they have taught? No, hardly, the parents will be responsible. . . .

And in a formal letter to the Board of Education, the same concern from another viewpoint:

> If the school assumes the responsibility of dealing in areas of personal values, morals . . . does the Board feel it is up to assuming the responsibility to these children that involves seeing them through the hurts and consequences of the decisions about morality? We as parents are suffering with them. . . .

One fear is that the nonjudgmental aspect of the program is *formal* only, that indirectly a type of value system is indeed being promoted—not just creativity, justice, and the like. In such a 'receptive' atmosphere, parents suspect, the 'is' of fashionable attitudes, professionally systematized and synchronized with an assimilated youth-culture, acquires artificially enhanced power; so that thought-structures deemed 'advanced' are substituted for the 'ought,' usurping the normative function of the latter.

Here, for example, is a key general definition used in the controversial course dealing with moral values: What is a "good decision"? The answer given by the DM Teacher's Manual: "the result of a decision is only good or bad in terms of the decision maker's own personal preferences."

And so, from the newspaper, a voice asks about the victims of such personal preference.

> If one chooses to steal, destroy, cheat—the list goes on—what about the victims? . . . *Thou Shalt Not Kill*. . . . On the other hand there is values clarification which teaches to "choose freely." I would say that Hitler fit into this category. As he chose freely to kill. . . . Bineum Heller, in reaction to the big lie of Hitler and its horrible consequences wrote this:
>> Perhaps part of the blame falls on me
>> Because I kept silent, uttered no cry
>> Fear froze my heart and confused my mind
>> And I did not resist the lie.
> So I have spoken up.

Thus a writer in the *Spencer Needle*.[6] The point must have been made elsewhere too, for in the Adirondack foothills, in a center of VC work, a reply for trainers and teachers has already been composed, and appears in a recent book:

> *Could a Person Use the Valuing Process and Become a Hitler?* This question repre-
> sents a classical test for the morality inherent in the valuing process. . . . *Short
> Answer*: No. A psychologically disturbed person is severely hampered in his or
> her ability to make free choices and to rationally examine a wide range of alter-
> natives and consequences. Hitler was clearly paranoid and could not effective-
> ly use the choosing process of valuing.

At Cornell, speaking on moral education, a professor recalls those
charges of old leveled at 'corrupters of youth'—and smiles. The audience,
knowingly, returns the smile. But vague resemblances, hurriedly glimpsed,
may conjure a distorted image. The quotations above, with their awesome
subject, suggest perhaps a truer resemblance—in the emphatic juxtaposition of a 'scientific' perspective to the concepts of 'good and evil.'

The protesting parents did not do particularly well in the school board
elections of June 1979—for reasons which will be touched upon. Though
they received a respectable portion of the vote, none of their three candidates was elected. (One, however, elected the previous year, remained on
the Board.) In September of that year, a new private Christian school
opened in North Spencer with a starting class of 23 pupils, and this year
there are 50.

It would be misleading to suggest that the values controversy alone
was responsible for the fact that children are now being withdrawn from
this public school system, though many pupils in the new school are children of the protesters. Scholastic achievement had been a major concern
before the values issue arose. Strangely enough, "Decision Making" was in
part developed to meet that concern—to combat apathy, and demonstrate
how valuable *thinking* could be in making life's crucial choices. And on that
basis it was defended.

To disenchanted parents, however, the relation between mental growth
and this type of values program is of an entirely different nature. It deepens their despair over the intellectual quicksands into which the schools
have now sunk, where each step taken to solve a problem seems to make
things worse. They say that in regard to rationality, the effect achieved by
methods like VC is the opposite of what is claimed; that the desire for
knowledge and delight in reasoning, naturally budding at that age, may be
channeled into something else: an endless sequence of games, of questions
and answers that place the trivial and the profound on equal footing. The
student is stupefied to a point where he is not likely to stop and think—or
object—when after innumerable other queries he comes across, ". . .
enjoy watching movies on T.V? . . . *enjoy going to church or syna-
gogue?* . . . enjoy going to a picnic?" (Emphasis added.)

The long-term intellectual effects of holistic elements in the curriculum
are always hardest to deal with, and in this area parents have the greatest
difficulty making themselves understood: "Do you mean that this innocent

little question will corrupt your child? You can't be serious." But these parents are serious. "Values clarification mocks the educational process."

Parent vs. School

For some families, no doubt, VC was the last straw, especially in view of the school's response to their considerable and initially hopeful efforts. This response—or rather, the response of the education system at its many levels—when faced with inquiry and criticism, is itself a matter that must receive special attention.

The first reaction was, unfortunately, a denial that any such thing as Values Clarification was included in the curriculum—an inauspicious beginning, immediately rupturing the trust between the school and the community. Unconvinced, some parents began their own education, gradually acquainting themselves with the basic texts of the VC movement. It was not long before they could place on the table the handbook by Simon, Howe, and Kirschenbaum, and beside it definitions, voting sheets, and tests that children brought home. From then on, the source of these lessons was never in doubt.

There followed months during which letters of protest and dissent appeared in almost every issue of the local weekly. Not all were restrained, or thoughtful, or distinguished by sensitivity to the teachers' unenviable position—letters to the editor being what they are. It is significant, however, that leaders of the protest group clearly set forth arguments paralleling those in the academic literature—a literature with which they were still unfamiliar. That scholarly criticism of values programs is not as easily found as the many volumes written by proponents was certainly a factor in these events.

The second phase of the school's reaction was to say, in effect: Well, yes, we use this 'technique,' but what of it? Many school systems do; it is state-evaluated, government funded, and approved by the best educators and universities. Then it was possible to quote Lawrence Kohlberg,[7] use a few psychological terms, mention the formidable name of Harvard, and so on—making it quite clear that behind the local course stands a vast professional world of *science*, prestige and official recognition, in comparison with which the doubting parent is a stumbler, at best. And to insure the final effect, there followed, at a public debate in 1978, this impressive counterthrust:

> Now I want to know who are *your* sources. . . . Where do you *get* these ideas? . . . You made very strong allegations against the school. . . . On what kind of *research* are these statements based, and *where* does this research come from?

Alas, at this point the spokeswoman for the parents' group could cite only the texts of VC itself—the books speak for themselves, don't they? How could anyone read this and *not* see what we see?

What happened instead was totally unexpected by the parents. In the third phase it was they who stood accused:

> We are witnessing an attack by an organized group . . . led by extremists from the far right who would like to see a little more hate than already exists in the world. The philosophy espoused in recent letters is embraced by the John Birch Society, the Heritage Foundation, the Conservative Caucus, the National Conservative Political Action Committee, and other ultra-conservative organizations.

This published outburst from a Spencer citizen became the keynote for much of what followed. The critics were described as anti-intellectual, simplistic, and opposed to independent critical thought. They were charged with 'lifting' ideas from outside sources, and their actions were blamed for 'dividing' the community. The insights of humanistic psychology and advantages of 'process methods' seemed far beyond their understanding. But that was hardly the worst of it: The teachers' spokesman—head of a recently formed "Academic Freedom Committee"—warned of possible "links with the new right coalition in the country, one aim of which is to do away with the public schools."[8]

Actually, the situation was quite different. In over two years of controversy, no evidence of any ties to an extremist organization was uncovered, though at least one Ithaca newspaper went to the trouble of "acting on local tips" and conducting its own investigation. "Are you now or have you ever been a member of the John Birch Society?" was asked of a member of the Board of Education.

In the beginning, lacking political experience, the letter writers did avail themselves of whatever materials seemed to indicate a shared concern—and these, it is true, do not all conform to the highest standards of scholarship. Still, such fragments as appeared in Spencer could not in seriousness be called extreme or 'ultra.' (Of 'outside sources' serving as a true inspiration here, C.S. Lewis's *The Abolition of Man* was one, and the Harvard speech of Alexander Solzhenitsyn another.) Thus the school critics of Spencer-Van Etten acquired a bad name: "We have become the target of much criticism, mockery, segregation . . . not only as a group, but as individuals as well. We are not enjoying this role." And a bad name is not easily shed.

Much later, when the relevant scholarly papers were discovered, when academics came to town and long discussions were held privately and publicly, and when highly respectable professional criticism was offered to the community, the labels still stuck. Repeatedly, the critics had to ask for attention to the substance of criticism, not its alleged sources:

> I am very concerned about being labeled "extreme right wing". . . . If you think what we are saying is wrong, then *talk* to us about those issues, don't call us names. . . .

And even when a scholar experienced with the problem (Professor Baer, of Cornell) came to the aid of the parents' cause—with detailed analyses, and lots of patience—that only discredited the scholar! "An avowed enemy of Sidney Simon," was the teachers' reaction. "Why has an outsider stepped into our local controversy?" asked the president of the Board of Education, "his remarks were . . . an affront to our local people."

Here the issue stands. Those parents who out of religious or moral conviction do not wish to have their child's values 'clarified' can with some effort avail themselves of a special procedure: The pupil is separated from his class and placed in study hall. This was the main concession brought about by the protests. Though parents point out the possible psychological harm, the injustice, and the educational waste of such a policy, no further accommodation has been offered. However, as people in this community continue their criticism—drawing in more detailed, credible support—the aura of unquestioned legitimacy surrounding the values program is dispelled. Since the election of June 1980, at least one third of the Board of Education is sympathetic to the case of the protesters.

Six Unsettling Questions

The controversy described obviously touches on issues of greater significance than a particular program used in one community. A critique of the Values Clarification method as such will not be given here. Rather, my aim is to reflect on a few of the more important problems brought to light by the Spencer-Van Etten story; and to pose, in the case of each problem, the basic question involved.

It is perhaps not accidental that the first difficulty encountered in Spencer-Van Etten, that which caused the earliest complaint and much bad feeling, had to do with the question of *truth*—not the problem of what to teach children about truth, but whether the school is telling the truth. Certainly access to fact was hard to obtain for those whom these facts affected most—the families of pupils, the 'consumers' of education. A controversial program was part of the required curriculum, but parents were told it was not. When they discovered otherwise, they were assured it played a minor role; when they perceived it was the basis of courses in decision making, the explanation was that no educator could be found to oppose the approach; and when they brought to town scholars who did oppose it—things turned nasty.

We have here a new kind of credibility gap, involving not merely the schools' shortcomings, but their very intentions. And this misinformation or lack of information regarding programs likely to be found objectionable is not unique to the towns we have focused on. It is reported of a district in Anne Arundel County in Maryland that the Board of Education itself "had no idea" of the use of VC in their school, and the revelation was "positively electrifying."[9] In a New Jersey course, Transcendental Meditation was apparently practiced with incense, chanting, and invocations to Vishnu, but that did not prevent school officials from blandly denying a religious content. "Inexcusable ignorance . . . bordering on deception," replied parents; a legal suit was needed to eliminate the program.

Did the response in Spencer-Van Etten also border on deception? Many people believe just this. Unfortunately, the VC handbooks lend support to that suspicion, since total frankness is not encouraged. What parents see, teachers are told, should be *carefully selected*. Trainers themselves prejudge the source and nature of objections in a way that corresponds exactly to what happened in our story.[10] The patronizing attitudes, so bitterly resented by parents in this case, are less surprising therefore.

National attention is now directed to problems of 'truth in lending,' 'truth in advertising,' the Freedom of Information Act, and so on. But 'truth in education'—in the same sense—seems to have been taken for granted. Thus, our first question:

1. *Is it acceptable that families of pupils in public schools be denied access to any information regarding curricula, school activities, or teacher training? Or that the standard of veracity for teachers and school officials be lower than that for auto makers, bankers, or the Department of Justice?*

The second point may also be seen in a consumer's perspective: Not only the purpose of a product has to be considered, but unintended results also. Cigarettes, energy use, and certain medicines have taught us that some side-effects are cumulative, long range, and hard to detect. Such experience supports the demand that on every new product or 'system' introduced into society there be imposed bounding conditions within which the designers must solve their problem. And to insure not only that positive claims are justified, but that these conditions too are met, *time* is required.

How do matters stand in education in this respect? When Values Clarification was introduced in Spencer-Van Etten in 1974, and the *Handbook of Practical Strategies* was beginning to be used there, this fundamental text of the movement was barely two years old. "Research" on the effectiveness of the approach was "far from conclusive"—by the authors' own admission—and side-effects were not even mentioned. In the following year, evaluators termed these studies "relatively unsophisticated";[11] and the first systematic appraisal of even this doubtful evidence—which did not appear until 1978—essentially refutes the positive claims of the proponents.[12] Thus, when the program was initiated in this school district, detailed professional criticism of Values Clarification had not yet begun. But advocacy by the developers was loud, their materials flooded the libraries, and their activities were gaining favor with the State and the teachers' union.

This situation gave rise to an especially instructive though predictable phenomenon. There was a time—some parents believe—early in the protest, when the protesters were being 'listened to' by their community, and indeed by a fair number of the teachers themselves. It was their inability *at that time* to produce unimpeachable professional critiques, supporting at least in part their own objections, that largely determined the ensuing course of sentiment. In the absence of scholarly judgment, the charge of 'right-wing sources' dominated the atmosphere. And once public positions were adopted, and commitments made, no amount of credible evidence could alter the outcome, for then a powerful new element had entered the picture—personal prestige.

Whether we take a 'consumerist' viewpoint, or that of scholarship and science, the implication is clear: Criticism *follows* innovation; a decent pause between proposal and full-scale use is therefore indispensable. The lack of thorough, many-sided evidence invites leadership by selected expertise and half-baked professionalism, and in such a context the ideal of local control of education through elected boards is hardly meaningful. Our second question, therefore, is:

> 2. *What checks and balances, if any, exist within the educational system to prevent questionable practices from being 'sold' to the school before competent evaluation has taken place? What mechanisms are there for communicating to the schools and their communities all aspects of such evaluation at the time they are needed?*

Now suppose a new program is in fact developed at the universities and widely implemented—and it does engender opposition. What then have we a right to require of the educational system?

Nothing was more evident, and more embarrassing to the teaching profession, than this fact: After a year of discussion and debate, the protesting parents were in possession of more knowledge, more documentation, and deeper understanding of the subject at issue than were the teachers defending the school's position. The parents, not the school, informed the community of the relevant literature and brought in scholars to speak. Teachers involved in the controversy seemed caught off-guard, frightened, and badly prepared to handle the intellectual substance of the protest.

Perhaps this explains why they never did come to grips with that substance, but concentrated instead on other replies.

"My intention here tonight is to get the facts out, and to demonstrate to you what we have found in our research." So began the teachers' representative in a public forum in Spencer-Van Etten. His posture is flawless. It is indeed the teacher's job to supply facts and dispel prejudice. What followed, however, was breathtaking: "I teach a college course in adolescent psychology and moral development . . . and the only place that I find the statements you people make about Values Clarification is in this right-wing literature." When this statement was made in December 1978, the *Moral Education Forum* had been reporting professional opposition for about three years, and a special issue of *Phi Delta Kappan* featured debates on the same kinds of objections local parents were raising.

The conclusion in Spencer was inescapable: Either school employees knowingly withheld critical opinion from the community, or they honestly did not know of it and could not find it even with effort. In any case, the professional training of these educators is called into question, and the responsibility of the academic world in such controversies reveals itself as greater than commonly assumed. Criticism of teacher education is nothing new, but the expansion of public schools into the realms of morals, sex, death, the self, and 'decisions' in general, raises additional questions.

Can the average teacher with the average training be a 'teacher'—or 'facilitator'—in such deep waters without comprehensive additional education in ethics, moral history, theory of knowledge, and the like? May we not doubt that these modern 'teachers of virtue' will themselves have had an education commensurate with the task? It does not inspire confidence to be lectured by an instructor in 'moral development' who does not even possess the basic skill of a 'library look-up' in his own field.

Unfortunately, most developers of these programs are far more engaged in stocking a veritable supermarket of materials—with films, tapes, games, numberless 'strategies' and 'moral dilemmas'—than the in-depth education of the teacher. Values Clarification, which recommends itself as "easy to get started," is especially culpable in this respect; to become a practitioner, the novice is assured, "a relatively short training program, as short as a few hours," will suffice! Summing up this side of the problem, then, we ask:

> *3. Do the universities—and does society—feel comfortable with a situation in which millions of children in thousands of schools receive systematic moral education from teachers whose study of the subject consists of no more than a one-week workshop and a handbook of strategies?*

Morality and Process

In discussions of whether moral education belongs in public schools at all, two questions always arise. Can you really avoid moral problems? And wouldn't prohibition violate a basic freedom?

Again the drama of Spencer-Van Etten casts the issue in a new light. The controversy there began in earnest not because an individual teacher made an isolated remark or gave a personal opinion, but when parents discovered exercises, questionnaires, and definitions, taken from *handbooks* and systematically applied in classes, and when they realized that this methodology projects a meta-ethic fundamentally different from the universal recognition of objective value and 'common human law'—what C.S. Lewis called the "Tao."

Teachers' freedom to express personal views on moral questions, when these questions arise naturally in subject matter or discussion, was *not* contested here. And of course such 'dealing' with moral issues cannot be avoided—nor is it. Nor does any serious person suggest it should be. The crucial question seems, rather:

> 4. *Is the issue of 'free speech,' when each individual teacher responds on the basis of his own unique background and convictions, the same issue as when a whole school, or district, adopts this or that developer's system of moral education, complete with hardware, software, 'trainers,' and gurus? In the latter case, may it not actually be a restriction on moral dialogue?*

The opposition in Spencer-Van Etten did indeed believe that a narrowing of moral vision had been imposed. Parents usually find the nature of this narrowing difficult to articulate in a form acceptable to educators, but in this community one school board member rendered the gist of it concisely: "We're not saying our precepts should be taught, but if ours are left out, then leave out all values."

The statement deserves attention. What is meant by 'our precepts'? It is not, as sometimes charged, 'our religion.' The Constitution's stricture in this regard is well known here and not challenged. We arrive therefore at a point where the pervasive effect of *meta*-ethics comes to the fore—the foundation of ethics, its origin, its categories. What is clearly felt by many parents is the exclusion—from the explicit methodology—of a whole realm of moral categories not specific to one religion, but part of that "Tao" encompassed by the heritage of both East and West: the *good*, the *true* (objectively), the *just, soul, faith, courage, moral rebirth*, (or 'turning,' or 'enlightenment'), etc. What they see instead is a new set, including such concepts as *self-image, life-style, rich experience, decision-model, risk strategy, maturity, self-actualization*, and *rationality* (in the hypothetico-deductive sense). The only element from the older list often kept is *justice*.

Granted that such a shift in ethical categories accords with trends in psychology and operations research, are there valid grounds for *excluding* from public education the more traditional set? Well . . . there may be. There may, indeed, be—*if* such language is in itself deemed to be religious, and its inclusion unconstitutional. But this raises perhaps the most fundamental question of all:

> 5. *If the systematic teaching of morals within the bounds of the First Amendment requires systematic exclusion of a whole realm of meta-ethics underlying moral thinking in our society, isn't this in itself a distortion of ethical discourse in the intellectual sense, and injustice in the social sense?*

Yet systematics is the main feature of the values teaching movement today; it meets the requirement that morals too be seen as *skill*, that here also a *science* legitimates. Let us look at the point from a humanist perspective.

The word 'humanism' is, of course, ambiguous. One usage,[13] relevant here, connotes by 'human' what is distinct from 'mechanical.' The 'knowledge factory' and 'mega-machine' embody its opposite. It is this antimechanistic, anti-rule-following inclination that led reformers to indict so-called traditional education, and to introduce a whole series of reforms to free the schools from needless stifling formalism. And this aim one can easily accept.

But today a curious reversal seems to be occurring. Humanist educators increasingly press to systematize the few realms that have so far escaped this fate. 'Traditional' education assumed certain values axiomatically—and for that very reason had no need to explicate them through a system. If a child cheated on a French test, for example, he lost face with his peers and his teachers—this is how the value was *supported* (not 'reinforced,' not taught!). Now, however, a pupil is asked whether, according to someone's criteria, that 'decision to cheat' is or is not a 'critical' one. Again, a child is tempted to "drop a friend who is being made fun of by others kids at school." In a traditional environment, chances are a youngster would *recognize* that act for what it is; but today in Spencer he is *taught*—to identify which of the four 'main risk-taking strategies' this decision represents.

Science, by definition, systematizes what it deals with, it formalizes its subject. And the formalization of decision making along the lines of management and computer models, if applied to morals, must obviously reduce that also to a system. It is fair at this juncture to ask, of humanists especially:

> 6. *Is it good 'strategy,' in the long run, to humanize our 'knowledge factories' by replacing the axioms and disciplines of one era with the systematics of another? Is rule-following in 'how to decide' really less mechanical than rule-*

following in regard to ends? In short, is an automatic 'value-driven decision system' the proper model for a more moral, more humane society?

A Warning, Ancient and Modern

The final point is not a question at all, but a reminder. The values education movement has justified its mission partly by the rapid social change occurring in our time—the 'collapse of the old morality.' True, the change is great. It lends urgency to the work of the developers; to do nothing seems dangerously complacent. That is one horn of the dilemma of values education. But in thinking about man, as about nature, there generally has been equal concern with the *invariants*, as science calls them—that which remains constant in the midst of all change. Compare now the dialogues from Tioga County that have occupied us here, with those of another, 'simpler,' age. In the invariant lies the second horn of the dilemma.

Protagoras, the Sophist, is come to town. And the young Hippocrates is all on fire—for this teacher, famous in all the cities, promises his student not the "drudgery of calculation, and astronomy, and geometry, and music," but "prudence in affairs private and public . . . to speak and act for the best." He has a special method, it is told, and all the youth and would-be teachers throng about him like a king's entourage.

But wait a minute, says Socrates—how do we know it's all true?

> When the soul is in question, which you hold to be of far more value than the body, and upon the good or evil of which depends the well-being of your all— about this you never consulted either with your father or with your brother or any one of us. . . .
>
> Surely . . . knowledge is the food of the soul; and we must take care, my friend . . . that the Sophist does not deceive us when he praises what he sells, like the dealers wholesale or retail who sell the food of the body. . . . In like manner, those who carry about the wares of knowledge, and make the round of the cities, and sell or retail them to any customer who is in need of them, praise them all alike; though I should not wonder, O my friend, if many of them were really ignorant of their effect on the soul. . . . If, therefore, you have understanding of what is good and evil, you may safely buy knowledge of Protagoras, or any one; but if not, then, O my friend, pause. . . . For there is far greater peril in buying knowledge than in buying meat and drink. . . . You cannot buy the wares of knowledge and carry them away in another vessel; when you have paid for them, you must receive them into your soul and go your way, either greatly harmed or greatly benefited. . . .

Plato, *Protagoras*

And today:

"You are what you eat," most of us have heard that expression.

There are those who are very concerned about "junk" being served through the school breakfast and lunch programs. They believe as adults we are responsible, and should be choosing nutritious food for school children. . . .

I have heard the phrase, "You are what you think." This also has clear meaning to me. What I think is eventually manifested in my character; the things I say and do. Shouldn't I be equally concerned with the nutritional value of what enters my mind? Shouldn't I be even more concerned as to my children's choice? . . .

Theresa Rimbey
Spencer Needle
August 10, 1978

[24]

A Tale of Two Controversies: Dissonance in the Theory and Practice of Rationality

Abstract: The relation between rationality in science and rationality in moral discourse is of interest to philosophers and sociologists of science, to educators and moral philosophers. Apparently conflicting conceptions of rationality can be detected at the core of two current socio-educational controversies: the creation/evolution controversy and that concerning moral education. This paper takes as its starting point the recorded views of participants in these controversies; exhibits the contradictions and their effect on the public; relates these contradictions to developments in the philosophy and history of science; and suggests, in a preliminary way, one approach for dealing with the problem. [—A.S.]

> *Education is the laboratory in which philosophic distinctions*
> *become concrete and are tested.*
> John Dewey

Introduction[1]

In the United States during the past two decades two educational controversies have become national issues: the creation /evolution controversy and the clashes concerning formal teaching of morals, values, and certain kinds of decision-making. The connection between these two problem areas has not been overlooked, but invariably it is the sociopolitical links that receive attention. First, there is simply the fact that conservatism in our culture includes people who favor creationism in the classroom *and*, at the

This paper appeared in *Zygon: Journal of Religion and Science*, 23, No. 3 (September 1988), 291–325; the same issue included comments on the paper by Mary Hesse, Abner Shimony, Thomas F. Green, Holmes Rolston III, and Daniel R. DeNicola, followed by Martin Eger's "Reply to Criticisms."

335

same time, oppose most values programs now in use. In addition, a larger group of *religious* conservatives believes that the teaching of evolution—by emphasizing man's continuity with nature while depicting an amoral universe—does indeed have something to do with declining morals, as William Jennings Bryan thought. Since the academic world largely attributes the reactions in both realms to educational backwardness or bad reasoning, and since these appear to threaten the integrity of teaching, many scientists and philosophers have been helping schools to ward off the 'attacks.' However, nothing in the complaints themselves is seen as worthy of serious discussion.

This is unfortunate because a real cognitive question does exist. Although religious and political motives are involved, these should not eclipse the underlying epistemic difficulties in educational philosophy, especially since the same concerns, in one form or another, lie also at the center of debates in philosophy of science and more generally in the theory of knowledge. One consequence of these difficulties is that even in the educational domain the positions of the academic community itself appear dissonant on a basic point—a point, moreover, that the public is now asked to take very seriously. It has to do with *rationality*.

True, the 'problem of rationality' has not, as such, captured the popular imagination. But just beneath the surface of these two controversies there lies the following profound question, at once practical and highly theoretical: *What form* of rationality shall be taught to the nation's children, and to their teachers, and to the public through state-supported schools?

The aim of this essay is two-fold: first, to sketch some aspects of a troubling situation, to tell a tale; and then to suggest that while we have been busy with symptoms, rather obvious causes have gone unexamined—causes that reach beyond the symptoms now in the forefront. I wish to show, contrary to prevailing opinion, that certain theories of education and the social disputes to which they give rise are related *not spuriously but logically*. Because this requires concrete examples, references to local conflicts that have actually occurred, I will draw in part on personal observation of one such series of events (Eger 1984 [1981b]).

The second half of the paper *begins* to show that the problem of dissonance is real in the sense that it cannot be easily disposed of; that it has depth, that aspects of it are manifest on several levels, and that the most obvious attempts to dismiss it fail. Along the way I indicate why the difficulty at the core is serious enough to exceed the bounds of current controversies. However, it should be understood throughout that the entire presentation is introductory, and no attempt is made to offer solutions or conclusions embracing the problem as a whole.

The stage is set for conflict when certain features of human reason are identified as essential, and curricula embodying these features are developed without regard for side effects or interactions with the total cognitive environment. In North America, in the area of morals, the elements singled out recently for special emphasis have been the 'critical attitude' and 'choice among alternatives': "The child must be encouraged to develop a critical attitude toward conventional right answers, rules and authority, whether they appear in the form of the Ten Commandments or of parental guidance" (Gow 1985, 77). "Many children at age 8 may be ready and eager to engage in the give-and-take of selecting among alternatives. . . . [There are] six basic steps in the valuing process: 1. Recognition of need for choice. . . . 4. Free choice among alternatives . . ." (The University of the State of New York 1976, 62, 89).

The first quotation refers to Clive Beck's 'reflective approach' (influential especially in Canada). The second is from a New York State teachers' guide, but the substance of both is found as well in the theory of Lawrence Kohlberg, in 'values clarification,' 'values inquiry,' 'values criticism,' and in other methods described in the literature (Kohlberg 1981; Chazan 1985; Morill 1980; Simon, Howe, and Kirschenbaum 1972). They exemplify what are widely regarded as the two necessary components of a rational attitude in ethics and in general: the Socratic, skeptical, questioning stance, and the almost tautological idea that *where choice is absent thought is impotent.*

Why negative reaction to these principles arises in some quarters is not hard to imagine. Again and again, parents of all ages have complained that whatever the motives, the schools are in fact "driving a wedge between the child and his family." Radical skepticism, they charge, is a strong solvent. Where, for one reason or another, families cannot provide the kinds of arguments that satisfy a teenager and make the moral demands placed upon him seem 'rational,' there, as one mother said, it "most certainly upsets the house" (see Eger 1984 [1981b], 213).

It is a mistake to believe that the local conflicts provoked by the application of this philosophy peaked in the 1970s and are now largely behind us. As implementation continues, opposition grows; and the dramas I witnessed years ago are reenacted—with a resemblance that is startling—in many communities today.[2] The question we must consider, therefore, is why, despite abundant criticism both popular and scholarly,[3] the movement as a whole has not substantially changed its direction. Even the proponents of these programs cannot be happy if in fact upset houses and torn families are among the outcomes of their work. Why, then, are such side effects not deemed strong enough to overrule, or at least substantially modify, strategies that (rightly or wrongly) are so often seen as morally and cognitively relativistic, even destructive?

This question is important. Without understanding the motive and claim to legitimacy behind contemporary forms of moral education, the problem which is the target of this essay cannot be fully appreciated. That motive is nothing less than a renewed faith in the power of *rationality*—in its active form—as the only available protection against the kinds of threats that have traumatized our society, from racism to Watergate, from Vietnam to street crime. Because "other methods have failed" so dismally, it is becoming credible again that Socrates, not Kierkegaard, was closer to the truth: Evil is largely error (see Kohlberg 1981, 30). Behind the moral disappointment of our time lies the failure of our *cognitive faculty*. From this it follows that formal education has much to offer, *provided* it takes seriously "its special connection with rational explanation and critical dialogue: with the enterprise of giving honest reasons and welcoming radical questions. . . . Free rational judgment by the student. This is what distinguishes teaching from propaganda or debating . . ." (Scheffler 1965, 11).

Here we reconnect with a major theme in the story of our 'coming of age'—an interest in the *manner* of thinking, especially the constraints that may be put upon it. The fear usually expressed is not of error, not of the propagation of falsehood; it is not in the first instance a question of content at all, but of "the structure of our ways of believing" (Green 1971, Ch. 3). In basic as well as in higher education, we seem to be attempting to implement certain insights of the previous century, when men like John Stuart Mill expounded this theme with even greater force: ". . . assuming that the true opinion abides in the mind, but abides as a prejudice, a belief independent of, and proof against, argument—this is not the way truth ought to be held by a rational being" (Mill 1961 [1859], 226). Then as now, the perceived dangers were *indoctrination, authoritarianism*, and *narrow prejudice*. As Mill maintained that "Whatever people believe . . . they ought to be able to defend against at least the common objections," so now, leading theorists insist that "this demand for *reasons . . .* is *essential* to the conversation of instruction" (Green 1971, 29).

Reason and Choice

Of course, couched in such language, it is hard to see how the policy could be opposed by anyone. Yet the critical principle—our first element of rationality—involves us in such a compelling logic that those who accept it very soon make the second move. It escapes few people in our time as in the past that all the 'radical questioning' and 'honest reasons' are by themselves insufficient if those answers and reasons never have to contend with *alternative explanations,* if genuinely different frameworks for viewing the problem are kept out of sight. Particularly with social and moral issues, whenever 'right answers' *only* are provided—*right reasons included*—there, it is widely believed, indoctrination is rearing its ugly head; for then, in effect, we are back to the old-fashioned, blinkered, "right-answer-inculcation-paradigm of knowledge" (Paul 1985, 2).

Michael Scriven has put the case most vividly: ". . . even if one believes that the old values are the best values, the reasons for—and against—them must be rehearsed by each generation or it will rebel against them" (Scriven 1985, 10); and this can be done only through frank encounter with *unorthodox* alternatives: "All the main examples should involve *highly controversial* issues of *considerable personal, social, or intellectual importance* that are *not seriously addressed in the regular curriculum.* Critical thinking is coping with controversy" (Scriven 1985, 12). Scriven is speaking of high schools here. He has in mind such matters as the decriminalization of marijuana, the criminalization of abortion, the effort at 'containing communist aggression' in Afghanistan or Central America, the rights of homosexuals and fascists to teach in schools, and the banning of atomic weapons and/or atomic power plants.

It is here, where 'reasonableness' demands the *radically* open mind, that educational strategy ignites its own fires. From 1978 through 1981 I observed at close range one such conflagration in a rural community in New York State. At the seventh and eighth grade levels, a school had put into effect programs embodying the approach advocated by Scriven (Eger 1984 [1981b]). Precisely the idea of 'choosing freely,' in the idiom of values clarification, or Israel Scheffler's 'free rational judgment,' caused not just dissent but incredulity—*incredulity that certain kinds of alternatives actually receive consideration inside the classrooms of a modern, civilized state, and thereby acquire a degree of legitimacy.*

In its defense, the board of education of this community invoked what the majority of its members clearly took as the established wisdom of the academic world. "To be a true value, it must be chosen freely," proclaimed an educator at a nearby college. The one difficulty, he said, is found in the example of a shoplifter who has undergone this process and still decides to shoplift. "The difficult part is that you have to respect that decision if they have reached it intelligently," he continued, "At least in this approach, you are respecting the person as a decision maker and you keep the lines of communication open."[4]

However, respect is just what many believe ought *not* be granted: "I don't want my son *even to think* of running away to Canada when he receives a draft notice; if his country calls it's his duty to go." That reaction from a father, incensed at some of the school exercises. Answer: To ask a boy to consider this option is not necessarily to recommend it. If upon serious reflection he rejects it, he will understand much better the reason for his behavior. Then, rhetorically, the president of the board of education asked "at what stage in an individual's life should he begin to learn to think for himself?"

Because they were aware of the prevailing academic opinion, because state agencies seemed to concur, local officials repeatedly invoked such arguments to shield themselves from charges of amorality. Richard L. Morill, for example, a college president, requires that students in a values program "be encouraged and enabled to assume the role of someone with a contrasting point of view" (Morill 1980, 245). Kohlberg and his collaborators assure us that "exposure to real or verbal moral conflict situations, not readily resolvable at the child's own stage, and to disagreements with and among significant others about such situations" bring about exactly that "conflict-induced-reorientation" which is needed if we are to ascend to a higher moral stage (Kohlberg 1981, 146).

While there are differences among the proponents of the several values education philosophies, it is important to keep in mind that concerning these two crucial elements—'criticism' and 'alternatives'—the consensus is rather close to unanimity. Despite serious opposition, educators have stood just as firm on the second point (the main bone of contention) as on the first; for this too is grounded in the illustrious tradition of western emancipatory thought. Mill's *On Liberty* is widely quoted with devastating effect as he lays bare precisely what is at stake: Even if the alternative "be an error, it may, and very commonly does, contain a portion of the truth; and since the general or prevailing opinion on any subject is rarely or never the whole truth, it is only by a collision of adverse opinions that the remainder of the truth has any chance of being supplied" (Mill 1961 [1859], 245).

In view of what is to follow, we will want to keep in mind every step in the logic of these classic passages. Evidently, 'collisions of adverse opinion' are to be promoted not to satisfy some questionable principle of fairness, but in order to attain as much of 'the truth' as may be available on any subject—the argument is cognitive. Moreover, something else is at issue besides that first attainment: "Even if the received opinion be not only true, but the whole truth; unless it is suffered to be, and actually is, rigorously and earnestly contested . . . the *meaning of the doctrine itself will be in danger of being lost, or enfeebled . . .*" (Mill 1961 [1859], 246, italics added). Here one sees, in a nutshell, the theory, the application, the protest, and the defense. One sees the antecedents, elaborately spelled out, of Scriven's need "for each generation to rehearse . . ." its views.

Little wonder, then, that backed by history and the finest thought of generations, leading educators are not inclined to compromise on this crucial matter. Somehow, the objections, the offenses (real or imagined) must and will be dealt with by the schools themselves; no doubt, the causes of whatever conflicts exist, and whatever complaints really are justified, lie in the local environment—misunderstandings, inadequately trained teachers, fearful administrators, all exacerbated by people whom rapid change has 'made insecure.'

This tempting response is heard frequently. It does contain a good deal of truth, and the venerable arguments on which it rests are still potent. Nonetheless, my purpose is to show that *because of the context* such a response is too narrowly conceived. Intellectually, it misses the thrust of the objections while underrating the opposition; socially, it fails to comprehend the genuine feelings of injury all this evokes. But to see that, and what I call here "*the* problem," we must now continue the tale by turning to the second of our two controversies.

In the wake of the Arkansas 'creationism trial'[5] of 1981–1982, a large number of books have appeared for teachers, students, and those who 'influence public policy,' explaining the position of the scientific-educational community.[6] The authors take great pains to show that the objection to discussing creationism in the classroom is based primarily *on scientific and educational grounds;* that it is valid not only against legislative interference and demands for 'equal time,' but against *creationist arguments in science courses under any circumstances:* They "have no proper place in the science classroom"[7]; "Creationism and evolution are mutually antagonistic and cannot survive in the same classroom" (Newell 1985, 14); "Like the belief that the earth is flat, creationism deserves *no* acknowledgment in the science classroom . . ." (Siegel 1981, 101). In fact, the most common state of mind among participants in this effort is (once again) *incredulity*— incredulity that 'in this day and age' such elementary principles should have to be explained yet another time, such old battles refought. There is also the fear, openly expressed, that creationism might be given "equal intellectual respectability with evolutionary doctrine" (Asimov 1984 [1981], 191). Thus, speaking as a philosopher of science, a witness for the State of California worries that "the court, by directing evolution to be taught as theory, not fact," awarded creationism an "implied legitimacy" (Siegel 1981, 101). The same anxiety was incorporated formally in a resolution of the American Anthropological Association.[8]

It is this standpoint that interests us, the justifications especially. The educational-philosophic reasons for excluding particular views *a priori* are familiar but as if from "the other side of the looking glass." Michael Ruse, philosopher of science, prolific writer on evolution, and major figure at the Arkansas trial, is somewhat more explicit than others, but his sentiments are widely shared. "One must offer children the best-sifted and most firmly grounded ideas that we have, together with the tools to move inquiry forward. . . . Unless we *exercise control over what we present*, the next generation will have no criteria by which to evaluate and advance knowledge . . . . All must agree that there has to come a time when *we have to cry 'finis' to the teaching of certain ideas. . . . It is an act of bad faith even to present such ideas as a possible basis of belief*" (Ruse 1982, 328–29, italics added; see also 1984).

What *would* happen if creationist views were discussed in schools? Analyzing the dispute as a philosopher, Phillip Kitcher describes the prospects:

> One may ask, why not let Creationists submit their case? Surely the truth will [win] out. Nor should we worry about a little wasted classroom time, *when a deeper understanding of the merits of evolutionary theory might be secured* by allowing students to think through the issues for themselves. *The argument is insidious.* . . . The previous chapters [of Kitcher's book] show that there is no genuine contest, no true comparison. What is in doubt is the possibility of a fair and complete presentation of the issues discussed above, in the context of the high school classroom. . . . There will be . . . much dredging up of misguided objections to evolutionary theory. The objections are spurious—but how is the teacher to reveal their errors to students who are at the beginning of their scientific studies? . . . What Creationists really propose is a situation in which people without scientific training—fourteen-year-old students, for example— are asked to decide a complex issue on partial evidence. (Kitcher 1983, 174–76, italics added)

Yet the educational policy here enunciated is addressed to the same population (including teachers) that reads Scheffler, Scriven, Kohlberg, Richard Paul—on 'critical thinking'—and through them also the words of Mill. Kitcher, as did Scriven before, reproduces Mill's argument with extraordinary faithfulness. But in one sphere that argument is gratefully embraced, in the other it is called 'insidious'—and this despite the fact that Kitcher too believes the "study of science provides important training in reasoning." To be sure, I have juxtaposed pronouncements that are aimed in one case at *science,* and in the other at something else. I will address that point, but for the moment let us continue the comparison.

In Kitcher's apologia, representative of a whole class of such arguments, an important question is raised: What age, level of maturity, and degree of relevant experience is required in order that a student profit by grappling with wrong or misleading or difficult arguments outside the domain of established ideas? We see a remarkable contrast when we put side by side the two answers widely accepted within academia—one dealing with human origins, the other with moral decisions. Regarding the former, Ruse supports Kitcher: "Exposing young minds to it, thinking that it passes for reasonable intellectual activity, reveals irresponsible behavior by the teachers. It is not simply mistaken: it is *corrosive.* Teaching scientific creationism will *stunt* abilities in all areas. . . . Thus, I say, 'keep it out of the schools'" (Ruse 1982, 328–29, italics added). On the other hand, speaking of social and moral concerns, Scriven gives quite a different view of 'young minds' and the need to protect them from 'corrosion':

It is sometimes said that the problem with critical discussion of sensitive matters in the school is that the students are not mature enough to handle them. The first fallacy with this argument . . . is the idea that the students are not already facing major decisions demanding every resource of the critical mind. . . . The second fallacy is the assumption that the students will mentally mature when they leave school in some mysterious way that will offset ignorance. . . . But still we avoid biting the bullet, we refuse to give time to the opposition, we refuse to meet the entry requirement for a school that wishes to engage in true critical thinking and teaching. . . . The shock of discovering that Karl Marx's *actual words* are being read by our own innocent children. It is the threat to our own beliefs that strikes fear into our hearts when we hear about alternatives to them being seriously considered in our children's schools. This is defensiveness, insecurity. . . . (Scriven 1985, 10–11)

A sweeping indictment, although many schools have in fact "bitten the bullet": A grant becomes available, a proposal is written—the program materializes. It *may* be successful, but sometimes it happens like this:

The following are 8th Grade Decision Making questions given to the children of our school.
   Q. Which would you prefer to give up if you had to? A. Economic freedom? B. Religious freedom? C. Political freedom?
   Q. Are you in love right now?
   Q. Do you think there are times when cheating is justified?
   Q. Do you think people should limit the size of their families to two children?
These are my questions to those questions: Who says one "has to" give up any of these freedoms? . . . Does an 8th grader really know if people should limit their families to two children? . . . What are they talking about? (see Eger 1984 [1981b], 212)

Let us reflect a moment on this outcry in a local newspaper, so jarring yet familiar. Imbedded in the blunt, excited phrases, one can easily discern the argument of Kitcher—translated only to fit the situation. Translating the other way, we recover the full power of the academic form: What *the school* really proposes is a situation in which people without *political and economic* training, without *exposure to the psychology of parenting*—thirteen-year-old students, for example—are asked to decide a complex issue on partial evidence.

This complaint is widespread, ongoing, and consistent. Recently, a more articulate young woman put it this way:

They have them debate on information that they don't give students the ability to research . . . to have time to reason it with their parents, their peers, someone else who knows more about the subject. . . . The students would learn about socialism; they would divide the class up and would vote on different aspects of it—if it was right or wrong. Well, the teacher, in his eyes, believes he is teaching children all sides of the story, that they have the right to choose their own values. I don't agree. . . . I believe in a high school American Government class, the student should be in a position to see the good points of *our nation*, that they should understand the *background* and *basis* of our nation.[9]

I cannot say whether the ridicule greeting such complaints, and the charges of anti-intellectualism, are due only to the fact that the complaints are seldom put in proper academic language. I do know, however, that the *dissonance* between the conventional justifications in the two realms is not missed. In the community I visited, the conflict centered on decision-making and values; creationism was not a part of the public debate. However, off-stage, protesting parents repeatedly called attention to that second issue—sometimes with confusion, sometimes with irony—confronting the interested listener with the following question: If it is a good thing for children to consider all sorts of alternatives in moral decision-making, no matter how repugnant—stealing, cheating, betraying one's friends—all for the sake of developing critical reason and autonomy, then why, suddenly, when we come to evolution, is it far more important to learn *right answers*[10] than to think critically? Why *just here* are certain alternatives taboo, even for the sake of discussion—despite the fact that being wrong about a scientific theory of remote origins can never have consequences as grievous as being wrong in one's moral outlook?

The most troubling aspect of these disputes is that while some parents see a glaring contradiction in the philosophies of education frequently expounded for their benefit, to the academic community it seems to be no issue at all. Little if any discussion of it can be found in print, and at times it does appear that such discussion is deliberately avoided. Consider the *National Forum*, one recent issue with 'critical thinking' as the featured theme: Scriven's comments above ("critical thinking is coping with controversy"), and those of Richard Paul, are taken from its pages; and there are other contributions by leaders of the movement. However, alongside their exhortations is an article by Ruse, "Creation Science: Enough is Enough is Too Much," in which the dictum again appears that "the teacher must sift and sort, giving the student the best," and that creationism in a science classroom would be as wrong as "to teach philosophy in a math classroom" (Ruse 1985, 37). It is striking that no article, not even a comment, dealing with the apparent contradiction is included. Yet if indeed the misunderstanding is all on the side of the public, then educators at least have a duty to *show why* the contradiction is in fact only apparent. If outsiders are troubled because they see the matter superficially, then they should be shown how a deeper analysis dispels the trouble.

One reason, perhaps, why this is not being done is that within the academic community the two controversies are not handled by the same group of people. Evolution and creationism involve biologists, philosophers of biology, and theologians. Decision-making and values programs attract the attention of psychologists, ethicists, and logicians. Possibly the right hand does not know what the left is doing. If so, the result is serious because while on the academic side there may be two different sets of speakers, on the public side there is often one and the same audience—comparing messages.

Sometimes the issue *seems* to be addressed. It is said that exclusion of creationist ideas from science classes is not a violation of 'fair play,' or democratic procedures, or the appealing notion of 'letting all sides be heard'—because science study must not be confused with debating, politics, or the judicial process; biology lies in a different sphere—it never was intended to be 'democratic.' However, this kind of answer misses the point. The dissonance does not arise in comparing classroom teaching with civic action outside the school; no one demands that political methods be imported into science. The context in which evolution is taught, and in which that teaching is viewed, is the scholastic environment itself: the various decision-making and critical thinking programs already in full swing and praised every day. The issue, in other words, is not fair play but *rationality*—*rationality* in the *search* for knowledge and rationality in *transmission* of knowledge.

Conventional Wisdom and His Problems

It is time now to bring into the picture the main objections. However, since many of these have already been heard in the current debates, I will summarize—or rather characterize—certain vital features of that opposition. Consider, therefore, the most common demurrers *as publicly voiced*—as phrased perhaps by a man I call 'Conventional Wisdom.' We shall see that the relation to politics has yet another aspect.

"Very clever," says Wisdom, "you have indeed shown us an ingenious juxtaposition, which, on first sight, seems to expose something serious." However, you are comparing oranges with apples. *On Liberty* is an eminently *political* treatise—political, that is, in the problems to which it is addressed. That is why the expressions Mill uses are appropriate: 'adverse opinion,' 'received opinion'—the key word is 'opinion.' There is a difference between *opinion* or *belief* on the one hand, and *scientific knowledge* or *fact* on the other. Mill's views apply very well to the former; for the latter they are almost irrelevant.

"True, Mill champions diversity of opinion as a *cognitive advantage,* not for the sake of some dubious intellectual egalitarianism. Yet it is *politics* to which his thought is directed. Though all fields of inquiry involve cognition, we must distinguish between the different types of questions asked. You might argue with your doctor about the *value* of his services or the *justice* of his fee, but you do not argue about whether your appendix should come out.[11] Medical theories are tested empirically and methodically. Karl Popper made this distinction when he introduced his criterion of falsifiability—and we forget it at our peril.[12] Yet Popper's criterion is only a way of making more precise what is obvious to common sense: We no longer debate whether the earth goes around the sun; we do not doubt that apples fall *down* from trees, rather than *up,* even when the theory of gravitation undergoes a revolution[13]; although we continue to argue about Aristotle and Marx, socialism and the categorical imperative.

"If parents here and there see in this a contradiction, the mistake is theirs. For such people, some appropriate educational literature does seem to be called for. However, to suggest that when schools promote unrestricted debate on *political and moral issues,* they are somehow at fault for not doing the same in *science classes,* is absurd. The implied, basic argument is a *non sequitur.*"

Conventional Wisdom may not be a philosopher of science, but neither is he a fool. Immediately, he zeroes in on the main point; he senses the power of the principle of insufficient reason: If there were *no difference* between science and morals as cognitive fields, then surely the burden of proof would fall on those who nevertheless object to using the same approach in both areas: "Why not?"

This point will occupy us in the next two sections. At the root of a socio-educational dispute, we find, strangely enough, a classic issue in the philosophy of science: a version of the problem of demarcation—not between science and pseudoscience, but between natural science and morals. No doubt science and morals do differ cognitively and in other ways. What is less obvious is whether these differences are relevant to critical thinking. Are the two fields different in the *structure* of their rationality? Einstein thought not: "Ethical directives can be made rational and coherent by logical thinking and empirical knowledge. . . . Premises play a similar role in ethics to that played by axioms in mathematics. . . . Ethical axioms are found and tested not very differently from the axioms of science. Truth is what stands the test of experience" (Einstein 1953 [1950], 779).

These words, provocative as educational policy some thirty-five years ago, could well be an epigraph for a current textbook. To the extent that the foundations of moral discourse continue to shift from a religious base to a rational one, to that extent does science, as *the* model of rationality, become *the* model for morals also. Yet with time, the trend becomes increasingly problematic. For if the core of rationality does include the two elements we have picked out—'criticism' and 'alternatives'—then why should not *the* model exhibit them in greater measure than all other fields? It is this that many people sense but often cannot express properly. In morality, in politics, in social problems—criticism and alternatives are the hallmark of *active reason*. And in natural science, *the pride of reason*—there not?

One more excerpt from Mill, not often quoted, is worth quoting at length:

> But, someone may say, "let them be *taught* the grounds of their opinions. . . . Persons who learn geometry do not simply commit the theorems to memory, but understand and learn likewise the demonstrations; and it would be absurd to say that they remain ignorant of the grounds of geometrical truths, because they never hear anyone deny, and attempt to disprove them." Undoubtedly: and such teaching suffices on a subject like mathematics, where there is nothing at all to be said on the wrong side of the question. . . . But on every subject on which difference of opinion is possible, the truth depends on a balance to be struck between two sets of conflicting reasons. *Even in natural philosophy, there is always some other explanation possible of the same facts; some geocentric theory instead of heliocentric, some phlogiston instead of oxygen, and it has to be shown why that other theory cannot be the true one:* and until this is shown, and until we know how it is shown, we do not understand the grounds of our opinion. (Mill 1961 [1859], 226, italics added)

Mill, writing with a judgment whetted by philosophy of science, reached this position more than 300 years after Copernicus, some 250 years after Galileo's discoveries.[14] Today, following his logic, we may well ask whether 'keeping creationism out of the schools' is not one reason why biology teachers are often *incompetent* to answer the classic challenges to evolution—never having faced these challenges themselves as did Darwin.[15] In New York City for example, teachers are advised ahead of time to redirect such questions to the experts, questions that any bright teenager might ask.

On the one hand, Einstein's remarks indicate that normative ethics can, or should, resemble science as a process; on the other, Mill insists that in regard to methodological pluralism, the educative aspect of natural science ought to resemble ethics. By now these hints have been developed substantially, as part of a trend; and that is the first, clearest, reason for the decreasing plausibility of the argument 'from unlikeness.' Although the story of this 'narrowing of the gap' is not new, its consequences for education and for a more general understanding of the role of science in society have yet to be assessed. It should be useful, therefore, to outline briefly the themes within that story directly relevant to the issue before us. Although the argument from unlikeness is broad enough to include several independent thrusts, we will focus on just one of these—however, a basic one. I want to show not only that it fails but that this failure leads to still other revealing questions.

Narrowing the Gap

Conventional wisdom continues to assert the relevance of the fact/value dichotomy, recent arguments to the contrary notwithstanding. Repeatedly, during these controversies, the public is told that science uses objective methods to establish *facts,* provides explanations based on *law;* and because of this all competent persons can agree on the outcome. Moral inquiry on the other hand depends strongly on the connotation of key concepts, which involve *judgment* and therefore *values* on which humans may justifiably differ. Even if the primary values are not in dispute, individuals assign to them their own priorities and weights; and if primary values *are* in dispute, then the very meanings of basic concepts—such as 'murder'—may not be understood the same way by everyone. From this difference in the nature of the object of knowledge we are asked to infer a difference in the nature of study.

The implication is that reason enters moral decision on a high level, but as a guiding principle, not as algorithm; while in science, alternative major theories simply do not have the same status as alternative ethical standpoints. Scientific rationality is best exemplified in the testing of low-level hypotheses, where it takes on the form of rule-governed procedure. Yes, science too has its uncertainties and disputes, but not interminable disputes; as data accumulate, a consensus emerges for one or another of the alternatives, and the frontier moves on. We know from experience that students need not retrace all the wrong steps of the past; and if they are to make progress, to reach the frontier, they *cannot* do so. Isaac Newton said it best: We see further because "we stand on the shoulders of giants." To study science is to climb up on those shoulders.

No one should underestimate the force of this line of thought. Its apparent self-evidence, coupled with the long-standing agreement of scientists and moralists to 'restrict' themselves to their own proper fields, accounts to a large extent for the continued strength of this view, although for some time we have known that the picture is much more complex. Recent work has emphasized the opposite idea, with the result that "restrictionism is dead as an intellectual option" (Graham 1981, 381).

It is a basic point of my argument that the simultaneous pressure of both the old insight and the new—respect for demarcation, and its decline—has much to do with some of the current strife. From the past, a deeply rooted, still viable, essentially positivistic concept of science seeks to keep the barriers of restrictionism effective (especially against penetration *into* science). At the same time, a number of developments in the sciences themselves, in their history, sociology, philosophy, *and* in other fields, tend to erode these barriers (see Laudan 1983).

I would like to pick from this latter category two such developments or movements to indicate how the breaching of the barriers from both sides intensifies the pressure on education. The first, associated with the work of Thomas Kuhn, takes as its raw material actual scientific practice, and is called sometimes "the new philosophy of science"[16] or "the new wave" (Laudan 1984, 13); the second, examining in a modern perspective the nature and logical basis of norms, provides a theoretical framework for the present revival of moral education.[17] Both trends are well known, but their combined impact carries an additional message. This message, while not decisive in itself, has to be taken seriously because it shapes indirectly the climate of opinion surrounding the kinds of controversies with which we are dealing.[18]

What Kuhn and other historians, sociologists, and historically based philosophers have put before us is, of course, more than delectable stories about how scientists 'fudge,' use 'intuition' and 'propaganda,' 'break every rule in the book,' and follow only in routine 'puzzle solving' that hypothetico-deductive recipe taught so solemnly in schools.[19] They find a serious qualitative difference between science as it really is and as it is pictured, especially in regard to the role of *method* and of *values*. Certainly this work has been criticized.[20] What has survived criticism is the recognition "that the appraisal of theory is in important respects closer to value judgment than it is to the rule-governed inference that the classic tradition in philosophy of science took for granted" (McMullin 1982, 8–9).

Whether this appraisal of theories takes place during brief periods of 'crisis' and 'revolution,' as Kuhn believes, or more frequently and evenly as some others insist, the point is that at such times the differences between scientists lie in the weights each attaches to the various desiderata: accuracy of prediction versus economy in assumptions, fruitfulness versus consistency, and so on.[21] Therefore, as one scientist explained recently in words amazingly close to Kuhn's, "science is an intensely personal enterprise. . . . In every real scientific problem I've seen, the evidence by itself never settled anything because two scientists of different outlook could both take the same evidence and reach entirely different conclusions. You eventually settle the differences, but not because of the evidence itself but because you develop a preference for one set of assumptions over the other."[22]

Nor is it just a matter of shared values *internal* to science itself; metaphysical bias, aesthetic preference, world views, even religious leanings all play a role.[23] The reason is not deviation from some ideal process but a feature increasingly recognized in the philosophy of science—that theories are *underdetermined* by empirical evidence. Something else *must* enter. Imre Lakatos, for example, after trying to save as much method as he could in reworking the Popperian philosophy, concluded that "The direction of science is determined primarily by human creative imagination and not by the universe of facts which surrounds us" (Lakatos 1970, 186–88).

Even allowing for a deliberately provocative style in some pronouncements, the meaning of these studies is unmistakable. If 'creative imagination' or 'sets of assumptions' have *such* weight, then at the very least the rationality of science is not bound to method as closely as hitherto pictured. When *judgment* based on *values* is the mode of decision-making at the highest level, then reasoning in the scientific realm looks much like that in the *politico-moral* realm.[24]

In just this way, discussing the legal status of marijuana as Scriven urges, high school students apply accepted values—health, personal freedom—each according to individual judgment, using individual relative weights. Many educators believe something valuable is to be learned by seeing how others do it, and by practicing this sort of judgment oneself. Thus, in light of the new philosophy of science, especially its indirect effects on language, on forms of thought, on public debate, we must not be surprised if some say "but if *science too* depends on judgment of this sort, why shouldn't one practice it when studying *that*?"—our old question. However, let us now note the trend on the other side.

Concurrent with this development in 'science-studies,' mainstream academic ethics seems to be flowing in more or less the opposite direction: away from relativism, decisionism, the extreme subjectivisms of earlier times (fostered by a positivism that made much of the is/ought dichotomy), and toward what might almost be called a *scientific approach*—the one indicated by Einstein. Today we hear frequently that "any science is significant for ethics on account of the ways in which it serves as an embodiment or exemplar of applied rationality" (Toulmin 1980, 59). Or, more boldly: "One can learn to think ethically much as one learns to think scientifically" (see Morill 1980, 46).

This should surprise no one. In the English-speaking world, with philosophical attention on deontological theories and on approaches that are increasingly analytical,[25] some papers have the structure of typical 'puzzle-solving' attempts of the mathematical sort. Ethics is losing its 'soft science' image. Meanwhile, in the public schools, there is growing pressure to 'do something' about the much-discussed moral vacuum, *and*, at the same time, to satisfy our religious and cultural pluralism—a feat that can be performed most easily by teaching morals with the barest of content! What theory promotes, practice invites, leading to a natural consensus—the consensus for *formalism*: ". . . those who object to a formalist definition of morality have no positive alternative to offer except (1) morality is what is in accord with my own system, or (2) morality is relative."[26] However, formalism in metaethics leads naturally to formalism in normative ethics and decision-making. A formalist conception of morality itself suggests formalist models and highly abstract thought experiments (dilemmas) to which, not surprisingly, formalisms such as game theory or decision theory may be applied, as well as insights gleaned from computerized 'value driven decision systems.'[27] All this, so characteristic of recent work, creates an environment in which any *physicist* can feel at home.

The influence on teaching, at least on its theory, takes place through a kind of two-stage process, the beginning of which is merely the idea that if moral education is to be serious then *rationality* must occupy the central place (see Bok 1976, 28). When we add an understanding of the 'moral point of view' as a formal structure of thought, then the first stage is complete. This in itself is a giant stride toward the methods of science even without attempts to emulate specific practice. The noteworthy feature of the second stage, direct emulation, is that the model of science used is wholly positivist and almost untouched by the "new philosophy." As a result, in many moral education programs the premium is on systematic procedure, rules for thinking, tight verbal argument, graphical schema, and technical jargon. If morality is recognized by its form, why not *teach directly the proper form,* the *process* of thought in six steps?[28]

Moreover, the continued existence of programs such as values clarification, widely criticized as too affective, in no way invalidates what has just been said. For in seeking to protect their legitimacy in the face of pressure for cognitive rigor, these programs too insert into their practice 'subroutines' that embody the sorts of features listed above (thereby becoming at once 'affective' and 'scientific'!).[29] The shift, then, is *from morality as synthesis involving judgment in the broad sense, toward morality as procedure, as skill.* Of course, I am speaking only of the direction of the shift; not all judgments in moral thinking can be streamlined, and not all procedure is algorithm.

Yet the antiparallel nature of the two movements is clear. While natural science is increasingly viewed by scholars as a 'human,' personal, and social endeavor, moral education is steadily taking on the aspect of a formal discipline. As *faith, metaphysics, conversion, discipleship, trust, dogma, anarchy, judgment* play a greater role in descriptions of physics, so in the teaching of values—even to thirteen-year-olds—the vocabulary now includes such terms as *equilibrium, rank ordering, decision model, decision point, risk strategy, utility function* and *maximin.*[30] The net effect of these two thrusts has been to narrow the gap between the image of the 'hard' sciences, long synonymous with method and objectivity, and that of the humanities, whose reputation (in our time) is the opposite. Especially noticeable is the fact that by directing attention to the *un*common—to crises, revolutions, and theory choice—the recent historical philosophies have uncovered the 'ought' in science. For at such times, the big questions are: Which theory *ought* we use, accept, support, or believe? In morals, however, since we now focus on the *common*—the universal and nonindoctrinating elements—naturally, the opposite effect occurs, and we find in the foreground *objective procedure*.

Note, however, that it is not just the extraordinary in science that shows a likeness to moral thought. The new philosophy has taught us also to appreciate the conservative side of scientific practice. Balancing Paul Feyerabend, there is Michael Polanyi; and Kuhn, of course, has a hold on both ends. When a paradigm is well in place, its function is to provide the practitioner with 'exemplars' of problem solving precisely because scientific knowledge is *not* wholly embedded in theory and rules. Theory often involves entities or processes that are not completely defined, causing problems of denotation. Sometimes this is because exhaustive definition would be too cumbersome; sometimes the concept is not yet clear. Theory also includes terms that can be understood only as part of a language, and thus there are problems of meaning.

Kuhn then sketches a picture of science education that has startled some people. The learner, it seems, is supplied with model solutions *showing* what the theory and rules really mean; then, forming new notions of similarity, he begins to *recognize* his own problem as one or another of the exemplars, and to solve it by *imitation* (see Kuhn 1970, 187). Because imitation is not just for beginners, but is practiced at the highest levels, Kuhn maintains that this reveals something very important: Stability in science, when it occurs, can be understood only on the basis of a practice in which 'tacit knowledge' embedded in the paradigm is passed from one generation to the next in a chain of imitations which, viewed as a whole, constitutes what can only be called a scientific *tradition*.

Here is something to ponder. In other sectors of education, students and teachers are accustomed to *oppose* tradition with the scientific method; hardly anyone believes this amounts to replacing one tradition with another, or that imitation counts. Imitation is important in religion, politics, or the raising of children, where fairy tales, folk heroes, histories, and saints' lives all have a well-known role. We can understand that the Ten Commandments—rules—are not enough for the learner because in the absence of exemplars, as Kuhn says in another context (see Kuhn 1970, 188), "the laws and theories he has previously learned would have little empirical content." The rule states: "Do not murder,"[31] but how are we to know what 'murder' is and how it differs from other kinds of killing until we are shown examples of murder and examples of non-murder, and after that, human models who distinguish murder from non-murder. In Kuhn's perception (see Kuhn 1970, 187), most of the time a law is not really a law but a 'law-sketch' or a 'law-schema' which must be interpreted "as the student moves from one situation to the next."

If as a description of physics this still seems strange, in the realm of morals it is only too familiar. After Raskolnikov kills the old woman in *Crime and Punishment,* he finds out it is murder, although at first he did not think so. By following his story, the reader too is enlightened about "Thou shalt not murder." One does not learn from Raskolnikov alone, or from King David, or Brutus, or the troubled and heroic figures of our own time, once and for all what is and is not murder; but taken together, thought about, invoked repeatedly, they do in fact fill in the meaning of a law which we also find 'schematized' in a few words.

Certainly this is a *traditional* approach to morality, not now at the center of scholarly or educational interest.[32] Yet precisely for that reason Kuhn's theory links science with morals in a particularly impressive way. For despite moderation, his 'sober middle ground' between the 'left' and 'right' wing of the new philosophy actually embraces the typical 'nonrational'[33] elements of both extremes. During rapid change, he points to pluralism in the application of values; in times of acceptance he calls attention to corporate authority. Both features intimate a logic closer in one way or another to the politico-moral than to the conventionally scientific.

Taking note of this unexpected turn toward 'subjectivism and relativism' (and shaking his head sadly), Alasdair MacIntyre believes we are witnessing in the philosophy of science a 'recapitulation' of the history of ethics and politics. Comparing the two realms man for man, he sees in Feyerabend a revival of Emerson, and in Polanyi a reincarnation of Edmund Burke (Popper, in this mapping, is Mill!).[34] Because Feyerabend (1988 [1975]) is so thoroughly convinced of the inefficacy of method, and because he views scientific theories as ideologies unbounded even by those minimal constraints imposed by Lakatos, his picture of science resembles the moral situation *in its modern, anarchic mode.* On the other end of the spectrum, we have Polanyi's master-apprentice relation (1958)—calling on trust, enduring by faith, providing the glue that holds the enterprise together. This too shows science and morals to be close cousins, but only if we view the latter in *traditional* terms.

Against this background, spanning both extremes with a two-phase process, Kuhn's scientific world resembles the moral *both* in the modern *and* in the traditional conception of the latter—or one could say that if morals *in the traditional sense* undergo periods of stability and of transition, then Kuhnian history of science reveals the correspondence at *every* phase. It reveals also the thrust of the movement[35] of which it is a part. Explicit, formally teachable, rule-governed method is downgraded all along the spectrum of views in both the static and dynamic aspects of science.

Let us pause here to take stock of the argument thus far. I have shown, first, that controversies over moral education and over the teaching of evolution are viewed by some people as linked in a way not acknowledged by the educational and scientific communities. I then suggested that *prima facie* this is indeed a logically valid linkage, centering on the concept of rationality; and the problem is that two different kinds of rationality are offered to the public and to students at all levels. In moral education, one form emphasizes methodological pluralism based on a philosophy of which Mill is the best exponent; in natural science, another emphatically rejects such pluralism, basing itself on the hypothetico-deductive method *within* the bounds of established paradigms. The problem is to justify this difference or in some other way to respond appropriately.

Because it appears that any successful justification would have to rest on a relevant distinction between the study of science and the study of morals, we considered a first order argument of this type, which calls for discussion because it is *socially and professionally operative*. It is used widely, is directed at the educated public (including teachers), and is made by responsible members of the academic community.[36] The argument rests on a cognitive demarcation in which science is characterized as a *logic,* operating on observation statements, by algorithm and explicit methods that lead to convergence of conclusions; while in morals, judgment is *expected* to result in *divergence* of conclusions. I then pointed out why, today, such an argument cannot convince those who have been made aware of a new philosophy of science and of what amounts to a new approach to moral education. Since these developments effectively erode certain criteria of demarcation, they surely weaken any argument based on those criteria.

Now we can go further. By examining the implicit interests and perspectives of the parties involved, the inadequacy of the conventional argument can be made more specific.

The Context of Education

One obvious question is this: Do any of the principals on the theoretical side—those responsible for narrowing the gap—have anything to say that bears directly on the educational problem? From their point of view, it might well seem that if methodological pluralism is to reign in moral education, then in science too it should be respected. However, on this point the testimony is sobering: Feyerabend does indeed advocate just such a course, but Kuhn, interestingly enough, comes to the opposite conclusion. The reason for this difference is suggestive.

Feyerabend, following unflinchingly the logic of his own analysis, does not shrink from coming out on the 'wrong side' of the creation/evolution debate. "A scientist," he writes, "must compare ideas with other ideas. . . . He will retain the theories of man and cosmos that are found in Genesis . . . to measure the success of evolution and other 'modern views.' He may then discover that the theory of evolution is not as good as is generally assumed and that it must be supplemented, or entirely replaced, by an improved version of Genesis."[37] Thus, when "some opponents of evolution in California in the seventies" applied the "counterforce of public action" to hedge and amend the teaching of evolution there,[38] this was a healthy sign, says Feyerabend (1988 [1975], 307), and we should "follow their example." In short, views like those expressed by Michael Ruse, quoted earlier in this paper, are but an instance of the 'chauvinism of science.'

According to Kuhn, however, such 'chauvinism' is not only 'normal' in his special sense of the word, it is beneficial. Science *requires* conformity in a profound way, as an approach—indeed, the only possible approach—to radical change. This Kuhn tries to show when he traces how, by a single-minded pursuit of the dominant paradigm, one reveals the limits of that paradigm, thus making way for the next paradigm (Kuhn 1970, Ch. 6). His is a *methodological* dogmatism. Accordingly, a training in the conservative mode of science is just what is needed; and this, speaking normatively, means nothing less than *respect for authority*—the authority of the reigning paradigm. One can only conclude that science education must be *different* in a basic, cognitive sense since today such a prescription for the humanities or social sciences invites only contempt.

Yet, on closer examination, we quickly discover that Kuhn and Feyerabend are not talking about education in the same sense or in the same context. Kuhn describes the *needs of a profession*,[39] freely admitting that in satisfying these needs there are drawbacks: "Of course this is a narrow and rigid education, probably more so than any other except in orthodox theology. But . . . *the loss due to rigidity accrues only to the individual*" (Kuhn 1970, 166, italics added).

Only to the individual! Well, yes. Since the goal of Kuhn's investigations is to understand the success of science as an institution, as an 'instrument' for puzzle-solving, the individual is secondary. Feyerabend, however, is interested precisely in that individual. He grants to professional training its prerogatives, but insists that "special standards which define special subjects and special professions must not be allowed to permeate *general* education and they must not be made the defining property of a 'well-educated man'" (Feyerabend 1988 [1975], 217).

Unfortunately, the term *general education* masks some important problems. Nevertheless, both Kuhn and Feyerabend point to the special role that science teaching has acquired even at the lowest levels. As a training and recruiting ground for the nation's scientific-technological infrastructure it is *expected* to be governed by interests other than those governing the rest of the curriculum—*but not by everyone*. And this brings us back to the controversies with which this paper began. Let us look at these again now, in light of the distinction made by Feyerabend explicitly, by Kuhn implicitly, and in passing by many others (see Boyer and Levine 1979, 41). However, this distinction itself requires a preliminary clarification.

Most subjects and certainly all sciences can be viewed and taught either in a context of education or in a context of application. Here *education* is intended to capture roughly what many call *general education*. Yet the latter term often denotes merely a subset of specialized or disciplinary topics made digestible for the outsider, while I mean by it simply an orientation toward *depicting* what the world is like and understanding its mysteries as far as possible. This can be done with any degree of rigor, detail, and depth. In contrast, *application* is intended to describe activities in which laws and explanations are seen as *tools* for the solution of predefined classes of problems.

Although the two contexts are not mutually exclusive, it is a fact that science courses today—with the exception of those in colleges labeled 'general education' and regarded as watered down—present natural science in the context of application. This is the education Kuhn discusses; it should be called *training*. Whether this orientation is inevitable and whether it represents a loss even as professional training has been debated. For our purpose it suffices to note that only in the context of education is there a conscious concern for the effect of subject matter on the learners' *orientation in the world,* on their actions in society.[40] In the other context any such effect is purely a by-product.

Let us see now what light this sheds on a phenomenon that so many have noticed in the creation/evolution debate at all levels: the lack of real communication due to the inability of either side to address the most vital concerns of the other. Anti-creationist writers routinely raise the fear of a scientific dark age if evolution is not taught within the framework of a totally dominant paradigm, if instead it is encumbered by distracting, useless comparisons with a theory long outworn. Often this warning is joined to a more general plea for keeping all our science at the forefront, lest we "inevitably recede into the backwater of civilization." "American science will wither. We will raise a generation of ignoramuses, ill-equipped to run the industry of tomorrow" (Asimov 1984 [1981], 193). "I'm convinced that if we fail to confront this issue squarely and publicly, we will have an American equivalent of the Lysenko affair. . . . The affair virtually killed genetics research in Russia until only a few years ago."[41]

In this type of argument (which is not the only one), all interest lies within the *context of application*. For it is *problems* that occupy center stage, while *students* are seen as potential problem-solvers. Even when industry and competition are not explicitly mentioned, but only the 'level' or 'quality' of science as an institution, the concern is still with the fate of that institution—the level of its members, the quality of its research.

By contrast, those who take seriously the creationist argument, or merely its critical side in relation to a magnified evolutionary vision, are oriented almost exclusively to issues within the context of education. They worry about what evolution *says,* or implies, or may imply to young minds, concerning the roots of human life. They think this important because beliefs about origin do contribute to our image of self, which, in turn, affects our human interaction.[42] And they have strong misgivings about a 'scientific world view' so pervasive and well articulated that it expels from consciousness all frameworks not based on an imperative of problem solving. True, this is a hermeneutic of suspicion, but it is one that many conservatives share with certain liberal[43] and Marxist thinkers.[44] The common concern is for a space free of natural science, where action-orienting self-understanding (Habermas) may be built on other foundations.

Granted that to recognize the role of different contexts in these conflicts is hardly to resolve the conflicts; nor, in the context of education, does creationism somehow become a better theory. But it is also clear that *the conventional argument for treating science differently fails.* It fails because the difference in object domains and the kinds of questions asked is *not sufficient reason* for the greatly diminished role of the critical stance in science study as compared with morals (or the enhanced role of criticism in moral discourse if science is the standard)—*provided the comparison is in the context of education.* Once the professional needs are set aside, it is easier to see that the distress voiced by Ruse, for example—the desire to protect young minds from 'corrosion' by pseudoscience—refers to particular arguments and particular contemporary authors. But if, in principle, Mill's analysis stands unrefuted, then the range of alternatives is much larger, and the danger cited must be weighed against cognitive benefits which were noted long ago.[45]

If, in other words, the controversy is addressed in the context in which it takes place (in application there is no such controversy), any gap that still remains between the two realms is far less relevant. In the context of education, 'knowing' the physical world appears more or less on an equal footing with 'knowing' the human world—morality included. Disputes notwithstanding, established propositions of great importance exist in both realms. Knowing some of these propositions, knowing how they were established, and knowing the disputes to which they gave rise, is part of knowing the world. However, if Ruse is right that we must carefully 'sift' and 'control' what we pass on to the next generation or it will have "no criteria by which to evaluate . . ." then it is hard to see why this applies only to science.

In the reverse direction the argument against demarcation is just as powerful. Ruse's impulse is the essence of conservatism: to save the best. Yet only if our moral experience has no accumulated knowledge worth saving, no giant's shoulders to climb up on for a better view, *is there sufficient reason* to reject here what is deemed so vital, there—*while at the same time* importing from 'there' the formal, structural elements. If in science, where criticism is vital, rationality still demands of the beginner exclusive immersion in the prevailing paradigm, then perhaps in morals too we might 'sift' carefully what is presented, and 'control' what is to be retained. If, even in science, exemplars are indispensable—cases and individuals—then why in morals should they now be out of place?

The objection that today we lack consensus in the moral realm is, perhaps, *not* sufficient reason. True, there is no consensus among *theoretical* experts on rules or principles, nor on frontier issues, from abortion to the 'right to die.' However, in regard to exemplars—history's moral heroes—the situation is quite different. When it comes to these *practical* experts,[46] a widely acceptable list can indeed be drawn up; and imitating exemplars, as Kuhn has shown, is at once surer and more flexible than acquaintance with rules. The implications are clear enough. It remains only to point out that in neither realm is imitation, rightly understood, the same as blind following. When, in certain respects, Martin Luther King imitated Gandhi, and Gandhi Tolstoy, they did so with as little blindness as when Einstein modeled himself on Max Planck, or Heisenberg on Einstein. But we cannot here develop this argument.[47]

The tacit assumption that in one case the issue is wholly between science and pseudoscience, and in the other wholly between autonomous thought and a no-longer-legitimate authority has obscured some serious questions about the *role of science in the cognitive domain,* its status in relation to other interests, its proper function in the schools; and about grounds for authority in any practice. By ignoring these concerns, the conventional argument appears to suppress them, and that too makes it inadequate. However, I do not suggest on this account that the defenders of some sort of demarcation in teaching are wholly wrong. For one thing, there are other ways of making distinctions than the one here discussed[48]; and, as some of my examples do show, application or general scientific excellence is not the only concern of those working to keep creationism out of the schools.

For these reasons the present treatment calling attention to context can only be preliminary. However, this preliminary overview is necessary because at present the two types of arguments are so intermixed that it is difficult to weigh opposing considerations meaningfully. It is not possible to compare the desire for rapid advance on the part of future researchers with the need for biology teachers capable of handling objections to accepted theory. It is not possible to weigh advanced technology against coherence in the images of rationality projected by schools. It is hard even to discuss seriously the role of authority and of exemplars in moral education if it is not realized that in science—the premier cognitive activity—these same features serve to enhance the social, institutional mode of problem solving.

Our discussion also suggests that to the extent that a desire for genuine dialogue exists on either side of the controversy, even if only for pragmatic reasons, to that extent the different cognitive contexts must be taken into account explicitly. There is no doubt that in science, as in morals, the interest in application is perfectly valid, and cannot be lightly brushed aside as was the tendency in the 1960s. However, we should remember that serious accommodations (in both colleges and high schools), accompanied by much good will and some success, were made in the 1960s and 1970s, precisely in response to criticisms in the context of education.[49] This alone should be ample warning that the issues are larger than the events described here. The pace of science makes it inevitable that other such conflicts are barely over the horizon.[50]

Some Preliminary Conclusions

Earlier I suggested that a basic problem for the schools is the collision between the older and the newer conceptions of science—the former largely shaped by the idea of demarcation, the latter less so; the older guiding the study of natural science, the newer exerting direct influence on educators, and evident also in periodic demands to include moral concerns or alternative theories within the teaching of science itself.[51] We see that this collision is exacerbated by the fact that at least two different orientations, both important, are possible in viewing natural science. People with a desire to maintain the dominance of the interest of application in the classroom tend to ignore the newer insights, retain a positivistic framework, and emphasize demarcation from all other fields. This is necessary because in the other fields—language, social studies, history, and so on—the context of education is still primary. On the other hand, those oriented toward the sciences in that latter context, for whatever reason, wish science study to be relevant beyond the professional horizon, and thereby more *educative*. Consequently, they soon notice that 'restrictionism is dead' and find "the new philosophy of science" more to their liking.

As part of this clash, everyone involved—students, teachers, and public—are offered simultaneously two differing and partly contradictory interpretations of rationality without serious explanation. This raises questions about the coherence, and therefore rationality, of the philosophy behind the entire educational effort in roughly the same way as occurred in the 1960s. At that time, radical students charged that while rationality was pursued *in the small* by individual instructors or subsystems of the institution, contradictions between different parts of the educational experience made that experience as a whole *incoherent*, made the academy an irrational place.[52] The problem was fragmentation, the desire was for integration. Today quite similar goals are embedded in carefully-thought-out educational philosophies: "A well ordered belief system is one which is internally consistent . . . and integrated rather than fragmented" (Strike 1982, 21).

No doubt, coherence in teaching can be considered a purely operational problem, a curricular matter internal to the profession. But that depends on what is at issue. Dissonance in regard to rationality is serious any time. It is all the more serious because in education, as we see, reason has been thrust into the spotlight lately. Kenneth Strike, for example, puts it in the form of a 'motto for schooling': "The central public function of schooling in a liberal state is the democratic distribution of rationality" (Strike 1982, 12). When the issue is so basic, so philosophical, and the dissonance so obtrusive on concerns outside the academy, the conflicts do become public and social. At such times, it may be quite appropriate to suggest a theoretical study of the causes of incoherence.

For those disciplines concerned seriously with the state of education, the problem discussed here is at the same time practical and theoretical. The practical part is: how to go beyond the *micro-rationality* embodied in small units of instruction to offer the student and the community an environment of *macro-rationality* as well. What good are 'critical thinking strategies' in one corner of the curriculum if that curriculum as a whole is *dis*orienting? How can the scientific, skeptical, critical attitude be an example for moral education and other studies *if,* for the sake of expediency or professional need, science courses themselves eschew it? And how can we require the next generation to accept on authority our scientific tradition if all other traditions are ignored or cheerfully dismembered? The more theoretical question is whether, despite the unlikelihood of universal agreement on the meaning of *rationality* any time soon, it may nonetheless be possible to develop a working conception that embraces *both* the teaching of morals and of science.[53]

A related point that also deserves attention is this: Does the label *science* now take on a semi-official, even official, status, entitling it to insert into the channels of communication filters that prevent possibly healthy contact with certain kinds of ideas? Do we really wish to subscribe to the principle that even outside the context of application a scientific idea can be confronted only by another scientific idea, and only one acceptable to *current* science?[54] This is not just a question of the 'dilution' of science, which many (rightly) fear; it is a question also of the *constraint of communication,* of *restriction of dialogue,* in cases where the subject partially, but not wholly, does belong in the context of application.[55]

In dealing with these problems, a study of the values-teaching and creationist controversies could be illuminating; that is why I am suggesting they be discussed within a wider horizon, and in an atmosphere of calmness. Perhaps, for the sake of overall coherence, it may be necessary to take seriously *selected aspects* of the critics' arguments[56] in regard to both science and morals: It may be that in present teaching *con*vergence of views and grounding (in experience) are overemphasized for natural science; *diver*gence and the lack of grounding overemphasized for morals. If so, the task would be to correct this imbalance without compromising progress in science or responsibility in morals. I suspect that a good deal can be learned about both problem areas by scrutinizing each *in light of the other, and* in light of recent philosophy of science.

The morals controversies warn us that a reluctance to communicate tradition, reinforced by overdependence on abstract, analytical skills, may reflect serious misunderstanding of what it means to be rational. Complementing this, the evolution/creation disputes raise to consciousness an aspect of science usually considered secondary, its 'communicative' (educative) side—the fact that it has a "social function . . . not as pure ontology, but as a mediation of man's views of himself in relation to nature" (Hesse 1980, xxi). Here claims to unconstrained discussion (unbounded by paradigm) confront our strongest conservative instinct. Perhaps in both realms, if reason is to be *one*—a culturally unifying force— it may not be identified with individual rediscovery any more than with accepted patterns of thought.

Both controversies suggest, in parallel with certain contemporary philosophical trends, that tradition and critique are related by interdependence as well as opposition (see MacIntyre 1981); that insofar as science has a communicative interest, to that extent must its teaching be subject to a communicative (not instrumental) rationality—which surely includes the critical mode[57]; and to the extent that morality embodies lived *experience,* to that extent must it be *transmitted,* not thought out anew. In a very general way, this two-fold view of reason has been advocated before. For example, Scheffler's carefully balanced position seems to address every aspect of our dilemma:

> In training our students to reason we train them to be critical. We encourage them to . . . seek and scrutinize alternatives. . . . *Such a direction in schooling is fraught with risk,* for it means entrusting our current conceptions to the judgment of our pupils. . . . Such risk *is central* to *scientific education.* . . .

[However,] scientific method can be learned only in and through its corpus of current materials. Reasonableness in science is an aspect or dimension of scientific tradition, and the body of the tradition is indispensable as a base for grasping this dimension. . . . *Analogously for the art of moral choice: the moral point of view is attained, if at all, by acquiring a tradition of practice,* embodied in rules and habits of conduct. (Scheffler 1973, 143, italics added)

But to offer this as pure theory, as a general ideal, far removed from specific conflicts, is one thing. It is quite another task to find appropriate ways (and the will) to take those risks Scheffler acknowledges in the face of a challenge such as the creationists have mounted. More difficult and problematic still is to criticize (with children!) ethical traditions at a time when crumbling traditions are sweeping away the most fundamental tradition of all—the family—*and* to do it in a manner that exhibits genuine rationality, not *rationalization* (of views popular in the academic world). All this is surely part of an unsolved problem of the *applied philosophies* of science and of morals.

[25]

Criticisms of "A Tale of Two Controversies"

'Rationality' in Science and Morals
By Mary Hesse

ABSTRACT: Martin Eger's comparison of controversies in science and morals is extended to a consideration of the nature of 'rationality' in each. Both theoretical science and moral philosophy are held to be relativist in social and historical terms, but science also has definitive non-relativist pragmatic criteria of truth. The problem for moral philosophy is to delineate its own appropriate types of social criteria of validity. [—M.H.]

Martin Eger has made a convincing case for the existence of a new controversy between currently received views of scientific and moral rationality, and of a new debate about the possibility of demarcating science from other forms of knowledge. He traces an antiparallelism in recent developments in philosophy of science and moral philosophy respectively, of which the main symptoms may be summarized as follows:

First, positivist philosophy characterized science as reliant on formal rules and methods yielding easily arrived at consensus, and as constituting a body of reliably established factual knowledge based on a sharp distinction between fact and value. Post-positivist philosophy, on the other hand, typified by Thomas Kuhn and the 'social constructivists,' characterizes science as theory- and value-laden, relying on imagination and creativeness rather than rules and methods, yielding a plurality of logically possible theories; consensus is achieved by extralogical and even extrascientific (social, moral, political) considerations. Thus, briefly, our view of science (not least

These three pieces appeared in *Zygon: Journal of Religion and Science*, 23, No. 3 (September 1988). They were among five comments on Martin Eger's paper, "A Tale of Two Controversies: Dissonance in the Theory and Practice of Rationality," which appeared in the same issue (291–325), followed by his "Reply to Criticisms" (363–68). The three criticisms he was particularly concerned to answer are reprinted in this chapter.

in education) has proceeded from a somewhat arrogant confidence in stable methods and results, to varying degrees of openness, skepticism, and relativism.

Second, moral philosophy, on the other hand, has proceeded in the opposite way. Until very recently moral education has presupposed a Millian liberality in which all rules and systems are open to critical examination, and there is an expectation of plurality of conclusions, none of which ought to be suppressed. More recent moral philosophy, typified by John Rawls and the general application of formal game- and decision-theory, has placed renewed emphasis on norms and rule-based calculation to the point where, as Eger nicely states it, "any *physicist* can feel at home" (Eger 1988b, this volume, 351). Briefly, moral philosophy has proceeded from critical openness and skepticism to something more dogmatic at least in terms of the formal structure of moral argument.

The antiparallelism of these developments takes them from controversy at one end to a mirror-image controversy at the other, and in both cases leaves education in science and morals in a state of confusion and contradiction. It is true that both sides appeal to the "giving of good reasons" and to "truth emerging from critical conflict," and one might naively expect that somewhere in the middle science and morals might be able to find common ground. However, Eger is right to argue that this does not take the matter deeply enough, because the questions of what are *good reasons* and what is *truth*, and whether these are uniform or multiform throughout *knowledge*, are not adequately addressed. The problem is that of the very nature of rationality, scientific and moral.

Eger makes a useful distinction between what he calls the contexts of 'education' and of 'application,' where application is concerned with practical problem-solving, and education with the learner's total orientation to the world. As an example he hints at the controversial suggestion that, while evolution theory is relevant to biological practice in a way creationism is not nevertheless creationists have a valid point in claiming that evolution theory has a deleterious effect on moral orientation, whereas the Genesis story is a valid moral myth. Creationists might not state it in those terms, but I use *myth* here in its proper anthropological sense of stories that have particular significance: "They are the stories that tell a society what is important for it to know, whether about its gods, its history, its laws, or its class structure" (Frye 1982, 32).

Expressed more generally, the context of education in Eger's sense is the context of *knowledge*, and the problem is shifted from the science/morals demarcation to the deeper level of discriminating theory and practice, and analyzing the conditions of validity for each. This is the question I shall briefly pursue here, and in doing so I hope to situate Eger's argument in relation to current discussions of realism and rationality in the realms of science and values.

It should be noticed that Eger's antiparallelism is not quite symmetrical because on both sides of the modern controversy we have, strictly speaking, forms of *relativism*. I shall argue this presently in the case of science, which is increasingly recognized to be underdetermined by data and partly dependent on various sorts of intellectual and social fashion. In the case of moral theory, Eger's account may not look like relativism, but there are two considerations that show that it is so.

First, moral theory is often explicitly *structural* in character, and is not intended to entail specific moral prescriptions. For example, Lawrence Kohlberg's influential theory describes the sixth and 'highest' stage of moral development as a moral *attitude* rather than a specific set of moral principles. It is formal principles of justice, reciprocity, and equality between individuals that are involved, not rules for particular moral decisions. If one asks what is the justification for these 'formal' norms, Kohlberg has no answer except the historical observation that moral philosophy in Western civilization has developed this way, buttressed by appeal to Jean Piaget's conclusion that children's moral sense shows the same sort of progressive development towards the ideal sixth stage.

Second, any application of formal decision-theoretic methods must presuppose some principles that are not formal. For example, in Rawlsian-type moral philosophy there are norms, which are taken to be self-evident in 'rational' and 'civilized' societies, and these norms have moral *content* as well as formal structure. They favor concepts of distributive justice, equality, and prudence which are patently ethnocentric to Western society, and in that sense relativist. Yet anthropologists have pioneered debates about the possibility of non-Western types of rationality, and it is no longer sufficient to take Western norms as needing no further justification. Good arguments for them there may be, but there are also good arguments, perhaps different kinds of arguments, for the more mythopoeic, authoritarian, and community-based ethics of traditional societies. In memorable rhetorical style, referring to the ultimate appeal to 'our' rationality, Hilary Putnam asked "Well, we should use *someone else's* conceptual system?" (Putnam 1975–1976, 192). Could there be a clearer expression of underlying relativism?

Relativism is in fact endemic in modern philosophy, both in science and morals. It sets the agenda for any discussion of rational justifications. Within these agenda, the question of justification of scientific theories has been around longer and has been discussed in more detail than in the case of moral theory, and so we may hope to use it as a model, both positively and negatively, for moral rationality. It is convenient to begin by considering the distinction between theory and practice in the analysis of scientific reason.

It is generally agreed that there *is* something special about scientific knowledge that characterizes its specific form of rationality. At least part of this specificity has to do with its practical success and its progressive extension of this success to wider and wider domains of natural phenomena. Given the complexity and social expense of modern experimentation it would be foolish to deny that science is at least partly justified by the fact that it delivers the goods—it is subject to control by empirical evidence, tested by its success or otherwise in empirical prediction. The objectivity of such tests is validated by the fact that they can *surprise* us, and eventually overthrow the most cherished theoretical structures. This is the hard edge of pragmatic objectivity within science that distinguishes it, in this sense, as a more cumulative form of knowledge than any other in the history of thought. In this sense everyone is a 'realist.'

The question that divides realists and antirealists in modern philosophy of science is the relative importance to be attached to this type of pragmatic or instrumental success, as against the aim of acquiring true, cumulative, and universal *theories* about nature. The view that science is in essence merely pragmatic appears to devalue its significance, and realists point out that good science has always required more than instrumental success. Other criteria come into play, including the aesthetic appeal of coherent, universal, and elegant theories, and the sheer delight of unraveling complex puzzles. The important question, however, is whether theory also gives us laws and descriptions of the real world, and whether these get progressively nearer to the truth. In the light of changing paradigms throughout the history of science, and of studies of the interactions among science, intellectual culture and social pressures, it is increasingly difficult to accept that scientific theory yields knowledge in this ultimate sense. The progressive character of science lies in its particular, localizable predictions, and their pragmatically successful outcomes. Its basic pictures of the world are *models* that evolve and change, sometimes very radically, and show no convergence to a unique theoretical truth. As an example one may think of the changing pictures of matter throughout physics: from atoms and the void, to continuous distributions of energy through space-time, to the current particle-field interpretations of quantum physics, where fundamental models come and go with bewildering rapidity. Mathematical structures may be preserved through such changes, but answers to the substantial question "What is matter?" are not.

This situation has two consequences for the relation between science and values. First, there is no doubt that value-implications are inseparable in practice from the theoretical models accepted in science at any given time. Atoms-and-the-void have traditionally been associated with a materialistic view of physical nature, where apparently regular behavior is in fact based on pure chance, and consequently with a mechanically reductionist view of human nature. On the other hand, at various periods astronomy and scientific cosmology have required some kind of 'design' in nature, from Isaac Newton's God who kept the planets in their orbits, to modern 'anthropic principles.' Such theoretical models appear more congenial to theistic or spiritual interpretations of the world. Whatever their implications may be from time to time, the crucial point for a relativist philosophy of science is that they are never conclusive nor permanent because they may be reversed by the next change in theoretical paradigm. It is therefore rash to try to derive value systems from scientific theory, even in the form of temporary apologetics. Any such conclusions are in constant danger of being undermined by theoretical developments whose objective rationale takes no account of their contingent value-implications.

Second, the pragmatic account of science poses difficulties for claims to knowledge that have no clear constraints of practical success to control their theorizing. The apparent symmetry of scientific and moral relativism breaks down at this point. For the sake of argument we may accept a sort of moral relativism which is parallel to theoretical relativism in science; that is, we may wish to regard moral systems as internal to the cultural and social pressures of a given society, subject to historical evolution and change. Even so, we have no explicit institutionalized constraints upon such moral change. We cannot test and compare the social and psychological outcomes of different moral systems against overriding and generally acceptable success-criteria, if only because the adoption of the success-criteria themselves are moral judgments. The outcome of easier divorce may be shown to be increased happiness for a certain proportion of divorcees, increased misery for others, and disruption of children's right to a stable family background. To weigh the positive and negative success-value of these outcomes is another moral exercise, and such problems are bound to ramify in the consideration of any serious moral issue.

To recognize these asymmetries, however, brings us back to the pragmatic criterion of science itself. Why did such a criterion ever become accepted as the touchstone of true knowledge in the first place, and how did every other knowledge-claim come to be seen beside it as subjective and arational? The answer lies primarily in the history of seventeenth-century Europe. For example, in the origins of the Royal Society it is clear that there was a conscious decision to adopt a mode of inquiry that was seen to have determinate results and to be relatively certain, one that evaded social and religious controversy, that was accessible and democratic (as Francis Bacon stated it, it "levels men's wits"), and that disposed of superstition, magic, astrology, and false (mainly Aristotelian) philosophy. In other words, the adoption of the distinctive style of scientific argument with its experimental tests was itself a largely *moral* decision, a decision for 'objective' fact uncontaminated by 'subjective' value, for plain linguistic prose in place of high-flown metaphor and rhetoric, for observable description in place of transcendental myth.

This whole complex of decisions was momentous for human history. That it was socially possible at all must be explained by a certain schizophrenia in the seventeenth-century mind—a Cartesian dualism between the natural and the mental and moral, and between science and theology. The social myths based on western theism were not seriously undermined for another century or two. Our current problems are the outcome of their eventual collapse, and unfortunately for us new social myths cannot be created to order in our kind of society. Perhaps the first step in trying to reconstruct our social and moral fabric is a negative one: to recognize the historically contingent and temporary character of our obsession with science as the norm of knowledge, to question its iconoclasm and restless search for progress and universality, and to recover a sense of the particularities and traditions of particular societies and their need for their own forms of social cohesiveness.

This conclusion is in close agreement with Eger's own, but as he remarks, its practical application remains to be worked out. We do indeed need to apply critical reason to tradition in morals as well as in science—this is an inescapable legacy of the 'rational scientific method.' However, philosophical critique should be directed as much to the presuppositions of different applications of rationality as to the content of scientific theories or moral norms. In the end society inevitably defends itself against its own disruption, and throws up its own moralities. Philosophers and educators can do little more than accept with realism the need for power and authority to be recognizably located within society, while exercising critical restraint on its excesses and those of the moralities it espouses.

On Martin Eger's "Tale of Two Controversies"
By Abner Shimony

Abstract: Criticisms are presented against Eger's challenge to the demarcation between the natural sciences and ethics. Arguments are given both against his endorsement of the 'new' philosophy of science and against his rejection of the fact-value dichotomy. However, his educational recommendations are reinforced rather than weakened by these criticisms. [—A.S.]

Introduction

Martin Eger's essay is extraordinarily rich in penetrating philosophical comments and in educational good sense. Nevertheless, I believe that there are serious errors in his fundamental philosophical theses, and much of this commentary will be devoted to exhibiting them. I shall then try to show that for the most part his educational recommendations are reinforced rather than weakened by my theoretical criticisms.

'New' and 'Old' Philosophy of Science

Eger challenges the demarcation of the natural sciences from the study of morals by questioning that they are different cognitively. The demarcation was clear and strict, he says, as long as an old, essentially positivist conception of the natural sciences was maintained and as long as the fact-value dichotomy was accepted. The 'new philosophy of science,' however, has profoundly criticized the old conception, and in spite of some reservations Eger on the whole assents to these criticisms. Likewise, he assents to arguments against the fact-value dichotomy, and indeed at one point he provides a very interesting argument of his own, which may be original while he makes no claim to this effect: "True, there is no consensus among theoretical experts on rules or principles, nor on frontier issues. . . . But in regard to exemplars—history's moral heroes—the situation is quite different. When it comes to these practical experts, a widely acceptable list can indeed be drawn up. And imitating exemplars, as Kuhn has shown, is at once surer and more flexible than acquaintance with rules" (Eger 1988b, this volume, 358).

Eger undervalues the 'old' philosophy of science partly because he uses this term to refer to only a part of a diverse and complicated collection of methodological and epistemological doctrines. Positivism—which maintains that the total content of a scientific theory lies in its implications for human experience—was indeed very influential during the first half of the twentieth century. During this period there were also influential realists, who followed the tradition of Galileo Galilei, Isaac Newton, and John Locke of attempting to infer from experience the properties of an objective world which has an existence independent of human beings. (For example, Bertrand Russell in 1914 was a positivist, but by 1927 Russell had been converted to realism.) Eger also attributes to the old philosophy of science the use of algorithms of inference, but I have no idea to whom he may be referring. The formulations of scientific methodology in the best of the old philosophers of science (my personal favorites being Charles S. Peirce and Harold Jeffreys) are complex and sophisticated syntheses of diverse intellectual elements: hypothetico-deductive reasoning, probability theory, decision theoretical arguments, appeals to evolutionary biology to justify human skill in hypothesis formation, and appeals to the history of science for a posteriori refinements of method (see Shimony 1970). It should not be surprising that an adequate scientific methodology is complicated, if one considers the ambitiousness of the scientific enterprise: namely, to obtain good approximations to the objective truth about the universe at large, on the basis of experience which is very limited in space and time and constrained by the peculiarities of human faculties for gathering and processing information. It is true both that "Facts are stubborn things" and that "Nature loves to hide," and an adequate methodology must do justice to each of these divergent dicta.

I do not wish to deny that the new philosophy of science has made some real contributions. It has emphasized the indispensability of the history of science for a rich philosophy of science. Some of the innovators (notably Michael Polanyi and Thomas Kuhn) have pointed out that skilled scientific practitioners typically have much more 'tacit knowledge' of their craft than they are able to articulate in explicit rules of scientific method. Some (notably Norwood Hanson) have drawn upon empirical psychology in order to carry out epistemological analyses of observation and other mental processes. For the most part, however, the great value of these insights has been debased by drawing from them relativistic and subjectivistic epistemological conclusions. The following are some of the important criticisms that should be made of the new philosophy of science.

First, the history of science need not be used as a surrogate for scientific methodology, as Kuhn maintains, with each historical epoch providing its paradigms which cannot be judged from a neutral standpoint. Instead, the history of science may be studied in order to provide a posteriori elements in scientific method, for there is no reason to believe that the human intellect is endowed a priori with all the methodological tools it needs for investigating the natural world. Experience is needed not only to learn substantive truths about nature, but also to learn how to learn (see Shimony 1976).

Second, the great virtue of the tacit knowledge of skilled investigators is not its tacitness but its knowledge. If methodologists eventually are able to articulate what these investigators know tacitly (as good athletic and musical coaches are able to do in their respective domains), then nothing is lost thereby and something is positively gained.

Third, the deployment of data from empirical psychology, especially from Gestalt psychology, in order to show that observations are 'theory laden,' provides prima facie evidence against the possibility of objective empirical assessments of competing scientific hypotheses. Yet a careful study of empirical psychology—such as the experiments on 'cognitive dissonance' by Jerome Bruner and Leo Postman—reverses this judgment, and shows that human beings are capable of switching between integrative and analytic strategies of perception, and the latter is strikingly liberated from theory-ladenness (see Shimony 1978).

Finally, the occurrence of scientific revolutions is an insufficient reason to deny the meaningfulness of the concept of objective truth and to recommend that it be replaced by a concept of historically relativized truth, as Kuhn recommends (1970, 171–73). The most that one can legitimately infer is the unlikelihood that human beings can ever achieve the goal of objective knowledge of the universe. However, even this concession to relativism is excessive, for it fails to pay due attention to the detailed history of those scientific revolutions which have occurred since the seventeenth century. Typically in these revolutions the displaced theory is a good approximation to the displacing theory, with regard to empirical predictions and in some respects with regard to conceptual structure. In the terminology of Niels Bohr, there is a "correspondence principle" governing the relation between the old and the new theory. Consequently, the appropriate moral to be drawn from the occurrence of scientific revolutions is not relativism but a doctrine of successive approximations to the truth.

In summary I find a pervasive slovenliness of reasoning in the new philosophy of science. Its advocates have failed to use its excellent insights constructively by exploiting them for the refinement of scientific methodology and epistemology. This constructive enterprise requires hard work, which is evaded by their relativism and subjectivism, and of these Eger is too tolerant.

The Fact-Value Dichotomy

The relation between facts and values is very subtle, and I make no pretense to professional expertise about it. However, I wish to present a few considerations which should make one resist any simple conflation of the natural sciences and ethics and to distrust the claim that rational criteria are the same in both domains.

At one level there is a universally admitted dichotomy: that what a person does and what a person ought to do in a given situation are not generally the same. What the person does is a fact, and perhaps what he or she ought to do is also a fact, but if so it is a fact of a different kind. The fact-value dichotomy would then be converted into a dichotomy within the domain of facts. But what, then, is the ontological status of the second type of fact, that is, of normative facts?

A possible answer is that desires are factual, and desires may or may not be achieved. Might not the ontological status of a normative fact be that of a desire? No, this suggestion is not sufficient for two reasons. Most people who are not morally nihilistic would make moral judgments not only about achievements but about desires themselves: that some desires are better than others. Secondly, in the same situation two different persons may have different desires as to what the actor should do. (The difference of opinion may depend upon whether the person judging is the same as the actor, but this is not the only crucial factor, for two people may disagree about what is desirable were their respective situations in the action to be exchanged.) Consequently, the identification of normative facts with desires would deprive normative facts of an objective or interpersonal ontological status.

It may be suggested that the ontological status of the normative fact is that of an authoritative prescription, with different versions of this point of view recognizing different authorities: God, society, the evolutionary history of the human race, and so on. I am skeptical that any of these appeals to authority can account for the motive force of a normative fact unless they endow the authority with a power of enforcement, to punish infractions and reward obedience. If the authority is so endowed, then its prescription is really a hypothetical imperative rather than a categorical imperative, for the motive force is effectively the desire of the subject to avoid pain or achieve gratification. In brief, despite the above-mentioned shortcomings of the identification of the normative fact with desire, it is unrealistic to neglect the role of desire in the analysis of value.

My own (tentative) position is to relate norms to desires, but to do so in the wise manner suggested by Aristotle's Ethics, which points out some remarkable facts about the structure of desire. He notes that all people agree verbally as to what is desirable: namely, happiness (Aristotle 1.4), but they have many different ideas as to what constitutes happiness. People also have subordinate desires—for example, for wealth or honor or learning—and the satisfaction of these may or may not lead to happiness. Furthermore, because a human being is neither a beast nor a god but a social animal, his or her happiness is bound up with the happiness of others, in the family or in the state (or—by a non-Aristotelian extrapolation—in the human race as a whole). It is a fact about human nature that character—and hence what is desired by an individual—is to some extent plastic, and is formed by habituation and education. It is also a fact that not all modes of forming character are equally conducive to that universally desired but vague end, happiness. Most of Aristotle's Ethics, after the preliminaries of the first book, is devoted to the investigation of the moral and intellectual virtues, which he considers to be the true way to happiness. Even if one has followed Aristotle this far, however, one may suspect that he was unavoidably ethnocentric, because of the limitations of his social experience. One may draw upon the mass of evidence accumulated by anthropologists to suggest that there is not one but a plurality of ways to human happiness (e.g., a contemplative life, a kinesthetic life, a ritually organized life). One is not thereby committed to a thoroughgoing cultural relativism, for not all cultures are equally satisfying to their practitioners in their own eyes; as Edward Sapir (1924) pointed out, some among the great variety of cultures are 'genuine' and some are 'spurious.'

It should be clear why I resist Eger's attempt to narrow the gulf between the natural sciences and ethics. With regard to the former, I have argued, albeit briefly, that there is a domain of entities independent of human experience which are endowed with definite properties, and a scientific proposition is objectively true if it correctly characterizes this domain. Whatever the difficulties may be for human beings to discover on the basis of their limited experience the objective truth, it is, so to speak, 'there' to be found out. In ethics, however, the ontological status of the normative facts is much more problematic. I do not wish to say that they are merely matters of convention or subjective opinion, for it is not the case that 'anything goes'; the constitution of the human psyche and the social character of human beings set limits upon the range of life styles which will permit the achievement of happiness by the standards of the subjects themselves. Yet it is by no means clear that these constraints uniquely determine an optimum life style, and with it a unique set of norms.

Pedagogy

After these criticisms of Eger's philosophy of science and ethics I shall turn to his discussion of pedagogy, with which I largely agree. He is properly outraged about an authoritarian handling of the creation/evolution controversy combined with the slackness of the program of moral education in the public schools.

The passages cited by Eger against teaching creationism express anxiety about its corrosive effect upon the intellectual faculties of young students. I agree, of course, that intellectual faculties should be cultivated, and a primary way to do so is to perform experiments and demonstrations concerning phenomena which are simple enough to permit a fairly complete exercise in scientific methodology, without gaps and without the intrusion of authority. It must be quite confusing, and perhaps even demoralizing, to thoughtful students to be presented with fragmentary and authority-adulterated applications of the scientific method unless the instructor is candid about the lacunae in the reasoning. How is the student to know whether his or her vagueness of understanding is due to the incompleteness of the reasoning itself or to personal intellectual shortcomings? With this consideration in mind, how is a rational account of the theory of evolution to be presented to young students? Here is the most outstanding case in the history of science of a great theory which is confirmed globally, by an immense variety of taxonomic, zoogeographic, embryological, and paleontological evidence falling into place, rather than by the prediction of striking, unexpected phenomena. The long history of resistance to the theory of evolution (see Mayr 1982, 510–570)—because of genuine conceptual difficulties, not just because of stubborn dogmatism—shows how ill-suited this theory is for elementary instruction. What then would be the danger of a good, open debate by clever students about the creation/evolution issue? It is hard to imagine a creationist positively persuading a classmate who is not antecedently convinced to accept the Biblical account. The worst that is likely to happen, from the standpoint of an advocate of evolution, is that the class will forcefully feel the lacunae in the standard textbook presentations, and is this such a bad thing for the cultivation of intellectual faculties? I suspect (partly because of introspection) that people who object to permitting creationism to be discussed in the public schools fear that it somehow will be the opening wedge of a general anti-intellectual, authoritarian, fundamentalist, and fascist seizure of political power. However, I believe that the real danger of such a catastrophe lies in racial and economic tensions, which will not be assuaged by the prohibition of a debate on creationism.

The case against a curriculum of moral education which emphasizes 'the critical attitude' and 'choice among alternatives' can be built solidly upon Aristotle's Ethics. A young person is not suited for lectures on ethics because of inexperience in the actions that occur in life (Aristotle 1.3). Furthermore, first principles in ethics (as opposed to those of the theoretical sciences, which are obtained by induction) are acquired by habituation (Aristotle 1.7), and inculcation of the moral virtues by habituation must precede the acquisition of the intellectual virtues by instruction (Aristotle 2.1). An important way in which the schools can contribute to this inculcation seems not to have been mentioned by Aristotle, namely, to capture the students' imaginations. For this purpose an exposure to the biographies of "history's moral heroes," in Eger's phrase, may be particularly efficacious.

At this point there is a major ideological conflict between Aristotle and the designers of the curriculum of moral education. Aristotle believes that ethics and politics are continuous, and that the state is responsible for the moral education of the child not just for the child's sake but for the good of the state as a whole. The designers of the curriculum of moral education, on the other hand, wish to develop the child's independence of judgment in order to be able to resist the authoritarian claims of the state.

How can one adjudicate this controversy? I would say, above all not a priori. Only on the basis of experience can one judge whether a person inculcated with moral virtues in childhood and only later exposed to ethical analysis is more self-confident, more judicious, more tolerant, and in general more rational than a person whose critical attitudes on moral matters is fostered in early childhood. Of course, the question is complex, and much depends upon the exact character of the early inculcation. Likewise, only on the basis of experience can it be decided whether the inculcation of a sense of responsibility to society is subversive of individual happiness. This question is also complex, and the answer depends crucially upon the mode of inculcation and upon the details of the relation between individuals and society. In appealing to empirical evidence, however, I do not mean to conflate these difficult ethical, political, and pedagogical questions with the problems of the natural sciences. Practical reason is not concerned with the aspects of the human mind which are genetically fixed, but rather with those which are plastic. Hence, the evidence which it must marshal has to be drawn from human history and from the experience of people who have struggled with the concrete problems of life.

A Tale of Two Controversies
By Thomas F. Green

ABSTRACT: The educational controversies that Martin Eger discusses regarding moral education and the teaching of 'creationism' arise from taking a single aspect of moral education and making it the whole, and from taking a single aspect of scientific work and assuming that it is the whole. The distinction between teaching science as application and teaching it as education is crucial in confronting these problems. [—T.F.G.]

Introduction

Martin Eger has presented a thorough and intricate account of what he conceives as two different presentations of rationality in the context of education: one drawn from prevailing views about the conduct of moral education in the schools and the other from controversies over the 'teaching of creationism.' "If it is a good thing," he writes, "for children to consider all sorts of alternatives in moral decision-making, no matter how repugnant—stealing, cheating, betraying one's friends—all for the sake of developing critical reason and autonomy, then why, suddenly, when we come to evolution, is it far more important to learn right answers than to think critically?" (Eger 1988b, this volume, 344). The reason that this stark contrast does not appear more transparently in public debate, he notes, is that the problems "are not handled by the same group of people" (Eger 1988b, this volume, 345). Scientists scream at the introduction of creationism in the schools and moralists and laypersons at the introduction of other forms of moral education.

In these brief comments, I wish to make only two points. First, the problems he describes inasmuch as they arise from the practices of moral education follow from an impoverished conception of what moral education is about, both on the part of professionals and on the part of laypersons; second, the conception of rationality represented in science arises from a similarity impoverished conception of what constitutes scientific education.

Stated simply, it is not a good idea for children, or for that matter anyone else, to consider all sorts of alternatives in moral decision making, nor is it good education for students to consider all sorts of answers in the study of science. The difficulties arise because a part of moral education has been taken in practice as the test of the whole and because a traditional and occasional part of scientific inquiry is assumed to provide the complete model of the educational process in science. The key lies in the important distinction Eger draws between the context of education and the context of application. Let us consider these items separately.

Moral Education

Certainly as mature moral persons we seek individuals who can and will think about their beliefs and reflect upon their behavior. In this sense we want persons who take responsibility for their own beliefs, and we recognize that this may imply that from time to time they will change their beliefs. We recognize that as they reach maturity they may have to alter their views about what is the useful, graceful, or fitting thing, even the efficient thing to do in this case or that or as a general rule. We expect them to acquire skills in human behavior, such as skills of foresight that often take years of experience to acquire and for which rules cannot be given in advance.

All this is part of what we include, or ought to include, under the general heading of autonomy and efforts to avoid the dangers of indoctrination. However, when we speak of the dangers of indoctrination we are not speaking simply of the early (or even late) inculcation of beliefs. Our concern with the dangers of indoctrination is the possibility that these eventual signs of autonomy—taking responsibility for one's beliefs and entertaining the possibility of changing one's mind, for example—will be hindered at a later time. We are concerned about a way of teaching that prevents later thought. All this is beautifully illustrated in an interview with David Wagner, administrator and principal of Abundant Life Christian School in Madison, Wisconsin. He is quoted as saying "Ideally, we would like to open up their minds and hearts and pour Christian values and Biblical concepts into them—and then close them up again" (Wagner 1988). This is almost a perfect statement of what we mean by indoctrination in the bad sense. It is something by no means limited to religious advocates. Often parents would like to do the same thing.

In any case, we ought to distinguish between moral training and moral indoctrination of this sort. Moral indoctrination is always wrong because it prohibits us from developing the signs of adult autonomy. It aims to close our minds. However, moral training is necessary, even necessary at all ages. Such moral training includes inculcating the idea that just because our choices are our own, that is, autonomous in that limited sense, it does not follow that they are also right or good choices. Having such a principle drummed in, as it were, may be indoctrination, but it is not bad indoctrination because it prevents no one from being free to think for themselves. Autonomy of choice is a good thing, but it cannot provide a model of the whole for moral education. Indeed, a person who does not understand that choices can be good or bad, or just plain stupid, is not a person whom we would admit had yet reached a stage of moral autonomy—no matter how 'autonomously' in some pedagogical sense such choices have emerged.

Where does this child exist who is or ought to be free to make any choices whatever, on the assumption that the more independently the choices are made, the better they are? It seems to me that no such child exists, but, as Eger points out, we do have pedagogies that seem to assume this is true of every child. What is it about moral beliefs that gives children the right to eventually make up their own minds about them, even to change their minds about what their teachers have taught them? Should a child be open to choose whether cruelty is wrong, for example, whether cruelty to animals is to be tolerated? Certainly not! Can there be an argument about this? It is not clear what the argument would be about. When we say that children have a right to make up their own minds about their moral beliefs, we have in mind the eventuality that they will need to be adaptable. It does not follow that moral training at the beginning or even as adults has no place in their education. If youth were to return later to their teachers suggesting seriously (not just as an academic exercise) that cruelty and dishonesty are acceptable, it is not even clear what the argument would be about. How would one trust the argument? Can we have such an argument if honesty is seriously in doubt? How would we know that the argument is serious? Would it include the claim that it is all right to torture cats, or the very different claim that it is cruelty to slaughter animals? How would we know what would even count as a point to be made and accepted in the course of such an argument? The point I am trying to stress is that the very idea of a moral argument of this sort in the context of teaching is an idea that is resident to a long and worthy educational tradition. It cannot be maintained independently of that tradition.

What is it that leads to this apparently morally vacuous approach to moral education in the schools and seems to excuse it as a form of rationality? The problem rests precisely where Eger places it, namely with the victory of formalism in the theory of moral education. "In the public schools," writes Eger citing Lawrence Kohlberg, "there is growing pressure to 'do something' about the much discussed moral vacuum, and, at the same time, to satisfy our religious and cultural pluralism—a feat that can be performed most easily by teaching morals with the barest of content!" (Eger 1988b, this volume, 350). The answer, then, is to find a form of pedagogy that allows us to assess not the moral conclusions or actions of children but the reasons they offer, and to do that on the basis of a moral theory about what constitute morally adequate reasons. These turn out to be reasons that are increasingly general in the cases they cover and increasingly universal in their logical form—a kind of neo-Kantianism. Furthermore, the pedagogy for eliciting these reasons typically involves confronting dilemmas or choices involving rank-ordering. In short we get a formalist moral pedagogy. It is this kind of pedagogy, typified by the followers of Kohlberg, that Eger points out seems to resemble a scientific kind of rationality. Indeed, this pedagogy intends to surmount the problems of pluralism and particularity for education in an institution that is supposed to be public and neutral and to give moral inquiry the rational appearance of scientific inquiry.

However, there is another point Eger makes that is absolutely essential to the full picture. He writes: "While there are differences between the proponents of the several values education philosophies, it is important to keep in mind that concerning these two critical elements—'criticism' and 'alternatives'—the consensus is rather close to unanimity" (Eger 1988b, this volume, 340). The rational picture of instruction that many of us had developed in the context of other views of instruction, this view of instruction as requiring a due respect for the student's sense of reason, is adopted in an entirely different context, namely to the rudiments of moral education, to produce a morally impoverished and, one might even say, a morally indefensible, view of moral education.

It amounts to the view that the moral life is a life that consists simply of making hard choices in dilemma situations (that fortunately occur rather rarely in life anyway) or a continual exercise in rank-ordering things.[1] Think of all that is deleted from this picture—how the virtues are formed, how the institutions within which we live, exist and change, how traditions persist but how they can be valued and also rejected. It is hard to imagine any view of moral education that does not include serious attention to prudence, to how things work, to the coming into possession of powers of foresight, and even acquiring standards of skill that involve caring not simply whether one does the right thing but whether one does good things well. It is equally implausible to suppose that a view of moral education can be adequate that does nothing to cultivate a social identity and hence a social memory. To suppose that we can have moral education independently of history is to suppose what even on the surface is hardly capable of defense.

Compared to the richness of what ought to be present in a philosophy or program of moral education, it is easy to see that the victory of the formalist pedagogy not only produces the social controversies that Eger describes, but also to see why one's sentiments are so easily aligned with the parents and not the professionals even in those cases where the parents' views can be described as bigoted. What we need is not so much an inadequate model of rationality that produces Eger's "Two Controversies" but rather an adequate understanding of what moral education involves and how it differs from, and perhaps how little it can gain from, moral philosophy.

Tradition

I am not a scientist nor am I as familiar with science teaching as Eger. However, I find it refreshing that he appeals to the fact that science is a tradition or has its traditions. This, it seems to me, is nowhere better stated than in Edward Shils's *Tradition*:

> Minds of the first order create new theories. In "normal science," in Professor Kuhn's popular distinction, second-order scientists work within a framework given by an accepted general theory or "paradigm." Paradigms are traditions of a limited life-span. The great scientist is in this respect like the founder of a great religion. Both are said to annul the tradition which has been presented to them. Both are aware of the inadequacy of what has been received and they aim to supplant the inadequate account by one which is fundamentally more adequate. In neither situation is the annulment of tradition complete. . . . The fruitfully productive scientist is thus not at war with tradition in general, insofar as he is attending to his business. In the field of his scientific work he is warily engaged in a complicated encounter with tradition. He cannot be oblivious to it, he cannot act without it, and he cannot just submit to it. (Shils 1981, 105–106)

These things that Shils says about tradition in science can also be said about tradition in moral philosophy and about traditions of rationality generally. I presume that creationism does not fall within the tradition of science and for that reason it should probably not be taught as a part of science. However, Eger draws another distinction of considerable importance which might give a different cast to the picture. He distinguishes between viewing sciences as taught within the context of education or taught within the context of application. When science is taught within the context of application, I understand, it is presented as activities in which "laws and explanations are seen as tools for the solution of predefined classes of problems" and not with a concern "for the effect of subject matter on the learners' orientation in the world, on their actions in society" (Eger 1988b, this volume, 356).

I wish he had elaborated the important aspects of his argument more at this point. I say this because the forms of rationality that are taught or can be taught within the schools cannot be construed as independent of traditions of rational inquiry (science or philosophy or religion) but neither can they be provided a place except within some distinction between application and education, which Eger depends upon. I am not sure how to elaborate the implications of this suggestion, and I do not suppose that in doing so the social disputes that he has so ably observed over the years would be calmed. Yet I am sure that in these respects he has pointed us in the right direction for an understanding that is fruitful, tolerant, and consistent with the values we seek to preserve in educational practice.

[26]

Reply to Criticisms

ABSTRACT: Comments on my essay, "A Tale of Two Controversies," were made by Daniel R. DeNicola, Thomas F. Green, Mary Hesse, Holmes Rolston III, and Abner Shimony. This reply focuses first on three issues: that very recently moral philosophy has taken a turn toward a more traditional, particularistic approach, which could mitigate the problems I described; second, that because creationism is essentially antiscientific, my more philosophical concerns miss the mark; third, that the relativism of the 'new philosophy of science' ought not be uncritically accepted. Finally, I compare Hesse's position with that of Shimony, indicating how the former implies a narrowing of distance between scientific *description* and moral *prescription*. [—M.E.]

I am of course gratified that so many outstanding scholars have engaged seriously with my essay. Their comments and criticisms clarify some important points, and in this brief response I would like to press further that clarification.

Thomas Green's belief (1988, this volume, 376–380) that an "impoverished conception" of moral education lies at the bottom of the controversies I describe is re-enforced by Daniel DeNicola, who reminds us that the formalist approach is not the latest word in moral philosophy. True, of the writers who have emphasized ethics of virtue, traditions, and contextualized life-stories, I mentioned only Alasdair MacIntyre. The reason is given by DeNicola himself: As yet, this work has had no influence on curricula. Nonetheless, I agree that my story is incomplete and that the situation would indeed be changed if this trend were translated into a new way

Martin Eger's paper, "A Tale of Two Controversies: Dissonance in the Theory and Practice of Rationality," appeared in *Zygon: Journal of Religion and Science*, 23, No. 3 (September 1988), 291–325, followed by comments by Mary Hesse, Abner Shimony, Thomas F. Green, Holmes Rolston III, and Daniel R. DeNicola. This is his response to their comments.

of teaching. However, the formalist approach has had an impact on education not merely because of its prominence but because it *is* formalist. Three advantages immediately accrue at the pre-college level: It appears scientific, it offends no tradition by favoring none, and it is 'teacher-proof.' The last point means that the designers provide *systems* that need only be understood before they can be administered.

In contrast, the nonformalist philosophies DeNicola mentions offer little that can be packaged and handed over in a summer workshop. For example, Green's application of traditional thinking of this kind, which he calls "education of conscience" (Green 1985), stresses conscience of craft, membership, sacrifice, rootedness, and story-telling. It is impressive, far richer than the programs I discussed. Yet precisely because it is so rich, his "curriculum of moral competence" is *not* a curriculum in the sense that a teacher can 'learn it and teach it.' It is a broad framework, an orientation, that an entire school would have to adopt, that teachers would have to acquire from the inside, slowly, over time—and there lies the difficulty. If we change our moral philosophy and the corresponding pedagogy every twenty years, there is not enough time for even one generation of teachers to have been themselves educated in such a tradition. However, someone may show how it can be done nevertheless.

DeNicola also believes that my unfocused portrayal of creationism pays too much attention to its more acceptable guises but fails to highlight its true viewpoint "in opposition to science itself" (DeNicola 1988, 359). This requires clarification indeed, but first I must decline the credit for conceding that creationists "reject the whole enterprise and vision of science." I do not think this is, in general, true. Rather, perceptions of the situation have been distorted by an indiscriminate use of the term *creationists* for millions of very different sorts of people, of whom only a tiny fraction are the 'creation scientists' or *creation activists* (as I prefer to call them) responsible for most of the news stories.

Within that larger population of creationists, active if at all only on the local level, I found—contrary to media impressions—that the prevailing attitude is surprisingly *respectful* of science, despite the fact that it is also suspicious of science. In this there is no contradiction. As American political conservatives are suspicious of government, but consider it a good thing when kept in bounds, so creationist parents suspect that when it comes to evolution something is being pressed on them in the name of science that actually *goes beyond science*. It is at this point that the question of the rationality of teaching enters. For many people, it becomes especially important to know what one *ought* to believe as a rational person, and what reason does *not* demand. These people do not wish their own views to be in conflict with science.

No doubt, as a maximum demand, most creation *activists* would like to see their beliefs studied in schools on equal footing. However, many *parents* would be content if, in their own district school, evolution were taught in what they regard as a 'less dogmatic' manner. This actually was the issue in the California controversy of the 1970s; and even Dorothy Nelkin, no friend of creationists, admits that some of the changes they proposed were in order (Nelkin 1982, 116). For another large group of parents and students, creationism is a starting point, a *preconception*[1]; evolution appears improbable, but there is room for discussion. Finally, there is also a small number of scientifically trained people among the creationists, who have tried to offer serious, technical critiques of theories of pre-biotic evolution (see Thaxton and Bradley 1984). Because creationism does come in many varieties, I would caution against the kind of language that needlessly places large populations 'out of court'—including intelligent, educated men and women who are open to dialogue. The sorts of people I describe, not members of the creation institutes, are the ones who actually interact with schools.

My own opinion is this: From the point of view of science, of philosophy, or of education, there is nothing *in principle* wrong with discussing creationist arguments alongside evolution. This is a minority position, but not just my own. Some philosophers and some evolutionists (see Shimony 1988; Alexander 1978) are essentially in agreement. Because such discussions ought to give each view the weight called for by the evidence, by history, and by its role in society today, creationism cannot receive equal weight. Moreover, if it is discussed at all, it must certainly be taken seriously *a priori*, not used merely as an example of 'antiscience' (a good analogy is Ptolemaic theory)—because it is a part of the history of science, because it is a widely held preconception, and because without such seriousness the discussion would be spurious. If questions are raised about extrapolation, about degrees of certainty, about differences between theory and fact, such questions are perfectly natural in this context and might offer opportunities for teaching something about the nature of science, although this may be more than the students (and teachers) can handle. However, as Abner Shimony comments, perhaps the whole theory of evolution is "ill suited . . . for elementary instruction" (Shimony 1988, this volume, 374). Therefore I agree completely with DeNicola that more specific educational (and social) considerations ought to decide *whether,* in a particular school or course, such discussion of creationism should be undertaken.

Concerning the philosophy of science, Shimony feels I am too tolerant of the relativism and of some other disturbing features in recent work. Since my purpose was not to analyze critically the various schools of philosophy, but only to describe the present situation in relation to teaching, Shimony's criticism must first be reformulated. The relevant question is, I think, whether my portrayal overemphasized the radical components in postpositivist philosophy of science, and thereby unduly narrowed the difference between science and morals. Fortunately, in this regard I am spared the need to amplify at length. This issue of *Zygon* includes very incisive but quite different evaluations of the role of the new philosophy by two prominent scholars in the field: Mary Hesse and Abner Shimony.

Shimony, who has criticized Kuhnian views deeply, sees the new elements as a contribution or amendment to the much weightier realist tradition which I included in my catch-all category 'older.' Therefore, to the extent that I did not discuss the continuing strength of that tradition, Shimony's comments make up the deficit. Hesse, on the other hand, as a contributor to the new, sees in it a major departure, resting on considerable historical and analytical support. In a sense, both these philosophers are talking about the same half-full glass: Shimony emphasizes that empirically *and* conceptually the scientific theories that succeed one another approximate one another *in most respects*. There is direction in the process, suggesting a convergence on *truth*. Hesse, however, wishes us to note well that in *some other important respects* the *concepts* of a new theory do *not* approximate or resemble the concepts they displace. The elaboration of this point, a major feature of the new philosophy, leads to the thesis of conceptual nonconvergence, and does encourage a certain degree of relativism. Some consequences of this were discussed by Hesse; but in regard to education and the science/ethics distinction, I would like to point out an additional implication.

Shimony believes that the gulf between science and ethics is still wide primarily because science attempts to describe an objective domain of entities which is "there to be found out" (Shimony 1988, this volume, 373). Hesse, however, distinguishes between a stable, growing core of successful, 'localizable predictions' on the one hand, and on the other the larger 'pictures of the world' which are models and which "show no convergence to a unique theoretical truth" (Hesse 1988b, this volume, 366). On this view, the localizable particulars resemble entities in that objective domain Shimony posits, but the high-level theories are indeed relative to the present moment in history, the present state of knowledge, ever subject to drastic (not just minor) change. True, the theories are far from arbitrary: They are constrained by lower-level particulars; but, to use language Shimony reserves for ethics, it is by no means clear that such constraints uniquely determine an optimum high-level theory.

These two differing pictures suggest different possibilities in our cognitive relation to science. Holmes Rolston points out that the idea of *responsibility* is often used to distinguish the ethical from the scientific: "People are responsible for their values as they are not for their science" (Rolston 1988, 351). Note that this way of putting things refers implicitly to an account such as Shimony's. If all science is a process of discovery of entities in an objective domain, then everyone is bound to accept the established results, and no one can be held responsible. We are not responsible for the temperature at which copper melts.

On the other hand, if Hesse's account is accepted, then the high-level theories are not part of that objective domain, which means there is a certain freedom in *the way we appropriate them* and the way we assimilate them in our self-understanding. We may choose to take these theories literally or we may take them metaphorically. We may regard them as good approximations or as purely methodological devices for giving form to the localizable particulars. We may place greater or smaller significance on the thought that the ultimate form may be radically different. Clearly such freedom implies responsibility; so in this sense we *are* responsible for 'our' science.

The appropriation of the intellectual products of science still differs from moral choice in one respect: Few individuals are in a position to exercise this freedom in the domain of scientific theories. The manner of appropriation is generally guided by societies, groups, institutions, and especially educational institutions—which brings us back again to the rationality of teaching and the role of authority. Even in science, teaching is rarely a pure transmission of content, free of all directives about sense or significance. In the quotation from the encyclopedia, for example, given by Rolston (1988, 352), it is made clear enough that Darwinism is to be understood as *contradicting* "scriptural legends." Evolution and the 'legends,' the reader must assume, refer to the same realm and should be taken literally.

A difference between science and ethics remains not only in Shimony's account but also in Hesse's, as it does in mine. However, I take it that in the newer accounts, such as Hesse's, this difference is reduced. *Some* high-level theories of natural science—in biology, brain research, cosmology—impinge more strongly on human self-understanding than others such as electromagnetism. At the same time, like all high-level pictures, these humanly more relevant theories are subject to choice regarding their role in the cognitive ecology.[2] We have the freedom, for example, to determine whether preconceptions are totally replaced by newly learned theories or whether they are moved and reconfigured, assuming new relations to the rest of that ecology. Therefore, in teaching *these* aspects of science to the next generation, we are in a position partially resembling the teaching of morals: We are *responsible,* that is, for the direct and indirect philosophical messages, transmitted along with the theories, that indicate to the learner *how* the content is to be understood.

Summing up, then, in regard to the question of the 'gulf' between natural science and moral thinking, it is a question of degree and of emphasis; few of us deny there is a difference. However, because of trends in the philosophy, history, and sociology of science, more scholars now claim that this difference has been exaggerated, that the resemblances deserve greater attention. Thus as Hilary Putnam says, "we tend to be too realistic about physics and too subjectivistic about ethics, and these are connected tendencies" (Putnam 1981, 143).

[27]

Kohlberg, MacIntyre, and Two Basic Problems of Education

ABSTRACT: Lawrence Kohlberg was a strong critic of ethical relativism, but paradoxically he was cited by New York State schools and teachers as a defender of relativism. The paradox is explained by his acceptance of the irreducibility of values to facts together with his proposal that the process of clarifying values can be dealt with rationally by open classroom discussion; and parents construed this openness—with some justification—as the acceptance of relativism, with consequent confusion and demoralization of the students. Alasdair MacIntyre offers an alternative theory of values which is more naturally and effectively adaptable to classroom instruction. Following Aristotle, MacIntyre maintains that morality is embedded in practice, because any systematic practice aims at certain kinds of excellence. In the course of teaching a skill—e.g., expository writing or scientific investigation—teachers can authoritatively, in their pedagogical role, explain and enforce the morality implicit in the mastery of those skills, such as truthfulness in writing and respect for data in scientific experimentation. Although MacIntyre does not regard these values embedded in practice as constituting the ultimate goal of a good human life, they may constitute the part of ethics that can be properly taught in the public schools of a pluralistic society, leaving the overarching values to be addressed by families and religious institutions. [—M.E.]

Three Puzzles and a Controversy

What I admire most in Lawrence Kohlberg's work is his unrelenting effort to provide solid ground for a theory of morality that refutes the major forms of relativism. Even more to his credit is the fact that theory alone does not satisfy him. The inability of the public schools to deal with moral education he sees as a scandal and a danger, and the frustration in this realm

---

This talk was presented in 1983 in Potsdam, New York, at the annual meeting of the New York State Sociological Association.

as a challenge; for he believes that the failure is due in large measure to the pervasiveness of relativistic attitudes. Consider, he says, this testimony from a junior high school teacher:

> My class deals with morality and right and wrong quite a bit. I don't expect them to agree with me; each has to satisfy himself according to his own convictions. . . . I often discuss cheating this way but I always get *defeated*, because they still argue cheating is all right. After you accept the idea that kids have the right to build a position with logical arguments, you have to accept what they come out with. . . .[1]

For me, this quotation has special significance—and I have no doubt that it is representative—because, for a period of several years, I had the opportunity to observe a controversy that revolved precisely about this point. It happened in the Spencer-Van Etten school district of New York State, from 1977 through 1982. *Values Clarification*[2] was taught in the 7th and 8th grades, and attacked by parents for its relativism. The furor grew and spread, and turned elections for the school board into major political events; and at the center of argument were just those philosophical issues that Kohlberg has made his special concern.

Kohlberg's views were cited in public. His name was on many lips. Strangely, however, his role in education was not exerting the kind of influence one might expect. It was the school, and the teachers' spokesmen, defending relativism explicitly, who cited Kohlberg as their champion; while dissenting parents, using arguments he might well admire, condemned him bitterly.

This was worth observing. Yet it was only the first of several puzzles I encountered in this community. Elsewhere, I have discussed in some detail particular issues and incidents—social, educational, political—especially the shortcomings of the school staff in dealing with the protest.[3] Here I will describe only certain conceptual problems, or puzzles. Regrettably, it will be seen that in spite of good intentions, Kohlberg's intervention in education is not solving the problem of moral relativism, that it is neither helping nor enlightening the junior high school teacher whom he quotes. It will also become apparent that his theory *cannot do so*, as it cannot account for the other puzzles I encountered. In the next section, I will suggest that an alternative theoretical formulation *can* account for the phenomena very well—and that quite possibly it is a superior formulation, not only in battle with relativism, but in regard to at least one other major problem of education.[4]

Yet had it been pure Kohlberg, the objections would have been the same, because the element in dispute was the open-ended class discussion where all decisions are made by the pupil. For this Kohlberg is no antidote. The difference in assumptions concerning morality may be relevant to teacher training, may perhaps result in some difference in emphasis; but to the parents I met such a difference would be trivial.

The second puzzle in Spencer was a certain troubling contradiction within the ranks of the protestors themselves. What would they have in place of relative morality? Immediate reply: "Absolute morality." Does that mean they would impose their absolutes on the children of others who do not agree? Oh no, everyone's traditions are to be respected. Well then, how is this to be done? Silence, then a second line of defense: "We're not saying our precepts should be taught, but if ours are left out, then leave out all values"—this actually a public statement by a key leader and member of the board of education. It does not remove the contradiction. For it was obvious on the basis of much interaction that the protestors did not want totally value-free schooling; largely for this reason, those who could sent their children to religious institutions.

From Kohlberg's point of view, the diagnosis is straightforward: These parents mistakenly believe that their 'bag of virtues' is absolute; they may genuinely dislike the thought of imposing it on others, but that is precisely what they would do if allowed. Moreover, their refusal to admit the contradiction shows a certain aversion to reasoned inquiry.[5]

It would have been easy, even natural, on the basis of reporters' news stories, to infer irrationality; we have read accounts of similar battles elsewhere, implying just that. In this case, however, such an assessment was ruled out—for I knew the leaders of the protests personally, having spent many hours with them discussing every aspect of the problem. Far from being irrational, their main line of reasoning in criticizing the school's program seemed to me flawless. I never found it difficult, in this group, to make a point or objection understood.

Kohlberg's own solution to the problems posed by moral education in a pluralistic society is to rely on formalism, and hope that the stage structure will take care of content. But a content morality, *seen as such*, is clearly what these parents want—in public schools too—or, as they say, none at all.

The third puzzle had to do with the relation between values and 'basics.' At countless meetings, in personal conversations, in newspaper items and public debates, hardly ever was the question of values discussed in isolation; scholastic achievement was inextricably intertwined. I would have found nothing unusual in this if the problem of 'skills' were merely concurrent with that of values—that "Johnny can't read" is, after all, a nearly universal refrain in the country. What impressed me, however, was the extremely strong correlation (in time and language) between arguments about values and arguments about 'skills.' Thus, in the middle of an internal discussion about the problematic values course, one leader waves his hand and says, "All this would disappear if we could just get the subject teaching back on track." On another occasion, as I listened to complaints about the quality of the literature textbook, I would suddenly hear the familiar arguments against relativism.

I did not then draw any particular consequences. In part, this was due to the explicit causal connection made by the protesting parents themselves: So much attention to values and 'decisions' takes time away from the basics; or, moral relativism causes discipline problems, which disrupt the academic program; or, relativism implies that pornography is as good as great literature, so the children are given junk to read. The links were many and varied and deeply felt, but a wealth of explanations is no explanation.

Concerning this issue, Kohlberg is almost irrelevant. For in his theory morality is *sui generis*; moral growth occurs as a result of *moral* interactions, preferably occasioned by *morally* problematic situations. The problem of reading or mathematics is another matter altogether.

At this point, the thought might recur that so much contradiction—first about indoctrination, then about causes and 'skills'—is mounting support for the nagging suspicion of irrationality on the part of the dissidents. In such cases we have to remind ourselves that surface contradictions are just that: It is a task for theory, not for observation, and certainly not for the subjects themselves, to provide coherent explanations. Irrationality in man does seem to exist; but in any situation, this circumstance can also serve to mask the fact that a better theory is not at hand. The point of my argument thus far is only that Kohlberg's approach by itself, and more generally a formalist approach, cannot even theoretically explain the events I witnessed; it certainly seems unable to cope with them socially. And these events are not unique.

Three problems remained, therefore, after I had left the scene: how to avoid moral relativism in actual practice; how to satisfy the requirements of religious and cultural pluralism, if a content morality were taught; and what, if any, connection there is between these two questions and the issue of competence.

MacIntyre and Values

There are obvious reasons why the framework recently advocated by Alasdair MacIntyre[6] may be more successful in dealing with these issues than that of Kohlberg. Following Aristotle rather than Kant, using a teleological approach from the outset, MacIntyre sees morality deeply imbedded in traditions of practice. Although here too we find stages, and a kind of development, the focus is on *historical* development; here too there is concern with logic, but with the logic of narrative, not of problem solving. To treat each moral issue in isolation, to apply reason alone in a timeless perspective—this he considers a project that has failed and must fail.

In view of these general features alone, such an outlook clearly has a better chance of comprehending the behavior of 'traditionalist' parents in situations of the kind I have described. But what MacIntyre suggests can, I believe, do more: It can provide the solution, or most of it, to the problem of relativism in education—a solution that is at once more natural, more compelling, and less likely to be socially divisive. And if it turns out that, without *ad hoc* tampering, this way of looking at things sheds light on another phenomenon, which it did not deliberately take into account, then, of course, it becomes more *credible* as well as more useful. Because MacIntyre's framework is broader than that of Kohlberg, his theory addresses itself not only to the junior high school teacher's dilemma, but to all three problems just raised; and it addresses these issues in such a way as to link them explicitly, bringing to light a relation often intuitively felt but not until now so convincingly developed.

The key feature in MacIntyre's analysis, and the means by which he bridges the logical gap between 'is' and 'ought,' is the concept of 'practice.'[7] Practice, in his sense, aims at the achievement of certain kinds of excellence which are "partially definitive of that form of activity." Within a practice, therefore, and in light of the standards of that excellence, particular circumstances may in themselves yield an imperative. Thus, a captain 'ought not' be among the first to leave a sinking ship, because to be a good captain is to command the ship as long as it floats and people are aboard. No higher principles are needed to reach this conclusion. A practice is not a skill or an institution, but a far more inclusive "form of coherent and complex, socially established cooperative human activity," which characteristically involves the realization of what MacIntyre calls "internal goods." To use his own examples, "bricklaying is not a practice; architecture is. Planting turnips is not a practice; farming is. So are the enquiries of physics, chemistry . . . painting and music." Games like chess too are practices: The external goods are winning and possibly winning a prize. Among the internal goods are certain kinds of analytic insights, imagination, and competitive intensity, which might result, for example, in a beautiful sacrifice play. A practice, then, includes skills and rules; but also values, standards of judgment, paradigms, traditions, and therefore life histories.

To delineate the genre more fully (which is not easy), MacIntyre gives a number of examples from favorite practices like chess and the art of portrait painting. I would like now to add to this list another example; one that will serve to illustrate not only the nature of 'practice' but alternative conceptions of values education. Again a teacher is involved—and I cite this deliberately, in contrast to the junior high school teacher quoted by Kohlberg. The following events took place in my own institution.

A physics instructor had become aware that some laboratory reports he was receiving were being copied from students of the previous year. One day, two boys submitted reports whose content revealed to the instructor, with 100% certainty, that the experiment had never been performed in his class. His response was immediate and unhesitating: After confronting the students with the evidence, he gave them a choice of failure or withdrawal from the course—to 'start clean' with someone else, as he put it. At the next meeting of the class, he delivered a three minute 'sermon' on cheating: "Cheating is the number one crime in science," he told them; "Two students who falsified reports are no longer in the class, and anyone else found doing anything like that will be treated similarly." He added one or two remarks on what he thought of cheating in a lab, and then finished by saying he did not wish to belabor the point, since at this level—college—such ideas, he would hope, were obvious enough.

In comparing the behavior of this instructor with that of the junior high school teacher, the first thing that strikes us is his speedy action and self-confidence as against her frustration. But note this important point: The physics instructor did not give his 'sermon' on cheating in general; he did not base his stance on some universal principle; he did not discuss whether cheating to save a life is justified; and he did not worry about the possibility that some students might still be at low stages of moral development while others were advanced. He spoke to all on the basis of the project in which they were involved: "Cheating is the number one crime in science." He exercised authority, but if we view physics as a practice, then he did so as a 'master' facing novices within the practice. He did what a coach does in suspending a player for breaking training rules, or what a thesis advisor might do if a Ph.D. candidate acquired data unethically.[8] In this perspective his self-assurance becomes rational and normal, but he gains it by limiting his teaching to the confines of a practice.

This instructor is not saying, "Because you cheated, you are immoral," but rather, "Because you falsified data you show that you have not yet understood the nature of this practice, *and* that you do not accept its rules even provisionally. But without that you cannot continue." What he assumes and teaches instinctively is what MacIntyre presents as a solution to the problem of the disjunction of 'is' from 'ought': Cheating *is* a taboo in physics. Therefore it is the duty of every physics teacher to 'teach' this taboo to his students as he teaches the methods of calibrating an ammeter. He should by all means try to make this taboo rationally comprehensible in terms of the practice and its aims, just as he tries to make the calibration procedure comprehensible in terms of the laws of electricity. But if a student is not quick enough to grasp this rationality at the very beginning, compliance is still required, as it is required that he accept and apply the calibration procedure whether or not he has thoroughly understood it—

otherwise, it is not possible to go on experimenting. One hopes that those who are slower will see the logic of it later on.

By contrast, a Kohlbergian dealing with issues of life and death, race, or social protest, must relegate himself to the role of 'facilitator'—a facilitator of moral development. The scope of his subject is tremendous, but his own position within it is dwarfed. He may ask for reasons, discuss dilemmas, suggest role-playing exercises, but he cannot, even in a circumscribed way, be a true 'master' or coach—since he cannot tell his students how "it ought to be done."

As soon as we notice that most of the activities students normally undertake—from first grade to graduate school—are indeed practices in MacIntyre's sense, then the relevance of his theory to education is manifest. And if so, the whole issue of values and morals teaching is seen in an entirely different light: In this light, it is not the connection between 'is' and 'ought' that has to be justified (as Kohlberg labors to do[9]) but their separation—a separation that now appears both irrational and artificial.

There is little doubt that the protesting parents in Spencer would have been delighted with our lab instructor. For this is essentially what they meant by 'absolute' values. The word as used in the larger community did not necessarily have the same connotation as it does in a religious context. In the school, values would be sufficiently 'absolute' for these parents if they were *not arbitrary* and not left to the 'decision' of the student. In this perspective, the contradiction noted earlier—between the desire to have 'absolutes' taught and the reluctance to impose on neighbors—disappears. We begin to discern some logic behind the attitudes of the dissenters, and some precision in the way they draw the line between the things they might be willing to relinquish to the school, and those they regard as 'better left for home and church': Roughly speaking, values internal to practices they would happily consider in the province of public education; values, or foundations of values, superordinate to practice, they would reserve for themselves.

Much too easy, one is inclined to say; something has to be wrong here. And, of course, by this time the objections are so obvious they must be taken up.

This 'derivation' of 'ought' from 'is' has been achieved at great cost, a formalist might well say: When the horizon of discourse shrinks to encompass nothing but a practice, a certain amount of authority and indoctrination is perhaps legitimized, but the 'ought' is thereby transformed to something very different from the true moral 'ought.' What purports to be a derivation is in fact a *reduction*. Such an 'ought' cannot be binding on those who do not care for the practice (as the example of the physics lab clearly shows). It is therefore a greatly impoverished morality that this physics teacher teaches—worlds away from the obligations emerging at Kohlberg's post-conventional level, for example. At best, it looks something like the conventional stage four (Kohlberg 1981), with special content, and thus a rather truncated stage four, for the conventions are not even cultural but subcultural (and we are back with a kind of relativism after all).

Such objections are not easily brushed away. MacIntyre himself admits the inadequacy of virtues based on practices alone. For him, this is only the initial stage in a logical development that proceeds first to abstract the common factor in all practices, then to formulate a conception of the good life—the overarching *telos*. Now this ultimate goal—the good life for man—was disappointing to a number of critics, as were the conclusions MacIntyre draws at the end.[10] However, it is entirely possible that whatever objections exist to the later stages of MacIntyre's argument are not necessarily relevant in the context of education. A hint of this was given already when I pointed out that values and foundations of values superordinate to practices are in any case something many parents do not like to see in public education. Thus, what appeared at first as an embarrassing break, or branch point, from the point of view of theory, could turn out to be quite an advantage in *this* application of the theory. Even if MacIntyre's 'social teleology' is not accepted as a framework for a complete account of morality, the earlier stages of his formulation may be both sufficient and appropriate for a comprehensive approach to the problems of the schools. And by this I do not mean merely an expedient approach, but one that is both reasonable and just.

Recall first that the natural sciences are not the only practices taught. This much the student should discover for himself (if all the 'masters' do their job): that plagiarism in a writing course is no less a 'crime' than false data in the lab, that teamwork is no less a virtue of the school band or newspaper or on stage than it is in a basketball game—and no more in a basketball game than in a complex physics experiment. In short, certain values and virtues are far more universal than a single practice alone reveals, and some are indispensable to the existence of practices as such. In the latter category, MacIntyre explicitly lists *truthfulness, courage, justice,* and *trust*. Does this realization lift truthfulness, for example, from the level of a mere instrument, inculcated within a practice, to that of a truly appropriated moral imperative? To a large extent—though not completely—the answer may be yes. And the argument supporting this contention can now be made stronger, since MacIntyre's reintroduction of the Greek conception of practice injects at least one important new point.

Although the classic Aristotelian idea of virtue through habituation should not be lightly dismissed, habituation is not at the focus of MacIntyre's account. The distinctive feature of practices is the large amount of rationality they involve, and the fact that here rationality is integrated with habits and values in an extremely natural way (as Polanyi and Kuhn have shown for the case of natural science). But it has always been possible to object that the values included in such practical contexts are highly instrumental, since the whole rationality of the sciences is instrumental; and instrumental reason is surely an odd foundation for a morality worthy of the name. (One has only to recall the writings of the Frankfurt school, and the fact that instrumental values are Kohlberg-stage-two.)

It is at this point that MacIntyre's distinction between internal and external goods is once again crucial: Yes, within physics truthfulness is supported as a means to a limited end; but it does seem to make an enormous difference whether that end is the external good of passing a course and getting a degree or the internal good of learning 'to do good science.' That is, having understood well what 'good science' is, wishing to do it truly and not just in appearance, a student acquires the habit of never presenting as fact what he knows to be untrue or doubtful or unestablished by credible means, *and* the conviction that to do otherwise is to be a bad researcher, or engineer, or technologist. Does it not seem that this is quite a bit beyond Kohlberg's stage four, and nowhere near the instrumental number two? And let us recall that with regard to educational intervention, Kohlberg now believes that just reaching stage four is a serious project. Finally, since several moral virtues do span the whole spectrum of human practices—or at least a large segment of it—the step, by induction, to consider these as 'human' virtues seems neither unreasonable nor counterintuitive.

That this is by no means the whole answer to moral relativism is once again admitted. Yet we cannot help noticing what even this much offers to the two teachers in our examples. To the physics instructor it says simply, you are right not to try to reach higher than your legitimate authority extends—but within those bounds, teach the practice, the *whole* practice, not just the skills. To the junior high school teacher it poses a series of questions: What subject are you teaching? Moral education in general? That does present problems. On morals, you cannot assume the role of a master merely because you are a teacher facing children—especially not in state controlled schools. Here we certainly agree with Kohlberg and many others. But if your subject is English, for example, then teach English from a position of mastery of your practice. Does a student plagiarize? Then deal with it as you would with a plagiarizing author if you were the editor of a respected magazine. Is it merely a class discussion? Then choose the appropriate literature: Instead of artificial dilemmas, let the student see how good writers treat the issue. And deal also with the question of *truth in literature*; here again you are the master of your practice. If you know how to make such things meaningful in the classroom, you are probably doing a lot for moral education.

As I mention literature, though, I recall vividly the incredulous voice of one of my conversation partners in Spencer: "What," said a young

mother, leaning across her living room table, "you want them to read
Solzhenitsyn? They can't even read a newspaper. Just give them a front
page article and ask what it says. You'll see—no comprehension."

Thus in the midst of social controversy, starting with the problem of
morals, one is brought face to face with the issue of skills and competence.

MacIntyre and Skills

The popular solution to such complaints, by now well known, was often
proclaimed in Spencer: 'Back to basics.'

But what is basic? Is not 'critical thinking' more 'basic' than a load of
facts soon forgotten? On the other hand, are not some facts so 'basic' that
critical thought without them is impossible?

In no other case does MacIntyre's set of categories make more difference than in regard to such questions, because it suggests strongly that the
whole 'back to basics' problem is misconstrued—the implied dichotomy is
false and probably pernicious. The slogan 'skills without frills'—whether
uttered by school administrators, critics, or discontented parents—is terri-
bly misleading.[11]

If there is such a thing as a 'basic' structure in education, then neither
skills nor facts, nor reasoning ability, nor values, and certainly not the disciplines, comprise that structure when taken alone. Nothing less can be
'basic' than the practice as a whole, because only within the practice is the
proportion of each of these elements to the others coherently defined, and
only within it can the proper balance between them be maintained. It is for
this reason that 'skills without frills' is not a solution to anything, but a
symptom of the problem—easily converted from a slogan against marginal and flabby courses to a rationale for wrenching skill out of its natural
environment and using it as means for some presumed goal in 'the real
world.' But one way that practices decay, MacIntyre points out, is when
external goods begin to dominate over internal goods; for then the whole
practice becomes instrumental, and the virtues associated with internal
goods lose credibility.

If a community really sees 'skills' as the only thing worth teaching, then
it is already moving down this road. But I believe that, on the whole, things
are not that bad. It is not *privatism*—as Kohlberg thinks[12]—but inability to
analyze a disturbing and very complex situation that is causing the stam-
pede to 'skills.' The overwhelming majority of people in the community I
observed undoubtedly do want values and morals and critical thinking and
all sorts of standards of practice to be part of the normal educational fare.
But since so many oppose relativism in values, concentration on the problem of 'skills' has become a kind of refuge, a way of "avoiding the worst"—

as one parent put it—when a better way of sorting things out is not in sight. In addition, some segments of the community discern an important relation.

What the discontented intuit is what MacIntyre's concept of practices implies about the *continuity* of values: If truth-telling is an internal good in the laboratory, then attention to detail, interest in error, getting the maximum possible accuracy from the available equipment, and all the skills this calls for, are means of attaining this good. Conversely, if all these disparate skills must be learned as part of the practice of physical science, then truth-telling in regard to the experiment is the obvious aim and unifying factor. But when the student perceives that care, accuracy, etc., are not valued in the classroom and laboratory, clearly truth is less valued. In that case, outright falsification of data is merely the final transgression of a standard already in neglect; if *then* the instructor reacts strongly, the student may well wonder what the 'fuss' is about—but more important, failing to see the link to practice, he is apt to regard the 'fuss' as authoritarian. And this can in no way be compensated for by discussion of dilemmas concerning cheating, role playing exercises, and so on. The whole environment is wrong (as, in regard to justice, Kohlberg has already perceived).

The situation therefore appears thus: First, we delete the most serious values from within the practices—either due to a spurious sense of the separation of 'is' from 'ought,' or because of misconceived regard for skill, or because external goods receive too much attention. Then we see a moral vacuum; something is clearly amiss. To fix that, we attempt to bring back the values as goods in themselves, using various theoretical formulations—via modules, units, and even separate courses. In the end, the same teachers who fail to do full justice to their own practice are called upon to develop and teach the values units and the modules on 'critical thinking,' where esoteric subjects such as topology are listed among the 'competences' to be imparted. It is easy to make the mistake of assuming that the strong negative reaction of many parents is not based on genuine insight into this process, because that insight does not fit an easily recognizable theoretical framework. If MacIntyre's viewpoint has any effect in education, however, this misunderstanding should abate.

The frequently seen connection between complaints regarding lack of values, or relativity of values, and complaints regarding scholastic achievement—of which Spencer-Van Etten was just one instance—is thus revealed as not only understandable but *inevitable*. Because certain values are internal goods of major practices in education, and some of these are moral values or closely related to moral values, either value relativism or neglect of values within the schools must of necessity involve a decline in the achievement within these practices. And vice versa, a decline in the skills that are components of practices must imply declining respect for those internal goods and values whose realization requires these skills. In retrospect, every one of the causal connections between skills and values that I heard in Spencer can be seen as different aspects of this underlying relation.

MacIntyre vs. Kohlberg

By way of final comparison, it is worthwhile to ask how Kohlberg treats the values of practice in relation to morality. His theory says almost nothing about it, but in dealing with major public controversies, as he often does, one cannot avoid the issue. Thus, when referring to Watergate, he usually points out that the President, as well as the advisors who went to jail, were *all lawyers*; and he expresses amazement at the 'moral ignorance' these lawyers show in their public pronouncements. Kohlberg would certainly like to bring moral discussion into legal education, into medical education, even into scientific education, and has made efforts to do so. How does this differ, in effect, from MacIntyre's approach?

The difference is vast. For MacIntyre, law and science *already are* projects that 'embody moral tasks' as internal goods. When many individuals show a lack of regard for such goods, this indicates the presence of decay somewhere in their practice. Whether the importation of moral discussion from the outside, as a kind of enrichment, can be of help depends on whether such discussion addresses the underlying sickness, and whether it leads to something more than discussion. If not, it may even do harm.

Unfortunately, many of the Kohlbergian dilemmas developed for the classroom fall into the latter category. I take one of these as my last example: It involves an ancient practice—competitive sports.

> A high school relay team of three boys and one girl is about to run the final event that will decide the championship. However, the girl is not the most suitable runner for this race; it is admitted that if a certain boy who has been alternating with her were to take her place, the team would win, otherwise it might lose. But the girl has been scheduled for this event long ago, because of an agreement resulting from a law that requires female participation in sports. Should she withdraw?[13]

What usually passes unnoticed is this: All the factors to be weighed, as presented in the dilemma story and its variations, are *external goods*. First, *winning* is opposed to the *agreement* occasioned by the law. If the class decide too quickly that the boy should run, the teacher is instructed to tip the scales by adding other goods to the picture: college scouts in the stands considering the girl for a sports *scholarship*; newspaper reporters on hand to do a *story* on the girl runner, and so on. Neither the coach, nor the girl, nor any of her teammates expresses any thought about what a *good athlete* should do in this situation—nor is the teacher-facilitator to raise any such question.

In the Kohlbergian view, the only relevant units are the individual, society, humanity. Jobs, professions, and activities exist as components of society and fields for individual achievement; but practices, as distinct structures, defining standards and virtues beyond skill, do not exist. A phrase of Robert Hutchins's sums it up well: We continue to educate "not *through* occupations but *for* them."

Obviously, the nature of this paper is that of exploratory thoughts and suggestions, whose main purpose is to provoke discussion. But were all these suggestions to be properly developed, and were they to pass the test of criticism, then the conclusions would be radical in relation to present trends. To state these concisely may be useful, and this I would now like to do Kohlbergian fashion—by means of an enumerated list, and with finality.

1. It is not true, as Kohlberg contends, that if moral relativism and arbitrary 'bags of virtues' are unacceptable then the only choice left is a formalist theory of morality.

2. Within the educational context, at least one alternative is conceivable: the first stages of MacIntyre's 'social teleology,' based on the concept of 'practices.'

3. At least in the context of education, this approach is not merely an alternative; it is superior, because it allows teachers to be more than 'facilitators,' and actually tells them how values—including some moral values—are to be taught.

4. This approach has the great merit of being completely acceptable to all sectors of a pluralistic society because it confines itself to moral values that are natural within practices that schools normally pursue.

5. It has the additional advantage of giving at least a partial account of morality that is not *sui generis* but continuous with the spectrum of other values. Therefore, the relation of morality to education's main practices is not one of optional superposition but one of necessity.

PART FOUR

Life, Death, and Dialogue

[28]

Letter to a Friend Regarding the Farm

You asked me to tell you how I spend my days when I'm not commuting to NYC. All right. I'll tell you about yesterday, a typical September Saturday for us, a 'nineteenth-century' day, so you'll know what I mean by this term.

About 5:30 A.M. we get up, a bit earlier than on most days—because of the cows. My neighbor's herd across the road had started bellowing (unusual, but sometimes it happens) and the windows are open to let in the frog-concert of the night. As of 5, we couldn't sleep, so might as well get up. It's not yet light outside and not obvious whether the day will be clear, but the temperature has been dropping lately, so by 6 am mist is rising out of the ponds, like smoke out of pots, drifting over the bound rolls of hay that have been left about in the fields. This scene out of the large dining room window is familiar yet different every time, and always it frames, by its sheer all-around presence, just about everything else on the horizon—the visual and the cognitive. We're anxious to start our morning routine: I put a scythe over my shoulder and head for the larger pond; it needs a lot of maintenance this time of the year. Since high school boys are not as ready to work now as they were ten years ago, even for high pay, I do it as part of morning exercise. Judith takes Wolf, our German shepherd, and heads for the higher-up paths through the woods, to keep the dog out of my neighborhood (once I nearly cut her legs off when she bounded up unexpected from behind while I was scything).

I get to work on the long grass already beginning to droop over the water. The object is to cut it low, heave it away from the shore, and prevent all that new bio-mass from filling in the pond. If you don't do that,

---

This letter was written in 1994 to a friend from Martin's graduate school years, who had tracked him down via the Internet after an interval of about thirty years.

the land will move in on the water and every year the pond will shrink. . . . Over a few years its effects can be big because this pond has been made deliberately shallow. It's a 'shallow-water area' I built seven years ago with some government financing, designed for wildlife preservation. Wetlands have been disappearing here in the East, so we're all eager to provide more habitat, especially for migratory birds. Well, it has certainly brought more wildlife to our land: deer, fox, raccoon, possum, large muskrat and turtle, many species of bird, and, most amazing to watch when he's fishing or hunting—the Great Blue Heron.

After ten minutes or so of scything, I stop, take an emery stone out of my back pocket and sharpen my blade. The sun is up no more than ten degrees over the far side of the pond, a red ball so faint through the mist you can look right at it. I take a rest for a while to listen to the silence. The silence here is a wonderful thing, due to the fact that this homestead is on a real 'back-in' road (as they say around here), with hardly ever a car going by. The only sounds now are those of a frog once in a while, leaping in—a plop here, a plop there, and silence.

When I stop like that and sharpen the blade, I can't help but think of all the farmers and peasants I've seen doing this all over Europe—in Norway, on Austrian hillsides, in Italy, Greece, on Crete—and everywhere I would stop the car, pull off the road and watch. It is a primordial scene—this scything—and if the weather, the position of the sun, and the background are right, it can be absolutely arresting when come upon suddenly after some turn in the road. But that was not the only reason I watched these men with scythes. There was something puzzling about the way they did their work. They would take five swings with the scythe, then stop to sharpen, another five or six swings, and again sharpen; and so on for hours. They seemed to take about as much time to sharpen as to cut. Why, in the twentieth century, do blades have to be so poor, I wondered. Why all this sharpening?

Even then, I think, I had an inkling of what was going on, but not until now do I really understand it. Of course no one has to stop that often—not for the blade's sake and not for the rest either. It is a gesture, and part of the ceremony of scything. By stopping, looking about, and letting the world around you impress itself upon you, you place your own consciousness at the center of this world, you interact with it as a subject. The cutting itself, with attention fixed on each swath, demands such a reduced subjectivity, such a detachment from the larger context, that it's intolerable for too long a time. So the peasant withdraws from his work, distances himself from it, and re-enters his real world. Whether working for himself or for someone else, he takes the time and sacrifices efficiency.

He is not one hundredth as productive as my friends here and tenants, who, single-handedly, or in teams of two, work hundreds of acres and dozens of cows, shipping every week enormous amounts of milk to market. But to watch my neighbors at harvest is to see something else altogether. I learned long ago never to stop to talk to anyone here when they're in the fields, especially in certain seasons, not even to wave from a distance. They would resent it. They move like men on a battlefield; every minute counts, and sometimes, far into the night, with tractor lights on, they're still cutting the crop—mindful of the coming rain, which will stop the tractors and cost tens of thousands of dollars. Here you have, in a nutshell, that contrast between two modes of human existence that has occasioned so much abstract theorizing, and so much distress in this sad twentieth century.

So now I stop cutting, sharpen, and look around, for more or less the same reasons men have always had for doing this. After a while, I resume and cut for another ten minutes perhaps, then clean off the scythe and start back (I have to take good care of it, by the way. A few years ago they stopped selling scythes around here; I'd have a hard time getting another blade). It's been a short workout, but we have to eat soon and head for Syracuse today to look at a truck.

We are buying a new truck. It will cost a lot, but it has to be the best, with all sorts of options, because it's absolutely fundamental around here and Judith's life will depend on it. It takes about a half hour's driving to get to Cornell from our place. In the warm seasons it's a magnificent ride overlooking valleys and grazing herds, but in the winter—which is long here, and severe—the weather conditions and the utterly reckless drivers combine to make the daily trip a real danger. One way to protect yourself is to be inside a sturdy four-wheel-drive truck. It will minimize slides, get you out of a snow bank even if you go off the road; and should you get hit by another car, the mass of the truck will shield you. In the winter, Judith drives the truck to her office; in other seasons, I use it for off-road work on the property. If you don't make the right choice on this purchase, you're asking for trouble for the next ten years.

We have our eggs and coffee by the window, watching the geese in the little pond just behind our back lawn, our talk this morning not quite as philosophical as it often gets since there are practical matters to be gone over. Beyond this pond, straight out, the fields stretch about a thousand feet to a hedgerow, while on the right they go down to the larger pond, then to hemlock woods and steeply rising slopes on the other side. The sun has come through enough now to throw a long shadow behind each of the hay-rolls, emphasizing the harvest season, the sheer blazing yellow of the scene—a true van Gogh picture, like the one I was looking at only a few days ago in a museum in Amsterdam.

This is why we came here. It is this daily living in nature that makes the difference now from what I could get as a mere visitor. It is the permanent, and more primordial context, to whatever else one may or may not do on this planet. In earlier, less 'sophisticated' times, all sorts of people understood the cleansing value, the necessity of distancing oneself from the 'doings of men.' Now, when even the poorest trailers have their VCR's and their 'dish' in the yard facing the satellites, 'distancing oneself' makes about as much sense as giving away a car. (A few years ago, when an old car we had was on its last legs, I gave it to a neighbor and friend here, a part-time mechanic, thinking he would appreciate the chance to make something out of it in his spare time. Since then, he hardly bothers to say hello to me, and looks upon me with mild amusement. Only a real fool would give away something for which he could have gotten maybe a couple of hundred dollars.)

But 'distancing' is not the only value of living and working on this land; there is something else it has taught and given me, of which I knew nothing before I came here. It's often said that we humans are 'thrown' into this world, strangers in an environment "we never made." Yet, when Judith and I look outside this window, what we see is by no means alien; it is by no means something "we never made"—quite the contrary. For it is we "who made it to be the way it is." When we came here, it was not so at all.

In May of '78, when I drove up for the first time on a windy afternoon, I was so unimpressed I told Judith over the phone there was hardly any reason for her to even look at it. The view in the back was squalid, the low area on the right was swampy and hardly passable; along the road, the homes on one side seemed much too close, fully visible from our windows, and the house itself was a relic, needing all sorts of work. Nevertheless, we bought it. The main reason was that we were not looking for the ideal country home. I was at that time almost driven to "get out of the city" at any cost. My money was limited, I had narrow constraints on location, and we had been looking for nearly two years.

Right after we bought it, I started working with my hands and learning, each year more, how big a difference could be made not just on a house but on the land itself. We transplanted trees by the hundreds, made wooded belts and groves so that our house area is now completely screened; I cleared paths through the woods to gain easier access, and laid out plans for ponds in the swampy areas. Then, a few years later, came the conceptual breakthrough, when I realized that even with modest amounts of money I could hire people and machines and change the very contours of the land: The cement remnants of the old burnt-out barn in the back were bulldozed down, along with the thick growth blocking the view to the main part of our property; the small pond was dug in a day, and the soil thus made available was used to reshape the whole back-area. A year later, the wildlife pond was dug by two giant machines working only a few days;

Judith started laying out flower gardens in the English style; and an additional room was built with a window and telescope overlooking that whole new scene.

And that is where we're sitting right now, this Saturday, looking out over the water at the four geese by the cattail reeds; they've been with us from the start, two pairs of Africans, with black beaks and knobs on their heads, now all lined up in a row—still, perfectly reflected in the pond's mirror surface. We're looking on a scene that's not alien. It is not 'a world I never made.' Of course, neither is it a world 'I made.' I intervene to dig large holes. Then, a whole sequence of processes takes over to change the shore line, the pond floor, and the life in the water. I act—nature reacts. Other times, nature acts and I react. And the scene outside is the outcome of this dialogue. It is not quite "nature" and not quite "man-made." It is something between nature and us—shared with nature, and among us.

As for the geese, they have pretty much of a paradise here during the summer. But in the winter, it's another world for them altogether, and every year I am again dumbfounded by what they can endure. They almost never go into the goose-house where they would at least have shelter from the wind, preferring the outside no matter how bad it gets. On dark February days, 20–30°F below zero, the yard completely filled in with big mounds of snow so that moving about is almost impossible, they stand— just stand—facing a howling wind, hour upon hour, day after day, sometimes sleeping on one leg with beaks tucked in under the wing.

We finish breakfast, drive up to Syracuse, take care of our business, and on the way back take a favorite route along the shore of Skaneateles, one of the Finger Lakes. And that's the way half our Saturday passes. I realize that I can't finish even this one day without making an already prolonged letter even longer. But in New York, my colleagues sometimes ask what I "do out there" in the country, don't I get bored?

You have other questions as well, about my work and so on. Another time I'll say something about that too. There are writings, and conferences and organizations, and ideas to which I have some attachment; but all that is, for me—how shall I put it?—perhaps not altogether secondary, but certainly derivative. I have not included with this first letter any reprint for you to read. I just cannot bring myself to "enclose my latest paper" as one routinely does with colleagues and people one meets at conferences. . . . For now what's needed, I think, is something like Alexei must have meant at the end of *The Brothers Karamazov*, when he said to the boys "certainly we will rise again, and see one another again, and gladly tell all that has happened to us." Is that possible even this side of eternity?

[29]

Dialogue Between Martin Eger and Abner Shimony

(October 2001–March 2002)

Martin Eger's Preface to a Dialogue with Abner Shimony

On Convergence, Reality, and Evolution (CRE)

In philosophy of science, "correspondence" and "convergence" have often been used as almost synonymous. I would like to distinguish between them as follows:

CORRESPONDENCE:

An essentially synchronic, theoretical procedure, showing that two realms governed by two apparently different theories can be made to 'coincide' in some limit ($c \to \infty$, $h \to 0$, small field, etc). I put aside here the question of various branches of the issue—whether only the experimental results coincide, or basic laws coincide, or all laws coincide, etc. Correspondence deals essentially with theoretical *continuity* throughout and between the two realms.

CONVERGENCE:

This I take to be descriptive of a certain kind of diachronic or *historical* sequence, using a rough analogy with the mathematical idea of convergence. The basic feature retained from the analogy is that in the case of convergence a small number of terms in the sequence contains essentially 'the whole' or 'the gist of the matter,' with all the rest being only a minor correction that may be left out without changing 'the picture.' This happens because each of the later terms is sufficiently close to its predecessor, thus

411

indicating a limit. If the sequence is *numerical* (e.g., results of experiments), then, possibly, a curve can be fitted to the points, and, with some assumptions and approximations, a quasi-mathematical conclusion may be reached as to whether or not the sequence converges. If it's a sequence of *theories*, there is little possibility of that because we have principles, laws, and concepts to compare. Nevertheless this does not mean that clear distinctions between known historical sequences cannot be discerned. Unfortunately, Peirce did not seem to have made this distinction in his famous formula of truth and reality as the "opinion fated to be ultimately agreed to by all who investigate." Otherwise he clearly had a convergent process in mind. But his word "ultimately" obscures that, and obscures the possibility of endless oscillation, at least in the conceptual sense (a defect that has not escaped your references to Peirce!).

EXAMPLES OF CONVERGENCE AND NONCONVERGENCE:[1]

1) The dynamics of the solar system from Kepler to Einstein appears convergent in both the numerical (instrumental) and conceptual senses. The corrections to orbits get smaller and smaller; *and* concurrently, the essential picture of approximate ellipses with the sun near one focus does not change. Nor do the laws governing the system. With Einstein, something new happens: The forces and energies are gone, the laws and concepts are different in a fundamental way, and we now often see pictures of the solar system—literally, in various kinds of publications—as imbedded in a warped space. Meanwhile, the numerical convergence continues, producing yet smaller corrections.

2) Atomic theory from Rutherford (1912) to Heisenberg (1925) also seems convergent: From Bohr's original circular orbits to Sommerfeld's ellipses with additional quantum numbers, and even some relativity, there is an increasingly better fit to spectral data, yet little change in the conceptual and visual picture—still electron orbits about a heavy nucleus. But the program runs into trouble with the He ionized molecule, excited He, and other things; and people like Born start calling for a "totally new" approach. We then have a sudden irruption of different forms of quantum mechanics, mathematically equivalent, but conceptually quite different from the 'old theory.' Gone is the famous picture of planetary-like orbits, or any kind of continuous path for the electron; gone the idea of a "law of motion" in the Newtonian sense. Again we see new kinds of pictures in textbooks—probability clouds, intriguing symmetries, etc. And again, the numerical convergence seems to continue: The Dirac equation brings more accuracy, but also more conceptual divergence—sea of electrons, antimatter, etc. Quantum electrodynamics allows for even smaller numerical corrections, but at the price of the picture of the atom now being full of appearing and disappearing particles with which the electron interacts, etc.

This, then, is where my first question regarding Bohr's atom comes in—in regard to convergence as I understand it: Are we *converging* from Bohr, to Schrödinger, to QED, in the full sense of the word—instrumentally and conceptually, at least in some plausible sense?[2]

INCOMMENSURABILITY:

As I understand it, this notion of Kuhn's and Feyerabend's consists of two features. First, they notice this "decoupling" of conceptual from instrumental convergence, and they attribute it to conceptual discontinuities. Second, they describe "paradigms" in the Wittgensteinian fashion, as languages in which scientists are apparently trapped without the possibility of real communication—"ships in the night," "old physicists must die out," etc.

The second feature, *incommunicability*, I take to be now largely rejected by most philosophers and scientists—as simply historically incorrect. But the first feature—conceptual nonconvergence—has continued to be a live issue, propounded by scholars such as Larry Laudan and Mary Hesse. And it is an issue that, at least for the time being, appears to be serious and unavoidable.

YOUR REPLY TO KUHN AND LAUDAN:

Your reply to this trend has, as I see it, two prongs: First you deny on general grounds, and in a historical context, Kuhn's thesis of the irrelevance of the search for truth ("Comments on Two Epistemological Theses of Thomas Kuhn"). This is based on the epistemological principle that the hypothetico-deductive argument ought to be as valid in philosophy as in science itself, which then leads you to propose a "hypothesis of verisimilitude." The latter, an idea of approximate approach to truth, seems Peircean, and closely related to the concept of convergence. You, however, discuss it in terms of the "correspondence" principle because, certainly, one can take any two consecutive terms in a series as given, and consider them in their theoretical approach in the appropriate limit. But then you claim that "also with regard to conceptual structure" continuity can be shown, and you list a number of such transitions as classical mechanics to relativistic mechanics, classical mechanics to quantum mechanics, etc. Certainly there is a great deal of correspondence and convergence in science history, and that is very important. Yet in some outstanding examples this seems to be questionable, as I've indicated above; and in the case of CM to QM I fail to see it in the conceptual realm. Nevertheless, the rejection of Kuhn's rejection of the search for truth still stands, I think, on intuitive and other grounds. But how do you square that with conceptual nonconvergence, if you concede that point in at least some theories dealing with the macro- or micro-cosmos?

The second prong of your reply is, I think, extremely interesting, provocative and fertile. Though it preserves the search for truth, it seems to point not only to an altered notion of 'truth' (as you admit), but also, possibly, to an altered conception of man in the world—much influenced by evolution.

In your reply to Laudan ("Reality, Causality, and Closing the Circle"), you partially accept conceptual nonconvergence, but only in regard to ontologies:

1) We must change our views of what counts ontologically. Is it the *visual* image or the causal, structural laws?

2) If we take the view that what the system "is" at any level is represented better by its causal structure than by visual images of essences or entities, *then,*

3) transition from one theory to the next is more continuous, or, at least, involves no contradictions *à la* Kuhn.

4) This implies a more Platonic turn: Only the forms exist, the mathematical forms are real, and these are quasi-invariant across theoretical revolutions (which seems closer to Heisenberg's later ideas regarding elementary particles).

THE NEW ISSUES RAISED

But this raises, on a new level, a whole host of issues—all the way from Schrödinger's *Anschaulichkeit*, which has to do, it seems to me, with the important question of *understanding*.

Feynman and Cushing are among those who have not yet given up the importance of visualizability of some sort. Of course, this does not have to be literal. Relatively simple causal relations and other abstract schemes can still be 'visualized,' i.e. in terms of models or maps. But when such visualizability breaks down substantially, then doesn't *understanding* (of what's going on) also break down? This lack of understanding may occur without any logical contradiction in the formulae, as many have pointed out. So how, in the larger picture of science, does one understand this consistent logic, instrumental efficacy, and lack of understanding? Does it not point essentially to instrumentalism (of one variety or another), and isn't instrumentalism uncomfortably close to magic?

SOME OLD ISSUES REVISITED

Historically, it's interesting that the Germans have taken much of this for granted for at least 100 years. Technical success has meant little to their view of the world, except to place upon it a dark cloud. This, for example,

from Spengler's *Decline,* way back in 1923: "It is *self evident* that no practical results and discoveries can prove anything as to the 'truth' of the *theory* of the *picture*." And this, more modern note: "By our analysis and synthesis, Nature is not merely asked or persuaded but forced. . . . Nature,
then, is an expression of culture, in each instance." And all this, of course,
is related to the 'will to command,' the Faustian spirit, and to Nietzsche's
larger views. I find it interesting that from the 1960s to the present much
of this has had a comeback seemingly on its own—and yet, of course, quite
continuously with the past. It is there today not only in Heidegger but in
Gadamer, Habermas, Apel, and even in old Herbert Marcuse, if anyone still
reads him (left and right, without prejudice!).

EVOLUTION

Bacon made too much, perhaps, of the possible corrections of the idols
of the tribe. Is it really a question of breaking idols and restoring some true
vision? Is it not rather a question of inherent limitations on the species?
The old idea of the great chain of being seems to be more relevant here:
Inherently, because of their neuro-sensory endowment, some species surely have greater access to the world than others—and this may hold (may
have its limit) even when all science possible to a particular species is taken
into account. Could it be, then, that your shifts from ontological continuities to causal structures support the idea that the 'truths' we can reasonably
hope to come by are projections of some sort, from a larger world, inaccessible to us even with science taken to it Peircean limit? In other words, we
may see, but through a glass darkly. Yet how could it be otherwise? That a
mind evolved by natural selection, on this infinitesimal region of space-
time, could reach 'truth' on the micro-, the macro-, and even the psychological scale, seems to me like the greatest miracle of all, greater certainly
than the old theologian's argument of the mind having been coordinated
to recognize the truth. That the evolutionary epistemologists have argued
realism out of this is understandable to me only in the limited realm of
direct human experience, and even there not so completely or unproblem-
atically. But this old argument goes all the way back to Darwin and Wallace.
Is it dead now by fashion or by reasoned conviction?

Thus, the lack of visualizability and understanding of modern science
may have to do with the same sort of situation as a dog's, hearing the
words we speak, maybe even recognizing some words, but being totally
and permanently in the dark as to what it all means—being outside 'our
world.'

Here, then, is the anti-transcendence argument entering again. The
theoretical or faith-based hope for full transcendence escapes me, though I
am by nature inclined toward it. Still, partial transcendence does not seem
ruled out (as I understand you to be saying). And then, the questions are:

How partial? How much? What is secure? And how does one make the relevant distinctions? A big program, but one that I think is something like what you have been doing. No?

These questions can also be posed in the language of CRE: In CRE a distinction is made between 'correspondence,' which is essentially a synchronic relation between theories in different realms, and 'convergence,' which is a property of a historical sequence of theories; Peirce takes this sequence to be ideally infinite, which introduces complications.

Finally, to come down to earth a bit, in this vast context, and to pinpoint one of its key features, I return now to the question of how much of Bohr's breakthrough (1913–1923) is still a permanent insight into the nature of the H atom and of atoms in general? If one can show that something permanent and truly enlightening was achieved already then, then it seems to me that the case for partial transcendence is stronger.

Dialogue Part I

A.S. This distinction between 'correspondence' and 'convergence' is correct and important. I have criticized Fritz Rohrlich for conflating these concepts. However, the concept of correspondence is sometimes useful in making a case for convergence, when a successor theory concerns a broader domain than a predecessor theory; but this is not the general situation.

M.E. The preface (CRE) raises several difficult questions concerning convergence: (1) whether a mathematical concept of convergence makes sense when applied to a sequence of theories differing in principles, laws, and concepts rather than in numerical predictions; (2) if this difficulty is overcome, can a Peircean justify neglecting the possibility of endless oscillation (or other kinds of divergence) in the sequence of theories?

A.S. As to (1), we must recognize that the concept of convergence, which in turn presupposes a concept of 'distance' between two theories, is used by Peirce and other methodologists metaphorically and impressionistically. A few methodologists, including Popper and Niiniluoto have tried to define distance between two theories by comparing sets of numerical predictions of the two theories. I suspect that this enterprise is scholastic, infected by inevitable arbitrariness, and sterile. What seems to me more promising is one or the other of the following, which also help to explicate 'correspondence': (a) the successor theory is more general than the predecessor, but under clearly stated conditions the former implies the approximate validity of the latter—e.g., the approximate validity of geometrical optics when appropriate conditions on the ratios of wavelengths to size of apertures are assumed in wave optics (Fresnel and Kirchhoff); (b) both predecessor and successor theories are formulated in a common mathematical language, which permits clear formulation of agreements and differences in structure (e.g., Cartan's use of modern geometrical concepts of affine, conformal, and metric structure to compare Newtonian gravitation theory with general relativity). Procedure (b) can be useful for carrying out procedure (a) with precision.

Peirce's program is indeed loose and conjectural. A serious attempt to clarify and justify it requires at least three major components.

(1) A historical investigation should be made of the sequence of respected theories in a given domain up to the present, because the claim that there will be convergence in the ideal infinite limit is clearly a vast extrapolation from the historically known finite sequence. The early part of the sequence (antique, medieval, pre-Newtonian) may very well not satisfy good relations of correspondence between successor and predecessor theories, because scientific methodology itself was in flux at that time, and standards for experimentation and inference on the basis of data were crude. If we were taking the concept of convergence literally in a mathematical sense, it would be trivial to neglect the early part of a sequence, since convergence only refers to the infinite tail of the sequence. But since our treatment of the infinite sequence is by logical extrapolation from a finite set of historical theories, it is nontrivial and important to give a criterion for bracketing a considerable part of the finite sequence up to the present. To establish an 'honorable retirement' of a predecessor theory in favor of its successor, it must be shown at least that the former agrees well with the experimental predictions of the latter in a well-formulated domain, which should coincide well with the domain in which the predecessor theory was well tested. Correspondence of concepts, in the sense discussed above, is desirable, but perhaps it is not necessary. The sequence of theories of heat from Carnot, to Clausius, to Gibbs's thermodynamics of phase equilibrium, to Tisza's statistical thermodynamics seems to me exemplary as a finite sequence which can serve as an evidential base for an extrapolation to the future. An equally good example is provided by the sequence of theories of light from Newton and Huygens through Fresnel, Kirchhoff, Maxwell, Planck and Einstein, to modern quantum optics.

(2) The extrapolation from a finite sequence of theories to the ideal infinite sequence of theories can be made reasonable only in the context of a full inductive logic, which treats, *inter alia*, the assignments of probabilities to universally quantified statements on the basis of a finite body of data. Popper thought that except for universal state-

ments that are analytic, Bayesian probability theory must assign zero probability to universally quantified statements. But Popper underestimates the flexibility of the Bayesian framework. Jeffreys and Wrinch, and then Jeffreys alone, provide outlines of procedures for assigning nonzero prior and posterior probabilities to generalizations. What these procedures are and how they are justified are discussed in my papers "Scientific Inference" and "Reconsiderations on Inductive Inference" in Vol. 1 of *Search for a Naturalistic World View*. I think much more work is needed along this line, but these papers are at least suggestive. I still think that a logic of extrapolation cannot dispense with a 'pragmatic justification,' ("Scientific Inference," Sect. 4), and that proper interplays of an empiricist and a rationalist component in induction, and of an *a priori* and an *a posteriori* component, are essential ("Reconsiderations").

(3) One possible objection to Peirce's vision, which was not mentioned in your paragraph on convergence in CRE, but which is pervasive in antiscientific philosophizing, is the radical gap between *any* theory and reality: Any theory is expressed in language, is articulated, and separates subject from predicate; reality is distinct from language, is not something articulated, and is unified, with all predicates inseparable from the subject (cf. Bradley in an older idealist tradition, I think Heidegger, and who else?). My reply is based upon Plato: Forms are essential to reality (a less idealistic statement than Plato's view that forms are pre-eminently real), and forms are somehow ingredient in the Receptacle. This role of forms ensures that articulation does not *ipso facto* falsify reality, though of course slovenly articulation can do so. The immense success of mathematical physics is *a posteriori* evidence that physicists collectively have found a very promising way to articulate physical reality. When Beerbohm's Enoch Soames says, "Life is web/ Nor warp nor woof is/ But web only," he is speaking very amusing gibberish.

Here are some further comments on CRE. First, a general comment, for the purpose of removing an obstacle created by ourselves: We are both believers in the progress of science, once high standards of experimentation and reasoning were established, and we both believe that Kuhn has overemphasized the discontinuities occurring in scientific revolutions. But surely a judicious balance is needed between an emphasis on continuity and a recognition of discontinuity. There is no doubt that quantum mechanics radically breaks with classical mechanics, even if a carefully formulated correspondence principle holds (Ehrenfest's theorem, approximate recovery of classical behavior for large quantum numbers), and even if geometric symmetries play similar roles in both classical and quantum mechanics.

In paragraph (1), regarding the relation between predecessor and successor theories, some rough criteria were given for the 'honorable retirement' of the former. These criteria seem to me to apply quite well to the cases you regard as problematic, such as the relation between Newtonian and general relativistic gravitation theory. In particular, using Cartan's geometry, a common formal framework can be formulated in which their agreements and differences can be formulated with precision. But you may object that there still is a difference in ontology between a fixed space-time and a space-time subject to dynamical modifications. At this point, I'm inclined to fall back on two propositions: first, the Platonic thesis of the importance of forms, and second, the thesis expressed in "Reality, Causality, and Closing the Circle" that ontology is inseparable from causal relations. Both of these theses are useful as antidotes to a fetishism regarding the 'stuff' of nature. The relation between classical and quantum mechanics I already talked about above.

You rightly regard incommunicability as a thesis largely passé. One can add some considerations. One was suggested by Howard Stein, that technicians can be trained to gather data in a laboratory without knowing any theory underlying the phenomena. Another is the historical example that Maxwell drew upon Stokes's measurements of viscosity of gases as a function of temperature to support his kinetic theory of gases, and Stokes's measurements were made without any commitment either to a continuous or a discrete model of gases. In my paper "Is Observation Theory Laden?" I conclude with a defense, based on psychological observations, of the logical positivists' view that a language of the ordinary world can be used as a metalanguage for reporting experiments in a way that is not biased by the theories being tested. You are right that incommensurability regarding concepts is a deeper matter than incommunicability, but I've tried at least partially to answer you above in my Platonic discussion of mathematical frameworks which can accommodate and permit comparison of theories which on the surface are utterly incomparable conceptually.

M.E. CRE defends the desirability of *Anschaulichkeit*, and cites eminent protagonists of it.

A.S. Yes, *Anschaulichkeit* is pleasant when we can have it, but it is not indispensable for scientific understanding. In fact, psychological investigations of our perceptual and conceptual faculties may exhibit fairly clear limits on *Anschaulichkeit*, well inside the bounds of scientific understanding. We have additional tools. One is mathematical structure; another is the hypothetico-deductive method, which usually tacitly assumes something about causal relations. A third tool is hardly ever discussed, because it sounds like cheating. It is opportunism in shuttling back and forth between intuitive, but manifestly oversimplified models, and mathematical descriptions which correct the oversimplification. I do this often.

M.E. How do we answer the (Germanic) claims that practical success has nothing to do with truth?

A.S. By a philosophical appeal to Bayesianism. H says that the practical success of electrodynamics in building dynamos and motors is due to the objective truth of electrodynamics. H′ says that the objective truth is irrelevant. Give to H and H′ prior probabilities of the same order of magnitude. The decision between them is then based on the posterior probabilities, which depend on likelihoods. If H′ is assumed, the successful design of a dynamo using electrodynamics is a coincidence—hence a negligibly small likelihood for a large sequence of successes. If H is assumed, a sequence of successes is not a coincidence, and hence the likelihood of a long sequence of successes is at least moderately large. Ergo. . . . This is crude, but it is on the right track and can be refined.

Dialogue Part II

M.E. CRE, section on Evolution, points out an apparent tension among several ingredients in your program: the adherence to an evolutionary account of human cognitive faculties, the acknowledgement of Bacon's analysis of the "idols of the tribe," and the Peircean optimism concerning convergence of science to the objective truth. Our faculties are the product of trial and error in the evolutionary laboratory, with selective pressure operating stochastically (not deterministically) towards sufficient (rather than optimum) aptitude for the practical matter of survival of the lineage. Evolutionary epistemologists like Donald Campbell have emphasized that a considerable degree of sensitivity to the objective facts about the environment is conducive to survival. But they are simplistic. Some traits contrary to objectivity may also be conducive to survival, such as myth-making that promotes group solidarity and raises morale. Furthermore, sufficing may have the consequence of consolidating perceptual and cognitive mechanisms which are less than optimum—which is an evolutionary explanation of "idols of the tribe" and "idols of the cave." Hence, Peircean optimism may be justified from an evolutionary point of view so long as progress towards the objective truth is consistent with the 'sufficiently good' structures of perceptual and cognitive faculties, but there is no basis for extending his optimism yet further. A corollary—returning to CRE—is an expectation that *Anschaulichkeit* will break down as science probes the very small and the very large.

A.S. I applaud your sophistication regarding the application of evolutionary theory to epistemology. The only way to answer your argument for a certain admission of pessimism regarding the long range process of science is further sophistication—some discriminations and distinctions, and more detailed attention to cognitive psychology. First of all, biological evolution, eventuating in the human genome, gives only a partial account of the development of human cognitive processes. There is a famous book (edited by James Spuhler) entitled *The Evolution of Man's Capacity for Culture*. The capacity for culture is to some extent 'hard-wired.' Certainly the capacity for language acquisition is an essential part of the capacity for culture, and if Chomsky is right, the capacity for language is quite hard-wired, and hence an important part of the capacity for culture is so. But the *content* of culture is most definitely not hard-wired, any more than the detailed syntax and lexicon of individual languages. Analogously, the detailed cognitive features of a normal individual in a culture surely depend on historical contingencies—the cultivation or non-cultivation of free debate, the attitude towards political and religious authorities, the encouragement of technical innovations, the procedures of education. The scientific revolutions of the Egyptians, the Greeks, the people of the Renaissance and the Enlightenment and their heirs were all cultural achievements. But the relative ease of the internationalization of science is quite good evidence that the hard-wiring of the human genome is amenable to the proliferation of these cultural achievements, especially when they are practically advantageous.

Can we put our finger on some hard-wired human perceptual and cognitive traits that permit, or even favor, the cultural achievement of science? Here are a few suggestions, which I hope can be extended: (1) the coordination of sensory modes, especially tactile-visual coordination—which infantile play seems to foster; (2) primitive visual geometry, susceptible to refinement; (3) the capacity for recursive procedures, essential for human syntax, but with the wonderful side-effects of permitting counting, arithmetic algorithms, and eventually the principle of mathematical induction; (4) some power of causal reasoning and counterfactual reasoning, which permitted the cultural development of hypothetico-deductive logic; (5) curiosity.

How should we assess Bacon's idols, in the light of these evolutionary and cultural considerations? Answer: somewhat as Bacon himself did—as regrettable flaws of individuals and groups, susceptible (culturally) to correction, by the invention of instruments like eyeglasses and metrological devices, and by the improvement of inductive logic. That Bacon's own inductive logic is faulty does not diminish his glory in realizing that the cultural procedures of investigation and education can correct faulty logic!

And how do these evolutionary and cultural considerations affect Peircean optimism? Answer: They indicate the need for caution, for it is not hard to write cultural scenarios that strengthen the idols—indeed, history has written some of the relevant scenarios. But if the cultivation of inquiry is not hampered for cultural reasons, is there anything in the hard-wired aspects of human cognitive faculties that undercuts Peircean optimism? With certain careful reservations, which I shall come to, the answer is no. As long as there is direct or indirect causal connectedness between a domain of reality (distant galaxies, physics below the Planck threshold, the human psyche, God), the hypothetico-deductive method carefully used can indirectly reveal properties of that domain. They may not be *anschaulich*—here there may be hard-wired barriers to one mode of knowledge—but as long as there are causal connections, there is no rigid barrier to indirect knowledge. Whether there can possibly be a domain that has no causal connection to the common sense world of human experience may be a good candidate for what the Vienna Circle called a *Scheinproblem*.

M.E. Might not a naturalistic epistemology admit the possibility of horizons beyond which there are existent domains which are inaccessible to human inference, without confronting a *Scheinproblem*?

A.S. Yes, this seems to me conceivable, in several ways. (1) It seems to me conceivable that the very distant parts of the universe, or the physical structures beneath the Planck threshold, may be plausibly inferred to exist, but the evidence necessary to raise the likelihood of one among the various theories conjectured about these domains is so attenuated by spatial distance or distance in scale that we cannot anticipate finding it with feasible apparatus. As a consequence there would have to be some retrenchment of Peircean optimism, but not a complete abandonment of it. Effective field theory is willing to sacrifice a precise characterization of the physics at very small Compton wavelengths, by showing that the predictions of the effective theory at longer wavelengths—perhaps those of nuclear lengths—are insensitive to the details at shorter wavelengths. Asserting an effective field theory is then tantamount to asserting a cluster or disjunction of more detailed theories, but the disjunction—though less informative than any of its disjunctive components—may nevertheless be a good approximation to an objectively true disjunction. (2) It is conceivable that the choice between alternative cosmologies can in principle be settled only by facts from an incredibly small interval after the Big Bang, facts whose implications are blurred as the universe ages and expands. If so, we may never be able to decide on the basis of humanly accessible data among a cluster of theories with very similar predictions for the developed universe. My subjective opinion is that this is not a worrisome consideration. We are not aiming at a chronicling of the entire universe in detail, but at its main principles. And it seems to me likely (subjectively) that a principle will have pervasive implications. (3) An entirely different consideration is psychic privacy. We have, I think, reason to believe that there are states of consciousness that cannot be exhaustively characterized by the physical state of the nervous system, and also reason to believe that other organisms than the self possess these states of consciousness. There certainly can be evidence and inferences to the effect that certain correspondences exist between the states of consciousness of one person and another, thereby accounting for apparent communication and coordination of behavior. But the detailed contents of consciousness of another person may lie beyond a horizon. I do not know whether the *quale* aroused in your sensorium by the physical stimuli from a patch of grass is the same as the *quale* aroused in mine, and I can't conceive of a way carrying out the comparison. Thus, there seems to be a horizon to my personal knowledge, or else raising the whole question is a *Scheinproblem*. . . . This is a problem in philosophy of mind which I find confusing. But is the existence of this horizon dangerous to Peirce's epistemological program? I think not, since he is mainly concerned with convergence of our theories to the objective truth, and theories are formulated more about structures than about *qualia* (though the generic existence of *qualia* may be asserted by a psychological theory). Peirce's discussion of firstness, secondness, and thirdness seems to agree with what I just now stated.

M.E. Let us return to the relation between practical success and truth discussed in Part I of our Dialogue. There are two ways in which the opponent of your Bayesian treatment might move. (i) The opponent might say that the opposition of H and H' is not the only way of drawing the pattern of opposition. Instead, H' can be regarded as an infinite disjunction of theories, all contradictory to H, but also to each other, and one of them may be as successful as H or better in accounting for evidence (including the successes based upon H). (ii) The opponent may dismiss as irrelevant the whole Bayesian machinery and indeed any other machinery which aims at inferring truth on the basis of evidence, because considerations of commitment, value, heroism, etc. override slavishness to objectivity.

A.S. As to move (i) I refer you to "Scientific Inference," where I defend a methodology that assigns non-negligible prior probabilities to all seriously proposed hypotheses concerning a matter under examination, but does not treat unarticulated hypotheses on the same footing as explicitly articulated hypotheses. The 'catch-all' hypothesis, asserting that something else is true other than the explicit articulated hypotheses, deserves non-negligible prior probability, but it is the disjunction of an infinite set of unarticulated hypotheses, and each of these does not deserve non-negligible prior probability. A pragmatic justification of induction demands treating articulated and unarticulated hypotheses differently. As to (ii) I'd say that someone who for any reason, good or bad, excludes the search for the objective truth from his or her values has withdrawn from the game of scientific methodology. Such a person can be answered by former Secretary of Agriculture Earl Butts's answer to the Pope: "He no playa the game, he no makea the rules." Can one argue with the ethics of such a person? I think yes, because dialogue, as Plato showed, can open irrational persons to the penetration of rationality. But in order to be an effective practitioner of such a dialogue, one needs to be as adroit as one's opponents, among whom are formidable intellects like Nietzsche and Heidegger.

M.E. After all these excursions into epistemology and methodology, how do you answer my question at the end of CRE about the permanence of Bohr's breakthrough of 1913–1923?

A.S. We need to look at the physics in some detail. A partial answer is honor to Bohr for posing the problem of combining a mechanism for discreteness with a mechanism for continuity, even if the way he does so is opportunistic and *ad hoc*. The honor is augmented by asserting that not only are both discreteness and continuity still combined in the most refined versions of quantum mechanics and quantum field theory, but the problematic character of the combination persists—specifically, in the measurement problem. Another answer is that certain parts of Bohr's opportunistic formalism have persisted, particularly the action integral. (You say that this does not appear in Bohr's first paper but in one soon afterwards. But since the question concerned the years 1913 to 1923, the action integral is still relevant.) And of course Feynman's formulation of the new quantum mechanics is essentially posed in terms of the action integral.

Dialogue Part III

M.E. I will comment on your most recent notes not in the order in which they were written but in the order of importance (in my view) so that, in case I run out of steam, what I leave out will be the less important.

Much of the remaining difference between us is on page 423 of Part II (M.E.) and on pages 423–24 of the same Part (A.S.): I say that an opponent of your Bayesian treatment makes two points labeled (i) and (ii). Then, in (A.S.) you reply to each of these in turn.

Consider (i) first. Implicitly, your reply makes clear that you are speaking from the methodological point of view, or within the methodological framework of concern (in regard to (ii) you make that even more explicit, and I'll get to that shortly). We have no difference whatsoever when we are discussing methodology. Within the framework of methodology, I consider your approach to be the best I know of—a most reasonable, flexible, open-minded, and historically sensitive naturalistic approach, surely the most likely to lead to progress in research (and possibly reveal truths about nature). Yet I would like to point out that what this involves is 'satisficing,' or choosing the 'best' from a finite set of articulated options (which, operationally, is all one needs), and leads therefore, in the first instance, to *acceptability*, not necessarily to 'truth.' As we know, a scientist will 'accept' a theory for all sorts of reasons, and in varying degrees, not necessarily because he thinks it true. The question of 'truth,' I want to reemphasize, would bring us back to the Peircean limit and the question of convergence: *Does the series of satisficing theories converge in all respects deemed important?* (CRE, page 411.)

However, the main issues I am raising are not *wholly* in the realm of methodology. That is to say, they are *not* addressed exclusively to the question of *how science should proceed to maximize progress and the chance of getting as close to the truth as humans can* (and this question is not the basis of the Continental 'critique' of natural science from Kant to Husserl to Heidegger, etc.). The questions I am focused on are: *How close have we in fact been getting? How close can we hope to get? Are we converging in, e.g., our theories of the H-atom? And,* inter alia, *what is 'truth' in this context?*

To pursue this further, I pass now to argument (ii). Here I cannot accept your formulation of what I am trying to say. It is not a matter of 'dismissing' the Bayesian or any other machinery based on evidence, in favor of commitment, heroism, etc. To 'dismiss' inference from evidence, where science is at issue, I take as simply foolish, and I would not wish to debate with anyone who took such a stance. Since I have not succeeded in making myself clear on this crucial point, let me introduce here, for the sake of discussion, *a parallel or analogy to science* (it is, I believe, a parallel that has some intrinsic merit, one that has intrigued me for years, and one that you touched on in your presidential address.[3] It is prompted by your quotation of Butts against the Pope, and your phrase "the game of scientific methodology" (on page 424).

Consider 'the game' of science and scientific methodology on the one hand, and the 'game of law' on the other. You and I agree that neither of these two games is *closed* in the sense that a 'play' or a 'success' in such a game has no value or meaning outside the game itself (as, e.g., in chess). Rather, there is something outside the game itself *for the sake of which* the game was developed: Truth is to the game of science as justice is to the game of law, and both are primordial, having had value for humans long before the corresponding games (in their modern forms) were begun. We know that in the realm of law (policing rules, trial rules, etc.) justice is not always attained, but we believe that a good game of this type is one that maximizes the probability of reaching the transcendent goal. Similarly, society and scientists in general assume that a good scientific methodology maximizes the probability of attaining truth. History shows also that sometimes contrasting methodologies compete with each other until one clearly proves itself superior, e.g., positivism and energetics vs. atomism at the time of Rutherford. We also know that there is a horizon problem in both cases: The justice we aim at in our courts is *human* justice, not some other kind, e.g., not God's; and, similarly, the truth we aim at in science can only be *human* truth, i.e., what is possible for humans in the long run, given their particular neuro-sensory endowment (this was the point of my evolution question regarding the uniqueness of man in being able to *attain*, not just strive for, the ultimate truth about the physical world).

Now, with the aid of the 'law game' analogy, it is possible for me to inject into our discussion the question of *standpoint* or *situation*. This is, of course, a key concept in 20th century hermeneutics, and now I believe that without it I have little chance of posing my questions in a more fruitful way. We all know that one can play the game of law as one plays a game of chess—and some do it just that way. They are technicians of their craft. Neither the individual lawyer, nor the law group he works for, nor his clients need to be in the least interested in the transcendent goal of the game—justice. In a similar way, many scientists, when asked why they do research, say simply "for the fun of it." Science is an enjoyable game. Few, in the recent past (in contrast perhaps to former times) have said that they do so to get at the truth about X or Y. As for the 'game of methodology,' as an academic subdiscipline of the philosophy of science, here I feel less qualified to speak because I have not taken this as my main interest and do not read many papers that are purely methodological in intent. I would assume that a *reflection* on the methods of science is, in general, closer to the question of truth than the *practice* of science, but I would also assume that, as methodology, it concerns itself mainly with the goals already mentioned—progress, satisficing, etc. Clearly these are the most pressing concerns for the 'players of the game,' the practicing lawyers and the practicing scientists (whether or not they care also about the transcendent goals). Are these not the concerns that define the *standpoint* or the situation of these 'insiders'? But regarding the game of law, it is certainly clear that a great many *other people* have a legitimate interest in the matter besides the lawyers, judges, etc., who play. Practically all other segments of society have a serious interest. Yet their interest is not essentially methodological: it is an interest in the presumed transcendent goal of the game—justice (I leave out of discussion, for the time being, the fact that *order* might also rank as a goal, not only justice, without, I hope, degrading the argument). Thus, in the case of the 'game of law' one would *not* say "he no playa the game, he no makea the rules." Many people who don't play the game are given a say about rules. Of course, the analogy is not perfect: In natural science, outsiders are generally not permitted to influence the actual rules of research. But *making* the rules is not the issue I raise; and for my issue the analogy is still good. The various other sectors of the public have a *legitimate* interest in the justice system in general, and in justice in a particular case. And sometimes, even though the rules of the law game have been followed properly, many people do not accept the legal outcome (e.g., the O.J. Simpson trial). Why? First, because they realize that the law game is *only* a game; it could yield the wrong result in any one case, *and* the rules have sufficient flexibility to allow bias—all sorts of human bias—to enter. But in addition—and this gets closer to your counterargument—there may indeed be a completely nonlegal component involved in the appraisal of the trial by particular individuals or groups. This could be faith, intuitive knowledge, and a more general consideration of life.

If the defendant is someone you know well as an honest man, and the evidence is clearly *against* him, *but not 100% against*, then, as a friend, you may think of the matter as follows: If I accept the weight of the evidence, I may be strictly rational, but I shall then have to change my mind about my friend (and perhaps others like him as well). As a result, I will probably change my orientation to life, my attitude towards people; e.g., I may become more cautious towards those who try to befriend me as has this defendant. If the evidence were perfect and complete, reason and objectivity would leave me no choice. But because the evidence falls short of 100%, it would *not be irrational* to consider the matter differently—to use the strategy of *Minimax* in a more general way than it is used in the law. Which way is it *worse* for *me* to be wrong? I will minimize the maximum loss. If he's guilty and I don't accept it, I will be fooled and may live to regret it. But if he's innocent and I go with the evidence, I will have done him a great injustice and betrayed a friendship. I may certainly judge this to be the greater loss. Therefore, I may make the choice 'against' evidence in a *perfectly rational way*; and this way, though partly subjective, does *not* leave out or ignore the evidence brought forth at the trial. That evidence is considered but weighed from a *standpoint different* from that of the law game. Had the evidence against my friend been, say, 95% (in my judgment) instead of only 80%, I might have found myself unable to remain on his side.

In this way, I try to emphasize the importance of *the standpoint* of the one who is *appraising* the significance of the results of the game. For those inside the game, the evidence and the vote of the jury terminate this particular play. For others, not necessarily.

This is in the tradition of the Continental 'critique' of science. All those who wish seriously to understand the significance and meaning of *the results* of science up to our time—the philosophers (non-methodologists), historians, theologians, journalists, policy makers, students, and all other interested outsiders—have a *different standpoint* from the methodologist. They have no intention of 'making the rules,' nor of 'excluding the search for objective truth.' They are content to leave all that to scientists and methodologists. But when it is done by these professionals, and the verdict is given, there still remains the freedom to *appraise* that verdict in a context larger than the game—*without ignoring* the outcome of the game or the evidence brought out by the game. Thus, here too, there is no intention of 'dismissing as irrelevant' the objective, scientific search for the best that humans can discover. Nor is there any implication here, in case an individual refuses to accept, or postpones acceptance of, an established scientific result, that this requires a change in the rules of scientific methodology. What reason compels in a limited context, it may not compel in a larger context. I admit that in the strongest and most objectionable 'critique,' some Continental philosophers suggest that the game of science has as its *attainable* (and even intended) goal—only *order*, not truth. Again, the analogy: for we often hear that social law serves mainly the interest in order, not justice. Some of the really unfortunate recent writing about science is of this sort; and this, of course, I don't accept. I pay attention only to those aspects of the critique that seem more cogent.

If one goes along with the argument above, then the old questions of validation, 'degree of certainty,' are back in focus, because the first thing the outsider wants to know about a particular scientific result, or a particular science, is "How certain is it?" We do not put skeptics about the big bang in the same category as skeptics in regard to the roundness of the earth. How strong the scientific evidence is determines how much room there is to *rationally* widen the context, how far outside the game of science one may go in *one's own* appraisal (science taken into account but not exclusively). As I read the historical record, this was the position taken by Pope Urban and Cardinal Bellarmino in the Galileo matter, at a time, be it understood, when everyone accepted religious belief by the people as an important and larger context. If the evidence were conclusive, said Bellarmino, we might have to reinterpret the Scriptures; but in the absence of such proof it is better, from the larger standpoint, to leave things as they are. It has always seemed to me that however wrong he may have been scientifically, it was neither irrational nor contemptuous of objective evidence to make this judgment, in so far as it was intended for his religious community, not for the community of scientists in their work. A major feature of the Continental critique of science (when cogent) is to ask whether or to what extent the *standpoint of science* is to be primary in life, in all contexts. Outside of purely personal matters, does the standpoint of scientific methodology, taken exclusively, always trump all other standpoints?

Thus, the methodological question is of course relevant, even in an enlarged context. For the methodologist, I take it, probes precisely the validity and procedural questions of science. But if the methodologist can only optimize and the scientist can only satisfice, then the results of these enterprises do not make obsolete all the other standpoints; rather, they *contribute* to them. It is in this context that, in my view, the evolutionary questions come in, raising, within the widest possible horizon, the question of the cognitive limitations on the human species, and the question of how to think about the world (and our science) not just from the human standpoint, but possibly from the standpoint of the evolutionary process as a whole.

Finally, I turn to page 424, where you respond to my question about Bohr. As you know, I asked about this mainly in regard to the section I am writing, as part of a *popular* essay, on his 1913 work. In other words, I wish my treatment to be meaningful for all those 'others,' not just for scientists. I asked, How much of the work on the hydrogen atom, 1913–1923, has survived as a permanent discovery or insight into nature, that can be communicated to the outsider as stable and trustworthy; and, by implication, *are we converging* to the truth about the H-atom? This is the connection with all those arguments above.

Your reply surprised me by its generality, but this may be because you took the question to refer to quantum mechanics, rather than to the H-atom specifically (and atoms, more generally). If the idea of combining discreteness with continuity, and of introducing the action integral as the key, is all that's left, what then is one to say to the outsider? Discreteness, continuity, and the action integral are highly significant to the physicist and the philosopher of science, but what could they possibly mean to others? This is not just an *Anschaulichkeit* problem. Outsiders are certainly capable of grasping some purely logical or conceptual schemes. It is a question of meaning of some sort, and understanding.

In regard to the H-atom, I thought you would grant that, later theories notwithstanding (and especially in light of 'decoupling' on the basis of energy realms), the following 'picture' still stands: the picture of an energy level scheme and the associated ideas of level transitions and probabilities of transition—this, together with the approximate spatial and compositional picture of a heavy nucleus distinct from a cloud of electrons (which determine most of the chemical properties). Not so?

So far, what I have written down (in my popular piece) emphasizes Bohr's instrumentalist moves in fitting a theory to the Balmer formula—the opportunism, *ad hoc*ness, etc., that you mentioned. Thus, I try to connect him historically with others who 'invented' rules to fit sets of numbers. But I would not wish to leave the reader with a picture of Bohr as merely an instrumentalist, because there were very interesting other motives at work. Bohr was thinking deeply and strategically long before he came upon the Balmer formula. He was fascinated by the problem of the stability and identity of matter; and he convinced himself that, in the context of the Rutherford atom, this could not be understood on the basis of classical theory. Such an atom could not possibly return to the original state after many collisions. In a hazy, groping, but nevertheless quite fruitful way, he was also thinking about the relation of the electrons to the periodic table. Is not Bohr a good example of that "sensitivity to truth" that you often ascribe to the scientific community in general? Doesn't this sensitivity have something to do with looking at the problem from many different sides and points of view, and seeing apparently different problems as related? Would it be too risky to say that, whatever the future may bring, we do not expect, as scientists, that this picture (within the appropriate energy realm), will be falsified or replaced qualitatively?

There are other things to comment on in the Dialogue you sent, but now I have run out of steam for the time being.

A.S. On page 425 the notion of 'game' is introduced, along with the distinction between 'open' and 'closed' and the examples of the 'game of science and of scientific methodology' and 'the game of law.' All of this is valuable for our discussion. The hermeneutic notion (or notions, are they the same or different?) of 'standpoint' or 'situation' is also valuable if proper precautions are taken. Two different insiders in the game of science may agree in being concerned with progress or with satisficing, but for different reasons. One may simply not be interested in the external standpoint of truth, but has simply internalized some norms of professional practice. The other may be vitally interested in the external standpoint of truth, but may regard progress and/or satisficing as criteria or indications of coming closer to the truth. We could also perform some depth analysis on the first group and say that science is 'fun' for members of this group just because they are playing for a high stake, the stake being truth; they may talk cynically or they may simply be inarticulate about their motivations, but in the backs of their minds is the adumbration of the value of truth and the belief that progress and satisficing are useful tools for the achievement or partial achievement of it. Thus, a case can be made that the distinction between the internal and the external standpoint is easily blurred and can hardly help being blurred.

Here is another consideration in favor of breaking down the distinction between internal and external standpoint. An articulate devotee of the external standpoint—e.g., a scientist who has brooded over Kant's transcendental philosophy but doesn't completely accept it— may claim that the hypothetico-deductive method is a good intellectual instrument not just within the phenomenal domain but within the noumenal domain. Peirce was just such a philosopher, and he regarded the hypothetico-deductive method as a tool of metaphysics. Gödel explicitly invokes the hypothetico-deductive method (e.g., in his essay "What Is Cantor's Continuum Hypothesis") as a proper tool of set theory and number theory. Query: Does the invocation of modes of inference characteristic of the natural sciences into the study of metaphysics and higher epistemology denigrate these 'higher' inquiries or does it 'elevate' the methods themselves? You know that I prefer the second answer, when I call the present age one of the "golden ages of metaphysics" just because of the fruitful use of scientific findings for sharpening traditional metaphysical questions and providing evidence for one answer or another.

In your paper "Language and the Double Hermeneutic in Natural Science" you multiply senses of 'standpoint' in a different way, in your discussion on pages 81–82 of 'interpretations' of scientific formalism. Here it seems to me that you are separating 'standpoint' from the value theoretical considerations implicit in the discussions of the hermeneuticists and are applying the concept within the domain of cognitive considerations: the roles of sense perception, integration of percepts, memory, imagination, language, mathematics, concept formation, inference. The dispute between Mach and Boltzmann, or between the energeticists and the atomists, did not seem to involve differences in values or cultures, but different epistemological assessments of the interplay of these various cognitive functions of the mind. To be sure, there were influential people who conflated cognitive standpoints with value-theoretical ones—notably Lenin, in his "Materialism and

Empirio-Criticism," but I regard this kind of conflation as seriously obscurantist, tendentious, political, and just plain bad!

I now turn to page 427 where you briefly discuss a well-known strategy of decision theory, the Minimax method. That method, like applications of Bayesianism to practical questions, makes a sharp distinction between probabilities, which are cognitive or epistemic in some sense (subjective, objective, etc.) and utilities, which are value-theoretical. Hence such methods really do involve the simultaneous deployment of two different kinds of standpoints. Now I am a dissident Bayesian, and I recommend something more 'existential' than evaluation of utilities. It is easy to envisage Kierkegaard's scorn for the evaluation of utilities. He was concerned with commitments as great as the wager of one's life, or Abraham's faith. To make tables of expected values of gains and losses, by multiplying utilities by probabilities at each entry in the matrix, is to cheapen the whole enterprise of living a serious life. Well, he is right. Such tables are sensible only on matters of tactics, once the constraints of one's commitments are acknowledged and articulated, but not for making the commitments. Otherwise one is a moral slob. There is still a role for Bayesian decision theory within the constraints of commitment, but I would characterize it as "Kierkegaardian Bayesianism" (almost an oxymoron, but not quite!).

Later, on page 428, you announce "the freedom to appraise the verdict in a context larger than the game." How is this to be done? Not without entering into the examination of one's life and/or the systematic study of ethics. I am not sure whether to use "and" or "or." I'm inclined to the latter, because I am acquainted personally with highly moral people who have surely examined their lives, though they may be inarticulate about their findings. On the other hand, I don't like to make a virtue of "tacitness" as Polanyi does. If one can articulate, then so much the better. Hence the Platonic philosophical examination— using all the information at our disposal, including the theory of evolution and the findings of psychoanalysis—may augment our self-knowledge and fine-tune our adoption of an ethical standpoint. This is the point of view I was urging in our debate in *Zygon*, years ago. With this view of meta-ethics I have a strong distaste for a set of moral commitments that condemns or disregards scientific evidence. To do so is deliberately to hold on to the comforts of immaturity. A wonderful defense of embracing the scientific method as an instrument for the "fixation of belief" is given in Peirce's essay with that title.

This consideration leads to an answer to your question on page 429, "whether or to what extent the standpoint of science (when cogent) is to be primary in life, in all contexts." That standpoint is not everything—it is not a substitute for the internalization of the outlook of fairness or empathy—but it is a valuable, and perhaps indispensable, complement to all other ingredients in the formation of a moral *Weltanschauung*.

On another matter, the phrase "being able to *attain,* not just strive for, the ultimate truth about the physical world," on page 426, needs some clarification before one can address it with clarity. What is 'ultimate'? Does it just mean 'literal,' so that the statement "Parameter x lies between 2.0 and 3.5" is literally true if the exact value of x is 2.9? Does it mean 'exhaustive'? Does it mean 'accurate' at the finest level that exists in nature? Whether we are able to attain the ultimate truth in some of these senses, e.g., the second, depends not only upon our faculties and the successful history of science in the future, but also on the structure of nature. For instance, if David Bohm's speculation (in his book *Causality and Chance in Modern Physics*) that there are infinitely many layers of physics of increasing fineness is correct, ultimate truth in the second or third sense would appear to be inaccessible to the human race. Also 'attain' is ambiguous. One might attain a true proposition in the sense of formulating it and preferring it, upon evidence, to all its well-formulated rivals, and nevertheless not be able to demonstrate its truth definitively. In *Naturalistic Epistemology: A Symposium of Two Decades*, I wrote, "Rather than agonize . . . if the scenario should turn out to be true, I recommend the philosophical strategy of Crossing That Bridge When We Come To It" (294).

Dialogue Part IV

A. Some Preliminary Clarifications: On science as a 'game' with internal and external standpoints (re: page 430)

M.E. I intended the completely internal and external standpoint to be *idealized* extremes. Certainly there are scientists who can be fully immersed in the game, once a 'play' is in progress, yet sit back, before and/or afterwards, and take the external standpoint, e.g., of the philosopher. How often such switching of standpoint occurs depends on the individual. (Of course, your case, that of an "experimental metaphysician," is in a class by itself.)

My point in making this distinction was not primarily a sociological one. Rather, it was to emphasize what follows from the idea of 'open games'—that being a successful 'player,' and insider, does not *necessarily* imply concern with the external goal. Nor is it *required* to be a player in order to concern oneself with such a goal. The only sociological element I can add is the minimal: I believe I can name individuals on both extremes. I do not claim that they are in the majority. But I do think you are too optimistic when you suggest that those who say that they practice a science because "it's fun" really mean that it is fun to play for the high stakes of *truth*. We all know people who play *closed* games, such as chess, very seriously; some play them professionally. Their stakes are the 'fun' of *actually playing* (intellectual enjoyment, "brainsport" Einstein called it), also the rise in their *rank*, and possibly money. Could it be that there are no such scientists?

Thus, I would not want to *blur* the distinction, but rather say that in any individual, or even group, the two standpoints can be *combined* in different ways. The underlying idea of such open games is that the *game as a whole* approaches the external goal, even if the players are mostly insensitive to that goal.[4] In science, then, the key question is whether *this* game as a whole (methodology included) gets closer to the *truth*.

B. STANDPOINTS, HERMENEUTICS, AND SUBJECTIVITY (RE: PART III, PAGES 430, 431)

M.E. First just a misunderstanding: I *did* cite the competition between energeticists and atomists as one between methodologies, not 'standpoints' (page 426).

But you also suggest that, in contrast to 'the hermeneuticists,' I am separating out (in my paper "Language and the Double Hermeneutic in Natural Science") the concept of 'standpoint' from the realm of values, or 'value-theoretical considerations,' and applying that concept to the cognitive realm only. It is certainly true that I deliberately confined myself to the purely cognitive realm in that paper. But to relate that paper to "the discussion of the hermeneuticists," I have to go afield a bit to disentangle several aspects of the very terms "hermeneutics" or "hermeneutic." (I hope you'll indulge me on this, not only because it's relevant to some key issues in this discussion, but because I want to make clear that I did not develop my interest in hermeneutics out of some distaste for objectivity or truth. Quite the contrary.)

First, hermeneutics sometimes designates the art or *methods of interpretation*; other times, it refers to the interpretation itself, especially if done with an awareness of hermeneutics in the sense of method or methodology (one of those unfortunate ambiguities of which Continental philosophy is so full). Hermeneutics seems to enter philosophy seriously in the early 19th century with Schleiermacher, whose interpretive interests ranged from Plato to theological writings. But the most general interest, from the start, was the problem of *dialogue*—dialogue in its widest sense, whether between two living individuals or

between a reader and an ancient author: How does one come to understand the *meaning* of the words or text one faces?

Second, an important change now has to be taken into account
between 'romantic' or 'classic' hermeneutics of the 19th century (which
continues to our time!) and the radically altered versions introduced
early in the 20th. In the earlier sense, the goal of interpretation was to
recover the *true meaning* intended by the author; and therefore,
progress in the interpretation of a particular text could be envisaged as
closer approach to that goal. The subjectivity of individual interpreters
is, of course, recognized, but that is what *method* is for—to purify oneself of various preconceptions, parochialisms, idols, etc., and then to
have one's work further improved by the next interpreter. This is not
only *a purely cognitive process*, but also one that is guided by *objectivity*,
in the sense that the true and usually unique meaning is *there* in the
text, independent of the reader-interpreter. When that meaning is not
obvious or clear, it may be necessary to go beyond the text to its appropriate context. That context, which then defines the *standpoint* of the
interpreter, can only be one—the standpoint of *the author*. To reach it,
the interpreter may be forced to suppress his own standpoint, but that
is only a further step in the self-cleansing or 'detachment' that is
required for objectivity. In recent times, this kind of hermeneutics is
represented by Emilio Betti and E.D. Hirsch, who have vigorously
debated the 'modern,' post-Heideggerian hermeneutics of Gadamer.

In the 'modern' version, the centrality and uniqueness of the
author's meaning are dropped or largely de-emphasized. It is assumed
that the author (of nontrivial texts) may have had multiple intentions
and/or an incomplete, even mistaken understanding of his own text.
Therefore, far from needing to be suppressed, the standpoint of the
reader-interpreter is now seen as a necessary element in reaching an
appropriate interpretation. Ideally, the reader should "fuse" his own
"horizon" (context) with that of the author, and from this higher
standpoint, try to find plausible or defensible interpretations, including
some that the author may not have had in mind. Thus, it can no longer
be said that one interpretation is closer to the true meaning than
another; it could still be 'better' by some other criteria, or it could also
be an alternative interpretation. In this version of hermeneutics, interpretations acquire a serious subjective component that cannot be completely eliminated by any process of self-correction, though all reasonable attempts of this sort should still be made. One way of looking at
this change is to consider the following question: *Where* is the message
(meaning) of the text? Is it exclusively in the mind of the author, with
language as the medium of transmission, and reader only as the 'receiving' destination? Is it fully in the text alone? Is it in the combination of

text and reader? The 'modern' view (Gadamer) is that there is a greater wholeness involved here; writer, text, and reader have to be taken as a unit, though sometimes, of course, the writer's full standpoint may be unavailable.

This, as I see it, was the crucial change in the 'modernization' of hermeneutics; and here, because of the inclusion of the *standpoint of the reader-interpreter*, there appears the *possibility* of the involvement of *values* in that standpoint. These may be aesthetic, cultural, etc., as well as purely cognitive values. Recently, of course, we have seen some absurd vulgarizations of this notion by those who claim, e.g., that 'women's physics' somehow differs from ordinary ('men's') physics, etc. But we need not spend time on such things.

Criticisms by people like Betti and Hirsch of the more serious ideas of hermeneutics were of this sort: The notion of the reader-participant is arrived at by *conflating* two different senses of 'meaning'—verbal meaning with significance. The former lies in the combination of author and text, the latter belongs to the standpoint of the reader. Meaning is determined only by author and text; the reader's *first* task is to recover it correctly. As a *second step*, the reader may then look for the *significance* of an already interpreted text in light of his own situation or concerns. The 'modern' reply is that this separation cannot always be carried out. And in science, I would point, for example, to Planck's initial 'derivation' of his radiation formula. Even to this day, historians argue about just how Planck understood his own energy quantization—how physically as opposed to merely mathematically, how necessary and irrevocable was it in his view, etc.? Didn't many physicists wave their hands at what 'Planck really meant' or what his 'derivation' really means, but took it as *they* saw it, as meaning what they thought it *should* mean, as it might be useful in their own work? And aren't there plenty of other such examples, e.g., Einstein and Bose?

A further change in the philosophical development of hermeneutics was its 'universalization.' Not only in texts of all sorts; not only in law, literary criticism, and theology; but in every aspect of human life, interpretation is present! Interpretation becomes part of the very *being* of a human. It acquires existential and ontological aspects. Nevertheless, since most of this was carried out in the context of European *Geisteswissenschaften*, natural science continued to be excluded from this universality. What happens if we try to include it? One possible approach is that of my paper "Language and the Double Hermeneutic in Natural Science."

And thus I return to your point about separating what I do from what the hermeneuticists seem constantly to emphasize—values and culture. You raise a cautioning flag not to conflate "cognitive stand-points with value-theoretical ones," as did Lenin (pages 431–32). In view of the paragraphs above, I think you see now the first part of my answer. In that paper, I tried to point out that hermeneutics is com-monplace in science (stage 0 and stage 1) in *its 'classic' version*. But then, in addition, I wanted to show that certain aspects of even its 'modern' version—in which the reader co-determines the meaning of a text—also occur in examples such as Hertz and Feynman. Thus, the scientific 'text'—a theory, a report on experiments, a proof, etc.—con-tains within itself *interpretive possibilities* that are not realized until a reader (interpreter) engages them in a genuine dialogue. I assume that physicists take all this for granted, but don't like to use words like 'interpretation' for fear of undermining the claim of objectivity. Outsiders, on the other hand, are often unaware of it, so for them it is perhaps worth saying. I deliberately stayed clear of nonscientific values or cultural differences to underline that 'hermeneutics' should *not* be associated with only its most radical or questionable forms, and cer-tainly not with its absurd forms, which, unfortunately are practiced by some high-profile academics like Stanley Fish.[5]

C. THE FACT-VALUE DICHOTOMY (ONE MORE TIME, BUT THIS TIME IN EARNEST!)

M.E. Your Criticisms:

On pages 431–32 you include several statements on the relation of the cognitive and scientific interest of the human being to his norma-tive, or valuing, interest (including commitments on a life-orienting scale). In this context, you refer again to your reply to my old *Zygon* article, in which there is a section with the same title as this Part C, above.

Big issue, of course; and I surely have no expertise in it, but per-haps that's just the point—the issue is too big to be left to the experts alone. We all get interested in it sooner or later, and can hardly help formulating something like a viewpoint on it. In response to your admonitions and criticisms, I will try below to give a coherent, if lim-ited, account of why I appear to you so often to be "trying to narrow the gulf" between these two types of concerns, what this 'narrowing' involves and does not involve, and what my 'standpoint' is in the mat-ter.

But first, I must point out my problem in *understanding what you have been saying* about it. In that criticism in *Zygon*, you include the fol-lowing introductory sentence: "It should be clear why I resist Eger's attempt to narrow the gulf between the natural sciences and ethics." The reason you give—in brief—is the difference between the "objective truth" of science, which is "there to be found out," and the "problematic" nature of the "ontological status of normative facts" (Vol. 1, 325). However, on page 432 you say "This is the point of view I was urging in our debate in *Zygon*. . . . I have a strong distaste for a set of moral commitments that condemns or disregards scientific evidence." Unless I miss something crucial here, this does appear as one kind of narrowing of the famous gulf. Yet, in the previous paragraph on the same page, you seem again to be trying to define the demarcation: "Such tables [of gains and losses] are sensible only on matters of tactics, once the constraints of one's commitments are acknowledged and articulated, but not for making the commitments. Otherwise one is a moral slob. There is still a role for Bayesian decision theory within the constraints of commitment, but I would characterize it as 'Kierkegaardian Bayesianism. . . .'"[6]

A.S In this paragraph it seems to me that you have confused two different philosophical problems—(1) the relation between 'is' and 'ought,' or alternatively, the ontological status of norms; and (2) the extent to which values can be expressed as utilities, which would render them susceptible to quantitative treatment in Bayesian or Minimax or some other version of quantitative decision theory. I think these are distinct matters, and I am not sure what I said to make you conflate them. Regarding (1) I think we both think there is some kind of bridge between 'is' and 'ought,' but that is a subtle one, resting on evolutionary, psychological, and social considerations, and we may differ regarding the constitution of this bridge. Problem (2) is entirely within value theory. I believe, borrowing ideas from both Aristotle and Kierkegaard, that a person's commitment to a kind of life is the overarching value judgment, which cannot be treated with anything approaching psychological adequacy in terms of utility; but that subsequent to such a commitment utilities can be calculated for the purpose of practical decisions such as military tactics and business investments. Von Neumann and Morgenstern have indeed accomplished something substantial in their *Theory of Games and Economic Analysis*, but it is not a surrogate for an ethics. I am confident that you agree, and also agree that this matter has little or nothing to do with the ontological status of norms.

M.E. I could certainly make a stab at *interpreting* all this in a way that harmonizes the apparent contradictions. For example, I could take your "Bayesian decision theory" as meaning strictly and only that, or, at most, some similar calculation machinery, but *not* a general process of rational inference from evidence. However, in an ongoing dialogue, it is best for me to leave that to you. I do want to understand clearly your position on this matter of the dichotomy because for me it is a big, big issue, and not just one of professional or philosophical or academic interest (although I do not intend to denigrate by any means the purely technical aspects of it). As far as the philosophical tradition goes, I can say here only, and without justification, that I have always been skeptical about the classic trichotomy of the cognitive, the moral, and the aesthetic. I think parts of it have done a good deal of harm in the modern world (e.g., in art, and cf. below). Of course, I do not quarrel with the claim that the 'ought' cannot be logically derived from the 'is'; nor do I have much sympathy with the various attempts to construct 'evolutionary ethics.' But I think that this first order disjunction, on the logical plane, tends to overshadow much of importance that still remains. As for the ontological approach (normative facts), I agree about the undecided, problematic status of the situation to which it has led; but I think we simply cannot afford to accept that situation as it stands. In other words, there may be something like a *moral imperative* for pursuing this issue with even greater effort than before.

D. A HISTORICAL STANDPOINT

M.E. To make this plausible, the standpoint from which I start now is not analytical but mainly historical. I find in human collective experience an indisputable fact of overwhelming, painful significance, of supreme moral importance, and even of current relevance. I take this fact to be a strong indicator of multiple underlying interconnections between the cognitive and the moral, some of which may well be so obvious that they are taken for granted by many—yet receive attention from few. I am referring here to the repeated instances of large-scale moral crimes preceded by, facilitated by, and indeed promoted by *large-scale untruths*. In the Jewish experience alone one can easily see the big lies at work, paving the way for the major slaughters; but it is evident too in the case of slavery in America, and numerous other examples. I do intend to move beyond the historical indicator; but as one example of how well all this is known and yet how easily shoved aside in politics and public life (due to intellectual neglect?), let me cite the interesting case of Shlomo Avineri, the Israeli academic and politician.

In a recent opinion piece, in the *Jerusalem Post*, Avineri related what he experienced long ago, when he was a delegate on a committee charged with the implementation of the cultural aspects of the Egyptian-Israeli peace treaty. This article was in the context of Arab 'incitement,' of which the Israelis are perhaps taking greater notice lately. Now, the significance of it in my view (though that was certainly not *his* thrust) is the fact of what *Avineri already knew* at the time of these committee meetings, because of *who* he is. As a historian, philosopher, expert on Hegel, and all-around Jewish intellectual, he surely knows more than I about the epoch of 'incitement' in the German-speaking world, during the century prior to the Holocaust. He knows well that it was not just the rabble, but, on the contrary, intellectuals at the highest level whose ponderous 'histories' and 'philosophies' changed the liberal, Enlightenment attitude toward the Jews precisely within those classes that have the most influence—the bourgeois, the professionals, the journalists, writers, and academics. He knew well that it was just these European intellectuals whose anti-Semitic theorizing (sometimes deceptively tempered by a few ambiguous appreciative remarks) *legitimized* on a vast scale the practice of the big lie. From Marx's accusation—the Jews have *corrupted* Europe by converting it to the worship of "their true God," money and huckster-ism—to Nietzsche's accusation that the Jews *corrupted* Europe by foisting on it, via Christianity, their baleful slave morality; to a whole slew of best sellers with titles like *The Victory of Judaism over Germanism* (Wilhelm Marr) or the pretentious *Foundations of the Nineteenth Century* by that famous English-German crony of Wagner's, Houston Stewart Chamberlain; to Göttingen Professor de Lagarde and Berlin Professor von Treitschke and many, many others—such men showed how the big lie can be constructed from a thousand *little ones* so ludicrous that your head spins at the thought of the lovers of Kant lapping it up respectfully even in recent decades! Chamberlain's book is still a 'classic' of German literature, sometimes mentioned matter-of-factly. Here's one nugget from it: "a fact occurs to me which I have received from various sources, viz., that very small children, especially girls, frequently have quite a marked instinct for race. It frequently happens that children who have no conception of what 'Jew' means . . . begin to cry as soon as a genuine Jew or Jewess comes near them!" (Vol.1, 537).

As I said, Avineri surely knew all this, *and more*. It was from one of his books (*The Making of Modern Zionism*) that I learned of the corresponding 'incitement' in the decades prior to the Russian pogroms of 1881, which led to the emigration of about three million Jews from that empire. He quotes Peretz Smolenskin as follows: "*Any intelligent person could have foreseen that it would not be long in coming*" (italics by Avineri). According to Smolenskin, for twenty years, Russian newspapers and intellectual journals were full of violent anti-Semitic writings . . . full of every possible accusation against the Jews—religious, moral, social . . ." (61).

Now quick shift to Egypt in the 1970's. Avineri, with his fellow Israelis, asks about the mutual correction of *textbooks* so that the school systems of each nation will exclude untruthful material that might tend to defame the other and build up prejudices between the two peoples.

But this changes the atmosphere immediately. His Egyptian counterpart becomes angry. He will not allow the Israelis to dictate policy in purely internal matters. Further discussion of this is unwonted. The Israelis gave it a try, made some arguments—about the treaty's stipulations on this, about the need for it, etc. No give. *So, finally, Avineri said to himself (according to his own account), "What does it matter what is in the schoolbooks, so long as people are not killed on the border."* And he let the matter drop. Now, however, when 'incitement' has born its fruits, even Avineri concedes that he did not quite understand the importance of this 'cultural' issue, which then appeared to him so secondary, compared to "people killed on the border."[7] But *Avineri knew all about it*!! He, better than most people, *knew* the historical precedents; he *knew* Smolenskin's view of what "any intelligent person" should have understood even long before the Holocaust. Yet "here we go again." It may be a *cliché* worthy of "the Gipper," but timelessly valid—"those who don't learn from history are condemned to repeat it."

E. The Cognitive and the Moral

M.E. Thus, from the macro-perspective of history, I am impressed by the hints at something more than a superficial connection between large-scale immorality and large-scale untruth. And that suggests, perhaps, connections between morality and truth in general. Well, as they say, "what else is new?" Is it accidental that in Christianity, "God is Truth" appears almost as often as "God is Love"? or that our Western justice system is elaborately entwined with the handling of *evidence*, rules for the presenting and questioning of evidence? Or that the word 'investigator' could refer to a police detective as well as a scientist?

Yet, a philosopher may respond as follows: Of course it is not news that *in social practice*, cognitive issues *contribute* to the just solution of conflicts and that injustices are aided by falsehood. But that should not excuse *conflating categorically* the moral with the cognitive. In particular, it does not follow that the process by means of which one *attains* to objective facts is similar to that by which one attains to moral commitments.

Yes and no. Perhaps that is also questionable. For I believe that the historical examples (but especially ones like the story of Avineri) lead us a bit beyond the truism that lies facilitate crimes. I want to approach the issue from several different directions. Let me begin by asking this: Is a *genuine* commitment to justice *possible without* a genuine and serious commitment to truth? I exclude here the person who *believes* himself committed to justice while showing little regard for truth. That's quite possible. I mean that the commitment would be judged genuine by you or me, were we to observe this person for a sufficiently long time, dealing with matters of justice.

As you know, in the past, quite a bit of metaphysics used to be done by asking oneself what could or could not be conceived of in isolation; what had invariably to be thought of together with something else. Well, when I avail myself of this venerable old way of philosophizing, I cannot help but give a negative answer to the above question. *I cannot* conceive of a commitment to justice without a commitment to truth. But this *brings together the moral and the cognitive* (including the scientific, in a loose sense). So even when we talk about the *attainment* of values, as distinguished from the attainment of facts, there is at least in this one important case a *restriction* on the famous demarcation: You *cannot attain* to justice as a value without *also* attaining truth as a value. Does it work from the other end as well? Is it possible to be committed to truth without a commitment to justice? We do still have that old notion that, in a sense, to seek or tell the truth about things is to 'do justice' to those things. But more convincing is the recollection that, often enough, in the *telling* of an untruth, an injustice is done to the hearers—for they are misled. The very word 'truth' itself seems to carry us into both realms. Are truth-telling and truth-finding so clearly disjoint? If you don't first look for it, how can you tell it? Isn't the *will to truth* both moral and cognitive?

F. Is a Sloppy Researcher/Reviewer in Medicine Merely a Bad Scientist or a Bad Man?

M.E. During the first phase of Judith's illness, her oncologist called our attention to a certain Dr. Werner Bezwoda, of South Africa. His was the only randomized trial, out of about four, that showed a documented advantage to the then-widely-practiced high-dose chemotherapy and stem cell transplant procedure for breast cancer. We made several calls to specialists; Judith was being urged to undergo this risky treatment. But when we read the Bezwoda paper—immediately, and independently, we were first puzzled, then had a decidedly negative reaction to it. The problem was that the control group in this trial did not 'look right.' They were dying too fast (10 weeks after progression of disease resumed, on average); their response to a presumably standard treatment was too poor. And the calculated advantage of the high-dose treatment was of course based on outcomes in this control group. I still recall how we sat at our dining room table, and I said to her, "Right, Judith, we will ignore Bezwoda."

It was several years later (February 2000) that we read in the newspapers about the outright fraud involved in that study, the confession and 'resignation' of Bezwoda, and the now-acknowledged worthlessness of that trial. My point about cognition and morality has little to do with the doctor who perpetrated this—such there will always be from time to time. But I do ask about *all the others involved*: Where were they? Where were the referees of the respected journal (*J. of Clinical Oncology*) in which this fraudulent paper was published? Where were the coworkers, the colleagues at the hospital, the institutional oversight boards? Perhaps they did not *know*; but was it not their cognitive-moral obligation to know? Take the example of one of the best-known breast cancer specialists in the country, a certain Dr. M. John Kennedy, then at Johns Hopkins. His detailed review of the Bezwoda paper was featured in October 1995; it was of great importance because hundreds of doctors all over the world were citing Bezwoda as part of their argument for this dangerous procedure. Dr. Kennedy, nobody's fool, immediately spotted the problems we spotted, and raised a note of caution. But then he went on to praise the design of the study, to endorse the results as "believable," and to "commend" Bezwoda for his work. Why? And when you know that most doctors don't read papers in full, the endorsement at the end is the bottom line.

What was the nature of Dr. Kennedy's failure? Was it cognitive or moral? True, one could not really know about the fraud if one did not go to Bezwoda's lab, as was done years later. But knowing the stakes, and seeing the flaws, why couldn't he write "there are too many questions here to accept this trial as evidence"? Well, someone might say, he had criticisms, but he wasn't sure; so he did not want to take a chance on destroying the credibility of what might, after all, be a useful contribution. Here is the crux of the matter—just here. He had well-formulated questions as befits an insider of this game. He could have picked up the phone and called Bezwoda himself before sending in his review. It appears he did not do this. Does this make Kennedy's failure all moral and not cognitive? Perhaps he thought a telephone conversation could not change the picture substantially. But on this he could have been wrong. Would that make his error cognitive rather than moral? Can a mere cognitive failure of this sort result in action that *should* be called immoral? Perhaps there were other things he could have done. Even some of our laws recognize this connection: People in certain situations are judged according to what they 'should have known,' even if they claim not to have known. At any rate, the effect was that for five years the medical community was misled. Who knows how many women would *not have died* in the transplant procedure, if Kennedy have given a clear warning signal.

As you know, the literature on ethics (including educational) is full of contrived moral problems, dilemmas, etc.—useful, it is believed, for forcing choice and making theoretical points. I dislike these. It is not just that they seem childish, but that they miss an important aspect of morality. Life is not a series of answers to textbook questions, and in 'practice,' one usually does not face a set of well-defined alternatives. Only when that is the case are probability estimates or decision theories even relevant. What actually happens is that the dilemma appears to us within a full context; the alternatives or possibilities are *not given* but become, in fact, the first phase of the *moral* task: Attention to the situation, to priorities, attention to the seriousness of the consequences, a will to find a way. And that must always be done by the individual himself, from within his particular context, or someone very close by. That is why I find more to ponder in real human experiences or in histories than in those set-piece ethical examples. In the case of Dr. Kennedy, I, as an outsider, could think of only one or two alternatives to what he did with that dilemma. But he himself might have known of more: He might have known, for example, that by calling the editor of the journal, the publication of the paper could be delayed, while a set of questions and requests for information would be sent to Dr. Bezwoda. Only an insider like Kennedy himself could know all or most of the realistic alternatives from which he had to choose.

But all this cognitive stuff—attention to possibilities, information-seeking—seems to be very closely associated with concern and responsibility. Is the former *merely a means to the latter*? No. Because only if Dr. Kennedy visualizes clearly the possibility of a totally worthless piece of research being used as a guide to life and death choices can he become concerned enough to trouble himself enough to make a responsible decision.

G. KNOWLEDGE AND COMMITMENT

M.E. Thus, I tend to see truth and justice as so intimately intertwined that one might even call the one a particular 'form' of the other. And, to some extent, this very idea is enshrined in language—"doing justice to the facts," "being true to oneself," "the truth shall make you free," "knowledge is good," etc. Historically, it would seem to me difficult to make a case against that; philosophically, there are still questions left, but let me continue.

Even if it is admitted that reaching justice as a value is inseparable from accepting truth as a value, this still has no bearing on the difference between how one attains values and how one attains facts. For in the discussion above, 'truth' was treated as a value, and as a value it appears inseparable from another value. All values, including the value of truth, are attained in a way that is different from the attainment of facts, even if 'truth' be a *cognitive* value in the sense that its actualization requires a cognitive, and therefore possibly a scientific, process. On the other hand, a Kiekegaardian commitment to God or a certain kind of life, or even to justice, may be an existential wager or leap (if it is not a matter of upbringing and socialization of some form). You can *adopt* 'truth' as your life's commitment; but then, when you carry out this commitment by becoming a scientist, for example, or a historical scholar, you will do so by practicing methods of rational inference from evidence, by means of which you could never reach your governing commitment. This gulf is surely there. The 'game' may be cognitive; the transcending goal, as such, is not.[8]

Again, I must say "yes and no." If we are talking about the highest, most general, or most fundamental commitments—God, justice, even friendship and honesty—the gulf seems real enough unless your values are purely utilitarian. But that much you would certainly expect, for otherwise what reason would there be for this gulf to be so prominent in philosophy. Although this is not my last word on the highest commitments, our experience is that most of life is not taken up with the problem of *whether* or *which* of them we should adopt. Nor are such commitments, in themselves, as subjective states, really *complete*. In practice, what we must decide over and over again is not whether "I value friendship" but whether this or that person is to be my friend. And if that friendship is to be more than the 'friendship' for a dog or cat, then it will have to be a cognitive decision, at least in part. I will have to *know* that person, as a person, before I can be committed to a friendship with him. Here, cognition cannot play its role just within the limits of commitment; it goes into the making and maintenance of the commitment; though of course, it will not usually take the form of a prescribed calculational scheme, Bayesian or otherwise. Nor is a *particular* friendship an *application* of the higher commitment to friendship in general. Friendships usually do not arise this way.

Explicitly or tacitly, cognition is right at the center of it if the friendship is to be a serious relation to *a person*. Yet, strangely enough, that 'person,' whom I think of as my friend, is, after all, a *construct* of my mind! How else could it be? I cannot get inside the being of the other person. His character, his deeper nature, the total *Gestalt* of his external and internal qualities, is something I must infer from what he says and does. Slowly, a *Gestalt* emerges of a worthy and attractive person; and I may notice that his attitude towards me is friendly as well. Consciously, I begin to think of him as a true friend; and the more I value this relation the more I will be committed to preserving it by doing what is appropriate to it under various circumstances. There's the commitment. This I take to be a *spiritual* (*geistlich*) construction of man in the sense of Cassirer, and also a *perception* of a spiritual *Gestalt* in the sense of Buber. I am not particularly a friend of this man's looks, his moods, or of most of the external visible traits. I am a friend mainly of the spiritual *Gestalt* embodied in this man, which I take to have communicated itself to my mind, hopefully growing and strengthening with time.

But is that the end of it? Is all further cognition within the *limits* of this new commitment? For most humans, I think not; nor is it clear that the highest morality demands it.

I may learn more. It may happen that despite my friend's admirable qualities, I notice from time to time certain jarring notes, inconsistent with what I take to be his character or 'real self.' Well, I wave my hand, of course; that's what friendship is about; you give your friend some elbow room, you distinguish between the important and the secondary, you don't insist on understanding everything, you have some *faith*. But what if, with time, the jarring notes repeat and elaborate themselves in such a way as to appear less isolated, less secondary, and more pattern-like, showing some coherence and getting closer to what I take to be the man's true character? No doubt, different individuals will react differently under those circumstances. But if this trend continues further; if as in my example, my friend is suddenly in court, accused of theft and the grossest dishonesty, of intrigue and injury to others, etc., and the evidence is overwhelmingly against him—then what? I am committed to being his friend, but that commitment is based on my *previous knowledge* of him as a person, a spirit. How unconditional can it be? A commitment to God may be absolutely unconditional (even that is not always so, and perhaps ought not to be), but here we are talking of a human being like ourselves. The experience on this, I think, is that a whole spectrum of responses is possible. I may, for example, say "I *know* him; he *couldn't* have done this no matter what the evidence appears to show." (*Brothers Karamazov: the evidence of the heart vs. the evidence of the court.*) But it could also happen that at some point in such a trend of new discoveries, if it goes far enough, my commitment will be damaged. I will not be *able to help it*; because the *Gestalt* of this man as a *person* will be altered in my mind and may crumble—not because of any cool-headed calculation of probabilities but because of the irresistible falling into place of new knowledge, into a *Gestalt* that is radically different from the old one (a "*Gestalt* switch," Kuhn would say, borrowing from psychology). Is it possible by sheer faith or commitment to maintain under the worst circumstances something like the original friendship? Is this what commitment calls for even in the realm of the mundane? By chance, we have just had an amazing example of such a thing, and I am compelled to mention it and marvel at it, though I think it goes beyond what most people are capable of.

In Texas, a trial is now taking place of a woman who has murdered five of her own children—all she had. She did so, according to her own account, methodically, with effort and much premeditation (a long story in the January 28, 2002 issue of *Time* magazine). The woman—well-educated, intelligent, middle-class—has a history of serious depression problems and medical treatment and is pleading innocent by virtue of insanity. Her husband, a sober, seriously Christian and scientific type—a software engineer with NASA—is left suddenly alone, without wife or any of the five children he loved so much. A tragedy of the highest magnitude. But still this man is fighting to save this wife from imprisonment or worse. He's all on her side; he says it would be grossly unfair to punish her further. Although she is not described as overwhelmed by remorse (she says only "I did not hate them"), he believes *he* has to be the one to stand up for her because he is her only friend!

In the face of such a thing, I can only be silent. But, in general, I would say that the process of cognitive build-up of a *Gestalt* to which one becomes committed, sometimes followed by cognitive disintegration of that *Gestalt* (*against all efforts to maintain it*) is a widespread and quite general human experience. Not only is this so with individual friendships and marriages, but with commitments to ideologies (*The God That Failed*), to groups, organizations, nations, professions, etc. Most commitments we make in life are inescapably cognitive and subject to cognitive erosion, with no exception, in principle, for even the highest commitment of all. I still have on my shelf a 1960s book by Richard Rubenstein, a Jewish theologian (*After Auschwitz*). His argument is that after the *evidence* of the Holocaust, the Jewish commitment to the God of the Bible cannot be maintained. If he exists in some sense, God is apparently not as the Jews and Christians have understood him for thousands of years. For those who demand, in this regard, unconditional faith, such 'rational reconsideration' may be contemptible. But to me it does not seem necessarily so. We are finite beings, condemned to live with infinitesimal knowledge. "Faith in things unseen" is one thing, but to "believe because it is absurd" (or "although it is absurd") is something else. There is a sense in which the latter may even be seen as a *loss of faith*—in a God who is *not* absurd (and not a "fool," as Kierkegaard so often emphasized).

H. OBJECTIVITY AND SYMMETRY

M.E. Of the various connections between facts and moral values, perhaps the most obvious is objectivity; again, not necessarily that specific values can be verified objectively, but that they involve objectivity. This seems an especially strong common factor if objectivity is taken *as detachment*. Objectivity in science—as detachment—and objectivity in ethics, as detachment, are major themes in countless philosophical works. But there is another interpretation of objectivity, ancient and modern, that is just as important in both realms, and perhaps more interesting—objectivity as *symmetry and invariance*. In physics it has of course been used abundantly, and in ethics my examples would be the ancient "Do unto others . . . ," Rawls's theory of justice (with its "original position," its "veil of ignorance"), and even Kant's categorical imperative. It seems the pre-Socratics had this idea already (Anaximander) when they brought justice into cosmology as the balance of "reparations" between opposites. All this may be such common knowledge that you may ask why I bother mentioning it. The answer is that I would like to probe somewhat the role of *cognitive "symmetry-breaking"* (pardon the allusion) in the dynamics of injustice.

But before we move on to that, I can again see my partner in dialogue fidgeting with impatience. Well, he might say, you keep elaborating merely the *connection* between the cognitive and the moral, but you avoid what everyone knows: On matters of fact, especially about nature, we can get "fixation of belief," intersubjective stability. On questions of value, moral values included, we have no such thing. There are no 'moral facts' in the sense of facts of nature. Even if everything you've said is granted, this gulf cannot be bridged.

Not really bridged, I agree. But recall that the argument is not about 'bridging' but about 'narrowing.' Not all values, surely, not even all moral values are of the same kind or order. Let us distinguish between the top tier and the rest. Is it not a fact that a certain fuzzy set of moral values—like courage, justice, friendship, truth-telling . . . overlaps to a large extent with several old versions of 'enduring' or 'true' or 'basic human' values, that are independent of religion and culture? To the Jews it is the code of Noah; C.S. Lewis called it "the Tao," and I would link these (loosely) with the 18th century version of "natural law." Most values may never reach the rank of "fixed opinion" in Peirce's sense, but some of them seem to be more fixed than others. Who argues that cowardice is desirable? Which culture promotes untruth as an aim? Some sort of quasi-objectivity seems to be involved here despite our inability to prove it. But then, mathematical truths also exist that we cannot prove, as Gödel and Penrose keep pointing out.

However *"un*-fixed" and culture-dependent most lower values might be, the highest and overarching set seems strongly intersubjective. If so, why not take this as a primary fact, a starting point, and *use* it, both theoretically and practically, for our benefit? In my *Zygon* paper, I suggested exemplars such as Socrates, Gandhi, Martin Luther

King, Janush Korczak precisely because their stature is tied to that highest tier of natural law. It is because of this, I think, that they receive widespread acceptance, and that might be another way of narrowing the gap: The *synthesis* of the *being* of an individual person is more powerful, psychologically at least, than a logical argument. The reduction, or explanation, of something like justice or friendship, in terms of a biologically acquired propensity for "reciprocal altruism" is simply not relevant here. That is a question of origins, not of the being of humans as they are. But if this quasi-objectivity is taken charitably, then it is *a kind of* objectivity, and objectivity is close to 'reality'—and that brings us back to those crucial words in your *Zygon* reply: "*The ontological status of normative facts" is "problematic*" (emphasis mine).

Problematic it may be, but this perhaps is the big issue at stake. For when "the Tao," or "natural law," being ontologically problematic, became unreal or *less real* than cognitive facts, that probably was when we Westerners embarked on the disastrous path of the 20th century— which is getting more disastrous every day. At this point, the thought occurs: Perhaps it is a *moral imperative* of our day to *restore the symmetry* by working as hard as we can to make the ontological status of normative facts *less* problematic—in other words, *to narrow the gulf*. I have no real suggestions as to how this might be done more effectively than in the past. For a start, though, just looking at the ontology of values as a 'glass half-full,' rather than a 'glass empty' or even 'half-empty,' may help. But to take seriously the possibility of real progress on this front, I rely on the faith of Imre Lakatos: *Do not underestimate*, he said, the ingenuity of bright theoreticians. If we can restore the symmetry between bosons and fermions, why not between facts and values?

I. Back to the Historical Standpoint

M.E. But what has all this fact/value business to do with Avineri, medical practice, and modern atrocities? My final point is just that—that it has a lot more to do with it than we acknowledge.

Why *did* Avineri accept the proposition that the contents of textbooks did not matter as long as people weren't killing each other? Of course, he could not force the Egyptians if they were unmovable. But neither did he have to drop it from mind, from print, from the awareness of the Israelis and the world. And it goes without saying that he was not the only one who had this view of things—I use him only as a symbol for so many others—in Israel, Europe, America. For decades, the media, the universities, the churches and synagogues, government people and intellectuals of all sorts *ignored* what was happening—first in the Arab world, then spreading to the larger Islamic world. It is not as though it had been hidden; it cannot be. Those who deal with these things professionally knew it; but it was swept under the rug, kept out of sight, or minimized, explained away, excused. Has there ever been a serious debate on the meaning of this 'incitement' phenomenon, say in some well-know university? How many scholarly papers have there been on this, coming out of our Middle East or Islamic studies departments? Where were the sociologists and social psychologists? When Mrs. Clinton hugged the wife of Arafat at a meeting in 'Palestine,' though the latter had accused Israel of deliberately poisoning Arab children, the media promptly made of it a Clinton issue—did Hillary do right or wrong? Not once in all the many articles I read on this, or the newscasts, did anyone ask what it means for the prospects of the 'peace process' when the wife of the Palestinian Chairman is a leader of such 'incitement' via the big lie. Only after 9/11 have some people finally begun to open their eyes slightly, to notice that as far as Jews and Israel go, the phenomenon is not only widespread, not only deeply rooted, not only strongest within the intelligentsia, but also that it is of a *kind*, in sheer audacity of the big lie, that outdoes even the Nazis (for they, after all, still had to take some account of their Western audience). And it has now had at least three decades to work on its people. Diplomats still keep saying in public that it is up to Arafat to do this, or agree to that, etc. But Arafat himself explained why he will not do these things (at the summit meeting with Clinton and Barak)—"do you want an invitation to my funeral?" On this, he surely was not lying. For he is only an icon, a 'front man.' He said clearly he doesn't speak for himself; he has 'responsibilities' to the Arab leaders, to the people, to Islam. In other words, *decades of unopposed 'incitement' on this scale cannot be turned around in a moment by anyone*. Of course, Arafat played his own part in it as a man with an unprecedented public voice—preaching jihad, "blood and sweat" to Arab audiences, praising the "martyrs," etc., etc.—still, he only contributed to the collective field (albeit as a large particle). The Egyptian newspapers, Arab T.V., the mosques and madrassas are even more powerful. Regarding Jews, the situation in the Islamic world now is certainly of the same order as that in Germany and central Europe in the 1930s. And, interestingly, in all three cases—Russia before 1881, Germany before 1933, and the Arab world for the past three decades at least—it is *intellectuals* who are spearheading the 'incitement.' So, as Mr. Avineri knows, "any intelligent person" should now be able to "foresee. . . ."

But what exactly should they "foresee"? That B follows A historically is not necessarily a causal connection. What do the various 'incitements' have to do with what followed? Isn't 'anti-Semitism' an 'irrational' or religiously motivated hatred that has 'always' been there? Yes and no. Anti-Semitism is not new, but when it (or something similar in regard to other people) rises to the level of dehumanization or delegitimization it radically *breaks the symmetry of the situation in the application of the principle of justice*.

Another example from contemporary life, less catastrophic, perhaps, but also pointing in the same direction: Some years ago, the popular singing group "2 Live Crew" recorded a song about men receiving pleasure from coitus by mutilating the vagina—all packaged in catchy tunes and familiar black street language. The state of Texas took exception to this kind of 'art,' and prosecuted. But this brought down on them an irresistible hue and cry about free speech—led by the distinguished professor from Harvard Henry Louis Gates, author of many books. These songs, he said, merely expressed the ways and values of a culture that whites did not understand. And so, in the face of such august expertise, the mostly white jury concurred; as they left the court house, they were heard humming the tune of the song they had been debating. Here too, in order to overcome the prompting of natural law, an asymmetry had to be created, again between black and white, but this time by subtly and acceptably dehumanizing whites— as being more repressed, less natural, more domineering, etc. And at this game, the universities have recently come into their own. Numerous studies claim to show no causal relation between word and deed, between value and fact, and yet we all know that this is mendacity on a colossal scale. Billions of dollars are spent by the advertising industry to get people to behave as desired; and these expenditures are frequently checked for efficacy, or the money would not continue to pour out. The trick is, of course, that the causality is statistical, and works mostly on a longer time scale.

But now something new has come up quickly, increasing the danger by orders of magnitude: In our ultra-communicative world, where 'facts' can be disseminated instantly to billions, when a single edited or even 'created' picture can mean more than thousands of words, would it perhaps not be in order to rethink from scratch the whole relation between word and deed, fact and value?

As a final objection, it may perhaps be said here that once again I am conflating something important—conflating *facts*, as such, well established, with mere 'information' which can, of course, be disinformation. And isn't the answer to this precisely more care in the distinction between the purely cognitive and the evaluative, whether the latter be individual, national, cultural, etc.? At the theoretical and methodological level, this objection is of course cogent; and that is another reason that the gulf cannot really be bridged but perhaps only narrowed. Yet even here, some things remain to be said. First, disinformation may not always be a lie; it may be simply an error at a particular stage of knowledge, e.g., the question of black inferiority in earlier centuries, the problem of the fetus now. Second, even if the lie is blatant, how is this to be avoided outside of the works of scholars and journalists of integrity? No one can directly find out everything for himself; we are forced to trust others who have looked it over more carefully. But if that trust keeps breaking down more and more—because the educational, journalistic, governmental, and social institutions fail to support it—then we are in a new age of trouble. As already mentioned, all these human failings might not be that new, but the *scale* on which they now operate is enormous.

So I have suggested that 'natural law' (loosely speaking) should be taken seriously again. I believe that, tacitly at least, it *is,* in the minds of large fractions of the various populations of the world. It is a natural predisposition based on symmetry, or the principle of insufficient reason: If 'he' and 'I' do not differ in any relevant way, then he and I are entitled to the same treatment—for if he got worse, then, under an exchange transformation (he and I change places), the same might happen to me. Even Nietzsche, in his odd way, supports this: Justice, he says, is possible only among equals. What he meant by "equals" were the strong and 'noble,' but Christianity, in its way, and the Age of Reason, in a different way, extended widely a fundamental *human* equality. This means that treating any group in a radically worse way requires a justification in terms of a sufficient and relevant *difference* that breaks the original symmetry. *That is the function and often the purpose of dehumanization.* That is what must be done by those who, consciously or tacitly, contemplate for another people what they would consider unjust for themselves.

In the case of Africans in America, dehumanization was achieved by pointing to cultural or anthropological 'proof' of their innate mental inferiority. Today, within the abortion debate, the same sort of thing is happening. One does not need to be a zealot on either side to notice that serious issues of dehumanization are involved: For if we grant that the union of a male and female germ cell is not yet human life, then at what stage is its humanity to be affirmed? And for those, at least, who do not take an extreme dogmatic view of this, doesn't the question become inextricably entwined with science? Below a certain stage, the 'fetus' is dehumanized. In the case of German Jews, there also were some 'stages'—*Ostjuden,* court Jews, assimilated Jews—but in the end, everywhere, this dehumanization has always taken the form of *immorality as evidenced by evil deeds.* Once the symmetry is broken, the 'I' can no longer think of itself as interchangeable with the 'he,' and justice therefore takes on a different form for 'him.' Justice itself is not repudiated today by anyone in the Islamic world; on the contrary, it is *just* in the eyes of many of them that Israel should disappear. But their ability to see this as justice rests on decades of 'information' about what happened and is happening now is 'Palestine'—and in this too they have been aided, and are today aided, by most of the Western media, intellectuals, etc. By the dissemination of thousands of 'little lies,' building up the big lie, and by means of dozens of journalistic tricks (easily spotted by those who watch carefully), many Arab accusations are supported—giving positive feedback and encouragement to the dehumanization program. Why is anyone surprised that tens of millions of Muslims 'believe' that the destruction of the twin towers was Israel's work? In their minds, Jews are the only people capable of a deed so demanding and at the same time so evil. Once again, Jewish 'character' has been elaborately pictured and defined on a very large scale—by the dissemination of 'facts.'

But if, as I claim, we in the West are all implicated in this dire development—why? Why *did* we (and still do) minimize, ignore, pay it little attention? Perhaps a clue can be found in the reaction to Hitler's big book, which spelled out his plans and *values* so clearly: Those on the other side waved their hand and said "just wait till he comes up against the *realities* of governing a nation, then he'll have to forget all this nonsense." The same was said about Arafat some years ago: "Wait till he has to govern like a statesman, provide his people with the goods they need, then economics will get the upper hand over ideology, and he'll forget all about terrorism." *There we see it—realities against values.* In our Western minds, the latter have no chance against the former. Values are associated primarily with words, articulated ideals; realities are events, facts. Kant and followers notwithstanding, the fact/value dichotomy (or duality) has shown itself capable of transmutation into the deed/word dichotomy—pragmatism. And from a pragmatist perspective it is easy to dismiss words in books, especially in textbooks,[9] in favor of 'facts on the ground.' That is what Hitler and Mussolini were admired for, but such bias is also supported from the left by the Marxist philosophy. And that is what, in our time, Avineri, many Jews, and most American media have been doing in regard to the third and most dangerous 'incitement' campaign against Jews within the past 150 years.

I do not suggest that the devaluation of values is the sufficient reason for the incitements and their consequences. Nor am I saying that a purely technical, philosophical distinction, as such, can have effects of this kind. I do suggest, however, that for us descendants of the Enlightenment, and especially those in the English-speaking world, this dichotomy, when taken as strict and wide, has in fact devalued values by de-ontologizing them, by placing them closer to opinion than fact. In short, it has provided a theoretical basis (quickly spreading as mass psychology) for the lessening of *attentiveness* and *diligence* in the face of a looming but not yet fully actualized danger.

The same sort of thinking seems to be evident in Dr. Kennedy's review of Bezwoda's paper: He wrote some cautionary words, and he wrote some words of praise. Only words, though they reflect his professional values. But since he was not *attentive* enough to visualize the *realities* (facts of life and death) that would ensue from these words, he did not place much importance on getting the words right. That in our Anglo-Saxon outlook, words don't count much, only deeds, has no doubt played a salutary role in our defense of freedom of speech. But here we see the reverse side of the issue: Kennedy only spoke. It was Bezwoda who had done the deed. Kennedy is innocent and respectable. But, of course, Dostoyevsky was well ahead of all this: Ivan only talked about his values; *then* Smerdiakov did the deed. Yet Ivan was not innocent.

This accounts, in part, for my suspicion of the dichotomies and trichotomies since at least Hume and Kant. It seems to me that the salt of life—in science as in morals—is primarily synthesis, not analysis. To be sure, analysis is the indispensable method. But in the 20th century, we have seen science flourish by means of a surprising balance between analysis and synthesis. In the arts, on the other hand, the extreme differentiation—the fads for 'purity' of the aesthetic, etc.—has brought an unprecedented decline; and something similar can probably be said about the differentiation between the cognitive and the moral. All of which brings us back to the need for interpretation—cognitive, moral, aesthetic—hermeneutic reinterpretation with a bit less respect for the most recent analyses. Where I seem to be closest to your viewpoint on the big questions, I suspect, is not in relation to Peirce but to Aristotle.

PERSPECTIVE

A.S. With installment IV, I no longer consider our exchange anything like a debate. There are very few propositions of yours that I take exception to. Many of your propositions are striking and inspiring, making me think about things more deeply and articulate more clearly, but not in opposition to what you say. The subject matter is quite close to our initial debate in *Zygon*, concerning the relation between facts and values, but much of the opposition of the original debate has been transformed by the evolution of our ideas since then.

I still have a reservation concerning your interest in hermeneutics. You clearly maintain your distance from what you call the "modern" version of hermeneutics on page 435ff, and you are explicit on page 434 that " I did not develop my interest in hermeneutics out of some distaste for objectivity or truth. Quite the contrary." But I cannot help wondering whether your attention and wide reading in all varieties of hermeneutics have not been a very indirect winding path to the matters that most concern you. Why not choose other guides in the philosophical literature, such as Aristotle whom you mention at the end, in conjunction with the immense (admittedly confused) body of relevant discoveries and discussions in psychology, anthropology, political science, and literature? But maybe that is an unfair question. After all, everyone has to use the material of one's past life and past reading to reach one's tentative world view. Even a deliberate break with one's past is a way of making use of it. In any case, I certainly admit that you made something fresh out of the materials you found in the literature of hermeneutics.

THE FACT-VALUE DICHOTOMY

A.S. This topic is the center of Part IV, and even though it revives the discussion in *Zygon*, it is much richer. On page 441 you reject "*conflating categorically* the moral with the cognitive," and on page 444 you give a reason for your rejection: "All values, including the value of truth, are attained in a way that is different from the attainment of facts, even if 'truth' be a *cognitive* value in the sense that its actualization requires a cognitive, and therefore possibly a scientific, process." I agree and applaud, but still feel the need for more explanation.

The basic reason for maintaining a dichotomy of facts and values was given by Hume: that "ought" cannot be defined in terms of "is." This can be rephrased, not so pithily, by saying that there is a recognized family of value terms—including 'ought,' 'valuable,' 'preferable,' 'desirable,' 'morally correct,' 'good,' 'bad,' 'right,' 'wrong'—any of which can be defined using the others, but none of which can be defined in a way that satisfies us conceptually by using only terms lying outside this family. This insight of Hume does justice to my own body of experience concerning values to the effect that I cannot judge one state of affairs to be more desirable than another without either a direct comparison of their desirability or else a rational exploration of the factual consequences of each, terminating with a comparison of the relative desirabilities of their respective consequences.

But human beings are so complicated that the dichotomy between fact and value is to a large extent bridged by their interactions. Your footnote 8 on page 504 is splendid: "I take Kant to be promoting not so much a 'dichotomy' between the is and ought, as a 'duality' (Cassirer), a combination of distinctions and interactions," and what results is the domain of practical reason. This remark is excellent, and I want to elaborate it. It will be useful to consider two classes of complications, (1) regarding the psychology of the individual human being in abstraction from society, and (2) regarding the social nature of human beings. This distinction will help the analysis even though, needless to say, (1) and (2) inevitably overlap.

(1) It is a fact that an individual human being has a multiplicity of desires, with interference between the conditions for satisfying them. It is a fact that the satisfaction of one desire has long-range consequences that may prove to be undesirable in due time, and it is also a fact that the cognitive powers of human beings are often sufficient without deliberate cultivation, and even more sufficient with cultivation, to anticipate these long-range consequences. Induction, the great tool for cognitive investigation of matters of fact, is also the great tool of hypothetical imperatives. A further complication: The early training of an individual human being can produce habits of self-control and deliberation that equip the mature individual with the power to compare the desirabilities of available choices, and to anticipate accurately the satisfactions that the various choices would provide, instead of being impelled to a choice randomly or erratically or by uncontrollable passion. That is why I call my version of Bayesianism "Kierkegaardian." This training is what we call the formation of character. I am, however, a Spinozist who holds that the person with strong character and the person with weak character are both infused with desire or appetite, rather than a Kantian who envisages the categorical imperative governed by rule rather than desire to be strictly distinct from a hypothetical imperative, which is only a wily way a satisfying desire. And I am enough of an Aristotelian to believe that the cultivation of virtues, which are the right kind of habits, is conducive to happiness (though by no means guaranteeing it, since ill fortune can bring misery to the virtuous man). From my naturalistic point of view—borrowing from Aristotle and Spinoza and from heroes of the Enlightenment like Diderot and from enlightened psychologists like Freud and from your choice of literary masters—the cultivation of virtues and the achievement of character convert the cacophony of competing and erratic desires into the harmony of a satisfying life, a life that is desirable not just in prospect but in retrospect and upon reflection. I am not very religious, and therefore what I desire in the long run is not "the peace that passeth understanding" but the peace achieved by understanding. Note that I am concentrating as much as possible upon the individual, though it was inevitable that a social consideration slipped in when I mentioned the cultivation of virtues, which Aristotle emphasizes is an unavoidably social contribution to the individual. Hence on to (2).

(2) Human beings are social animals in several different ways. As a matter of fact, given our physical and mental constitutions and the characteristics of our natural environment, we cannot reliably satisfy our most primitive desires without human cooperation. Now this fact could be the basis of a hypothetical imperative: Cooperate to the extent necessary to obtain the reciprocal aid in satisfying personal desires, but no further. In other words, no altruism that does not maximize utility for me personally. But it is another fact that on the whole we are not constituted to live by this rule. Some altruism seems to be genetic, at least the concern for one's offspring. And it may also be that there is genetic altruism that could not be accounted for by the 'selfish gene,' but somehow has been accomplished through the evolution of unselfish and empathetic genes. (I don't know whether this is true, but I doubt that the evidence is definitive on the matter or even that the question is well posed.) Another body of relevant facts is connected with the social nature of man—namely cultural facts, which have the peculiarity that generically they are indispensable but specifically they are variable. There is an important book of anthropological theory, *The Evolution of Man's Capacity for Culture*, edited by James Spuhler, whose title expresses a profound truth. The various social groups into which a child is born—the nuclear family, extended family, the clan, the caste, the city, the nation, etc.—care for the children in various ways, and in return, to some extent consciously and to some extent unconsciously, they inculcate in the child some feeling of loyalty and responsibility to those social groups. They shape the child's desires, which are the stuff of value, by teaching it to desire recognition, approbation, and reliable acceptance by the group; and these desires in turn imply hypothetical imperatives, the desire to acquire those talents and modes of behavior which guarantee social acceptance. A corollary is the enculturation of the 'ought' of the social group: what behavior is right, what dress and ornament are beautiful, what skills are admirable, etc. These differ from society to society, but the coherence of a social group—upon which depends its ability to enhance the satisfaction of the rudimentary biological desires—requires some kind or another of enculturation.

On page 445 you argue well that fundamental commitments are usually remote from practical questions—e.g., a commitment to friendship does not help one very much in choosing friends. Aristotle anticipates your point, when he says near the beginning of the *Ethics* that in practical reason the first principles are particulars, in contrast to the situation for theoretical reason, where they are universal. (Of course, Aristotle has a whole apparatus of essences and universals, which you can borrow from when convenient and disregard when not. But as usual there is wisdom in what he says: Small children are trained with particular instances of encouragement and correction, rather than with general maxims. And your thesis in the *Zygon* article about centering moral education around moral heroes or practical experts rather than around rules has an Aristotelian ring.) But in the first sentence above I qualified with "usually." Certainly among adults who have chosen an intellectual lifestyle, by proclivity or training, rules really make a difference. I remember reading the statement of a professional philosopher who remained in Germany throughout the Nazi period without capitulating to the ideology, and he credited Kant's rule of practical reason—so act that the maxim of your action can be universalized—for preserving his decency through the terrible period. (Do you remember who this person was?)

There is another aspect of the problem of universality that is generated by the undeniable influence of culture on norms. My exposure to anthropology by talking to my wife, Annemarie, did have the effect of making me sympathetic with pluralism or at least of being disturbed by it. Here is one of the most troublesome cases. Her specialty was the Iroquois, and I came to know a number of living Iroquois personally and to hear about historical heroes and models. They had norms—at least prior to recent acculturation and conversion—in strong opposition to my own, particularly glorification of war and approbation of extreme cruelty to prisoners of war. Each Iroquois man had a private song—the "ado'wa" song—which was to be sung on his deathbed, and the Iroquois warriors in particular were supposed to sing this song while being tortured after capture in battle. Now to me both the experiencing and inflicting of torture are ultimate horrors. What then do I make of a culture in which torture was condoned as a way of giving a warrior a chance to show his courage? Of course, I am thoroughly repelled. Nevertheless, there is something about this culture that I can understand. It values a commitment to courage and steadfastness, which of course can be manifested in many different situations, in battle and elsewhere, but which is subjected to the extreme test only under torture. One cannot make this commitment the highest norm, at least for the warrior class, without condoning the occurrence of torture. A few of the old Iroquois chiefs whom I happen to have met seem to me to have been men of extraordinarily strong character. Furthermore, I am always moved by photographs of the famous Indian chiefs of various tribes who were involved in the wars with white men—how strong, serene, and self-confident they look. What universality of norms, at the highest level of generality, would subsume simultaneously the norms of these fierce chiefs and of tender me? Well, a universality is discernible, if one is willing to generalize enough: namely, character, that quality that I was talking about in (1) above, which some people achieve when they cultivate successfully a hierarchy and a harmony of values. At that level of generality, the Indian chief and Socrates are admirable by a common highest norm. Yes, but. . . . Isn't it the case that gradation is still possible? Aren't some achievements of firm character better than other? Isn't Socrates's courage in his willingness to accept great discomfort during his career as a soldier and to accept death with serenity, for reasons which he was able to articulate even when they opposed some of the norms of Athenian society, more admirable than the hardness of the Iroquois warrior towards himself and towards others? Isn't the subjection of one's life to examination, which Socrates preached and practiced, an additional universal that must be considered along with character when one evaluates norms? This question brings me to the brink of your fine discussion of objectivity and symmetry, but I am reminded of something else. A few days ago I read in the Boston Globe the obituary of Richard M. Hare, "an advocate of prescriptivism . . . that moral beliefs can be defended through objective reasoning. He outlined his ideas . . . in books that included *The Language of Morals* (1952) and *Freedom and Reason* (1963)." I have never read anything by Hare, but these few remarks are impressive. We probably should include him in our Dialogue.

OBJECTIVITY AND SYMMETRY

A.S. Symmetry—in its various forms of Hillel's and Jesus's maxims, Kant's categorical imperative, Rawls's theory of justice and (though I never would have thought of it) Anaximander's epigram—does seem to be a candidate for a supreme norm, under which all other can be subsumed or by which they can be evaluated. Using symmetry it is easy to resolve my uncertainty about judging the ethos of the Iroquois warrior. Standing where I do—after early indoctrination, a professional commitment to rationality, reactions to the grotesque ethnocentrisms of our time, inculcation of empathy by literary masters and peacenik propagandists—the rule of reciprocity comes as close to seeming irresistible and even synthetic *a priori* as anything in the literature of morality. Why not stop here, then, at this comfortable intellectual resting point?

    Answers to this question: (i) Our sophistication as scientists makes the breaking of symmetry seem as fundamental in the universe as the maintenance of symmetry. Both are there, and hence one needs to have definite reasons in any situation that the one and not the other holds. (ii) Acknowledgement of the cultural origin of norms and the diversity of cultures makes one examine the anthropological literature for evidence for and against the universal acceptance of the rule of reciprocity, and what does one find? Unless one twists and sanitizes the data contrary to our intellectual consciences, we simply don't find cultural universality of the rule of reciprocity. (iii) We can write a plausible evolutionary scenario for the creation of asymmetry—the dehumanization or at least debasement of the out-group. The scenario is that the psychology of identification of the ends of the relevant social group (family or clan or city or nation) with one's personal ends requires paying a price: namely, the debasement of the out-group. Even a philosopher with the astonishing scope of Aristotle endorses the idea of natural slaves! And Plato endorsed the "golden lie"—that all citizens of the city have a common ancestry!

What kind of 'proof' can be offered for universal symmetry against the foregoing scenario to the contrary? And what good is a proof if one can be invented? Well, I remember one, in an inspiring lecture that I heard at the University of Chicago around 1949 by Chancellor Robert Maynard Hutchins at the Thomas Aquinas Society. He summarized Aquinas's version of the Aristotelian argument that man is a social animal, whose individual ethics cannot be distinguished from concern with the good of his relevant social group. And then he added a modern twist. The economic interconnection of everyone in the world and the potentialities of modern military technology are so great that the relevant society for every individual is the whole human race! Hence, on prudential grounds, it would be unwise to exclude anyone and any group from the rule of reciprocity. I loved his lecture and was moved by it. But is Hutchins's argument definitive? I regret that I think not, because of the peculiarities and the variations of human psychology. The prudential argument is slave morality according to the followers of Nietzsche. What if one's system of desires is systematized into a Will for Power? Then even if one prudentially calculates that the probability of winning the Great War is small (which was in fact not the case in 1940!), the attraction of winning is so great that the expectation value, relative to that system of values, of the outcome of a strategy of aiming for world conquest is overwhelmingly higher than the expectation value of compromise and acceptance of the rule of reciprocity. And with such a psychology one may either with complete conviction or with complete cynicism (wasn't Goebbels a bizarre combination of both?) proceed with a campaign of debasing and dehumanizing one's opponents. And then the 'incitement' which you eloquently summarize and correctly fear becomes prudentially a reasonable strategy. And suppose you try to answer this gloomy analysis by taking the high road, as one may hear in sermons and moral discourses, and say, "What sane person would want to be a Hitler, even a victorious one?"—would this help? I don't think so, because I am sure that many people would want to be a victorious Hitler. And how do you convict them of insanity without the circular argument of applying the rule of reciprocity universally?

What course remains open? It seems to me that I must to some extent retreat to Kiekegaardian commitment, though with some extra armaments of psychology and pugnacity. Here I am, fashioned as I am by background and circumstances of my life, with my entire set of values systematized by the universal rule of reciprocity, even though in no sense of proof known to me can I show that I am right in maintaining this rule. I therefore join organizations that support the universal rule of reciprocity and encourage it (even if one characterizes this encouragement as propaganda). And I encourage the literature, plays, and films that cultivate empathy with people in other ethnic groups. I agree with your criticism of Avineri for tolerating incitement against Jews in the textbooks of the Egyptian educational system, as if debasement has no consequences. And I support legislation that criminalizes overt hate literature, all in the hope that my commitment to the universal rule of reciprocity will spread and spread. Finally, I am willing to fight for this commitment. I continue to regret that my service in the army was in the Korean War, partially but not completely justified, and not in World War II, when I would have been fully committed to the fight.

Looking back over what I've been moved to say, it seems strange to me—and certainly it will to you—that my commitment to the most reasonable and rational general rule of morality has such an emotional basis. But this is where my reflections upon the human condition lead, and it seems to me to be a position one can live with well. But if Richard Hare or any other moral theorist can show me more, I shall welcome the argument.

FINAL THOUGHTS

A.S. In our most recent exchange we do seem to have come very close to each other on two fundamental points: the 'yes' is that there are many connections ('narrowing' of differences, as you say on page 448) between the cognitive and the moral; the 'no' is (page 441) that one must not categorically conflate the cognitive and the moral. Because of agreement on these points there is no sharpness in our debate, but only an exploration of refinements and elaborations of the theses that we share. For instance, you did not comment on my (Humean) proposition that there is a family of value-theoretical terms which can be defined with reference to each other but not with reference to terms outside the family. I can see lines of division on this matter. An obvious one is between terms concerning moral value and those concerning aesthetic value. Two great philosophers whom we both respect disagree on this. Kant treats moral value in the second *Critique* and aesthetic value in the third *Critique*. In Vol. 1 of the *Collected Papers*, where Peirce presents his fascinating hierarchy of disciplines, he subordinates ethics to aesthetics, in that the latter studies what is valuable in itself, whereas the former takes the results of aesthetics as settled and studies, in light of the complexity of human desires, how they are to be achieved. (Note: I am paraphrasing from memory. Peirce surely says things differently and better.) I suspect that you would lean more towards Kant than towards Peirce (in view of your paragraph on overarching values and intersubjectivity on pages 448–49), and I the other way, but I am not sure even about myself and certainly not about you.

Where the most refinement of our substantial but not complete agreement is needed is on the status of the "highest and overarching set" of values (your page 448). I think we agree that the defining condition for this set is 'fairness' or 'symmetry' or 'empathy' or "seeing oneself in the other person's position." But I am most uncomfortable about the scope of "other persons." I find no way of proving in a rationally definitive way that the scope is *all* of humanity, and I become even more confused when I consider animal rights. It was at this point that the gulf between the cognitive and the moral seemed hardest to bridge. In the absence of a satisfactory bridge I proposed Kierkegaardian subjective commitment. What difference does it make to my conduct? I'm not sure that it makes any difference. There may be no pragmatic difference between acting on the basis of a firm commitment that gives structure to my whole set of values and acting on a cognitive basis that is attentive to the constitution of human beings and to their place in nature. I confess to discomfort at having to resort to Kierkegaardian commitment on moral questions when my professional commitment as a scientist is to objectivity and naturalism, and I certainly would be gratified by an ethical argument that avoids this duality. But if we, and nature, and the realm of concepts are so constituted that the duality is uneliminable, then respect for objectivity says, "So be it; I'll live with the duality." Do you agree with me, or do you have a different elaboration of the theses which we hold in common?

FINAL THOUGHTS

M.E. My argument is that for a particular value—such as "this man is my valued friend"—the way in which the moral value (the friendship) is attained is not disjoint from the cognitive.

Practically the whole of Part IV is devoted to elaborating on the 'no' in the 'yes and no' that I frequently give to the arguments for the classic dichotomy such as you and many others might give. The culmination of this argument of mine is given in such passages as that on page 449, where I talk about a moral imperative of narrowing the gulf between fact and value.

There are other points I would like to take up but this is all I can do for now.

Down, but not yet out,

Martin

[30]

Chronology of Martin Eger's Published and Unpublished Papers

Publications of Martin Eger

"Point Transformations and the Many Body Problem" (with E.P. Gross), *Annals of Physics* 24, 63–88 (1963).

"Point Transformations and Scattering by a Hard Core" (with E.P. Gross), *Nuovo Cimento* 34, 1225–241 (1964).

"Spatially Inhomogeneous States of Many-Body Systems" (with E.P. Gross), *Journal of Mathematical Physics* 6, 891–901 (1965).

"Point Transformations and the Hard Sphere Bose Gas" (with E.P. Gross), *Journal of Mathematical Physics* 7, 578–582 (1966).

"Simulation of Classical Many-Body Systems," *American Journal of Physics* 38, 1475–76 (1970).

"Science and Liberation," *Richmond Times, The Student News-Magazine of Richmond College* (October 5, 1970), 1.

"Science and Society at Richmond," guest editorial, *Richmond Times, The Student News-Magazine of Richmond College* (January 6, 1972), 4.

"Physics and Philosophy: A Problem for Education Today," *American Journal of Physics* 40, 404–415 (1972). This volume, 167–185.

Review of *Physics and the Physical Universe* by J.B. Marion, *American Journal of Physics* 40, 1052 (1972).

All of Martin Eger's writings and professional papers—published and unpublished papers; work in progress at his death; notes, files, and correspondence—will be placed in the Archives of Scientific Philosophy at the University of Pittsburgh, University Library System, Special Collections Department, 363 Hillman Library, Pittsburgh, PA 15260. Tel: 412–648–8190.

"Positivists v. Positivism," *The Center Magazine* (Invited: the Center for the Study of Democratic Institutions), VI, 4, 61–66, (July/August 1973). This volume, 149–159.

"Philosophy in Physics and Physics in Philosophy" (1976), forthcoming in a volume to be edited by Calvin S. Kalman. This volume, 187–190.

"The Price of Collaboration," *The Bulletin of the Atomic Scientists* (December 1978), 55–56. This volume, 297–99.

Review of *Knowledge and Wonder* by Victor Weisskopf, *American Journal of Physics* 49, 605 (1981). This volume, 191–93.

"The Conflict in Moral Education: An Informal Case Study," *The Public Interest*, 63, 62–80 (spring 1981). Reprinted in Nathan Glazer, ed., *The Public Interest in Education* (Cambridge, MA: Abt Books, 1984, 209–227). Reprinted in *Annual Editions – Education* (Guilford, CT: Dushkin, 1982/3, 1983/4). This volume, 317–334.

"Simulation of the Boltzmann Process: An Energy Space Model" (with M. Kress), *American Journal of Physics* 50, 120–24 (1982).

"The Philosophy of Science in Teacher Education," in J. Novak, ed., *Proceedings of the Second International Seminar: Misconceptions and Educational Strategies in Science and Mathematics*, 163–176 (1987). This volume, 195–210.

"A Tale of Two Controversies: Dissonance in the Theory and Practice of Rationality," *Zygon* 23, No. 3, 291–325 (1988). Five contemporary essays on this paper appear in the same issue of *Zygon*. Reprinted in M.R. Matthews, ed., *History, Philosophy, and Science Teaching: Selected Readings* (Toronto and New York: Teachers College Press and OISE Press, 1991). This volume, 335–362.

"Reply to Criticisms," *Zygon* 23, No. 3, 363–68 (1988). This volume, 383–88.

"The 'Interests' of Science and the Problems of Education," *Synthese* 80, 81–106 (1989). This volume, 211–230.

"Rationality and Objectivity in a Historical Approach: A Response to Harvey Siegel," plenary talk, in Don E. Herget, ed., *The History and Philosophy of Science in Science Teaching: Proceedings of the First International Conference*, 143–153 (1989). This volume, 121–131.

"Hermeneutics and Science Education: an Introduction," *Science and Education* 1, 337–348 (1992). This volume, 231–241.

"Hermeneutics and the New Epic of Science," in M.W. McRae, ed., *The Literature of Science* (Athens, GA, and London: University of Georgia Press, 1993, 186–209). This volume, 261–279.

"Hermeneutics as an Approach to Science—Part I," *Science and Education* 2, 1–29 (1993). This volume, 3–27.

"Hermeneutics as an Approach to Science—Part II," *Science and Education* 2, 303–328 (1993). This volume, 29–51.

An overview of Martin Eger's writings on teaching evolution appears in the section "Responding to Creationism" in the anthology *Evolution Extended: Biological Debates on the Meaning of Life*, C. Barlow, ed. (Cambridge, MA: MIT Press, 1994, 274–75).

"Alternative Interpretations, History, and Experiment: Reply to Cushing, Crease, Bevilacqua and Gianetto," *Science and Education* 4, 173–188 (1995). This volume, 133–147.

"Achievements of the Hermeneutic-Phenomenological Approach to Natural Science: A Comparison with Constructivist Sociology," *Man and World* 30, 343–367 (1997). This volume, 53–71.

"Language and the Double Hermeneutic in Natural Science," in M. Feher, O. Kiss, and L. Ropolyi, eds., *Hermeneutics and Science* (Dordrecht, Netherlands: Kluwer Academic Publishers, 1999, 265–280). This volume, 73–88.

Presented Talks and Circulated Writings

"Position Paper on the CUNY B.A. Degree" (1970), a response to a proposed new university-wide baccalaureate program permitting students to design individualized programs of study and take courses at any CUNY campus, thus allowing the closing of some academic departments in 'unpopular' disciplines (like Philosophy). The Position Paper, a parable about a young Brahmin who searches for 'true knowledge,' was circulated among Martin Eger's colleagues. The new degree option went into effect in 1971.

"Philosophy and the Physical Sciences in Undergraduate Education," presented at the annual meeting of the American Philosophical Society, Boston, MA (1976). This volume, 243–259.

"Kohlberg, MacIntyre, and Two Basic Problems of Education," presented to the New York State Sociological Association, Potsdam, NY (1983). This volume, 389–401.

"Rationality of Science and Social Controversy," presented at a joint meeting of the History of Science Society, the Philosophy of Science Association, the Society for the History of Technology, and the Society for the Social Studies of Science (1986). This volume, 301–316.

"The New Epic of Science and the Problem of Communication," presented to the Society for Literature and Science, Albany, NY (1988). This volume, 281–294.

"Hermeneutics in Science and Literature," presented at a conference on Cybernetics: Its Evolution and Praxis, Amherst, MA (1991). This volume, 115–120.

"Meaning and Contexts in Physics: A Case Study," presented at the Universities of Auckland and Waikato, New Zealand (1992). This volume, 101–114.

"Natural Science and Self-Understanding," presented at the European Association for the Study of Science and Technology—Hermeneutics Section; Budapest, Hungary (1994). This volume, 89–99.

"Evolutionary Cosmology at the Turn of the Century: The Case of George Darwin," presented in New Orleans at the annual meeting of the History of Science Society, the Philosophy of Science Association, and the Society for the Social Studies of Science (1994).

"Academic Dialogues and the Post-Modern Mood," memorandum to colleagues at City University of New York's College of Staten Island (1995). This volume, 161–64.

"Gene Therapy: New Rules," response to a National Institutes of Health request for comments on proposed new rules for gene therapy trials (2001). The statement was shared with the NIH Human Protections Committee of the NIH Director's Council of Public Representatives and elicited substantive replies from several members, to which Martin Eger wrote "Further comments on rules governing clinical trials" (2001).

"Reflections on the CLL gene therapy story, Part I" and "Reflections on the CLL gene therapy story, Part II" (2001). These essays, concerning the current state of research and government involvement in gene therapy for chronic lymphocytic leukemia, were posted on the CLL online discussion site.

Work in Progress at Martin Eger's Death

The major active project was a book for the serious general reader on the history, scope, and implications of the realism/instrumentalism controversy in science. It was partly narrative in form, interspersed with substantive discussions and historical illustrations. Full drafts exist of the following chapters:

"How I Invented the Laws of Arithmetic (A Story-Essay on the Problem of Knowledge")): a 15-page account of Martin Eger's arrival in the United States as a boy and his experience attending public school while knowing no English. Without understanding the meaning or purpose of his actions, he learned to 'solve' problems in fractions by watching the teacher's operations and fitting arbitrary rules to them. The teacher was happy; he was not. Comparison is made with the Rhind Papyrus and Egyptian mathematics.

"World Systems in Old Europe": a 24-page essay that begins with a narrative description of an encounter in 1624 between Galileo and his old friend Maffeo Barberini, who had just become Pope Urban VIII. This meeting leads Galileo to write, and to publish in 1632, the *Dialogue Concerning the Two Chief World Systems*, which in turn leads to his trial and recantation in 1633. The conflicting views of the Pope and the scientist are discussed in terms of realism and instrumentalism; a connection is made with the "discovery of the laws of arithmetic" in the previous essay.

"Quantum Mechanics at M.I.T.": a 15-page discussion of instrumentalism in relation to quantum mechanics as it was taught to Martin Eger in the 1950s. It includes a narrative on the discovery of the "Balmer series."

"Bohr's Rules": an 18-page narrative and discussion depicting Bohr in 1914, "those last months of the old European world," and the change from classical to quantum physics "with the guidance provided by the Balmer formula" (Pais). *"From Copernicus to Balmer to Bohr, we are still in the game of fitting rules to numbers."* Bohr's rules – stationary states; quantization; the radiation rule – are then discussed.

There follows a "Plan for the remaining sections":

An aside on 'understanding.' What does it mean to understand? Parts and wholes. Attempts to understand the Balmer formula by means of classical physical models.

A section on the Bohr model as an 'explanation' of the Balmer formula. Yet the Bohr model is another set of rules! One set of rules explaining another set of rules? Of course it's more than that if one accepts the physical picture.

A section on the Schrödinger equation as an explanation of Bohr's rules, but now the physical picture changes. The problem of interpretation of the formalism (rules). Discussion of the original question – why q. m., and much of science, is regarded instrumentally (*de facto*). The more we advance, the more we shed everything but the mathematics – the rules.

The strange story of E. Husserl in the 1930s, the Vienna lecture and the *Crisis*. The fear of realism. Husserl's great charge: science mistakes what is "just a method" for reality itself, that's what's causing the crisis of Europe—1936! Husserl was an instrumentalist in science, like Urban VIII. Like Urban, he thought the issue had world-importance.

Beyond instrumentalism: constructivism today, the heirs of Husserl.

What I make of it all, why I am no longer an instrumentalist: When the ship lists too much to the port side, I hear a voice from some invisible loudspeaker above, saying "Go to starboard."

Other files of notes for future writing include:

For a work on the intersection of art and science: Notes taken in the Val Gardena (in the Italian Dolomites), a community of hereditary, apprenticeship-trained wood sculptors, where the Egers spent two summers living in the home of a sculptor.

Also relating to the intersection of art and science: Notes concerning a family friend and major personage in Martin Eger's boyhood, the Hungarian aristocrat Paul Fejos, a physician, distinguished silent-film director, and later an anthropologist (the founding director of the Wenner-Gren Foundation and the initiator of Libby's development of carbon dating).

Notes concerning the 'creationism/evolution' trials in the United States.

Notes for a paper provisionally titled "What Is Scientism?"

Notes for a paper or papers on 'objectivity.'

Notes

Introduction

1. Leming, James. "Historical and Ideological Perspectives on Teaching Moral and Civic Virtue," *International Journal of Social Education* 16, no. 1, Spr/Summ 2001, 5. This article is a useful brief survey of the history of moral education in New York State and more generally in the United States, with a substantial bibliography. See also Thomas Lickona, *Educating for Character* (New York: Bantam Books, 1991).

PART ONE

HERMENEUTICS AND NATURAL SCIENCE

Chapter One
Hermeneutics as an Approach to Science:
Part I

1. Hermeneutics is the theme of practically all of Gadamer's writings, but the major work is *Truth and Method* (1975). On Habermas's use of hermeneutics, see his (1971) and (1978), and McCarthy (1978). Hesse's formulations are given in her (1980) and in Arbib and Hesse (1986). Excellent overviews of all these approaches are contained also in Bernstein (1983).

2. This separation pervades the work of both these authors (see Gadamer 1975, 409–412; Habermas 1971, 308–309; 1984, 109–110; 1988, 300). But the idea is widespread; it follows from the assumption that the sole purpose or 'interest' of natural science is 'prediction and control' (see von Glasersfeld 1988, 89). However, Habermas especially has emphasized that in expositions of science *beyond* the professional community, hermeneutics *is* relevant (see his 1985, 89; 1988, 299).

3. Among the few who have applied hermeneutics to natural science explicitly,

though in different ways, are Hesse (1980), Arbib and Hesse (1986), Stent (1985), Heelan (1983; 1988a), Ihde (1991). Polanyi (1958; 1969) and Kockelmans (1966) have done so implicitly. A physicist, Bruce Gregory (1988), has used the interpretive-linguistic approach in a recent popularization.

4. A classic argument for the separation, given specifically to refute Hesse on this point, is contained in Habermas (1984, 109–110).

5. In his widely read *The Postmodern Condition* (1984), Jean-Francois Lyotard described how the development of communications technology makes of science, and knowledge in general, a "commodity" whose worth is measured not by its truth value but by its "performativity," its use—a development that he sees as leading to the end of the "age of the professor" and of the university as we know it (1984, 53).

6. Reduction of the objective claims of science to rhetoric or institutional practice exist in both explicit and implicit form. Implicitly, this occurs when the focus on the rhetorical, "discursive," or social aspects of science is so strong that the most serious cognitive claims of science are overshadowed—as in the well-known work of Michael Foucault (1972; 1980, 85, 130–33). Less sweeping analyses in which science is also treated as rhetoric include, for example, Bazerman (1988). On the other hand, from Harry Collins, a sociologist, we have this kind of explicit reduction: "The natural world has a *small or nonexistent* role in the construction of scientific knowledge" (Collins 1981, original emphasis). A more restrained form of sociological downsizing comes from leaders of the "strong program" such as David Bloor (1991). Out of philosophy proper, the best known figure to participate in this general trend is Richard Rorty (1980; 1988), but see also Harding (1991). The widely accepted view among critics, that science is 'prediction and control' is not seen by them as in conflict with the picture of science as rhetoric.

7. Unfortunately, the word 'objectivism' is used very inconsistently. Usually, such criticisms refer to the supposition that science is well separated from the objects of its study, that the objects have an independent existence, and that science, nevertheless, gives true descriptions of these objects. 'Foundationalist philosophies,' like the realist school in the philosophy of science, are opposed by those who follow Rorty (1980) in his skepticism that *any* foundation exists for any branch of learning that would raise its cognitive status above that of others. Lyotard's "grand narratives" (1984) refers to social or metaphysical justifications for the pursuit of science and learning, e.g., the 'story' of the 'emancipation' of humanity through reason.

8. The charge of 'domination' (a translation of the German *Herrschaft*) leveled at science, and used in this open-ended way, is typical of European philosophers but has now become popular in many parts of the world. It is made possible by the idea of science as 'prediction and control.' See Marcuse (1964, 158), Habermas (1970, 112), Gadamer (1975, 433); for recent connections to 'Western' culture and objectivity, see Keller (1985, 96–97) and Harding (1991, 267 and Ch. 6).

9. See Oliver and Greshman (1989), Hostetler (1991), Young (1990). In presenting these critiques as 'problems' for science education, I do not by any means deny their value as critiques—a point of view that should be increasingly apparent, since hermeneutics too is a critique, having much in common with those now receiving attention.

10. This seems to be the upshot of Habermas's well-known theory of "knowledge-constitutive interests" (1971, Appendix), which I have criticized (Eger 1989b). See also notes 3 and 9.

11. Polanyi (1958), Hanson (1965), Kuhn (1970; 1977), Feyerabend (1988 [1975]), Lakatos (1970), Laudan (1977), Hesse (1980), Piaget (1970; 1971).

12. That many students, including future *teachers,* 'hate' science is well known and well documented (Stinner 1992; Bettencourt 1992). Deeper roots of the problem were much discussed in early twentieth century continental philosophy (see Forman 1971).

13. Arbib and Hesse (1986, 181), Heelan (1983, 193–97). See also the references in note 4. Mary Hesse has given general arguments for a "continuum" theory, which assigns to all disciplines, from literature to physics, varying degrees of hermeneutical concern. See her (1980, xxi, 185–86). For an extended discussion of construction and hermeneutics see Arbib and Hesse (1986, 171–196).

14. I have reformulated somewhat the description of the 'fore-structure' to clarify the skimpy or abstract definitions often given. The original statement is in Heidegger (1962, 191). But see also Connolly and Keutner (1988, 18), Bernstein (1983, 136), and Jones (1989, 126, 138, 146).

15. As a general idea, this is of course not new; but to Kant's *innate* forms of understanding, many other, contingent ones are now added. For basic expositions on interpretation, "the circle," pre-understanding, etc., see Mueller-Vollmer (1988), Gadamer (1976), and Dreyfus (1985). For Heidegger's views on natural science, see his (1977, 247–318) which has his analysis of Galileo and Newton, including mathematization as the pre-understanding of modern physics (1977, 268), and his essay on technology (1977, 283).

16. See references in note 11.

17. Eger (1992b) includes a brief discussion of preconceptions in science education and further references.

18. This feature is embedded in Gadamer's concept of "effective history" (Gadamer 1975, 267). Different ways in which the interpreter becomes part of the interpretations in science are described by Forman (1971) and Feuer (1982). See also Cushing (1994) and note 28.

19. This basic contemporary idea is a major point for Kuhn (1970, 206) and for Arbib and Hesse (1986, 181), and was treated in detail by Laudan (1981).

20. The critical literature on Kuhn is vast, but Lakatos and Musgrave (1970) and Gutting (1980) are two indispensable collections. In addition, see Scheffler (1967) and Shimony (1976).

21. That is, *as a mode,* the hermeneutic circle is prior. In particular ("regional") concerns, it may follow some previous construction.

22. This appears to be in sharp contrast to Popper's "world 3" concept, in which ideas coded in texts exist complete in a world of their own (Popper 1972, 106–152). However, although Popper classifies himself as an 'objectivist,' and is criticized as a 'positivist,' he has been open to the arguments of hermeneutics, and has himself offered a theory of understanding based on the dialogue between "world 2" and "world 3" (Popper 1972, Ch. 4). He denies that hermeneutics is characteristic of the humanities rather than the sciences, or that a deep form of 'understanding' can be used to distinguish between the two (1972, 183–85).

23. The performing arts play an important and unique role in the form of hermeneutics developed by Gadamer (1975, 91–152).

24. As a philosophical concept, "horizon" originates in phenomenology. But care has to be taken because it is not always used in the same way (see Husserl 1970, 162, 358–59; Heelan 1983, 8–12; Ihde 1977, 60–66).

25. In this paper I use the word "phenomenological" mostly in the general sense (the scientist's meaning rather than philosopher's), as something that avoids delving into theories or mechanisms by staying close to the perceived phenomenon. Whenever specifically Husserlian phenomenology is intended, I try to indicate that, the latter being highly relevant to the form of hermeneutics discussed here.

26. It is recognized that this contextual or 'horizonal' quality of the hermeneutic dialogue resembles the situation in quantum mechanics, especially as formulated by Niels Bohr (Heelan 1983, 208–209).

27. The complex relations among interpretation, hermeneutics and embodiment cannot be discussed here in depth; nor are they the same for all authors. I take all perception, including that which is instrumentally mediated, to involve interpretation. Embodiment, by incorporating instruments, internalizes some interpretation, which otherwise would be external, overt, subject to reflection, and therefore hermeneutic. In this sense, then, every embodiment reduces hermeneutics and brings the object 'closer.' Ihde (1979, Ch. 1) gives detailed discussion of varying degrees of embodiment. However, the act of successive, improved embodiments, as a whole, is always hermeneutical in the sense of "the circle." The relation of the embodiment concept to realism in regard to the findings of science is discussed at length by Heelan (1983) and Ihde (1991).

28. Polanyi refers to Wilhelm Dilthey's concept of "empathy" or "transference." For an overview of Dilthey's theory of understanding, of hermeneutics, and of his distinction between the two types of sciences, see Habermas (1971, Chs. 7 and 8).

29. In Gadamer's philosophy, the subject is embodied in a historically constituted forestructure, which he calls "effective history." See his (1975, 267ff).

30. On the connection between conceptual change in the individual and in science, see especially Piaget and Garcia (1989), and Franco and Colinvaux-de-Dominguez (1992).

31. It has to be kept in mind that constructivist approaches exist in several varieties. In education, the most familiar is probably that of Piaget (1970; 1971) and von Glasersfeld (1988; 1989), which are based on psychological and linguistic studies. Social constructivism is represented, for example, by Bloor (1991). In science history, Kuhn's (1970) is the best known, and in philosophy, Van Fraassen's "constructive empiricism" has had strong impact. For criticism of an extreme form by one who is herself a constructivist, see Hesse (1988a). Criticism in the context of science education, and further references, are given by Matthews (contained in Alexander [1972]).

32. For realist replies to constructivism (van Fraassen's in particular) and to relativism, see Churchland and Hooker (1985), Cushing et al. (1984), and Hollis and Lukes (1982).

33. This 'root' of constructivism is attributed to Giambattista Vico (von Glasersfeld 1989, 123). For other formulations of the same idea, and its relation to Piaget, see von Glasersfeld (1988, 105).

34. Hacking (1983) in effect describes instrumental observation as perception. Theoretical entities become "real" when "observed" by means of different technologies and used instrumentally for further observation. His detailed discussion of microscopy and its history (Ch. 11) is particularly relevant to McClintock's experience.

35. The degree and nature of 'interaction with the environment' (or with persons) in Piaget, for example, is problematic according to von Glasersfeld (1988, Ch. 12), and "does not include an exchange of 'meaning,' 'knowledge,' or 'information'" (1988, 253). This is a difficult point, but one on which a decision is not necessary for a hermeneutical approach.

36. It is an interesting hermeneutic question whether Ibsen's play, *The Master Builder*, is an allegory on precisely the 'Western,' scientific man as he was seen at the turn of the century.

37. The universe as a "black box," von Glasersfeld's metaphor (1988, 109), embodies the conviction that what the scientist investigates can never be directly perceived. This metaphor exemplifies well the double-edged nature of constructivism in education: As a precaution against the seductive power of the big stories about how things are and how they got this way, as a reminder that it might all 'be' radically different, the "black box" is highly effective; but regarding human beings trying to understand the world, the image is one of dismal isolation, of the student as hopeless *outsider*. Because "equilibration" is envisaged by Piaget (1971) as a type of cybernetic feedback mechanism of trial and error, it bears a suggestive resemblance to the hermeneutic circle and reflects once again a certain parallelism of ideas about knowledge. That the parallel is only partial should be clear at least from the ontological nature of hermeneutics—which explains also why the "black box" is not a hermeneutical concept.

38. See note 5.

Chapter Two
Hermeneutics as an Approach to Science: Part II

1. Eger 1993a (this volume, 3–27) will be referred to in the text and in the notes as "Part I" (of this two-part paper).

2. Other versions of this argument can be found in Taylor (1971), and Dreyfus (1985). Rouse (1987, Ch. 6) criticizes both these arguments in detail but in a manner different from that offered here.

3. Were we to accept Galileo's view that nature *already has* its own language, which we must try to understand, it would follow immediately that in physics there is also a "double hermeneutic," more or less of the same kind as in anthropology; and this, of course, was the point of Galileo's philosophical debate with Cardinal Bellarmino. For a modern view of science as language, see Gregory (1988).

4. The problems of language use in a laboratory have been discussed, for example, by Polanyi (1958, 101) and Schilk et al. (1989).

5. When I use the adjectives "focal" and "peripheral" (or "subsidiary") as modifiers of the noun "attention," I understand them in the technical sense of Polanyi (1958, 59), since it is my aim to show here the confluence or coherence of several distinct lines of phenomenological description.

6. See Popper (1972, Chs. 3–4). It may appear at first sight as though the two-book metaphor assumes a realism—the first book being the 'real.' This is not so (although that view is not excluded thus far). While the second book, nature in the language of science, is essentially Popper's "world 3," the first book is not necessarily his "world 1" (independent physical reality). It could be simply the *phenomena* of nature as given to the researcher. See further in the text.

7. See Part I, note 7. The new organization, which includes more than the "book of science," is the Society for Literature and Science, initiated by faculty at Worcester Polytechnic Institute and the Rensselaer Polytechnic Institute.

8. To document here all these debates, well-known to historians of physics, would be inappropriate. As an example, regarding classical electromagnetic theory, see Bevilacqua (1983; 1984).

9. Feynman actually used the 2nd (1935) edition of Dirac's book and a paper (Dirac 1933) devoted to this subject. The 4th edition, to which I refer, includes the topic at issue in Sect. 32. The relation in question is $<q'_t|q''>$ "analogous to" $\exp\{iLt/\hbar\}$ where t is an *infinitesimal* time interval; the left side of the relation is the quantum probability amplitude for a particle transition from q' at time 0, to q'' at time t; L is the classical Lagrangian at the earlier time, i is the imaginary, and $\hbar$ is Planck's constant divided by 2π (Dirac 1958, 128).

10. See Part I (this volume, 3–27) for a discussion of the "circle." Some authors (e.g., Dreyfus 1985) distinguish between a "theoretical circle," such as is found in science, and a "hermeneutic circle" that involves pre-scientific, even non-cognitive preconceptions (related to "stage 0" of Giddens and Habermas). I do not make such distinctions here because I wish precisely to call attention to the structural similarity of these circles. In our example, Feynman is struggling with *the already existing language of quantum mechanics,* not with natural phenomena as such.

11. Although the reinterpretation of quantum mechanics was essentially achieved in the early 1940s, it was not published until after the Second World War (Feynman 1948). For a historical account, see Schweber (1986).

12. Feynman says "equal" here (1966, 37), but he means "equal to within a multiplicative constant."

13. This volume, 3–27.

14. Throughout this paper, the term "ontological" refers to questions concerning *modes* of being in the world and their meaning, and is thus to be distinguished from "ontic." See, for example Dreyfus (1991, 19–21).

15. For an example of 'losing oneself' in the 'game' of science, see the discussion of Barbara McClintock in Part I (this volume, 3–27).

16. Ingarden includes coherence, or "harmony," not only with the core but with the other interpretations (1989, 113–15).

17. This is an old idea, developed at length, for example, by Ernst Cassirer (1953). See also Heelan (1991).

18. Popper (1972, 164) allows for something like a circular or iterative process of problem solving, but problems themselves already exist in world 3, though not necessarily explicitly. See note 6.

19. As a result possibly of the work of the historians of experiment (e.g., Hacking 1983; Galison 1987), of sociologists of science (e.g., Latour 1987), and of historians of science focusing on differing interpretations (e.g., Bevilacqua 1983).

Bevilacqua (1986) is particularly effective in using his own historical investigations to argue that the traditional subject/object distinction is questionable even in physics. Other developments also point toward the presentation/performance aspect of science. One is the emergence of *visualization* (in experiments and computer simulations) as a research field combining objective and perceptual aspects. Another is the proliferation of science museums such as the Exploratorium in San Francisco and the Ontario Science Centre in Toronto. The most explicit example, no doubt, is the new Biohistory Research Hall in Takatsuki, Japan, where "scientists will 'perform' for the public in glass-walled labs" (*Science* 260, 1993, 391).

20. Williams (1965, 326).

21. In his book on relativity, Bohm is quite explicit on this matter: "The difficulty was not that the Lorentz theory disagreed with experiment. On the contrary, it was in accord with all that had been observed at the time of Lorentz, and in fact, it is also in accord with all that has been observed since then" (1965, 42. See also note 25).

22. It is now well known that, personally, Einstein did not start from Lorentz's equations to arrive at the crucial results of his 1905 paper. Not having actually seen Lorentz's important work of the year before (Lorentz 1904), he re-derived these equations himself. However, from the point of view of the physics community, his derivation appeared as a reinterpretation. This is precisely what was implied in one of the most important histories of the subject (Whittaker 1910). See Holton (1960).

23. K.-O. Apel (1968, 322) is one of the few philosophers who recognizes the role of hermeneutics in natural science and has explicitly called attention to this aspect of Einstein's theory.

24. For example, Weinstock (1965) and Penrose and Rindler (1965). One relatively recent reinterpretation using the *ether* has been developed by H. E. Ives, following Lorentz, and was discussed historically and analytically by H. Erlichson (1973). Erlichson's conclusion is that so far there has been no experimental proposal that would distinguish between "the special theory of relativity and this Rod Contraction-Clock Retardation Ether Theory" (1973, 1076). See this paper for further references.

25. Bondi (1964). See also Bohm (1965, Ch. 26).

26. Pais (1982, 152).

27. Such reinterpretations as Hertz's, and Feynman's non-relativistic quantum mechanics, do not fit into Kuhn's picture of science either (Kuhn 1970). They were not part of any revolution, and they were surely not an exercise in normal 'puzzle-solving.'

28. See Gadamer (1975, 85, 441–43), Heelan (1983, Chs. 10 and 11) and Ihde (1991). See also Part I regarding objectivity and the dialogical character of interpretation (this volume, 3–27).

29. Collins (1981).

Chapter Three
Achievements of the Hermeneutic-Phenomenological
Approach to Natural Science: A Comparison
with Constructivist Sociology

1. See, for example, Gyorgy Markus (1987). For replies to Markus see Heelan (1989b) and Eger (1993a, this volume, 3–27). Of the major theorists, H.-G. Gadamer is reputed to confine hermeneutics strictly to the *Geisteswissenschaften*. Were this so, however, it would contradict his basic thesis of the universality of hermeneutics. Actually, he can he interpreted both ways, especially in view of his belated recognition (prompted by Thomas Kuhn) that "there is something like a hermeneutical problematic in the natural sciences too" (Gadamer 1985, 179). Apel's position also contains ambiguities: On the one hand, he counterposes hermeneutics to "scientistics" and considers them complementary; on the other, he describes hermeneutics as a "precondition" for science (Apel 1972, 28–29; 1988, 330–32). That the "precondition" is often an overt process within science itself is something he seems to deny, though that is arguable.

2. In order to retain the more direct and personal style of this talk, as given at the conference, I have refrained from converting its phraseology to that of a 'paper.' However, since it was originally given to a group of scholars fairly familiar with each other's work, I thought it advisable for a larger readership to supplement the original with more extensive explanatory notes. I also wish to thank John Bahcall, Harry Collins and Trevor Pinch for reading an earlier draft of this paper and offering their critical comments.

3. The term "objectivism" will be used here always in the sense of Husserl (1970, Sect. 14) and as generally understood in phenomenological philosophy, and is not to be confused with the ordinary use of the word "objectivity" in science. The phenomenological approach to natural science seeks *to understand* and to articulate more explicitly the nature and the varieties of "objectivity" in each science, as, for example in Heelan (1967).

4. Habermas (1984, 110), Giddens (1976, 158).

5. Eger (1993a, 305ff, 317ff), Føllesdal (1993). At the 1993 conference in Veszprém, Hungary, I argued that the literature of a scientific field constitutes a chapter in what might be called the "book of science" (as distinguished from the "book of nature"), and that it must be studied to gain access to the conceptual language of that field. That this is not the *same* as studying the language *used* by the objects of study, as in anthropology, is clear. But in both situations, at the stage of language acquisition, interpretation is needed; and in this sense the two are comparable. Regarding stage 2 interpretations such as those of quantum mechanics, my point is that they are more frequent in science than is generally realized by anyone except the historians, that they often have direct cultural significance, and that they deserve more attention.

6. Crease's formulation builds on Heelan's analogy between a scientific phenomenon and a perceptual object (see note 19). In this view, an experimenter "perceives" a profile of the phenomenon by means of apparatus, and "recognizes" that phenomenon as an *invariance* (under transformation of standpoints or perspectives,

in the Husserlian sense) which structures still other "profiles" (Crease 1993, Ch. 6). Thus also, the word "phenomenon" will be used here not in the Kantian sense but as an *extension* to natural science of its Husserlian meaning, as in Crease's book. Heelan, however, sometimes calls it the "experimental object" (Heelan 1989a).

7. Polanyi (1958; 1969; 1975), Kuhn (1970).

8. The contributors to whom I refer are Kockelmans (1968; 1985; 1993), Kisiel (1973; 1980; 1983), Heelan (1977; 1983; 1989b; 1991), Ihde (1977; 1979; 1991), and Crease (1993).

9. STS stands for the numerous "Science and Technology Studies" centers on such campuses as Cornell, Stanford, Wesleyan, Virginia Polytechnic Institute and State University, and to such British "Science Studies Units" as those at Bath and the University of Edinburgh. For typical descriptions of "negotiation" in natural science see Latour and Woolgar (1979, 134, 156–57), Collins (1985, 143ff), Shapin and Schaffer (1985, 226ff), and Rouse (1987, 125–26).

10. Collins and Pinch (1993). The chapter on solar neutrinos is based on an earlier, more extensive study of this subject by Pinch (1986).

11. Collins and Pinch (1993, 139). The emphasis on careerism as a driving force of science is a distinctive mark of the writings of the constructivists and is part of the meaning of negotiation. How far such bias can go is shown by the following "speculations": "Although we should be wary of simple-minded models of scientists as rational calculators who always try to promote what is in their best career interests, we can nevertheless speculate as to what rationale Bahcall might have had for his dramatic change of position" on whether experiment agreed with theory (Collins and Pinch 1993, 133–34). On the one hand, we are told, there was the "pressure" of theorists like Iben, whose findings made Bahcall's position "tenuous"; on the other hand, the famous Richard Feynman "advised the young Bahcall that . . . if there was a contradiction this made the result *more* rather than less important." No mention here of any new evidence, no *scientific* reason is given for the change of mind. However, in his own earlier work on this, in a section entitled "Bahcall Changes His Mind," Pinch had given two technical reasons—changes in numerical values of the solar opacity and in the cross-section for the reaction $Be^7 + H \rightarrow B^8 + \gamma$ (two key parameters in solar models). This raised the predicted neutrino capture rate considerably, and made it much higher than the increasingly more accurate and *lower* results of Davis (Pinch 1986, 144–46). In the earlier work, after reporting this scientific evidence, Pinch rejected it as insufficient for the change of mind, and offered instead his own careerist explanation. But in *The Golem,* the reader is not even allowed to know that this evidence was in the picture. In addition, "rationality" undergoes a significant change of meaning: No longer is it possible for a reader to disregard Pinch's interpretation and to see in this history an example of rationality in the classic sense—an inference from evidence to a scientific conclusion. Now a scientist is described as "rational," if at all, only in the *instrumental* sense of using reason in the service of career—with scientific conclusions falling out as they may. In a comprehensive review of *The Golem,* David Mermin called this type of bias the "sustaining myth" of constructivist sociology (Mermin 1996).

12. The analogy with art, especially performing art, is a major feature of contemporary hermeneutics of science. It is the theme of Crease's (1993), based on Heelan's (1988b, 522; 1989a, 302–303). My (1993a, this volume, 3–27) places

somewhat more emphasis on theory than does Heelan, and is based on Gadamer's and Ingarden's phenomenologies of art (Gadamer 1975, First Part, II; Ingarden 1989). In all these treatments the concept of interpretive "performance" is crucial.

13. The shift of the subject/object cut in physics, for example, takes place, according to Heelan, in analogy with what happens when an archaeologist learns to read an ancient text: At first, before the language is deciphered, the archaeologist makes the words themselves the objects of attention; later, he or she sees *through* the words, as the *meaning* becomes the object. Here the subject/object cut has shifted from a position between the archaeologist and the words to a position between the words and the meaning (Heelan 1977, 12; 1983, 201–207). Michael Polanyi said something very similar when he made his distinction between "focal" and "subsidiary" awareness (Polanyi 1958, 87–95; Polanyi and Prosch 1975, 33–35).

14. Refer to note 6 regarding my use here of the word "phenomenon."

15. In one of the earliest technical papers on the subject, Bahcall used the phrase "only neutrinos . . . can enable us to *see into the interior of a star*" (Bahcall 1964, 300, his emphasis).

16. Heelan (1977, 31–35; 1983, 197–213), Ihde (1977; 1991, 97ff). The idea of embodiment is closely related to that of the "shift" of the subject/object cut (note 13) and to Polanyi's "indwelling" (Polanyi 1958, 58–63; 1969, 148). Note that not every use of an instrument can be subsumed under the idea of embodiment; only that which has become unproblematic, allowing a "seeing" of the object so well controlled and reliable that the analogy makes sense. Dudley Shapere, an analytic rather than a hermeneutic philosopher, asking the same question about this issue of "seeing" into the sun, reaches a similar conclusion: If by using telescopes and counting photons we observe the exterior of stars, then using other instruments and counting neutrinos we can "observe" their cores (Shapere 1984, Ch. 16).

17. Latour and Woolgar (1979, note 11, 150), Latour (1987, 2, 131).

18. Bahcall (1996a) describes, in addition, a third problem resulting from experiments using gallium instead of chlorine, which further deepens the mystery of the missing beryllium neutrinos.

19. Conversions or "oscillations" of one neutrino "flavor" to another are discussed in Wolfenstein and Beier (1989). A mass of 3 milli-electron-volts would actually be too small to have a significant effect on the cosmic mass; but, in general, the problem of neutrino mass relates this solar research to cosmology. For the latest on independent attempts to measure the neutrino mass, see Glanz (1996).

20. Gadamer(1975, 235–38, 261). Heidegger (1962, Sect. 32).

21. See note 13.

22. Dreyfus (1985, 228–235). See also Rouse (1987, 50–68).

23. *Fore-conception, fore-sight,* and *fore-having* are part of what Gadamer calls the fore-structure of understanding (Gadamer 1975, 235ff; Heidegger 1962, 191). See also Dreyfus (1985, 233–34).

24. Perhaps the earliest effort to call attention to the scientist's embeddedness in the problem situation was that of Kisiel (1973; 1979). A later effort to adapt to laboratory research the Heideggerian concept of "circumspection" has been carried out by Rouse (1987, Ch. 4). The point here is to show that the background tacit knowledge of scientists is not just their 'scientific world-picture' but also their pro-

fessional life-world understanding of their own situation in relation to equipment, the state of the art, their financial means, the work of others, etc.

25. See note 11.

26. The new generation of very large detectors may lead to one of several 'smoking gun' indications of something happening to the neutrinos *en route,* such as a distortion of the shape of the neutrino energy spectrum (Bahcall 1996a and personal communication). But I am specifically referring to planned experiments like the one in which a neutrino beam emanating from Fermilab in Batavia, Illinois, will cut through a slice of the curving Earth, emerging 460 miles away at a large detector in an iron mine in Duluth, Minnesota. The aim will be to see what fraction of this known beam's neutrinos convert during that long trip. Collins would object here: But how will you know whether the results of *this* experiment are right? Reply to Collins: If the results do not cohere with the solar experiments, we will indeed still be in the dark; but if they do, then—closure. Because then, an even larger set of experiments and theories will all validate each other simultaneously like the last pieces in a jig-saw puzzle.

27. However, a most interesting discussion of the possibility of shrinking ("closing") the large circle between knowledge of the natural world and the nature of the knowing subject (ontology and epistemology) is given by Abner Shimony (1993).

28. One example of "negotiation" regarding performance is given by the philosopher Ronald Giere (1988, 14). Two experimenters argue over a slight malfunction of the instrument (a cyclotron at Indiana University) because it is not clear whether the trouble will affect the results. Is it better to correct the malfunction, playing it safe at the price of time lost, or to take more data at the risk of it all proving useless in the end? But this is *not* negotiation. There is no trading on interests here. Both scientists have the same interest—to acquire reliable data. They differ in judgment on which path is more likely to lead to that goal and their arguments are wholly scientific. Giere's other "negotiations" (274) are of the usual sort—over whose paper is to be published first, who should have privileged use of a facility, etc. That any of this might have influence over the final scientific result is, in effect, repudiated by Giere, though by discussing it he seems to be "paying his respects" to a powerful school of thought.

29. *Agon* (the joust) is Lyotard's metaphor for the language game scientists "play against one another" while nature stands by "mute" (Lyotard 1984, 10, 17, 57). This view has been enthusiastically adopted by Latour and Woolgar (1979, 237), by Collins and Yearley (1992, 382), and by others. Latour's view of scientists as "entrepreneur-generals . . . waging war" has been criticized, for example, by Amsterdamska (1990). See Fujimura (1992, 170–71).

30. A well-conducted dialogue, in Gadamer's view, is like a game in which the players forget themselves and allow the thing at issue to "take over" so that the players are in fact "played." The players here are not the subjects, contemplating the object of their conversation; rather, the object "reaches presentation" through the dialogue of the players (Gadamer 1975, 91–92, 345ff, 341ff). Similarly, Kisiel (1979, 406) describes the dialogue between a scientist and the problem situation as one in which "the situation 'asks.'" In view of the point made by Evelyn Fox Keller about Barbara McClintock's non-dominating way of doing biology (Keller 1983, 117–125), and of the protracted discussions of alleged "dominating" attitudes

among the founders of Western science (Keller 1985), it is worth noting that "giving oneself" in Gadamer's sense is a totally genderless concept. Gadamer has such examples in mind as men playing soccer or chess, men and women acting in a theatrical performance, and so on. It is in this sense that I use the expression here.

Chapter Four
Language and the Double Hermeneutic in Natural Science

1. Hesse (1980), Bernstein (1983), Grene (1985), Arbib and Hesse (1986).

2. Habermas (1984, 110). Other versions of this argument can be found in Taylor (1971), and Dreyfus (1985). Rouse (1987, Ch. 6) criticizes both these arguments in detail but in a manner different from that offered here.

3. For a direct exchange between Hesse and Habermas, see Hesse (1980, Ch. 7) and Habermas (1984, 109–111). H.-G. Gadamer, for the most part, shares the objectivist view of science (e.g., Gadamer 1975, 251–53), but more recently, under the influence of Kuhn and the "new philosophy of science," he has admitted that in natural science too "there is something like a hermeneutical problematique" (Gadamer 1976, 179). Of all the leading continental theorists of hermeneutics, K.-O. Apel, with his "principle of complementarity" (Apel 1988, 330–31), and his direct approach to the issue, seems to be the most accommodating to the idea of a hermeneutical aspect of natural science. Nevertheless, he too takes an essentially positivistic view by assimilating natural science to objectivity or instrumental action or technological praxis.

4. Habermas's general discussion of "cultural value spheres," from which the notions of "expert cultures" and "cultural expert systems" are derived, is contained in his (1984).

5. Polanyi (1958), Polanyi and Prosch (1975), Heelan (1977; 1983; 1991), Ihde (1979; 1991).

6. Latour and Woolgar (1986), Latour (1987), Traweek (1988).

7. Latour and Woolgar (1986, 128). Clearly, this distancing posture is itself problematic *as social science,* because it does not pursue the *meanings meant* by the people studied; it raises a kind of double-issue concerning the role of hermeneutics in a social science whose object is natural science. But this is not our issue here. See Taylor (1971).

8. Edge and Mulkay (1976), Latour and Woolgar (1986 [1979]), Knorr-Cetina and Mulkay (1983), Collins (1985), Pinch (1986), Galison (1987), Latour (1987), Pickering (1984; 1992).

9. See Popper (1972, Chs. 3–4). It may appear at first sight as though the two-books metaphor assumes a realism—the first book being the "real." This is not so (although that view is not excluded here). While the second book, nature in the language of science, is essentially Popper's "world 3," the first book is not necessarily his "world 1" (independent physical reality). It could be simply the *phenomena* of nature as given to the researcher, or 'inscriptions' on instruments. See further in the text.

10. To document here all these debates, well-known to historians of physics, would be inappropriate. As an example, regarding classical electromagnetic theory, see Bevilacqua (1983; 1984). For a recent historical-philosophical study of the

debate on the "Copenhagen interpretation" of quantum mechanics and its alternatives, see Cushing (1994).

11. Feynman actually used the 2nd (1935) edition of Dirac's book and a paper (Dirac 1933) devoted to this subject. The 4th edition, to which I refer, includes the topic at issue in Section 32. The relation in question is $<q'_t|q''>$ "analogous to" $\exp\{iLt/\hbar\}$ where t is an *infinitesimal* time interval; the left side of the relation is the quantum probability amplitude for a particle transition from q' at time 0, to q'' at time t; L is the classical Lagrangian at the earlier time, i is the imaginary, and $\hbar$ is Planck's constant divided by 2π (Dirac 1958, 128).

12. Although the re-interpretation of quantum mechanics was essentially achieved in the early 1940s, it was not published until after the Second World War (Feynman 1948).

13. Feynman actually says "equal" here, rather than "proportional," but he means "equal to within a multiplicative constant" (Feynman 1966, 37).

14. For example, Gadamer (1975, 274).

15. One who has made this attempt is Joseph Rouse. As an example, he suggests that the use of "master molecule" theories in genetics (in contrast to collective mode theories) is an expression of a "political" self-understanding of our agency as scientists, reflecting our own powers of one-way intervention in natural phenomena (Rouse 1987, 188–191). This is speculative, at best, and narrowly aimed; and it supports, I would suggest, the need for more serious studies of self-understanding in natural science.

Chapter Five
Natural Science and Self-Understanding

1. This volume, 335–362.
2. See, for example, Segerstrale (1986).
3. Graubard (1988); Dreyfus and Dreyfus (1986).
4. Grünbaum (1990, 1991); Russell et al. (1988); Peacocke (1981).
5. See also von Wright (1974) and Hoyle (1982).
6. See, for example, Russell et al. (1988); Peacocke (1981).
7. See, for example, Hoyle (1950).

Chapter Six
Meaning and Contexts in Physics: A Case Study

1. Lyotard, J.-F. (1984 [1979]), *The Postmodern Condition: A Report on Knowledge*, University of Minnesota Press, Minneapolis, 4–5.

2. Hide, R. and Dickey, J. O. (1991), "Earth's Variable Rotation," *Science* 253, 629–637.

3. Darwin, G. (1879), "On the Precession of a Viscous Spheroid and on the Remote History of the Earth," *Phil. Trans. Roy. Soc.*, Part II, 170, 447–530. *Scientific Papers*, Vol. II, 104.

4. See Burchfield, J. D. (1975), *Lord Kelvin and the Age of the Earth*, University of Chicago Press, Chicago, 35–36.

5. Op. cit., 36–37.

6. Darwin, G. (1880), "On the Secular Changes in the Elements of the Orbits of a Satellite Revolving about Tidally Distorted Planet," *Phil. Trans. Roy. Soc.*, Part II, 171: *Scientific Papers*, II, 208–382. See also Burchfield, 115.

7. Darwin, G. (1905), "Cosmical Evolution," (Pres. Address, Br. Assoc. South Africa), *Papers*, Vol. IV., 524.

8. Darwin, G. (1905), "Cosmical Evolution," 533.

9. See Chandrasekhar, S. (1969), *Ellipsoidal Figures of Equilibrium*, Yale University Press, New Haven.

10. Boss, A.P. and Peal, S.J. (1986), "Dynamical Constraints on the Origin of the Moon," in *Origin of the Moon*, ed. W.K. Hartmann, R.J. Phillips, and G.J. Taylor, Lunar and Planetary Institute, Houston, Texas, 71. George Darwin worked on the problem of 'stable' fission from the Poincaré figure, but Lyapunov later proved the Poincaré figure to be unstable.

11. "Cosmical Evolution" (op. cit.), 539.

12. Ibid., 543.

13. Ruse, M. (1982), Darwinism Defended, Addison-Wesley, Reading, Massachusetts.

14. Papers were published in Hartmann, W.K., Phillips, R.J., and Taylor, G.J. (Eds.) (1986), Origin of the Moon, Lunar and Planetary Institute, Houston, Texas.

Chapter Seven
Hermeneutics in Science and Literature

1. Heidegger (1977, 247–282) has excerpts from *What Is a Thing* (1962), specifically on science, and his analysis of Galileo, including mathematization as the pre-understanding of modern physics (268). See all the excerpts from *Being and Time* in Mueller-Vollmer (1988, 214ff), for basic expositions on interpretation, the circle, the pre-understanding, etc. See also Hubert Dreyfus, "Holism and Hermeneutics" in Hollinger (1985, 227–247).

2. Gadamer (1984); Gadamer (1976). See also the introduction to Connoly and Keutner (1988), and Gadamer's essay there, "On the Circle of Understanding." In addition, Hollinger (1985) has good critical essays on Heidegger's and Gadamer's version of hermeneutics. Bernstein (1985) has an excellent chapter on Gadamer, and a good introduction to the larger problem.

3. Gadamer (1984, 236).

4. Gadamer (1984, 261).

5. Hesse (1980, 185–186, 225). The very useful introduction to this book also contains discussion of the continuum idea (xxi). See Arbib and Hesse (1986, 171–196), for an extended discussion of "construction" and hermeneutics. For a brief summary of her position, see the introduction to Bernstein (1985, 31–33).

6. Stent (1985, 219).

7. Habermas (1984, 110).

8. *Proceedings of the Second International Seminar: Misconceptions and Educational Strategies in Science and Mathematics* (1987).

9. Gadamer (1984, 85).

Chapter Eight
Rationality and Objectivity in a Historical Approach:
A Response to Harvey Siegel

1. Siegel treats rationality as the answers to two separate questions: Q1) In what does the rationality of science consist? and Q2) What is to count as evidence or good reason for hypotheses or procedures? (1989, 12). He takes this distinction as crucial, charging critics (and most philosophers of science) with failing to understand it.

2. Siegel offers the following explanation: Only "good reasons . . . which actually have convicting force" count. "To be *appropriately* moved is to be moved to just the extent that the reasons in question warrant" (1989, footnote 14).

3. All this, according to Siegel, involves Q2, not rationality itself (Q1). I agree that the distinction, when it can be made, is not trivial: In education, at certain early levels, emphasis on reasons distinguishes rationality from sheer whim, conformity, or blind authority. But a fair fraction of *twelve-year-olds* have already understood this, and take reasons for granted. For them, as for Hoyle, the real questions are: *What sort* of reasons? How to evaluate them?

4. About one proton per liter per billion centuries must be created *ex nihilo*. For the sake of simplicity, I have paraphrased Hoyle's argument (1980, 52).

5. Laudan (1977, 132), for example, does accept the possible rationality of philosophical, religious and moral considerations in science; others do not. This, however, is not merely an argument over Q2, or the confusion of Q2 with Q1. It suggests that in many situations the distinction itself is difficult or inadvisable.

6. Of course Hoyle's struggle was waged by means of *counterevidence*, which proves his commitment to evidence. But his *motive* for seeking counterevidence came also from commitments of a different sort. "Commitment," then, is another hazy concept in need of clarification.

7. The prior commitments of scientists are freely admitted by Siegel on the first page of his paper (1989), but thereafter, when mentioned at all, they are treated as deviations. When he speaks approvingly of the scientist's passion, it is a passion for *reasons* (Siegel 1989, 25–26), a very different thing from commitment to *one side* of the issue.

8. One far-reaching criticism has been wholly omitted from this paper because that subject is treated in my (1989b). Siegel's underlying assumption, completely undiscussed, is that the *process* of science (method) is more important than the *content* or product (Siegel 1989, 41). Although this position has been prominent in American reform movements for decades, it now appears to me that there are good reasons for questioning it.

9. I am simplifying somewhat. A crucial contribution at this point was made also by Salpeter (1952). For a concise description of these events, and their relation to the *anthropic principle* and the "fine-tuning" of the universe, see Barrow and Tipler (1986, 250–254).

10. I am simplifying again: Other evidence was of course also involved; see Hoyle (1980). For a concise history of the standard model, see Penzias (1979). However, in a more complex form, the steady-state framework is not dead. Even the three degree background radiation can be included in it. See Hoyle (1974), Ellis et al. (1978), and Barrow and Tipler (1986, 601–608).

11. More recently, Tipler does regard the theory as refuted on theoretical grounds (personal communication).

12. Note that the strong version of objectivity does not preclude periods of systematic protection of young theories, as discussed earlier. Protection is effected at the higher level where the impact of evidence and other considerations are weighed. In a sense, objectivity as defined here might be thought weaker, not stronger, than Siegel's view of the critical thinker; but it *answers* the question about what it is to be "appropriately moved." The only kind of motion required is that *all* the evidence be *developed and revealed*. On final judgment it is silent.

13. The French philosopher Jean-François Lyotard goes so far as to suggest that western society has now changed to the point where all speech should be viewed not primarily as communication but as competition, as part of agonistics

Chapter Nine
Alternative Interpretations, History, and Experiment: Reply to Cushing, Crease, Bevilacqua, and Giannetto

1. However, see Patrick Heelan's paper (Heelan 1995) for a contrasting picture. Heelan sees the standard Bohr-Heisenberg interpretation as "more right than wrong."

2. In addition to Cushing (1994), see also his (1991).

3. Cushing (1994) and Cushing and McMullin (1989).

4. Feynman (1985, Ch. 2), Feynman et al. (1963, Vol. 2, Ch. 26). See also Feynman (1967) for elementary discussions of alternative interpretations of Newton's Laws.

5. Rouse (1987, Ch. 3) gives an extended analysis of this distinction under the heading "What Is Interpretation: Two Approaches to Universal Hermeneutics." But see also Dreyfus (1985, 233ff) and Ginev (1995). For my own discussion of this point and further references, see Eger (1993a, this volume, 3–27, and notes 14, 15; 1993b, this volume, 29–51, and note 9).

6. The first of these is "Meaning and Contexts in Physics: A Case Study" (of George Darwin), a lecture delivered at the Universities of Auckland and of Waikato, New Zealand, October, 1992 (this volume, 101–114).

7. "Fusion of horizons," a concept used by H.-G. Gadamer (Gadamer 1975; Eger 1993a), does not mean *overlap* of languages, or translatability. It means that partially, at least, one language can be *interpreted* by the other.

8. In this book, Bohm gives a fairly standard presentation, based on Bohr, though a highly perceptive one. His 1952 deterministic reinterpretation is not in evidence here.

9. Simon (1977), quoted in Turkle (1984, 351).

10. It may be objected that meaning is being confused here with significance; the former being the permanent message embodied in a 'text,' the latter a variable extrapolation related to readers' contexts. See the extensive discussion of this and related questions in Hirsch (1967), especially the appendix "Gadamer's Theory of Interpretation." Hirsch criticizes Gadamer on this very point, that in the latter's approach (and in my own, by extension), the *locus* of meaning and the degree of its permanence is not as clear as in previous theories. This, of course, is true, and now

generally acknowledged, but three decades of debate have not resolved the issue. It has occurred, therefore, to some people that this situation may reflect real continuities in the matter at hand, the matter of 'text' and its range of meanings. In the case of a 'text-equivalent,' this is even more likely to be true, which explains, for example, why Charles Taylor (1971) also uses "meaning" in the enlarged sense, as I do. The point is that to explore and map these continuities in particular situations now seems a better course than to try to eliminate them by definition. Thus, I have set aside the dichotomy of permanent 'meaning' and variable 'significance' in favor of the image of a 'core' of meaning in the 'text,' surrounded by more fluid regions of 'potential interpretations,' forming a kind of cloud (resembling perhaps that of a "probability cloud" of the electron in an atom). In this image, the text may be seen as providing a "force field" to *constrain* interpretations mostly within the core, to a lesser extent outside but close to the core, and occasionally farther from the core. This substitutes a *continuum,* in the degree of attachment of meaning to 'text,' for the sharp cut of the dichotomy (Eger 1993a, this volume, 3–27). But even this substitute should be seen only as a working preconception; for the whole point of hermeneutics is to *explore* the subject area—with examples, analogies, histories—and thus to clarify ideas, not to introduce hard boundary lines at the start. The theatrical analogy, with its "cascade of interpretations," is one device for visualizing within science precisely this spectrum of relations of meaning/significance to 'text' (see Eger 1993b, this volume, 29–51).

11. As for the "reduction" part of the process, likened by Bevilacqua and Giannetto to a Husserlian suspension or "bracketing" of meaning *(epoche),* I can perhaps see its role in dealing with misconceptions; but even there, the "fusion of horizons" seems to me a better approach, for it is not always necessary to 'put out of play' a troublesome preconception. As I have emphasized, preconceptions, even misconceptions, are *needed* to arrive at alternative interpretations.

12. The fear of fusions, and of 'flattening' the distinction, is often related to the fear of a 'colonization' or 'domination' of the 'life-world' by the 'science-world' (Eger 1992b, this volume, 231–241; 1993a, note 9), as exemplified by the student of computers quoted above. The same fear is expressed in a different way in Heidegger's writings on technology (see his "Question Concerning Technology" in *Basic Writings,* 1977). This is an enormous problem of modern times, with large educational significance, but a problem that cannot be solved by maintaining theoretical distinctions in the traditional manner of continental hermeneutic thinkers. The 'fusions' I have been discussing are not of the colonizing kind—that is why they are *fusions,* not displacements; stemming from a hermeneutic awareness, they make possible an *understanding* of the colonizations that already exist or those that are coming.

13. Some educators oppose repeating classic experiments in which the "outcome is known in advance." I think this misses the point, and the theatrical analogy is just what is needed to make that clear: When we go to see *Oedipus Rex* for the third or fourth time our enjoyment is not diminished but enhanced by the fact that the outcome is known in advance. It all depends on how the re-enactment is *presented.*

PART TWO
PHILOSOPHY AND EDUCATION

Chapter Twelve
Physics and Philosophy: A Problem for Education Today

1. P. Morrison, *American Journal of Physics* 32, 441 (1964); *The Proceedings of the Boulder Conference on Physics for Nonscience Majors*, edited by M. Correll and A. Strassenburg (College Park, MD: Commission on College Physics, 1965). See also E.A. Wood, *American Journal of Physics* 34, 891 (1966); G. Holton, *Physics Today* 19, 31 (March 1967); S.P. Heims, *American Journal of Physics* 37, 319 (1969); F. Reif, *Science* 164, 1032 (1969); E.H. Marston, *American Journal of Physics* 38, 1244 (1970).

2. H. Marcuse, "Remarks on Redefinition of Culture" in *Science and Culture*, edited by G. Holton (Boston, MA: Beacon Press, 1965), 233.

3. Concepts such as "paradigm" and "normal science" are used here in the sense of Thomas Kuhn. See note 8.

4. R. Dubos, "Science and Man's Nature" in G. Holton, note 2, 254.

5. H. Marcuse, *One-Dimensional Man* (Boston, MA: Beacon Press, 1964), 156–157.

6. See, for example, *Tai Chi—A Way of Centering and I Ching*, translated by Gra-Fu Feng and Jerome Kirk (London: Collier-Macmillan, 1970), 4.

7. D. Easton, *American Political Science Review* 63, 1051 (1969).

8. T. Kuhn, *The Structure of Scientific Revolutions* (Chicago, IL: University of Chicago, 1962).

9. S. Wolin, *American Political Science Review* 63, 1062 (1969).

10. A. Baker, *Physics Today* 23, 34 (Mar. 1970).

11. A. Baker, *Modern Physics and Antiphysics* (Reading, MA: Addison-Wesley, 1970), 118.

12. J. M. Fowler, *American Journal of Physics* 37, 1193, (1969).

13. P. W. Bridgman, *The Logic of Modern Physics* (New York: Macmillan, 1927), 31.

14. H. L. Davis, *Physics Today* 23, 80 (Dec. 1970).

15. L. Mumford, *The Myth of the Machine—The Pentagon of Power* (New York: Harcourt, Brace, Jovanovich, 1970), 88.

16. R. Dubos, note 4, 264.

17. B. F. Skinner, *Science and Human Behavior* (New York: Free Press, 1965), 5.

18. H. Arendt, *The Human Condition* (Garden City, NY: Doubleday, 1959), Ch. VI.

19. H. Marcuse, note 5, Ch. 6.

Chapter Fifteen
Philosophy of Science in Teacher Education

1. Gould (1983, 253-262).

2. Gould (1977, 160-167).

3. Nagel (1979, Ch. 5).

4. Hugh Helm, "Do Students Understand the Nature and Role of Scientific Concepts?" in Helm and Novak (1983).

5. Novak and Gowin (1984).

6. Butts (1968, 20).

7. It should be noted, though, that Whewell actually used brackets instead of arrows, to emphasize the generalizing and subsuming aspects of science.

8. Popper (1968, 53).

9. The key work in the historico-philosophic school is still Thomas Kuhn (1970). But useful also are Lakatos and Musgrave (1970); Brush (1974). On history of evolution, there are, among many others: Gillespie (1979); Hull (1973). For representative statements of the 'constructivist' position in the sociology of science, we have Barnes (1974); and Barry Barnes and David Bloor, "Relativism, Rationalism, and the Sociology of Knowledge" in Hollis and Lukes (1982).

10. Thomas Kuhn, "Replies to Critics," in Lakatos and Musgrave (1970, 239).

11. Feyerabend (1975, 181).

12. Butts (1968, 275).

13. Ibid., 161.

14. Rohrlich (1983, 1251-1255).

15. One of the more scientifically responsible in this category is Price et al (1986).

Chapter Sixteen
The 'Interests' of Science and the Problems of Education

1. But Markus (1987) argues that this demarcation remains because it is *institutionally enforced*.

2. This was a basic motivation for the development of my own course in philosophy of science for high school teachers, taught since 1985. The basic approach is to raise to consciousness the philosophical problems in the teacher's own disciplines (biology, physics, geology, etc.); then to show how historic and contemporary theories of science answer these questions. A detailed discussion of the course is given in my (1987).

3. Two hedging remarks must be made here. Many courses in the category 'physics for poets' do exist, especially at large American universities. Closer examination, however, reveals that these are not regarded as serious by faculty or students. In part, the reason for their existence is to provide researchers with teaching duties at a time of shrinking enrollments (Fowler and West 1970, 31); in part it is to propagate at large a favorable image of scientific activity, thereby insuring future public support (Priestley 1971, 1500). Nonetheless, there are some well-known exceptions: I.B. Cohen's course at Harvard, for example, and Gerald Holton's books (see Holton and Roller, 1958) did indeed offer real alternatives. Unfortunately, these are rare exceptions.

4. I have discussed the influence of this literature at some length in my (1993c).

5. For an especially succinct description of the post-empiricist position, see Hesse (1980, 172ff).

6. For the effect on the evolution/creation controversy, see Eger (1988b); influence on theology can be seen in Küng (1980, 643-647) and in many recent issues of the journal *Zygon;* the impact on philosophy of mind, via such figures as Dennett (1981) and Churchland (1984), is perhaps better known than any other.

7. Throughout, I use the term "cosmology" in the ancient sense of an all-encompassing order. Today, that means from the very largest structures science treats to the very smallest; from the most abstract laws of physics to the observable features of insect behavior.

8. In Habermas's theory, the relation between "interests" and "actions" is close; hence, the same terms are sometimes used for both. Nonetheless, the distinction is useful in a larger context that includes sociological issues. Because Hesse does not explicitly make this distinction, what I here call "actions" will correspond to her "interests." In this paper, I use a nomenclature that seems appropriate to the educational problems at hand. Anyone following the work of Hesse, however, should not have any trouble in seeing the intended correspondences.

9. Goals that are often mentioned include these: leading students to understand "physics as a dynamic enterprise of unique coherence . . ." (Nedelsky 1973, 364) and to "enable them to understand scientists and their activities" (Fowler 1969, 1198).

10. However, if A.N. Whitehead was right, then for each of us and for every new field, learning involves three stages: a "stage of romance," a "stage of precision," and a "stage of generalization," in that order. To some extent, my 'cosmological interest' overlaps his first and third stages. See Whitehead (1955).

11. The "evolutionary epic," as E.O. Wilson describes it, is "probably the best myth . . . far more awesome than the first chapter of Genesis or the Ninevite epic of Gilgamesh" (1978, 209).

12. The evolution controversies in California are described by Nelkin (1982, Ch. 7). The trial in Arkansas is well documented in La Follette (1983). The concerns with ethology in the classroom surfaced in regard to the MACOS program (Nelkin 1982, Ch. 8).

13. Mario Savio was a leader of the Berkeley student protest movement in the mid-1960s. On the role of fragmentation of knowledge in this movement, see his introduction to Draper (1965).

14. Richard Baer, Jr. (1983), for example, has analyzed in detail the extra-scientific elements in Carl Sagan's film, *Cosmos,* which has been used in schools.

15. For an account of how, in one community, a far-reaching controversy broke out over that restriction of viewpoint (or its flattening out) for which the term "secular humanism" is intended, see Eger (1984 [1981b]).

16. For example: "A scientist is interested neither in proving nor in disproving. What he is interested in is discovering what the facts are" (Gallant 1984, 302, quoting Ashley Montagu).

17. See note 2.

18. I found many of the teachers familiar with such 'writers on science' as I. Velikovsky and E. Von Deniken, but not with the authors in my sample of the 'third genre.' An important aspect of this course is to introduce people to appropriate readings, understandable yet challenging.

19. See Hodson (1988, 21), for his views and further references.

20. Hodson (1988), for example, reports that the various efforts to shift attention to 'process-oriented' teaching have failed.

Chapter Seventeen
Hermeneutics and Science Education: An Introduction

1. For the general problem of meaning in education, see Ausbel (1963); conference papers on preconceptions are collected in Novak (1987); for constructivism see for example von Glasersfeld (1989) and Mestre (1991); the *National Forum,* Winter, 1985, features critical thinking; and an *Education Week Special Report* (October 9, 1991) reviews several related approaches of cognitive science.

2. Eger, M. (1993a, b), "Hermeneutics as an Approach to Science."

3. These major postempiricist developments can be found, for example, in Polanyi (1958; 1969), Hanson (1965), Kuhn (1970). Richard Bernstein is one of the few philosophers who sees in these developments a meaningful parallel to hermeneutic theories in the social sciences. See especially the introduction and 'PART III' of his (1983).

4. Hesse (1980) and Arbib and Hesse (1986) include hermeneutics within a general schema theory; Heelan (1983; 1988a) treats science specifically within a phenomenological-hermeneutic approach, with special emphasis on space perception.

5. The contrast is between natural science as a 'monologue' and the humanities as dialogue. See, for example, Habermas (1988 [1971], 291–301).

6. I leave out, for special treatment at a later time, the role of experiment and demonstration in science education. On this, Heelan (1983) will be especially relevant.

7. By focusing on the study of science rather than of nature directly, I do not imply that the *goal* is to understand science *rather* than nature. The difference is of course a basic philosophical problem that we do not address here. That there stands something 'real' beyond the 'language of science' is neither neglected nor affirmed in this first-order phenomenological description. A more detailed treatment is given in the reference of note 2.

8. The linguistic aspect of science has, of course, been widely noted, in education as well as in philosophy. See for example Schilk et al. (1989, 326), Thompson and de Zengotita (1989), and Champagne et al. (1980).

9. Mestre (1991) gives the transcript of a student-teacher dialogue on this point, which must have lasted ten to fifteen minutes. At the end, the students were no closer to the 'right' answer than at the beginning.

10. The concept of "horizon," crucial to hermeneutics, is used by Gadamer (1975 [1960] 269) to indicate the entire field of objects and events that may be seen from a particular standpoint or orientation to the world.

11. One who has dealt with this problem extensively, though not specifically within education, is Michael Polanyi (1958).

12. My direct experience is with ABET (American Board for Engineering and Technology), which demands a greater role for 'design problems' or 'projects' in the standard curriculum, and insists that such problems must have more than one answer.

13. Underdetermination of theory by observation was treated at length by Duhem (1954 [1914]). Kuhn (1970), Laudan (1977), Hesse (1980), and others have based their theories on this point.

14. That is why, for example, Kuhn (1970, 187-89) calls Newton's second law a "law sketch."

Chapter Eighteen
Philosophy and the Physical Sciences in Undergraduate Education

1. Weisskopf (1976, 23).
2. Brush (1974, 1174).
3. Habermas (1968, Ch. 4).
4. Habermas (1968, 308).
5. See, for example, Maslow (1966); and Wolin (1969, 1069).
6. Kuhn (1970, 194).
7. Polanyi (1958, 88).
8. Polanyi (1969, 199–203).
9. Kuhn (1970, 88).
10. Habermas (1968, 52).
11. For example, Thomas Kuhn, in Lakatos and Musgrave (1970, 23).
12. Popper (1970, 53).
13. A. Einstein, quoted in Frank (1957, xi).
14. Habermas (1968, 309–310).
15. Frankel (1974, 44–45).
16. An idea and phrase much used by Polyani. See, for example, Polanyi (1969, 141); and Kuhn (1970, 191).
17. It should of course be clear from the context that the *choice* referred to here is qualitatively different from the 'decision' point of even the most sophisticated computer programs thus far envisaged. And, moreover, that this 'essentially algorithmic' character of science study is unfortunately *not* confined to elementary courses only.

Chapter Nineteen
Hermeneutics and the New Epic of Science

1. Yet Stephen Gould, for example, sees no essential difference between his strictly scientific writing and his 'popular' essays—only the technical arguments being omitted from the latter (lecture at Worcester Polytechnic Institute, Worcester, MA, 8 October 1987).

2. Particularly worth noting are Yantsch (1981), Davies (1983), Sperry (1985), Chaisson (1981; 1987), Margulis and Sagan (1986), and Penrose (1989). These works fit my description in varying degrees. Chaisson's books are far less demanding than *Gödel, Escher, Bach*, for example, while Yantsch's contribution is a collection of papers presented at a meeting of the American Association for the Advancement of Science.

3. E.O. Wilson, for example, mentions "the Huxleys, Waddington, Monod, Pauli, Dobzhansky, Cattell" (1978, 214).

4. On Bohr's and Pauli's views, see Mayr (1982, 428–29). On Wittgenstein, see Putnam (1981, 108–109). Some of Popper's questions are found in his *Objective Knowledge* (1972, 267–270).

5. Haeckel, for example, gives many qualitative discussions of developments that have only recently turned quantitative; for example, the size of proteins ('plasm molecules'), prebiotic evolution, and what is now known as the Gaia hypothesis—the idea that the whole earth should be regarded as one giant organism (1905, 36).

6. Many scientists object to Wilson's and others' use of the word "myth" in this way. From the viewpoint of the philosophy of science, I prefer "metaphor" or "high-level metaphor."

7. Further evidence of the effects of this literature can be seen in *Zygon* 24 (2) (1989), where the authors in my sample are much quoted in a debate on science and the theology of Wolfhart Pannenberg.

8. Julian Huxley's view of the larger sociomoral role of science is very close to that of Monod and Wilson. See, for example, J. Huxley (1957), especially the essays "Man's Place and Role in Nature" and "Evolutionary Humanism."

9. For a modern version of living in 'harmony with nature,' see Toulmin (1982, 262).

10. The idea of the 'differentiation of values spheres' originates with Max Weber and is now used extensively by Habermas, as in his *Theory of Communicative Action* (1984, 233–247).

11. Proponents of hermeneutics in natural science are in the minority. But Hesse (1980), Arbib and Hesse (1986), Grene (1985), and Stent (1985) represent this position well. Anthony Giddens and Habermas have countered that nevertheless, social science is different because it has an *additional* need for hermeneutics: Scholars must first use interpretive understanding to gain access to a language already in use, prior to any theory construction (Habermas 1984, 109–111); only after that do their problems begin to resemble those faced by natural scientists. Such considerations do not undermine my argument here that in natural science at least one important role for hermeneutics exists. However, one could go further. The outsider, the nonscientist, the student, and sometimes even the researcher may have need for the 'double hermeneutic' as well: first, to 'gain access' to the logic of a science *as a whole* (through history, philosophy), and only then to *properly* understand particular theories in light of the evidence. Many scientists would not accept this, and to argue it here would be a digression. But the recent attention given by biologists to the history and philosophy of evolution, as this theory undergoes yet another reappraisal, is some indication that my point has support.

12. In 1989, Philippe Rushton, a professor at the University of Western Ontario, was the target of protests and of investigations by Ontario police for publishing scientific papers that some people describe as racist and reactionary. These papers suggest that certain behavioral differences between major ethnic groups can be traced to different evolutionary strategies adopted by these groups. See Gross (1990).

Chapter Twenty
The New Epic of Science and the Problem of Communication

1. My talk at the 1987 SLS meeting, "Literature and the New Epic of Science: Why Habermas Was Wrong and Wilson Is Right," included a brief reference to the

subject of this paper. The questions and comments elicited by that reference in Worcester convinced me to devote to it a separate, expanded treatment.

2. For a brief review of this movement, see Randall (1977, Ch. 3).

3. As may be expected, criticism of such cosmic extension of science appeared in the 19th century as it does today. Among the best known critics of Spencer was Chauncy Wright (1865). Recently, Mary Midgley (1985) has made similar indictments: Extension beyond a certain point is not science, not intellectually legitimate.

4. As examples of the *biologization* of language, consider the transition in our time from 'men' and 'women' to *males*, *females*, from 'boy' to 'male juvenile,' and from 'making love' to *having sex*.

5. Lest it be thought that Wilson's models are intended as scientific methodology only, not as a way to better understanding in general, his direct remarks on this should be taken seriously: "One is tempted to . . . accept the purest elements of altruism as simply the better side of human nature. . . . a transcendental quality . . . But . . . it is precisely through the *deeper analysis* of altruism that sociobiology seems best prepared at this time to make a novel contribution" (Wilson 1978, 156–57, emphasis added).

6. Pure or near-pure types can actually be listed: For example, in addition to Weinberg (1979), there are Chaisson (1981), Gribbin (1981), and Jastrow (1967). Books that are 'standard' only in part, but project or refer to a complete standard narration, include Weisskopf (1979), Rose (1976), Monod (1971).

7. An American-made supercomputer.

8. See, for example, Collins (1985). See also criticisms of this position in Hesse (1988a).

PART THREE

MORALITY

Chapter Twenty-One
The Price of Collaboration

1. New York Times, August 1, 1978.

2. Cablegram from Philip Handler to M.V. Keldysh, *Science*, 181 (September 21, 1973).

3. Cablegram from Robert Marshak to A. Logunov, May 19, 1978, quoted in *Physics Today*, 31 (July 1978) 63.

4. Private correspondence between members of the Federation of American Scientists and its director, April 1978.

Chapter Twenty-Two
Rationality of Science and Social Controversy

1. Kuhn (1970). For more recent clarifications and changes, see his (1977).

2. Popper (1970). See also Gutting (1980).

3. Eger (1988b).

4. Mill (1961, 245).

5. Ibid., 246.

6. Kohlberg (1981, 146).

7. See, for example, Morill (1980).

8. For the influence of Rawls's work on Kohlberg, see Kohlberg (1981, Ch. 5), and Boyd (1980).

9. For Judge Overton's "Opinion," pre-trial briefs and commentary, see La Follette (1983). Among the many books that give the academic view from various perspectives are: Nelkin (1982), Ruse (1982), Godfrey (1983), Kitcher (1983), Wilson (1983), Montagu (1984), and Newell (1985).

10. Ruse (1982, 328–329).

11. Resolution of the Board of Directors of the Iowa Academy of Sciences, quoted in Wilson (1983, 196).

12. An idea of the 'dangers' of real science history in science teaching is given by Brush (1974).

13. The "close connection between the general concept of *reasonableness* underlying the moral and the scientific points of view" is stressed, for example, by Israel Scheffler (1973, Ch. 2).

14. Hesse (1980, xxiv).

15. Eger (1984 [1981b]) and (1988b).

16. Scriven (1985, 10–12) (the sequence of some phrases has been changed).

17. See, for example, Gallant (1984, 290).

18. Kitcher (1983, 174–76), emphasis added

19. Mill (1961, 226).

20. Ibid., 172. In fairness to Kitcher, it must be added that his main worry appears to be about situations where "responsible challenges to neo-Darwinian orthodoxy may be left to dangle with the fallacies that the creationist offers—and when the teacher exits without explaining the difference." The fears are understandable but *our problem here is a comparative one*: The same reservations could be voiced (and *are* voiced!) about debates on capitalism and Marxism. 'Fallacies' and 'distortions' are not scarce in the political arena, but there Scriven fears not for the child. That this very point is part of the *social conflict* is shown explicitly in my "A Tale of Two Controversies . . ." (1988b, this volume, 335–362).

21. See note 3.

22. For a recent stimulating overview, see Bernstein (1983), especially parts I and II. See also the "Epilogue" in Barnes (1974).

23. In addition to the 'maverick' position of Paul Feyerabend, the more limited criticisms include Laudan (1983b), and Quinn (1984).

24. Most important is his (1984), but see also McCarthy (1978), Bernstein (1985), and Thompson and Held (1982).

25. Bernstein (1985, 20–21).

26. See, for example, Jürgen Habermas (1985b, 206). In his more "Weberian" mode, Habermas now speaks of *three* 'spheres'; but the third is the aesthetic, which does not concern us here.

27. Marcuse (1964, 166).

28. Habermas (1985b, 209).

29. Frankel (1974, 44–45).

30. Mayr (1982, 77).

31. For a nontechnical presentation, see Davies (1982, Ch. 5). But even before the anthropic principle, physics had already brought the 'observer' into quantum mechanics; and in Bohr's 'principle of complementarity' the influence of William James and Kierkegaard have been detected. See Holton, "The Roots of Complementarity" in his (1973).

32. Habermas (1985a, 89).

33. For example, Baer (1977).

34. Habermas (1973, 270).

35. Habermas (1968a, 105).

36. Horkheimer (1974, 13).

37. That professional logic should sometimes yield to that of education is taken seriously by only a handful of scientist-educators. Gerald Holton is among these few: See his (1973, Chs. 14–15).

38. Feyerabend (1988, 307).

39. Habermas (1968a, 8–9).

40. For example, McCarthy (1985).

41. Hesse (1980).

42. Ibid., 225.

43. Ibid., 186. emphasis added.

44. The distinction between the empirical and theoretical objects of natural science is of course no longer unproblematic. But even the theory-ladenness of observations comes in degrees, and so the distinction here is not to be construed as a disjunction but as a variable in the continuum.

45. Habermas's direct answer (1984, 109–110) to this move by Hesse is revealing. He acknowledges a hermeneutic rational component in confrontations with theories of natural science, but insists that the human sciences then involve a *double* hermeneutic—that due to theories *about* its objects (social facts), which he calls 'first' level, and in addition, that required to *discover* the objects, which have a meaning constituted in human communication (the 'zero' level). This is not the place to argue that physics too is full of objects (sometimes vaguely understood) constituted by 'consensus in communication.' But even if some difference of this kind is granted, we have still done away with the sharp boundary between realms of hermeneutics and realms of no-hermeneutics. Once again, it becomes a matter of levels and degrees.

46. Hesse (1980, Ch. 4).

47. On account of such statements as "rational norms and true beliefs are just as much explananda of the sociology of science as are non-rationality and error" [quoted in Hollis and Lukes (1982, 13)], the work of Mary Hesse is sometimes classed with that of sociologists of science espousing relativism. However, it is not the relativist aspect of her treatment that I incorporate here, but only her ascription of hermeneutics to natural science.

48. See the discussion in Bernstein (1983), especially 31ff.

49. In a private communication, Professor Philip Quinn has suggested to me that this should be understood as a *political accommodation*: There exist *experts* in both realms; on questions of biological origins, these are scientists; on questions of morals, the experts are not psychologists, decision theorists, or philosophers—for

this *is* a practical matter—but those who 'live well' (which can be established from their biographies). That we do not have this latter group of people deciding, or at least advising, on moral education curricula is due essentially to the political infeasibility of such a thing. Thus, in Quinn's view, we compromise by trying merely to get young people to be thoughtful about controversies involving values. I find this argument stimulating in a number of ways, and have therefore taken it into account to some extent in the second part of the present paper. If one takes seriously the idea explicitly introduced by Hesse (but accepted also by many others) of a hermeneutic and communicative interest in natural science, it becomes debatable whether the research scientist is the *only* expert on science curricula in general education (as distinguished from professional education). In particular, his priorities (as Kuhn has shown) may well differ from those of the *educator*. For the latter, we are now told, the following problem should be high on the agenda: how to convey a *core of reason*, the *same* core, in all cognitive teaching. However, Quinn's point cannot be treated fully within the scope of this paper.

Chapter Twenty-Three
The Conflict in Moral Education: An Informal Case Study

1. *Goals of Elementary, Secondary and Continuing Education in New York State* (New York: The University of the State of New York, 1974).

2. For a general introduction and critique, see Bennett and Delattre (1978).

3. See Simon et al. (1972). This is a widely used text.

4. For example: Stewart (1975), and other articles in that issue. Lockwood (1975); and Lockwood (1977).

5. Professor Richard A. Baer, Jr. (Cornell University). See also his (1977).

6. *Spencer Needle* (*September* 14, 1978, 7). Most of the local quotations given in this article appeared in the *Spencer Needle* during 1978 and 1979.

7. Lawrence Kohlberg has criticized Values Clarification for some of the same shortcomings as have the parents of Spencer-Van Etten. But since he has defended the use of VC in public schools, his name was used here in support of the controversial course. See Kohlberg and Simon (1973).

8. *Ithaca Journal Magazine* (January 27, 1979, 1).

9. Private communication. See also *Maryland Gazette*, July 30, 1979, 1, and *Sunday Sun* (Baltimore), May 27, 1979, A–1. In Anne Arundel County, parents apparently won some legal restrictions on the use of VC.

10. Howard Kirschenbaum (1977, Chs. 4 and 13).

11. See Superka et al. (1975).

12. See Alan Lockwood (1978).

13. But "humanism" also connotes an opposition to theism: in our culture, Judaism/Christianity. The confusion of the two meanings of the word greatly exacerbated the conflict in this community, and elsewhere.

Chapter Twenty-Four
A Tale of Two Controversies: Dissonance in the
Theory and Practice of Rationality

1. The author wishes to thank Nathan Glazer, Thomas Green, Mary Hesse, Larry Laudan, Alasdair MacIntyre, Larry Nachman, Philip Quinn, Israel Scheffler, Diane Ravitch, Abner Shimony, and Kenneth Strike for their helpful comments and criticisms, and Holmes Rolston III for valuable stylistic suggestions.

2. See Holden (1987) and Lewin (1987). Some of the recent court cases are: *Mozert et al. v. Hawkins County Public Schools et al.* (United States District Court for the Eastern District of Tennessee); *Douglas T. Smith et al. v. Board of School Commissioners of Mobile County et al.*, in the United States District Court for the Southern District of Alabama (popularly known as the 'Alabama Secular Humanism Trial'). For more general testimony on various aspects of these issues, see Department of Education of the United States 1984.

3. Scholarly criticism includes: Lockwood (1978), Oldenquist (1979), Munsey (1980), Chazan (1985), and Baer (1977).

4. Newspaper article quoting Professor John Kaufhold. (Elmira *Star Gazette*, 1 March 1979, 4.)

5. *The United States District Court Eastern District of Arkansas, Western Division: McLean et al. v. The Arkansas Board of Education, et al.* For the *Opinion* of Judge William R. Overton, and excerpts from pretrial briefs, see La Follette (1983).

6. Nelkin (1982) is an overall introduction. For detailed scientific, philosophical and educational arguments, see Kitcher (1983), Godfrey (1983), Montagu (1984), Futuyma (1982), Ruse (1982), Newell (1985), Eldredge (1982), and David B. Wilson (1983).

7. Resolution of the Board of Directors of the Iowa Academy of Sciences, quoted by David B. Wilson (1983, 196).

8. Passed by the Council of the American Anthropological Association at the 79th annual meeting, (December 1980, 3–7).

9. WMHR-Syracuse, *Focus on the Family*, 3 September 1985. The form of education desired by this speaker is termed by Kohlberg, "indoctrination of fourth stage values" (1980, 64).

10. This complaint is the more striking in view of the recent academic movement to redirect precollege teaching away from 'right answers,' toward the *process* of science (see Ravitch 1983, 242–243). Compare with this: "In indoctrination, we are concerned primarily with *what* people believe. . . . In teaching, however, we are concerned primarily with *how* persons believe . . ." (Green 1972, 44).

11. This argument was used by Dorothy Nelkin in a public debate against creationist Kelly Segraves at the Chautauqua Institution (Chautauqua, NY, 10 July 1985).

12. Falsifiability (see Popper, 1968) was a major criterion at the Arkansas trial, used by the judge to disqualify creationism as a science. However, it has also been used to criticize evolution, especially by Popper (1982, 167–172). But see also the exchange of letters concerning this criticism in *Science* 212 (22 May 1981, 873) and

(26 June 1981, 1446). A related discussion appears in *Nature* 290 (12 March 1981, 75–76, 82).

13. It is a common claim in these debates that "evolution is a fact" of the same type as that "the Earth is round," or that apples fall rather than rise from trees, or that "I have a heart" (though I have not seen it). See Gould (1983, 254–55), Eldredge (1982, 29), and Ruse (1982, 58).

14. In view of the fact that the Ptolemaic system is routinely 'taught' in astronomy, and even caloric and phlogiston have a place in certain physics courses, the total ban on creationism is less justified than most people think. Creationism, after all, bears the same relation to evolution as does the Ptolemaic system to post-Newtonian astronomy. To what extent it might have *value* in biology study is a question of educational policy; but it is *historical* creationism, not the writings of a small group of contemporaries, that should be the main reference for any academic discussion. That creationism *was* an important paradigm in science is well established. See Neal C. Gillespie (1979), Hull (1973), Rudwick (1985), and C.C. Gillespie (1951), among others.

15. This is true at least of those in-service teachers taking my philosophy of science course.

16. This term has been widely used, but see McMullin (1983, 3) and Shapere (1966, 50).

17. Sociobiology will not be discussed here because I focus first on method, or approach, not on findings. The potential impact is great. Concerning the influence of ethology on morals and moral education, see Nelkin (1982, 47–51), Conlan (1975), and Dow (1975). On the combined influence of biology and computers, see Pugh (1977).

18. Among moral philosophers the response to the 'new philosophy of science' is varied: Some still dismiss it with hardly more than a note, maintaining the traditional 'autonomy of moral discourse.' See, for example, Gewirth (1978, 4). On the other side, MacIntyre (1980 and 1981, Ch. 15) has found in the work of Imre Lakatos independent support for his own idea that science *and* morals "can only become intelligible to us" through their history.

19. The key work in the historico-philosophic school is still Kuhn's (1970). But see also Lakatos and Musgrave (1970), Feyerabend (1988), Laudan (1977). For representative statements of the 'constructivist' position in the sociology of science, see Barnes (1974) and Barnes and Bloor (1982).

20. Examples of criticism from an objectivist position appear in Scheffler (1967) and Shapere (1966). For more recent appraisal of Kuhn's influence, see Cohen and Feyerabend (1976) and Gutting (1980). Some of Kuhn's revised views are given in his (1977).

21. In Kuhn's (1977), see especially the chapter, "Objectivity, Value Judgment, and Theory Choice." See also his (1970, 199).

22. An Apollo scientist quoted in Mitroff (1981 [1974], 173).

23. See, among others: Agassi (1965), Lakatos (1970), Laudan (1977), and McMullin (1982).

24. This feature has been noted widely: "Kuhnian analysis . . . seems to put science on the same level as ethics, aesthetics, and literary criticism" (Hull 1973, 451).

25. For example, see Rawls (1971) and the subsequent debates.

26. Kohlberg (1981, 173). A formalist definition of morality involves such criteria as impersonality, universalizability, and reversibility. For the relation of Kohlberg's formalism to Rawls and Hare, see Kohlberg (1981, Ch. 5) and Boyd (1980).

27. For value-driven decision systems see Pugh (1977). The "basic outline of a decision system" he gives on page 54 can be found with minor modification in various teachers' guides. See The University of the State of New York (1976, 89). I also found it in the *Teacher's Manual, Vocational Decision Making 7* of the Spencer-Van Etten Central School, New York.

28. See note 25 and The University of the State of New York (1976).

29. See note 27 and Kirschenbaum (1977, 10).

30. The *Teacher's Manual* in note 27 is an example.

31. I am using the Hebrew-English edition of the translation by the Jewish Publication Society of America (Philadelphia, 1955). Some English translations that use the word 'kill' also mention in annotation that the meaning is 'murder.'

32. Modeling oneself on human exemplars (Martin Luther King, Socrates, Jesus) is expressly ruled out by most formalists, e.g., John Wilson (1972, 20–30). However, on the importance of story-telling see MacIntyre (1981).

33. The charge of "irrationality," denied by Kuhn, continues to be made by some, e.g., Alasdair MacIntyre, although others regard subsequent reformulations as moves back toward the 'received view.' See Musgrave (1980 [1971]).

34. MacIntyre (1978) and (1980). But see note 18. MacIntyre rejects the subjectivism of the "new philosophy of science" although he heartily welcomes its seriousness about history and shares its attitude toward method.

35. Some developments of this direction are today far beyond Kuhn, e.g., Laudan (1977, Ch. 3). Laudan includes as possibly legitimate factors moral attitudes and world views of all sorts.

36. See notes 11 through 13. The same criticism of the public argument is made in Laudan (1983a). Among the views offered the public authoritatively, some are far more outdated than the conventional wisdom, e.g., "a scientist is interested neither in proving nor in disproving. What he is interested in is discovering what the facts are" (Gallant [1984, 302] quoting Ashley Montagu).

37. Feyerabend (1978, 30). For a more detailed discussion of Feyerabend's view of radical alternatives, see his (1983 [1965]).

38. To see exactly how creationists wished to amend the California *Science Framework* concerning the teaching of evolution, see Nelkin (1982, Appendix 3).

39. Although this seems obvious, it is not always acknowledged in the literature. Harvey Siegel (1979), for example, arguing that science need not be taught as Kuhn describes, offers "as counter-example to Kuhn's educational directives," the treatment of alternatives (Ptolemaic system) in the Project Physics Course. But Project Physics was aimed at art and humanities students—an unusual effort in which the concern is certainly within the 'context of education.'

40. The two contexts require more discussion. They partially overlap Habermas's "practical" (or communicative) and "technical" interests. But while Habermas at first used this dichotomy to differentiate natural from human sciences, it is becoming more apparent that these interests do not correspond to different modes of

knowledge, that the communicative interest inheres also in natural science. See Hesse (1980, 167–186).

41. Wayne A. Moyer, executive director of the National Association of Biology Teachers, quoted in Gallant (1984, 289).

42. This is evident in nearly all creationist pronouncements, e.g., Morris (1974, 1–2).

43. One recent protest against the *tacit* expansion of 'the scientific world view' into the religious realm is given by Baer (1983).

44. An example is Habermas (1980, 310–15). However, Habermas has tried to reintroduce, on nonpositivistic grounds, a radical distinction between the 'empirical analytic' sciences and other disciplines. Discussion of this type of distinction must be postponed.

45. Ruse (1982, 328) does feel the need to respond to Mill's position, but, significantly, he does so only in regard to general tolerance of dissent. Mill's cognitive argument is not dealt with seriously.

46. For calling my attention to the importance, in the moral realm, of distinguishing between theoretical and practical experts, I am indebted to Professor Philip Quinn.

47. See note 32. But recently, Kohlberg too acknowledged the value of exemplars. In addition to Socrates and Martin Luther King, whom he cited for years, he also recommends the life of Janusz Korczak (1981, Epilogue).

48. See note 44.

49. For one example, and further references, see Eger (1972).

50. See notes 17 and 27.

51. That suggestions for a radically alternative science, one in harmony with philosophic and political views, emanate also from the left, can be seen in Marcuse (1964, 166–69).

52. See Mario Savio's introduction to Draper (1965).

53. The present much used idea does not go far enough: 'Reason-giving' may be sufficient to tell the difference between enlightened instruction and old-fashioned rote-learning, but it does not succeed in drawing a clear line between rationality and even the cruder forms of rationalization or propaganda—which excel precisely in the art of providing the 'right sort' of reasons. To his credit, Thomas Green (1971, 51) is one theorist of education who acknowledges the difficulty of making the distinction externally and objectively.

54. In regard to sociobiology, some scientists are obviously violating this principle. See Ann Arbor Science for the People Editorial Collective (1977). For an in-depth study of the interaction between scientific, philosophical, and moral concerns in the debate between E.O. Wilson and R. Lewontin, see Segerstrale (1986).

55. Despite his emphasis on rationality, Kenneth Strike (1982a) gives detailed arguments in defense of such restriction of dialogue in science education. There, he maintains, the point is to "internalize the standards and procedures of a field" and to advance "the goals of a discipline" (1982a, 139). 'Liberty' is rightly restricted to the community of 'experts.' In political discussion, however, liberty of dialogue should be based not on Mill's consequentialist cognitive arguments but on Kant's principle of respect for persons as ends. Strike assumes that even on the nonprofessional or pre-professional level, disciplinary goals should dominate. This in itself is

problematic; it is more problematic in sciences that affect self-understanding, where the distinction between the student as 'person' and the student as 'novice' is not so clear-cut.

56. For one of the few papers that, cautiously, suggest some role for creationism in a biology course, see Anderson and Kilbourn (1983).

57. See Hesse (1980, Ch. 7) and Arbib and Hesse (1986, Ch. 8), although the inference here is my own.

Chapter Twenty-Five
Comments on Eger's 'A Tale of Two Controversies', by M. Hesse, A. Shimony, and T. Green

1. See Pincoffs (1986) for a discussion of how modern moral theory has been constructed around the image of the moral life as a life filled with quandaries and how this has falsified especially the tasks of developing an adequate philosophy of moral education.

Chapter Twenty-Six
Reply to Criticisms

1. Recent research has revealed that students do not come to science without any notion at all of how bodies move, what heat is, and so on. Instead they have preconceptions based on intuition or other sources, which are at variance with present-day science yet persist even after the relevant science course has been taken. See Helm and Novak (1983) for many papers on this subject. An interesting point is that a number of studies have identified some of these preconceptions as Aristotelian, that is, resembling the science of Aristotle.

2. Cognitive or conceptual ecology is a term adapted to education by Kenneth Strike (1982, 54–61).

Chapter Twenty-Seven
Kohlberg, MacIntyre, and Two Basic Problems of Education

1. Kohlberg (1981, 7).

2. Simon et al. (1972) and Raths et al. (1966).

3. Eger (1981b).

4. MacIntyre does not claim that the categories of his analysis (e.g. practices, virtues) are directly applicable to formal education. The responsibility for this is mine.

5. Kohlberg (1981, 12); see also his (1973).

6. MacIntyre (1981) and (1978).

7. MacIntyre (1981, Ch. 14).

8. However, professional ethics are not the same thing as the moral values imbedded in practices, though they are often closely related. See below.

9. See the revised version of his "From Is to Ought: How to Commit the Naturalistic Fallacy and Get Away with It in the Study of Moral Development," in Kohlberg (1981).

10. For example, see Callahan (1982).

11. For a discussion of the "Competence" movement in education, see Grant et al. (1979).

12. See Kohlberg, "Educating for a Just Society: An Updated and Revised Statement," in Munsey (1980).

13. Galbraith and Jones (1976, 118-122).

PART FOUR

LIFE, DEATH, AND DIALOGUE

Chapter Twenty-Nine
Dialogue between Martin Eger and Abner Shimony
(October 2001–March 2002)

1. The examples here are given very roughly, leaving out numbers, dates, and specifics which can be supplied from the literature.

2. More on this question later.

3. *Philosophy of Science* 64, S1-S14 (1967).

4. That is the dominant view, as I understand it: Peer review, repetition, mutual criticism, etc.—these professional or communal features are the key ingredients. However, for some players success in the game is not everything. The atmosphere, the types of conversations, the intellectual life mean a lot. In this respect, the 'progress' of the 20th century has made the game seem more successful, while the personal experience less so. An important question perhaps is whether open games like law and science can survive with integrity if the fraction of pure game players becomes too big. But that is a large sociological issue.

5. "In my model, the reader was freed from the tyranny of the text and given the central role in the production of meaning" (Fish, *Is There a Text in This Class?*, Harvard U. Press., 1980, 7). Of course, he's talking about literary texts, but even there it is silly, and he had to abandon all this years later (without embarrassment or career damage).

6. Your reference to value tables seems to be in response to my hasty example of a friend on trial (pages 427–28). This shows the pitfalls of hasty argumentation—all sorts of implications can be drawn that were not intended. Though I did use the phrase "minimize the maximum loss," the gist of the matter was rather in the phrase "which way is it *worse* for *me* to be wrong?" This question does not necessarily require any sort of table, nor does it necessarily cheapen life by reducing commitments to calculation. I will make this more clear in the sequel; here I only want to add that tables are not what I had in mind. The purpose of the example was much more limited—to emphasize that the outcome of a 'game,' however legitimately and well played, is not necessarily decisive for an outsider situated in a larg-

er context. I should not have used the word 'Minimax' because that does bring to mind the whole quantitative apparatus of game theory.

7. Jerusalem Post, Nov. 23, 2001.

8. I take for granted here, without discussion, the philosophies maintaining that values or ends can be reached by reason. But reason applies to both valuation and cognition, and I take Kant to be promoting not so much a 'dichotomy' between the is and ought, as a 'duality' (Cassirer), a combination of distinctions and interactions.

9. Part of the present constructivist movement is to ignore scientific texts of all kinds as façades, to go directly into the labs, and to interview researchers. This seems to have the imprimatur of Einstein, who famously said: If you want to know how science is done, don't listen to what scientists say; watch what they do.

References

Agassi, J. (1965). "The Nature of Scientific Problems and Their Roots in Metaphysics." Contained in M. Bunge (ed.), *The Critical Approach to Science and Philosophy* (New York: The Free Press of Glencoe, 1965), 189–211.

Alexander, H. (ed.) (1992). Philosophy of Education 1992, Proceedings of the Forty-Eighth Annual Meeting of the Philosophy of Education Society. Urbana, IL: Philosophy of Education Society.

Alexander, R.D. (1978). "Evolution, Creation and Biology Teaching." *The American Biology Teacher* (February), 91–104.

Amsterdamska, O. (1990). "Surely You Are Joking, Monsieur Latour!" *Science, Technology and Human Values* 15, 495–504.

Anderson, T. and Kilbourn, B. (1983). "Creation, Evolution, and Curriculum." *Science Education* 67, 45–55.

Ann Arbor Science for the People Editorial Collective. (1977). *Biology as a Social Weapon*. Minneapolis, MN: Burgess Publishing.

Apel, K.-O. (1968). "Perspectives for a General Hermeneutic Theory." Contained in Mueller-Vollmer, (ed.) (1988), 320–345.

_____. (1972). "The A Priori of Communication and the Foundation of the Humanities." *Man and World* 5(1), 3–37. Also contained in Dallmayr and McCarthy (1977), 292–315.

_____. (1988 [1968]). "Scientistics, Hermeneutics, Critique of Ideology: An Outline of a Theory of Science from an Epistemological-Anthropological Point of View." Contained in K. Mueller-Vollmer, (ed.) (1988), 321–345.

Arbib, A.M. and Hesse, M.B. (1986). *The Construction of Reality*. Cambridge, England: Cambridge University Press.

Aristotle. *Nicomachean Ethics*. Contained in many editions including McKeon, Richard. (1941). *The Basic Works of Aristotle*. New York: Random House, 927–1112.

Asimov, I. (1984 [1981]). "The 'Threat' of Creationism." Contained in Montagu (1984), 182–193.

Ausbel, D.P. (1963). *The Psychology of Meaningful Verbal Learning*. New York: Grune and Stratton.

Baer, R.A., Jr. (1977). "Values Clarification as Indoctrination." *Educational Forum* 41 (No. 2), 155–165.

Baer, R.A., Jr. (1983). "'Cosmos', Cosmologies, and the Public Schools." *This World* 5, (Spring/Summer), 5–17.

Bahcall, J.N. (1964). "Solar Neutrinos I. Theoretical." *Phys. Rev. Lett.* 12, 300–302.

_____. (1996a). "Solar Neutrinos: Where We Are, Where We Are Going." *Astrophysics Journal* 467, 475–484.

_____. (1996b). "Ray Davis: The Scientist and the Man." *Nuclear Physics B (Proc. Suppl.)* 48, 281–283.

Baigrie, B.S. (1988). "Siegel on the Rationality of Science." *Philosophy of Science* 55, 435–441.

Barnes, B. and Bloor, D. (1982). "Relativism, Rationalism, and the Sociology of Knowledge." Contained in M. Hollis and S. Lukes (eds.), *Rationality and Relativism* (Cambridge, MA: MIT Press, 1982).

Barnes, B. (1974). *Scientific Knowledge and Sociological Theory*. London: Routledge and Kegan Paul.

Barrow, J.D. and Tipler, F.J. (1986). *The Anthropic Cosmological Principle*. Oxford, England: Clarendon.

Baynes, K., Bohman, J., and McCarthy, T. (eds.) (1987). *After Philosophy: End or Transformation?* Cambridge, MA: MIT Press.

Bazerman, C. (1988). *Shaping Written Knowledge*. Madison, WI: University of Wisconsin Press.

Ben David, J. (1981). "Sociology of Scientific Knowledge." Contained in J.F. Short, Jr. (ed.), *The State of Sociology* (Beverly Hills, CA: Sage, 1981), 40–59.

Bennett, W.J. and Delattre, E.J. (1978). "Moral Education in the Schools." *The Public Interest* 50 (Winter).

Bernstein, R.J. (1983). *Beyond Objectivism and Relativism*. Philadelphia, PA: University of Pennsylvania Press.

_____. (ed.) (1985). *Habermas and Modernity*. Cambridge, MA: MIT Press.

Bettencourt, A. (1992). "They Don't Come Easy to Us: On the 'Alienness' of the Natural Sciences and Their 'Appropriation.'" *History and Philosophy of Science in Science Education* (Proceedings of the Second International Conference on the History and Philosophy of Science and Science Teaching). Contained in Hills (1982), Vol. 1, 87–104.

Bevilacqua, F. and Giannetto, E. (1995). "Hermeneutics and Science Education: The Role of History of Science," *Science & Education* 4.

Bevilacqua, F. (1983). "'Textbooks' Physics Versus History of Physics: The Case of Classical Electromagnetic Theory." in *The Principle of Conservation of Energy and the History of Classical Electromagnetic Theory*. (Pavia: La Goliardica Pavese), 1–34, 231–244.

_____. (1984). "H. Hertz's Experiments and the Shift Towards Contiguous Propagation in the Early Nineties." *Rev. Stor. Sci.* 1, 239–256.

_____. (1986). "'Enchantment' and the Post Industrial Society." in P. Thomsen (ed.), *Science Education and the History of Physics* (Aarhus, Denmark: Physics Dept., University of Aarhus, 1986), 42–54.

Bloor, D. (1991 [1976]). *Knowledge and Social Imagery,* 2nd ed. Chicago, IL: University of Chicago Press.

Bohm, D. (1951). *Quantum Theory.* New York: Dover.

_____. (1965). *The Special Theory of Relativity.* New York: W. A. Benjamin.

_____. and Peat, F.D. (1987). *Science, Order, and Creativity.* New York: Bantam Books.

Bok, D.C. (1976). "Can Ethics be Taught?" *Change* (October), 26–30.

Bondi, H. (1960). *Cosmology.* Cambridge: Cambridge University Press.

_____. (1964). *Relativity and Common Sense: A New Approach to Einstein.* Garden City, NY: Doubleday.

Born, M. (1950 [1935]). *Atomic Physics.* New York: Hafner.

Boyd, D. (1980). "The Rawls Connection." Contained in Munsey (1980), 185–213.

Boyer, E.L. and Levine, A. (1979). *A Quest for Common Learning.* Washington, D.C.: The Carnegie Foundation for the Advancement of Teaching.

Bruner, J.S. and Postman, L. (1949). "On the Perception of Incongruity: A Paradigm." *Journal of Personality* 18, 206–223.

Brush, S.G. (1974). "Should the History of Science be Rated X?" *Science* 183, 1164–1172.

Buchdal, G. (1992). *Kant and the Dynamics of Reason.* Oxford, England: Blackwell.

Butts, R.E. (ed.) (1968). *William Whewell's Theory of Scientific Method.* Pittsburgh, PA: University of Pittsburgh Press.

Callahan, D. (1982). "Tradition and the Moral Life." *The Hasting Center Report* (December 1982).

Cartwright, N. (1983). *How the Laws of Physics Lie.* New York: Oxford University Press.

Cassirer, E. (1953). *The Philosophy of Symbolic Forms.* New Haven, CT: Yale University Press.

Chaisson, E. (1981). *Cosmic Dawn.* Boston, MA: Little, Brown & Co.

_____. (1987). *The Life Era.* New York: Wiley and Putnam.

Chaisson, E.J. (1988). "Credo: Our Cosmic Heritage." *Zygon* 23, 469–480.

Chambers, R. (1845). *Vestiges of the Natural History of Creation,* 2nd ed. (From the 3d London ed.) New York: Wiley and Putnam.

Champagne, A.B., Leopold, E.K., and Anderson, J.H. (1980). "Factors Influencing the Learning of Classical Mechanics." *American Journal of Physics* 48, 1047–1079.

Chazan, B. (1985). *Contemporary Approaches to Moral Education.* New York: Teachers College Press.

Churchland, P.M. and Hooker, C.A. (eds.) (1985). *Images of Science: Essays on Realism and Empiricism.* Chicago, IL: University of Chicago Press.

Churchland, P.M. (1984). *Matter and Consciousness.* Cambridge, MA: MIT Press.

Cohen, R.S., Feyerabend, P.K., and Wartofsky, M.W. (eds.) (1976). *Essays in Memory of Imre Lakatos.* Dordrecht-Holland: D. Reidel.

Collins, H.M. and Pinch, T. (1993). *The Golem.* Cambridge, England: Cambridge University Press.

Collins, H.M. and Yearley, S. (1992). "Journey into Space" in Pickering (1992), 369–389.

Collins, H.M. (1981). "Stages in the Empirical Programme of Relativism." *Social Studies of Science* 11, 3–10.

_____. (1985). *Changing Order*. Chicago, IL: University of Chicago Press.

Conlan, J.B. (1975). "MACOS: The Push for a Uniform National Curriculum" *Social Education* (October), 389–392.

Connolly, J.M. and Keutner, T. (eds.) (1988). *Hermeneutics Versus Science?* Notre Dame, IN: University of Notre Dame Press.

Crease, R.P. (1993). *The Play of Nature: Experiment as Performance*. Bloomington, IN: Indiana University Press.

Cushing, J.T., Delaney, C.F., and Gutting, G.M. (eds.) (1984). *Science and Reality: Recent Work in the Philosophy of Science*. Notre Dame, IN: University of Notre Dame Press.

Cushing J.T. and McMullin, E. (eds.) (1989). *Philosophical Consequences of Quantum Theory*. Notre Dame, IN: University of Notre Dame Press.

Cushing, J.T. (1991). "Quantum Theory and Explanatory Discourse: End-Game for Understanding?" *Philosophy of Science* 58, 337–358.

_____. (1994). Quantum Mechanics, Historical Contingency, and the Copenhagen Hegemony. Chicago, IL: University of Chicago Press.

_____. (1995). "Hermeneutics, Underdetermination and Quantum Mechanics." *Science & Education* 4, 137–146.

Dallmayr, F.R. and McCarthy, T.A. (eds.) (1977). *Understanding and Social Inquiry*. Notre Dame, IN: University of Notre Dame Press.

Davies, P.C.W. (1982). *The Accidental Universe*. Cambridge, England: Cambridge University Press.

_____. (1983). *God and the New Physics*. New York: Simon and Schuster.

Davis, H L. (1970). "Are Physicists to Blame?" *Physics Today*, December, 80.

DeNicola, D.R. (1988). "Comments on Eger's 'A Tale of Two Controversies." *Zygon: Journal of Religion and Science* 23, 357–361.

Dennett, D.C. (1981). *Brainstorms*. Cambridge, MA: MIT Press.

Department of Education of the United States. (1984). *Official Transcript of Proceedings Before the U.S. Department of Education. In the Matter of: Proposed Regulations to Implement the Protection of Pupil Rights Amendment Section 439 of the GEPA, Also Known as the Hatch Amendment*. Washington, D.C.

Depew, D.J. and Weber, B.H., (eds.) (1985). *Evolution at a Crossroads: The New Biology and the New Philosophy of Science*. Cambridge, MA: MIT Press.

Dirac, P.A.M. (1933). "On the Lagrangian in Quantum Mechanics." *Physikalische Zeitschrift der Sovietunion* 3, 64.

_____. (1958 [1930]). *The Principles of Quantum Mechanics,* 4th ed. Oxford, England: The Clarendon Press.

Dow, Peter B. (1975). "MACOS Revisited: A Commentary on the Most Frequently Asked Questions about *Man: A Course of Study*." *Social Education* (October), 389, 393–96.

Draper, H. (1965). *Berkeley: The New Student Revolt*. New York: Grove Press.

Dreyfus, H.L. and Dreyfus, S.E. (1986). *Mind Over Machine*. New York: The Free Press.

Dreyfus, H.L. (1985). "Holism and Hermeneutics." Contained in Hollinger, (ed.) (1985), 227–247.

_____. (1991). Being in the World: A Commentary on Heidegger's Being and Time, Division I. Cambridge, MA: MIT Press.

Duhem, P. (1954 [1914]). *The Aim and Structure of Physical Theory* (translated by P. P. Wiener from the 2nd ed. in French). New York: Princeton University Press.

Dunbar, D.N.F., Pixley, R.E., Wenzel, W.A. and Whaling, W. (1953). "The 7.68-Mev State in C^{12}." *Physical Review* 92, 649–650.

Edge, D. O. and Mulkay, M. J. (1976). *Astronomy Transformed*. New York: John Wiley.

Eger, M. (1972). "Physics and Philosophy: A Problem for Education Today." *American Journal of Physics* 40, (March), 404–415, (this volume, 167–185).

_____. (1984 [1981b]). "The Conflict in Moral Education: An Informal Case Study." Contained in Nathan Glazer (ed.), *The Public Interest on Education* (Cambridge, MA: Abt Books, 1984), 209–227, (this volume, 317–334).

_____. (1987). "Philosophy of Science in Teacher Education." Contained in J.D. Novak (ed.), *Proceedings of the Second International Seminar: Misconceptions and Educational Strategies in Science and Mathematics* (Ithaca, NY: Cornell University, 1987), 163–176, (this volume, 195–210).

_____. (1988a). "The New Epic of Science and the Problem of Communication." (This volume, 281–294.)

_____. (1988b). "A Tale of Two Controversies: Dissonance in the Theory and Practice of Rationality." *Zygon: Journal of Religion and Science* 23 (September 1988), 291–325, (this volume, 335–362).

_____. (1989a). "Rationality and Objectivity in a Historical Approach: A Response to Harvey Seigel." in *The History and Philosophy of Science in Science Teaching*, D.E. Herget, ed., Tallahassee, FL, 1989: 143–153, (this volume, 121–131).

_____. (1989b). "The 'Interests' of Science and the Problems of Education." *Synthese* 80, 81–106, (this volume, 211–230).

_____. (1992b). "Hermeneutics and Science Education: An Introduction," *Science & Education* 1, 337–348, (this volume, 231–241).

_____. (1993a). "Hermeneutics as an Approach to Science: Part I." *Science and Education* 2, 1–29, (this volume, 3–27).

_____. (1993b). Hermeneutics as an Approach to Science: Part II. *Science and Education: Contributions from the History, Philosophy and Sociology of Science* 2, 303–328, (this volume, 29–51).

_____. (1993c). "Hermeneutics and the New Epic of Science." in M.W. McRae (ed.), *The Literature of Science: Perspectives on Popular Scientific Writing*. (Athens, GA: University of Georgia Press, 1993), 187–209, (this volume, 261–279).

_____. (1994a). "Natural Science and Self-Understanding." (This volume, 89–99.)

_____. (1994b). "Letter to a Friend Regarding the Farm." (This volume, 405–409.)

_____. (1999). "Language and the Double Hermeneutic in Natural Science." Contained in Fehér et al. (1999), 289–305, (this volume, 73–88).

Eigen, M. and Winkler, R. (1981 [1975]). *Laws of the Game* (translated by R. Kimber and R. Kimber). New York: Harper & Row.

Einstein, A., Lorentz, H.A., Weyl, H., and Minkowski, H. (1952). *The Principle of Relativity: A Collection of Original Papers on the Special and General Theory of Relativity* (translated by W. Perrett and G.B. Jeffery). New York: Dover.

Einstein A. (1949). "Autobiographical Notes." Contained in P.A. Schilpp (ed.), *Albert Einstein: Philosopher-Scientist* (New York: Harper Torchbooks, 1949), 1–96.

———. (1953 [1950]). "The Laws of Science and the Laws of Ethics." Contained in H. Feigl and M. Brodbeck (ed.), *Readings in the Philosophy of Science* (New York: Appleton-Century-Crofts, 1953), 779–780.

Eldredge, N. (1982). *The Monkey Business.* New York: Washington Square Press.

Ellis, G.F.R., Maartens, R., and Nel, S. (1978). *Monthly Notices of the Royal Astronomical Society* 184, 439.

Erlichson, H. (1973). "The Rod Contraction-Clock Retardation Ether Theory and the Special Theory of Relativity." *American Journal of Physics* 41, 1068–1077.

Fehér, M., Kiss, O., and Ropolyi, L. (eds.) (1999). *Hermeneutics and Science: Proceedings of the First Conference of the International Society for Hermeneutics and Science, Boston Studies in the Philosophy of Science*, Dordrecht: Kluwer Publishing.

Feuer, L.S. (1982). *Einstein and the Generations of Science*, 2nd ed. New Brunswick, NJ: Transaction Publishers.

Feyerabend, P. (1983 [1965]). "Problems of Empiricism." Contained in Robert G. Colodny (ed.), *Beyond the Edge of Certainty* (Lanham, MD: Univiversity Press of America, 1983), 145–260.

———. (1988 [1975]). *Against Method.* Thetford, Norfolk: Thetford Press.

Feynman, R.P., Leighton, R.B. and Sands, M. (1963). *The Feynman Lectures on Physics.* Reading, MA: Addison-Wesley, 3 Vols.

Feynman, R.P., Leighton, R B., and Sands, M. (1963). *The Feynman Lectures on Physics*, Vol. 1. Reading, MA: Addison-Wesley.

Feynman, R.P. (1948). "Space-Time Approach to Non-Relativistic Quantum Mechanics." *Review of Modern Physics* 20, 367–387.

———.(1966). "The Development of the Space-Time View of Quantum Electrodynamics" (Nobel lecture). *Physics Today,* August, 31–44.

———. (1967). *The Character of Physical Law* (Messenger Lectures, Cornell, 1964). Cambridge, MA: MIT Press.

———. (1985). *QED.* Princeton, NJ: Princeton University Press.

Fiske, J. (1874). *Outlines of Cosmic Philosophy.* London: Macmillan & Co.

———. (1902 [1874]). *Outlines of the Cosmic Philosophy*, 4 vols. Contained in John Fiske, *The Miscellaneous Writings of John Fiske* (Boston, MA: Houghton Mifflin, 1902).

Føllesdal, D. (1993). "Hermeneutics and Natural Science." Lecture, First International Conference on Hermeneutics and Science, Veszprém, Hungary, 6–9 September, 1993.

Forman, P. (1971). "Weimar Culture, Causality and Quantum Theory, 1918–1927: Adaptation by German Physicists and Mathematicians to a Hostile Intellectual Environment." *Historical Studies in the Physical Sciences* 3, 1–115.

Foucault, M. (1972 [1969]). *The Archaeology of Knowledge* (translated by A. M. S. Smith). New York: Pantheon Books.

_____. (1980). *Power/Knowledge: Selected Interviews and Other Writings*. New York: Pantheon Books.

Fowler, J. and West, R. (1971). "What Our Left Hand Has Been Doing." *Physics Today,* March 1970, 24–41.

Fowler, J. (1969). "Content and Process in Physics Teaching." *American Journal of Physics* 37, 1194–1200.

Fowler, W.A. (1984). "The Quest for the Origin of the Elements." *Science* 226, 922–936.

Franco, C. and Colinvaux-de-Dominguez, D. (1992). "Genetic Epistemology, History of Science, and Science Education." *Science & Education* 1, 255–272.

Frank, P. (1957). *Philosophy of Science*. Englewood Cliffs, NJ: Prentice Hall.

Frankel, B. (1974). "Habermas Talking: An Interview." *Theory and Society* 1, 37–58.

Franklin, A. (1986). *The Neglect of Experiment*. Cambridge, England: Cambridge University Press.

Frye, N. (1982). *The Great Code*. London: Routledge & Kegan Paul.

Fujimura, J.H. (1992). "Crafting Science: Standardized Packages, Boundary Objects, and 'Translation.'" Contained in Pickering (ed.) (1992), 168–211.

Futuyma, Douglas J. (1982). *Science on Trial*. New York: Pantheon Books.

Gadamer, H.-G. (1975 [1960]). *Truth and Method* (translated by G. Barden and J. Cumming, from the 2nd [1965] ed.). New York: Crossroads.

_____. (1976 [1966]). *Philosophical Hermeneutics*. Berkeley, CA: University of California Press.

_____. (1984 [1960]). *Truth and Method* (translated by G. Barden and J. Cumming, from the 2nd [1965] ed). New York: Crossroads.

_____. (1985 [1977]). *Philosophical Apprenticeships*. Cambridge, MA: MIT Press.

_____. (1988 [1959]). "On the Circle of Understanding." Contained in Connolly and Keutner (1988), 68–78.

Galbraith, R.E. and Jones, T.M. (1976) *Moral Reasoning: A Teaching Handbook for Adapting Kohlberg to the Classroom*. Minneapolis, MN: Greenhaven Press.

Galison, P. (1987). *How Experiments End*. Chicago, IL: University of Chicago Press.

Gallant, R.A. (1984). "To Hell with Evolution." Contained in Montagu (1984), 282–305.

Garrison, J.W. and Bentley, M.L. (1990). "Science Education, Conceptual Change and Breaking with Everyday Experience." *Studies in Philosophy and Education* 10, 19–35.

Gell-Mann, M. (1989). "Dick Feynman, the Guy in the Office Down the Hall." *Physics Today,* February, 50–54.

Gewirth, A. (1978). *Reason and Morality*. Chicago, IL: University of Chicago Press.

Giddens, A. (1976). *New Rules of Sociological Method*. New York: Basic Books.

Giere, R.N. (1988). *Explaining Science: A Cognitive Approach*. Chicago, IL: University of Chicago Press.

Gillespie, C.C. (1951). *Genesis and Geology*. Cambridge, MA: Harvard University Press.

Gillespie, N.C. (1979). *Charles Darwin and the Problem of Creation*. Chicago, IL: University of Chicago Press.

Ginev, D. (1995). "Between Epistemology and Hermeneutics," *Science & Education* 4, 1995, 147–159.

Glanz, J. (1996). "Added Weight for Neutrino Mass Claim." *Science* 272, 812.

Gleick, J. (1992). Genius: The Life and Science of Richard Feynman. New York: Pantheon.

Godfrey, L.R. (ed.) (1983). *Scientists Confront Creationism*. New York: W. W. Norton.

Gould, S.J. (1977). *Ever Since Darwin*. New York: W. W. Norton.

_____. (1981). "Evolution as Fact and Theory." in Gould (1983), 153–262.

_____. (1983). *Hen's Teeth and Horse's Toes*. New York: W. W. Norton.

Gow, K.M. (1985). *Yes Virginia, There Is Right and Wrong*. Wheaton, IL: Tyndale House.

Graham, L.R. (1981). *Between Science and Values*. New York: Columbia University Press.

Grant, G. et al. (1980). *On Competence*. San Francisco, CA: Jossey-Bass.

Graubard, S.R. (1988). *The Artificial Intelligence Debate*. Cambridge, MA: MIT Press.

Green, T.F. (1971). *The Activities of Teaching*. New York: McGraw Hill.

_____. (1972). "Indoctrination and Beliefs." Contained in L. A. Snook (ed.), *Concepts of Indoctrination* (London: Routledge & Kegan Paul, 1972), 25–46.

_____. (1985). "The Formation of Conscience in an Age of Technology." *American Journal of Education* (November), 1–38.

_____. (1988). "A Tale of Two Controversies: Comment." *Zygon: Journal of Religion and Science* 23, 341–46, (this volume, 376–381).

Gregory, B. (1988). *Inventing Reality: Physics as Language*. New York: John Wiley.

Grene, M. (1985). "Perception, Interpretation, and the Sciences: Toward a New Philosophy of Science." Contained in Depew and Weber (1985). 1–20.

Gribbin, J. (1981). *Genesis: The Origins of Man and the Universe*. New York: Delacorte Press/Fride.

Gross, B. (1990). "The Case of Philippe Rushton." *Academic Questions* 3, 24–46.

Grünbaum, A. (1990). "The Pseudo-Problem of Creation in Physical Cosmology." Contained in J. Leslie, (ed.), *Physical Cosmology and Philosophy* (New York: Macmillan).

_____. (1991). "Creation as a Pseudo-Explanation in Current Physical Cosmology." *Erkentniss* 35, 233–254.

Gutting, G. (ed.) (1980). *Paradigms and Revolutions: Applications and Appraisals of Thomas Kuhn's Philosophy of Science*. Notre Dame, IN: University of Notre Dame Press.

Habermas, J. (1968a). *Toward a Rational Society*. Boston, MA: Beacon Press

_____. (1968b). *Knowledge and Human Interests*. Boston, MA: Beacon Press.

_____. (1970 [1968]). *Toward a Rational Society: Student Protest, Science and Politics*. (translated by J. Shapiro). Boston, MA: Beacon Press.

_____. (1971[1968]). *Knowledge and Human Interests* (translated by J. Shapiro). Boston, MA: Beacon Press.

_____. (1973). *Theory and Practice* (translated by J. Viertel). Boston, MA: Beacon Press.

_____. (1980). *Knowledge and Human Interests*. Boston, MA: Beacon Press.

_____. (1981). *The Theory of Communicative Action* (translated by T. McCarthy). Boston, MA: Beacon Press.

_____. (1982). "A Reply to My Critics." Contained in Held and Thompson (1982), 219–283.

_____. (1984 [1981]). *The Theory of Communicative Action,* Vol. 1 (translated by T. McCarthy). Boston, MA: Beacon Press.

_____. (1985a [1983]). "Neoconservative Culture Criticism in the United States and West Germany: An Intellectual Movement in Two Political Cultures." Contained in Bernstein (1985), 78–94

_____. (1985b) "Question and Counterquestions." Contained in Bernstein (1985), 192–216.

_____. (1988 [1971]). "On Hermeneutics' Claim to Universality." Contained in Mueller-Vollmer (1988), 294–319.

Hacking, I. (1983). *Representing and Intervening.* Cambridge, England: Cambridge University Press.

Haeckel, E. (1900). *The Riddle of the Universe.* New York: Harper & Row.

_____. (1905). *The Wonders of Life* (translated by J. McCabe). New York and London: Harper.

Hanson, N.R. (1965 [1958]). *Patterns of Discovery.* Cambridge, England: Cambridge University Press.

Haraway, D. (1989). *Primate Visions: Gender, Race and Nature in the World of Modern Science.* New York: Routledge.

Harding, S. (1991). *Whose Science? Whose Knowledge?* Ithaca, NY: Cornell University Press.

Heelan, P.A. (1967). "Horizon, Objectivity and Reality in the Physical Sciences." *International Philosophical Quarterly* 7, 375–412.

_____. (1972). "Toward a Hermeneutic of Natural Science." *Journal British Society for Phenomenology* 3, 252–260.

_____. (1977). "Hermeneutics of Experimental Science in the Context of the Life-World." Contained in Ihde and Zaner (1977), 7–50.

_____. (1983). *Space Perception and Philosophy of Science*, University of California Press, Berkeley.

_____. (1988a). "A Heideggerian Meditation on Science and Art." Contained in Kockelmans (1988), 257–275.

_____. (1988b). "Experiment and Theory: Constitution and Reality." *Journal of Philosophy* 85, 515–524.

_____. (1989a). "After Experiment: Realism and Research." *American Philosophical Quarterly* 26, 297–308.

_____. (1989b). "Yes! There is a Hermeneutic Philosophy of Natural Science: Rejoinder to Markus." *Science in Context* 3, 469–480.

_____. (1991). "Hermeneutical Phenomenology and the Philosophy of Science." Contained in Silverman (1991), 213–228.

_____. (1995). "Quantum Mechanics and the Social Sciences: After Hermeneutics," *Science & Education* 4.

Heidegger, M. (1962 [1926]). *Being and Time* (translated by J. Macquarrie and E. Robinson). New York: Harper & Row.

_____. (1977). *Basic Writings* (edited and translated by D. F. Krell). New York: Harper & Row.

Held, D. and Thompson, B. (eds.) (1982). *Habermas: Critical Debates*. Boston, MA: MIT Press.

Helm, H. and Novack, J.D. (eds.) (1983). *Misconceptions in Science and Mathematics*. Proceedings of the International Seminar, June 20–22. Ithaca, NY: Cornell University.

Helm, H. (1983). "Do Students Understand the Nature and Role of Scientific Concepts?" Contained in Helm and Novack (1983).

Herbert, N. (1985). *Quantum Reality*. New York: Doubleday (Anchor).

Herget, D.E. (ed.) (1989). *The History and Philosophy of Science in Science Teaching* (Proceedings of the First International Conference). Tallahassee, FL: Science Education and Department of Philosophy, Florida State University.

Hertz, H. (1956 [1894]). The Principles of Mechanics Presented in a New Form. New York: Dover.

Hesse, M.B. (1980). *Revolutions and Reconstructions in the Philosophy of Science*, Bloomington, IN: Indiana University Press.

_____. (1988a). "Socializing Epistemology." Contained in McMullin (1988), 97–122.

_____. (1988b). "'Rationality' in Science and Morals." *Zygon: Journal of Religion and Science* 23, 327–333, (this volume, 363–68).

Hills, S. (ed.) (1992). *History and Philosophy of Science in Science Education* (Proceedings of the Second International Conference on the History and Philosophy of Science and Science Teaching), Vol. 1. Kingston, Ontario: The Mathematics, Science, Technology and Teacher Education Group and The Faculty of Education, Queen's University.

Hirsch, E.D., Jr. (1967). *Validity in Interpretation*. New Haven, CT: Yale University Press.

Hodson, D. (1988). "Toward a Philosophically More Valid Science Curriculum." *Science Education* 30, 19–40.

Hofstadter, D.R. (1979). *Gödel, Escher, Bach: An Eternal Golden Braid*. New York: Vintage Books.

Holden, C. (1987). "Textbook Controversy Intensifies Nationwide." *Science* 235 (2 January), 9–21.

Hollinger, R. (ed.) (1985). *Hermeneutics and Praxis*. Notre Dame, IN: University of Notre Dame Press.

Hollis, M. and Lukes, S. (eds.) (1982). *Rationality and Relativism*. Cambridge, MA: MIT Press.

Holton, G. and Roller, D.H.D. (1958). *Foundations of Modern Physical Science*. Reading, MA: Addison-Wesley.

Holton, G. (1960). "Origins of the Special Theory of Relativity." *American Journal of Physics* 28, 627–636.

_____. (1967). "Harvard Project Physics." *Physics Today*, March 1967, 31–34.

_____. (1973). *Thematic Origins of Scientific Thought*. Cambridge, MA: Harvard University Press.

Horkheimer, M. (1974 [1967]). *Critique of Instrumental Reason*. New York: Continuum Publishing Corporation.

Hostetler, K. (1991). "Community and Neutrality in Critical Thought: A Nonobjectivist View on the Conduct and Teaching of Critical Thinking." *Educational Theory* 41, 1–12.

Hoyle, F. and Tayler, R.J. (1964). "The Mystery of the Cosmic Helium Abundance." *Nature* 203, 1108–110.

Hoyle, F. (1950). *The Nature of the Universe*. New York: New American Library.

_____. (1960 [1950]). *The Nature of the Universe*. (revised) New York: Harper..

_____. (1974). "On the Origin of the Microwave Background." *The Astrophysical Journal* 96, 661–670.

_____. (1980). "Steady State Cosmology Revisited." Contained in Terzian and Bilson (1985), 17–57.

_____. (1982). "The World According to Hoyle." *The Sciences* 22, 9–13.

Hull, D.L. (1973). *Darwin and His Critics*. Chicago, IL: University of Chicago Press.

Husserl, E. (1962 [1913]). *Ideas: A General Introduction to Pure Phenomenology* (translated by W.R. Boyce Gibson). London: Collier-McMillan Ltd.

_____. (1970 [1936]). *The Crisis of the European Sciences and Transcendental Phenomenology*. Evanston, IL: Northwestern University Press.

Huxley, A. (1963). *Literature and Science*. New York: Harper and Row.

Huxley, J. (1957). *Knowledge, Morality and Destiny*. New York: Mentor Books.

Ihde, D. and Zaner, R.M. (eds.) (1977). *Interdisciplinary Phenomenology*. The Hague, Holland: M. Nijhoff.

Ihde, D. (1977). *Experimental Phenomenology*. New York: Putnam.

_____. (1979). *Technics and Praxis: A Philosophy of Technology*. Dordrecht, Holland: Reidel.

_____. (1991). *Instrumental Realism: The Interface Between Philosophy of Science and Philosophy of Technology*. Bloomington, IN and Indianapolis, IN: Indiana University Press.

Ingarden, R. (1973 [1937]). *The Cognition of the Literary Work of Art* (translated from the German by R. A. Crowley and K. R. Olson). Evanston, IL: Northwestern University Press.

_____. (1989 [1928]). *Ontology of the Work of Art* (translated by R. Meyer with J. T. Goldthwait, from the German edition of 1961). Athens, OH: Ohio University Press.

Ithaca Journal Magazine. (1979, January 27).

Jastrow, R. (1967). *Red Giants and White Dwarfs*. New York: Harper & Row.

Jeffreys, H. (1948). *Theory of Probability*, 2nd ed. Oxford, England: Oxford University Press.

Jensen, U.J. and Harré, R. (eds.) (1981). *The Philosophy of Evolution*. New York: St. Martin's Press.

Jones, E. (1989). *Reading the Book of Nature*. Athens, OH: Ohio University Press.

Keller, E.F. (1983). *A Feeling for the Organism*. New York: W. H. Freeman.

_____. (1985). *Reflections on Gender and Science*. New Haven, CT: Yale University Press.

Kierkegaard, S. (1954). *Fear and Trembling and Sickness unto Death*. Garden City, NY: Doubleday.

Kirschenbaum, H. and Simon, S. (eds.). (1973). *Readings in Values Clarification*. Minneapolis, MN: Winston.

Kirschenbaum, H. (1977). *Advanced Values Clarification*. La Jolla, CA: University Associates.

Kisiel, T. (1973). "Scientific Discovery: Logical, Psychological or Hermeneutical?" Contained in David Carr and Edward S. Casey (eds.), *Explorations in Phenomenology* (The Hague, Holland: Nijhoff, 1973).

_____. (1979). "The Rationality of Scientific Discovery." Contained in Theodore F. Geraets (ed.), *Rationality Today/Rationalité Aujourd'hui* (Ottawa: University of Ottawa Press, 1979).

_____. (1980). "Ars Inveniendi: A Classical Source for Contemporary Philosophy of Science." *Revue Internationale de Philosophie* 131–132, 130–154.

_____. (1983). "Scientific Discovery: The Larger Problem Situation." *New Ideas in Psychology* 1, 99–109.

Kiss, O. (1995). "Hermeneutic Roads to Commensurability." 3rd International Conference on Hermeneutics and Science, Leusden, Holland, 12–15 July, 1995.

Kitcher, P. (1983). *Abusing Science*. Cambridge, MA: MIT Press.

Knorr-Cetina, K. and Mulkay, M.J. (eds.) (1983). *Science Observed*. London: Sage.

Kockelmans, J.J. and Kisiel, T.J. (1970). *Phenomenology and the Natural Sciences*. Evanston, IL: Northwestern University Press.

_____. (1958). *Hermeneutic Phenomenology: Lectures and Essays*. Washington, D. C.: University Press of America.

_____. (1966). *Phenomenology and Physical Science*. Pittsburgh, PA: Duquesne University Press.

_____. (1985). *Heidegger and Science*. Washington, D.C.: Center for Advanced Research in Phenomenology and University Press of America.

_____. (1988). *Hermeneutic Phenomenology: Lectures and Essays*. Washington, DC: University Press of America.

_____. (1989). *Heidegger's "Being and Time."* Washington, DC: Center for Advanced Research in Phenomenology and University Press of America.

_____. (1993). Ideas for a Hermeneutic Phenomenology of the Natural Sciences. Dordrecht, Holland: Kluwer.

Kohlberg, L. and Simon, S. (1973). "An Exchange of Opinion between Kohlberg and Simon." Contained in Kirschenbaum and Simon (1973).

Kohlberg, L. (1980). "Educating for a Just Society: An Updated and Revised Statement." Contained in Munsey (1980), 455–470.

_____. (1981). *The Philosophy of Moral Development*. San Francisco, CA: Harper & Row.

Kuhn, T. (1970 [1962]). *The Structure of Scientific Revolutions*, 2nd ed. Chicago, IL: University of Chicago Press.

_____. (1977). *The Essential Tension*. Chicago, IL: University of Chicago Press.

_____. (1983). "Commensurability, Comparability and Communicability." *PSA 1982 Volume Two*, (ed. P.D. Asquith and T. Nickles). East Lansing, MI: Philosophy of Science Association.

Küng, H. (1980). *Does God Exist? An Answer for Today*. (translated by E. Quinn). Garden City, NY: Doubleday.

La Follette, M.C. (ed.) (1983). *Creationism, Science, and the Law*. Cambridge, MA: MIT Press.

Lakatos, I. and Musgrave, A. (eds.) (1970). *Criticism and the Growth of Knowledge*. Cambridge, England: Cambridge University Press.

Lakatos, I. (1970). "Falsification and the Methodology of Scientific Research Programs." Contained in Lakatos and Musgrave (eds.) (1970), 91–196.

_____. (1976). *Proofs and Refutations*. Cambridge, England: Cambridge University Press.

Latour, B. and Woolgar, S. (1979). *Laboratory Life*. Princeton, NJ: Princeton University Press.

_____. (1986 [1979]). *Laboratory Life*. Princeton, NJ: Princeton University Press.

Latour, B. (1987). *Science in Action*. Cambridge, MA: Harvard University Press.

Laudan, L. (1977). *Progress and Its Problems*. Berkeley, CA: University of California Press.

_____. (1981). "A Refutation of Convergent Realism." Contained in Jensen and Harré (1981), 232–268.

_____. (1983a). "Commentary on Ruse: Science at the Bar—Cause for Concern." Contained in La Follette, M.C. (ed.) (1983),161–66.

_____. (1983b). "The Demise of the Demarcation Problem." Contained in R.S. Cohen and L. Laudan (eds.), *Physics, Philosophy and Psychoanalysis* (Dordrecht-Holland: D. Reidel, 1983), 111–127.

_____. (1984). *Science and Values*. Berkeley, CA: University of California Press.

Layman, J.W. (1983). "Overview of the Problem." *Physics Today*, September 1983, 26–30.

Leiss, W. (1972). *The Domination of Nature*. New York: George Braziller.

Lewin, R. (1987). "Creationist Case Argued Before Supreme Court." *Science* 235 (2 January), 22–23.

Linge, D.E. (1976). "Editor's Introduction." Contained in Gadamer (1976), xi–lviii.

Lockwood, A. (1975). "A Critical View of Values Clarification." *Teachers College Record* 77 (No. 1).

_____. (1977). "Values Education and the Right to Privacy," *Journal of Moral Education* 7 (No. 1).

_____. (1978). "The Effects of Values Clarification and Moral Development Curricula on School-Age Subjects: A Critical Review of Recent Research." *Review of Educational Research* 48 (no. 3, Summer), 325–364.

Lorentz, H.A. (1904). "Electromagnetic Phenomena in a System Moving With Any Velocity Less Than That of Light." Contained in Einstein et al. (1952), 11–34.

Lyotard, J.F. (1984 [1979]). *The Postmodern Condition* (translated by G. Bennington and B. Massumi). Minneapolis, MN: University of Minnesota Press.

MacIntyre, A. (1978). "Objectivity in Morality and Objectivity in Science." Contained in H.T. Engelhardt, Jr. and D. Callahan (eds.), *Morals, Science and Society* (Hastings-on-Hudson, NY: Institute of Society, Ethics and Life Sciences, 1978), 21–39.

_____. (1980). "Epistemological Crises, Dramatic Narrative, and the Philosophy of Science." Contained in Gutting (1980), 54–74.

_____. (1981). *After Virtue*. Notre Dame, IN: University of Notre Dame Press.

Markus, G. (1987). "Why Is There No Hermeneutics of Natural Science? Some Preliminary Theses." *Science in Context* 1, 5–51.

Marcuse, H. (1964). *One-Dimensional Man*. Boston, MA: Beacon Press.

Margulis, L. and Sagan, D. (1986). *Microcosmos*. New York: Simon and Schuster.

Marston, E.H. (1970). "A Course on the Physics of Urban and Environmental Problems." *American Journal of Physics* 38, 1244–47.

Martin, M. (1972). *Concepts of Science Education*. Glenview, IL: Scott, Foresman.

Maryland Gazette. (1979, July 30).

Maslow, A.H. (1966). *The Psychology of Science*. Chicago, IL: H. Regnery Co.

Matson, F.W. (1966). *The Broken Image*. Garden City, NY: Doubleday (Anchor).

Matthews, M.R. "Old Wine in New Bottles: A Problem with Constructivist Epistemology." Contained in Alexander (1972), 303–311.

Mayr, E. (1982). *The Growth of Biological Thought*. Cambridge, MA: Harvard University Press.

McCarthy, T. (1978). *The Critical Theory of Jürgen Habermas*. Cambridge, MA: MIT Press.

_____. (1985). "Reflections on Rationalization in the Theory of Communicative Action." Contained in Bernstein (1985).

McMullin, E. (1986). "The Shaping of Scientific Rationality." A conference paper presented at the University of Notre Dame.

_____. (ed) (1988). *Construction and Constraint*. Notre Dame, IN: University of Notre Dame Press.

_____. (1982). "Values in Science." Contained in *PSA 1982: Proceedings of the 1982 Biennial Meeting of the Philosophy of Science Association* 2. East Lansing, MI: Philosophy of Science Association, 3–25.

Merleau-Ponty, M. (1962). *The Phenomenology of Perception* (translated by Colin Smith). London: Routledge & Kegan Paul.

Mermin, D. (1996). "What's Wrong with This Sustaining Myth?" *Physics Today*, March issue.

Mestre, J.P. (1991). "Learning and Instruction in Pre-College Physical Science." *Physics Today*, September, 56–62.

Midgley, M. (1985). *Evolution as a Religion*. London: Methuen.

Mill, J.S. (1951). *Utilitarianism, Liberty, and Representative Government*. New York: Dutton.

_____. (1961 [1859]). "On Liberty." Contained in M. Cohen (ed.), *The Philosophy of John Stuart Mill* (New York: Random House, 1961).

Minkowski, H. (1908). "Space and Time." Contained in Einstein et al. (1952), 75–91.

Mitroff, I.I. (1981 [1974]). "Scientists and Confirmation Bias." Contained in R.D. Tweney, M.E. Doherty, and C.R. Mynatt (eds.), *On Scientific Thinking* (New York: Columbia University Press, 1981), 170–75.

Monod, J. (1971). *Chance and Necessity* (translated by A. Wainhouse). New York: Random House.

Montagu, A. (ed.) (1984). *Science and Creationism*. Oxford, England: Oxford University Press.

Morill, R.L. (1980). *Teaching Values in College*. San Francisco, CA: Jossey-Bass.

Morris, H.M. (1974). *Scientific Creationism* (public school ed.). San Diego, CA: CLP Publishers.

Mueller-Vollmer, K. (ed.) (1988). *The Hermeneutics Reader*. New York: Continuum.

Munsey, B. (ed.) (1980). *Moral Development, Moral Education, and Kohlberg: Basic Ideas in Philosophy, Psychology, Religion and Education*. Birmingham, AL: Religious Education Press.

Musgrave, A. (1980[1971]). "Kuhn's Second Thoughts." Contained in Gutting (1980), 39–53.

Nagel, E. (1979). *The Structure of Science*. Indianapolis, IN: Hackett.

Nedelsky, L. (1973). "Physics Taught as a Liberal Art." *American Journal of Physics* 41, 364–67.

Nelkin, D. (1982). *The Creation Controversy*. New York: W.W. Norton.

Nersessian, N.J. (1989). "Conceptual Change in Science and in Science Education." *Synthese* 80, 163–184.

Newell, N.D. (1985). *Creation and Evolution*. New York: Praeger.

Novack, J.D. and Gowin, D.B. (1984). *Learning How to Learn*. Cambridge, England: Cambridge University Press.

Novak, J.D. (1977). *A Theory of Education*. Ithaca, NY: Cornell University Press.

_____. (ed.) (1987). *Proceedings of the Second International Seminar: Misconceptions and Educational Strategies in Science and Mathematics*, June 26–29, 1987. Ithaca, NY: Cornell University.

_____. (1983). "Can Metalearning and Metaknowledge Strategies to Help Students Learn How to Learn Serve as a Basis For Overcoming Misconceptions?" Contained in Helm and Novak (1983).

Nussbaum, J. (1983). "Classroom Conceptual Change: The Lesson to be Learned from the History of Science." Contained in Helm and Novak (1983).

Oldenquist, A. (1979). "Moral Education Without Moral Education." *Harvard Education Review* 49 (May), 240–47.

Oliver, W.O. and Greshman, K.W. (1989). *Education, Modernity, and Fractured Meaning*. Albany, NY: State University of New York Press.

Pais, A. (1982). *Subtle Is the Lord: The Science and the Life of Albert Einstein*. Oxford, England: Oxford University Press.

Paul, R.W. (1985). "The Critical Thinking Movement." *National Forum* (Winter), 2–3, 32.

Peacocke, A.R. (1981). *The Sciences and Theology in the Twentieth Century*. Notre Dame, IN: University of Notre Dame Press.

Peirce, C.S. (1932). *Collected Papers*, Vol. 2 (ed. C. Hartshorne and P. Weiss). Cambridge, MA: Harvard University Press.

Penrose, R. and Rindler, W. (1965). "Energy Conservation as the Basis of Relativistic Mechanics." *American Journal of Physics* 33, 55.

Penrose, R. (1989). *The Emperor's New Mind*. New York: Oxford University Press.

Penzias, A.A. (1979). "The Origin of the Elements." *Science* 205, 549–554.

Piaget, J. and Garcia, R. (1989). *Psychogenesis and the History of Science*. New York: Columbia University Press.

Piaget, J. (1970). *Genetic Epistemology*. New York: Columbia University Press.

_____. (1971). *Biology and Knowledge*. Chicago, IL: University of Chicago Press.

Pickering, A. (1984). *Constructing Quarks*. Chicago, IL: University of Chicago Press.

_____. (ed.) (1992). *Science as Practice and Culture*. Chicago, IL: University of Chicago Press.

Pinch, T. (1986). Confronting Nature: The Sociology of Solar-Neutrino Detection. Dordrecht, Holland: Reidel.

Pincoffs, E. (1986). *Quandaries and Virtues*. Lawrence, KS: University of Kansas Press.

Polanyi, M. and Prosch, H. (1975). *Meaning*. Chicago, IL: University of Chicago Press.

Polanyi, M. (1958). *Personal Knowledge*. Chicago, IL: University of Chicago Press.

_____. (1967). *The Tacit Dimension*. London: Routledge.

_____. (1969). *Knowing and Being* (ed. M. Grene). Chicago, IL: University of Chicago Press.

Popper, K.R. (1968). *Conjectures and Refutations: The Growth of Scientific Knowledge*. New York: Harper and Row.

_____. (1970). "Normal Science and Its Dangers." Contained in Lakatos and Musgrave (1970).

_____. (1972). *Objective Knowledge: An Evolutionary Approach*. Oxford, England: Clarendon Press.

_____. (1982). *Unended Quest*. LaSalle, IL: Open Court.

Price, D., Wiester, J.L. et al. (1986). *Teaching Science in a Climate of Controversy*. Ipswich, MA: The American Scientific Affiliation.

Priestley, H. (1971). "Science For the Nonscience Major—Problem and Partial Prescription." *American Journal of Physics* 39, 1498–1503.

Prigogine, I. and Stengers, I. (1984). *Order Out of Chaos: Man's New Dialogue with Nature*. New York: Bantam Books.

Proceedings of the Second International Seminar. (1987). *Misconceptions and Educational Strategies in Science and Mathematics*. June 26–29. Ithaca, NY: Cornell University.

Pugh, G.E. (1977). *The Biological Origin of Human Values*. New York: Basic Books.

Putnam, H. (1975–1976). "What Is Realism?" *Proceedings of the Aristotelian Society* 7, 177–194.

_____. (1981). *Reason, Truth and History*. Cambridge, England: Cambridge University Press.

Quinn, P.L. (1984). "The Philosopher of Science as Expert Witness." Contained in Cushing et al. (1984).

Raghavan, R.S. (1995). "Solar Neutrinos—From Puzzle to Paradox," *Science* 267, 45–50.

Randall, J.H., Jr. (1977). *Philosophy After Darwin*. New York: Columbia University Press.

Raths, L., Harmin, M., and Simon, S. (1966). *Values and Teaching*. Columbus, OH: C.E. Merrill Books.

Ravitch, D. (1985). *The Troubled Crusade: American Education 1945–1980*. New York: Basic Books.

Rawls, J. (1971). *A Theory of Justice*. Cambridge, MA: Harvard University Press.

Rigden, J.S. (1979). "What Is Our Rationale?" *American Journal of Physics* 47, 667.

_____. (1981). "What Is Scientific Literacy?" *American Journal of Physics* 49, 107.

_____. (1987). *Rabi: Scientist and Citizen*. New York: Basic Books.

_____. (1988). "A Call For a Marshall Plan to Create a New Ecology for Education." *American Journal of Physics* 56, 203.

Rohrlich, F. (1983). "Facing Quantum Mechanical Reality." *Science* 221, 1251–55.

Rolston, H. (1988). "Scientific Education and Moral Education." *Zygon: Journal of Religion and Science* 23, 347–355.

Romer, R.H. (1976). *Energy, An Introduction to Physics*. San Francisco, CA: W. H. Freeman.

Rorty, R. (1980). *Philosophy and the Mirror of Nature*. Oxford, England: Basil Blackwell.

_____. (1987). "Pragmatism and Philosophy." Contained in Baynes, et al. (1987), 26–66.

_____. (1988). "Is Science a Natural Kind?" Contained in McMullin (1988), 49–74.

Rose, S. (1976). *The Conscious Brain*. New York: Vintage Books.

Rouse, J. (1987). *Knowledge and Power: Toward a Political Philosophy of Science*. Ithaca, NY: Cornell University Press.

Rowell, J.A. (1989). "Piagetian Epistemology: Equilibration and the Teaching of Science." *Synthese* 80, 141–162.

Rudwick, M.J.S. (1985). *The Meaning of Fossils*, 2nd ed. Chicago, IL: University of Chicago Press.

Ruse, M. (1982). *Darwinism Defended*. Reading, MA: Addison-Wesley.

_____. (1984). "A Philosopher's Day in Court." Contained in Montagu (1984), 311–342.

_____. (1985). "Creation Science: Enough Is Enough Is Too Much." *National Forum* (Winter), 37–38, 48.

Russell, B. (1914). *Our Knowledge of the External World*. London: Allen & Unwin.

_____. (1927). *The Analysis of Matter*. London: Allen & Unwin.

Russell, R.J., Stoeger, W., Jr., S.J., and Coyne, C.V., (eds.) (1988). *Physics, Philosophy, Theology*. Vatican City State: Vatican Observatory.

Salpeter, E.E. (1952). *Astrophysics* 115, 326.

Saperstein, A.M. (1975). *Physics: Energy in the Environment*. Boston, MA: Little Brown.

Sapir, E. (1924). "Culture, Genuine and Spurious." *American Journal of Sociology* 29, 401–429.

Scheffler, I. (1965). *Conditions of Knowledge*. Glenview, IL: Scott, Foresman & Co.

_____. (1967). *Science and Subjectivity*. Indianapolis, IN: Bobbs-Merrill.

_____. (1973). *Reason and Teaching*. Indianapolis, IN: Bobbs-Merrill.

Schilk, J.M., Driscoll, S.E. and Carter, C.S. (1989). "Problem Solving and Construction of Scientific Knowledge: A Case Study in Epistemology." Contained in Herget (1989), 322–331.

Schweber, S.S. (1986). "Feynman and the Visualization of Space-Time Processes." *Review of Modern Physics* 58, 449–508.

Schwinger, J. (1989). "A Path to Quantum Electrodynamics." *Physics Today,* February, 42–48.

Scriven, M. (1985). "Critical for Survival." *National Forum* (Winter), 9–12.

Segerstrale, U. (1986). "Colleagues in Conflict: An 'In Vivo' Analysis of the Sociobiology Controversy." *Biology and Philosophy* 1, 53–87.

Shapere, D. (1984). *Reason and the Search for Knowledge.* Dordrecht, Holland: Reidel.

_____. (1966). "Meaning and Scientific Change." Contained in R.G. Colodny (ed.), *Mind and Cosmos* (Pittsburgh, PA: University of Pittsburgh Press, 1966), 41–85.

Shapin, S. and Schaffer, S. (1985). *Leviathan and the Air-Pump.* Princeton, NJ: Princeton University Press.

Shils, E. (1981). *Tradition.* Chicago, IL: University of Chicago Press.

Shimony, A. (1970). "Scientific Inference." Contained in R.G. Colodny, *The Nature and Function of Scientific Theories* (Pittsburgh, PA: Pittsburgh University Press, 1970).

_____. (1976). "Comments on Two Epistemological Theses of Thomas Kuhn." Contained in Cohen et al. (1976), 569–588.

_____. (1978). "Is Observation Theory-Laden? A Problem in Naturalistic Epistemology." Contained in R.G. Colodny (ed.), *Logic, Laws, and Life: Some Philosophical Complications* (Pittsburgh, PA: University of Pittsburgh Press, 1978).

_____. (1988). "On Martin Eger's 'A Tale of Two Controversies.'" *Zygon: Journal of Religion and Science* 23, 333–340, (this volume, 369–375).

_____. (1993). "Reality, Causality, and Closing the Circle." Contained in his *Search for a Naturalistic World View,* Vol. I (Cambridge, England: Cambridge University Press, 1993).

Siegel, H.. (1979). "On the distortion of the History of Science in Science Education." *Science Education* 63 (1), 111–122.

_____. (1981). "Creation, Evolution, and Education: The California Fiasco." *Phi Delta Kappan* 63 (October), 95–101.

_____. (1985). "What Is the Question Concerning the Rationality of Science?" *Philosophy of Science* 52, 517–537.

_____. (1988). "Rationality and Anemia (Response to Baigrie)." *Philosophy of Science* 55, 442–47.

_____. (1989). "The Rationality of Science, Critical Thinking, and Science Education." *Synthese* 81, 9–41.

Silverman, H.J. (ed.) (1991). *Gadamer and Hermeneutics.* New York: Routledge.

Simon, H. (1977). "What Computers Mean to Man and Society." *Science* 195, 1186–1191.

Simon, S.B., Howe, L.W. and Kirschenbaum, H. (1972). *Values Clarification: A Handbook of Practical Strategies for Teachers and Students.* New York: Hart.

Simpson, G.G. (1949). *The Meaning of Evolution.* New Haven, CT: Yale University Press.

Spencer Needle. (1978, 1979).

Spencer, H. (1958 [1880]). *First Principles*, 4th ed. [2nd ed. 1867]. De Witt, NY: Revolving Fund.

Sperry, R. (1985 [1983]). *Science and the Moral Priority*. 2nd ed. New York: Praeger.

Stent, G. (1985). "Hermeneutics and the Analysis of Complex Biological Systems." Contained in Depew and Weber (1985), 209–225.

Stewart, J.S. (1975). "Clarifying Values Clarification: A Critique." *Phi Delta Kappan* 56 (No. 10).

Stinner, A. (1992). "Science Textbooks and Science Teaching: From Logic to Evidence." *Science Education* 76, 1–16.

Strike, K. (1982a). *Educational Policy and the Just Society*. Urbana, IL: University of Illinois Press.

_____. (1982b). *Liberty and Learning*. New York: St. Martin's.

Sunday Sun (Baltimore). (1979, May 27).

Superka, D., Johnson, P.L., and Ahrens, C. (1975). *Values Education: Approaches and Materials*. Boulder, CO: Social Science Education Consortium Inc.

Taylor, C. (1971). "Interpretation and the Sciences of Man." Contained in Dallmayr and McCarthy, eds. (1977), 101–131. Original publication in *Review of Metaphysics* 25, 3–34.

Terzian, Y. and Bilson, E.M. (eds.) (1982). *Cosmology and Astrophysics: Essays in Honor of Thomas Gold*. Ithaca, NY: Cornell University Press.

Thaxton, C.B., Bradley, W.L. and Olson, R.L. (1984). *The Mystery of Life's Origin*. New York: Philosophical Library.

The University of the State of New York. (1974). Goals of Elementary, Secondary and Continuing Education in New York State. New York.

_____. (1976). *Valuing, A Discussion Guide*. Albany, NY: University of the State of New York.

Thompson, J.B. and Held, D. (eds.) (1982). *Habermas: Critical Debates*. Cambridge, MA: MIT Press.

Thompson, M. and de Zengotita, T. (1989). "Science Literacy and the Language Game of Science." Contained in Herget (1989), 344–357.

Tipler, F.J. (1983). "Black Holes and the Nature of Quantum Gravity." *General Relativity and Gravitation Journal* 15, 1147–49.

_____. (1985). "How to Construct a Falsifiable Theory in Which the Universe Came into Being Several Thousand Years Ago." Contained in P. D. Asquith and P. Kitcher (eds.), *PSA 1984*, Vol. 2 (Ann Arbor, MI: The Philosophy of Science Association), 873–902.

Toulmin, S. (1980). "How Can We Reconnect the Sciences with the Foundations of Ethics?" Contained in H.T. Engelhardt, Jr. and D. Callahan (ed.) *Knowing and Valuing, The Search for Common Roots* (Hastings-on-Hudson, NY: Institute of Society, Ethics and Life Sciences, 1980), 44–64.

_____. (1982). *The Return of Cosmology*. Berkeley, CA: University of California Press.

Traweek, S. (1988). *Beamtimes and Lifetimes*. Cambridge, MA: Harvard University Press.

Turkle, S. (1984). *The Second Self*. New York: Simon and Schuster.

Van Fraassen, B.C. (1980). *The Scientific Image*. Oxford, England: Clarendon Press.

Van Hise, Y A. and Nelson, J. (1988). "AAPT's Physics Teaching Resource Agent Program." *Physics Today,* March, 47–50.

Von Glasersfeld, E. (1988). *The Construction of Knowledge: Contributions to Conceptual Semantics.* Salinas, CA: Intersystems Publications.

_____. (1989). "Cognition, Construction of Knowledge, and Teaching." *Synthese* 80, 121–140.

Von Wright, G. H. (1974). *Causality and Determinism.* New York: Columbia University Press.

Wagner, D. (1988). Interview in News Letter of the National Center on Effective Secondary Schools 3 (Spring).

Wagner, P.A. (1983). "The Nature of Paradigm Shifts and the Goals of Science Education." *Science Education* 67, 605–613.

Weinberg, S. (1979 [1977]). *The First Three Minutes.* New York: Bantam.

Weinstock, R. (1965). "New Approach to Special Relativity." *American Journal of Physics* 33, 640–45.

Weisskopf, V. (1976). "Is Science Human?" *Physics Today,* June 1976, 23.

_____. (1979). *Knowledge and Wonder.* 2nd ed. Cambridge, MA: MIT Press.

Weizenbaum, J. (1976). *Computer Power and Human Reason.* San Francisco, CA: Freeman.

Westfall, R.S. (1980). *Never at Rest: A Biography of Isaac Newton.* Cambridge, England: Cambridge University Press.

Westheimer, F.H. (1987). "Are Our Universities Rotten to the Core?" *Science* 246, 1165–66.

Whitehead, A.N. (1955). *The Aims of Education.* New York: Mentor Books.

Whittaker, E.T. (1910). A History of the Theories of Aether and Electricity: From the Age of Descartes to the Close of the Nineteenth Century. London: Longmans, Green & Co.

Williams, L.P. (1965). *Michael Faraday.* New York: Da Capo Press.

Wilson, D.B. (ed.) (1983). *Did the Devil Make Darwin Do It?* Ames, IA: The Iowa State University Press.

Wilson, E.O. (1978). *On Human Nature.* New York: Bantam Books.

Wilson, J. (1972). *Practical Methods of Moral Education.* London: Heinemann Educational.

Wolfenstein, L. and Beier, E.W. (1989). "Neutrino Oscillations and Solar Neutrinos." *Physics Today,* July 1989 issue.

Wolin, S. (1969). "Political Theory as a Vocation." *American Political Science Review.* 63, 1062–1082.

Wright, C. (1877). "The Philosophy of Herbert Spencer." Contained in *Philosophical Discussions*, ed. posth., C. E. Norton, New York, 1877, 43–96.

Yantsch, E. (ed.) (1981). *The Evolutionary Vision.* Boulder, CO: Westview.

Young, R.M. (1985). *Darwin's Metaphor.* Cambridge, England: Cambridge University Press.

Young, R.E. (1990). "Habermas' Ontology of Learning: Reconstructing Dewey." *Educational Theory* 40, 471–482.

Index